Microbial Technologies for Wastewater Recycling and Management

This book introduces the innovative and emerging microbial technologies for the treatment, recycling, and management of industrial, domestic, and municipal water and other wastewater in an environment-friendly and cost-effective manner. It discusses existing methods and technologies, up-gradation of existing technologies, and new technologies. It also highlights opportunities in the existing technologies along with industrial practices and real-life case studies.

Vineet Kumar
Department of Basic and Applied Sciences,
School of Engineering and Sciences,
G D Goenka University, Gurugram, Haryana, India

Joginder Singh
Department of Microbiology
School of Bioengineering and Biosciences,
Lovely Professional University, Phagwara, Punjab

Microbial Technologies for Wastewater Recycling and Management

Recent Trends, Challenges, and Perspectives

Edited by
Vineet Kumar and Joginder Singh

CRC Press
Taylor & Francis Group
Boca Raton London New York

CRC Press is an imprint of the
Taylor & Francis Group, an **informa** business

Designed cover image: © Shutterstock

First edition published 2023
by CRC Press
6000 Broken Sound Parkway NW, Suite 300, Boca Raton, FL 33487-2742

and by CRC Press
4 Park Square, Milton Park, Abingdon, Oxon, OX14 4RN

CRC Press is an imprint of Taylor & Francis Group, LLC

ISBN: 9781032137582 (hbk)
ISBN: 9781032139715 (pbk)
ISBN: 9781003231738 (ebk)

DOI: 10.1201/9781003231738

Typeset in Times
by codeMantra

Dedicated to our teachers and mentors, from whom I continue to learn, and

to my family for their support, blessings, motivation, and love.

—Vineet Kumar

This book is dedicated to my children and my wife. Their patience and understanding

have given me the time and inspiration to research and to edit this project.

—Joginder Singh

Contents

Preface

The safe disposal of wastewater from different sectors such as municipal, domestic, and industrial activities has become a global threat due to its elevated content of organometallic materials, nutrients, pathogenic microorganisms, and many toxic and refractory chemical elements including heavy metals and endocrine-disrupting chemicals. Due to rampant industrialization, population, and urbanization, it is not possible to stop the discharge of wastewater into the ecosystem. Hence, adequate management and reclamation of wastewater streams for their recycling and reuse as well as resource recovery of value-added materials have become a priority of environmental engineers and biotechnologists for sustainable development. Although several conventional biological, physical, and physicochemical methods dealing with a wide array of chemicals discharged in wastewater have been suggested or tested widely by the scientific community, these treatment approaches are associated with high operational cost, excess use of chemicals, less effective decolorization rate, and a huge amount of toxic sludge generation with subsequent disposal problems. These methods only change the form of the problem and fail to remediate the pollutants from wastewater thoroughly. Therefore, sufficient treatment is essential before the wastewater is disposed into the open environment. In the current scenario, considerable research efforts in government institutions, industrial, and academic, have been looked at on the development of new and sustainable technology for the adequate reclamation and/or detoxification of chemical contaminants of wastewater discharged in wastewater that can recover and reuse wastewater for sustainable development. In recent years, microbial biotechnologies have introduced a myriad of novel microbes and/or their enzyme systems that can have promising outcomes in environmental clean-up and remediation. Microbe-based biotechnological methods offer clear advantages over traditional individuals' methods for clean-up of wastewater. Research and development studies conducted around the globe established that microbial technologies play a promising role in the reclamation of wastewater contaminated with refractory organic, inorganic, and organometallic pollutants and enable them to be safely disposed into the environment without posing a serious danger to human and environmental health and subsequently recycle and reuse them for various applications. Hence, microbial technologies play a vital role in the development of a circular economy in wastewater recycling and management by integrating resource recovery and bioenergy generation for clean water production. Due to the increased importance of the latter and to the need of finding innovative, low-cost, low-energy, sustainable, and eco-friendly solutions for wastewater treatment, the proposed book is timely and appropriate. Editors' effort is to provide complete and advanced knowledge of wastewater reclamation, recycling, reuse and resource microbial recovery technologies available across the globe, which will be a boon for the concern broad range of readers and industry professionals to their minas of sustainable development for pollution prevention and economic growth of the country.

This book, *Microbial Technologies for Wastewater Recycling and Management: Recent Trends, Challenges, and Perspectives,* introduces the innovative and emerging microbial technologies for the treatment, recycling, and management of industrial, domestic, and municipal wastewaters in a cost-effective and environmentally friendly way. The whole book is spread over 21 chapters, contributed by leading workers, drawn from the world over, in their field, and provides the latest research and development in microbial wastewater treatment technologies. In this book, expert authors critically review the recent advances and development in this exciting field and further address different topics related to different existing microbes-based technologies widely used for wastewater remediation, recycling, and reuse for environmental sustainability and future development. This book "highlights latest advanced and updated knowledge on various developed individual and/or combine microbial technologies/approaches such as constructed wetland-microbial fuel cell, anammox, vermifiltration, bioaugmentation, innovative integrated treatment approaches for biodiesel production and wastewater treatment by oleaginous microorganisms and many more, which is presently restricted only in the specific journals, review articles, and research papers conference proceeding. Moreover, the present book provides information on the upgradation of existing microbial technologies and focus areas for further research and developments along with industrial practices and real-life case studies. All chapters in this book are comprehensible straight forward, with relevant graphical and photographic illustrations to simplify the topic and provide additional help to clarify. This book is intended to appeal to students, researchers, and scientists a wide range of professionals responsible for regulating, monitoring, and designing industrial wastewater treatment and management facilities who aspire to work on the remediation of wastewater or wastewater pollutants for sustainable development. Thus, in the editor's opinion, this book will be a valuable resource for engineering consultants, industrial waste managers, educational and research institutes, policy-makers, environmental engineers, government implementers, nongoverment organizations, and others concerned with industrial wastewater reclamation, recycling and management towards sustainable development.

Acknowledgments

In this endeavour, we, the editors, were not alone but assisted by many people, want to acknowledge the contributions of the numerous people who have made this book possible.

First, we greatly appreciate the efforts of all the authors who have given their valuable contributions and shared their knowledge with us to make this primer a reality.

Dr. Vineet Kumar most sincerely thankful to Dr. Sunil Kumar, Senior Principal Scientist and Head in the Waste Re-processing Division (WRD) at CSIR-National Environmental Engineering Research Institute (NEERI), Nagpur, Maharashtra, India, for his dedicated help, moral support, and outstanding and invaluable advice on how to construct an effective book. Moreover, Dr. Sunil Kumar had provided a fantastic facility in his laboratory to complete this task. They helped us in editing some of the manuscripts for value addition, without him this book would not have been possible. We are also thankful to all the faculty and staff members of our division for their help and support during our course of study.

We are grateful to Dr. Ram Chandra, Professor & Head of the Department of Environmental Microbiology at Babasaheb Bhimrao Ambedkar (A Central) University, Lucknow, Uttar Pradesh, India, for the tremendous academic support that they provided to Dr. Vineet Kumar during his doctoral studies and, most prominently in the unique field of metagenomics, phytoremediation, bioremediation, and biodegradation of industrial wastewater pollutants.

Each of us owes a special debt of gratitude to our colleagues, from whose insights and suggestions we have benefited greatly and borrowed freely. We would like to thank those colleagues who took the time to read over individual chapters of this book and those who reviewed the entire manuscript. Their comments have been gratefully received, and in some cases, spared us from the embarrassment of seeing my mistakes perpetuated in print.

We wish to acknowledge our students, past and present, who provide the stimulation for the continuation of this work, that make it possible. We want to thank our teachers/mentors – who, each in his own way, have true passion for science and an insistence that the right experiments are done.

We would also like to thank CRC Press (Taylor and Francis Group) for giving us the opportunity to accomplish the project and share the knowledge with the scientific and academic fraternity. We are particularly indebted to Dr. Renu Upadhyay, Senior Acquisition/Commissioning Editor (Life Science), for the execution of the publishing agreement, motivation, trust, encouragement, support, and her valuable suggestions, which is simply the best. We want to acknowledge Ms. Jyotsna Jangra, Editorial Assistant (Life Science) at CRC Press (Taylor & Francis Group), India, for her constructive criticism, sound advice, and invaluable support throughout the project. Dr. Kumar has also had the good fortune to work again with two of the same professionals who did a masterful job with the earlier two books.

We are grateful to the literary Jenny Guildford in his role as Senior Production Editor at CRC Press (Taylor & Francis Books) for ensuring smooth production of this book. Without his prodding, this book would never have been completed. Thanks are also due to the production editor at Codemantra, Assunta Petrone, for ensuring the smooth production of this book.

We are also deeply indebted to the many publishing professionals whose consistent encouragement, hard work, and careful attention to detail contributed much to the clarity of both the text and the art. A special thanks are owed Assunta Petrone, Project Manager at codeMantra, and team who skilfully laid out the pages for each chapter, and reviewed the material and provided editorial comments.

Last but not least, Dr. Vineet Kumar would like to acknowledge his family members with love and affection, in particular his Parents (Mr. Niranjan Singh and Mrs. Pawan Devi), younger brother, Rohit Chowdhary, and Sister, Ms. Khushboo).

Finally, we would like to apologize in advance for any errors that may occur in the text and express my heartfelt embarrassment

We should be pleased to receive any comments on the content and style of Microbial Technologies for Wastewater Recycling and Management: Recent Trends, Challenges, and Perspectives, from students, professionals, environmentalists, environmental engineers, scientists, and policymakers, all of which will be given serious consideration for inclusion in any further editions.

Vineet Kumar
Haryana, India

Joginder Singh
Punjab, India

Editors

Dr. Vineet Kumar is presently working as an Assistant Professor in the Department of Basic and Applied Sciences, School of Engineering and Sciences at G D Goenka University, Gurugram, Haryana, India. Prior to joining at G D Goenka University, Dr. Kumar served in various reputed institutions in India like CSIR-National Environmental Engineering Research Institute (NEERI), Maharashtra, India; Guru Ghasidas Vishwavidyalaya (A Central University), Chhatisgarh, India; Jawaharlal Nehru University, Delhi; Dr. Shakuntala Misra National Rehabilitation University, Uttar Pradesh, India, etc. He received his MSc and MPhil degree in Microbiology from Ch. Charan Singh University, Meerut, India. Subsequently, he earned his PhD (2018) in Environmental Microbiology from Babasaheb Bhimrao Ambedkar (A Central) University, Lucknow, India. Dr. Kumar's research interest includes: Bioremediation, Phytoremediation, Metagenomics, Wastewater Treatment, Environmental Monitoring, Waste Management; Bioenergy, and Biofuel Production. Currently, his research mainly focuses on the development of integrated and sustainable treatment techniques that can help in minimizing or eliminating hazardous waste in the environment. He has published 32 research and review articles in reputed International Journals, written 4 Proceeding Papers, 45 book chapters, and authored/edited over 15 books on the different aspects of Phytoremediation, Bioremediation, Wastewater Treatment, Omics, Genomics, and Metagenomics published by CRC Press (Taylor & Francis Group), Elsevier Science, Springer Nature, Wiley, Royal Society of Chemistry. His recently published book is '*Recent Advances in Distillery Waste Management for Environmental Safety*' (From CRC Press; Taylor & Francis Group, USA). He has presented several papers relevant to his research areas at national and international conferences. He is an active member of numerous scientific societies including the Microbiology Society, UK, and the Indian Science Congress Association, India. Dr. Kumar has been serving as a guest editor and reviewer in many prestigious International Journals, including Frontiers in Microbiology, Environmental Research, Chemosphere, Journal of Basic Microbiology; International Journal of Environmental Science and Technology; CLEAN-Soil, Air, Water, etc. He is the founder of the Society for Green Environment, India (website: www.sgeindia.org). He can be reached at drvineet.micro@gmail.com

Dr. Joginder Singh is presently working as Professor at the Department of Microbiology, Lovely Professional University, Punjab, India. Previously, he worked as a Young Scientist at Microbial Biotechnology and Biofertilizer Laboratory, Department of Botany, Jai Narain Vyas University in Department of Science and Technology, Govt. of India. He is an active member of various scientific societies and organizations including, the Association of Microbiologists of India, the Indian Society of Salinity Research Scientists, the Indian Society for Radiation Biology, and the European Federation of Biotechnology. He has more than 130 research and review articles in peer-reviewed journals, edited 15 books published by Springer International Publishing, Elsevier Science Publishing, and authored/co-authored 65 chapters in edited books. He serves as a reviewer for many prestigious journals, including Current Research in Engineering, Science and Technology, Journal of Cleaner Production, Science of the Total Environment, Environmental Monitoring and Assessment, Pedosphere, Soil and Sediment Contamination, Symbiosis, International Journal of Phytoremediation, Ecotoxicology and Environmental Safety, Annals of Agricultural Sciences, Annals of the Brazilian Academy of Sciences and many more. Dr Singh attended several International and National Seminars, Symposia, and Conferences, chaired technical sessions, and presented papers. Despite teaching aptitude, his experience in planning and execution of R&D facilities at the scientist level and knowledge of microbial implications for resolving social problems have been well recognized in terms of research projects he is handling and has worked upon. His training and research experience prepared him to work with microbial consortium formulation and its ecology. His interest is to explore efficient strategies towards bioremediation and phytoremediation of pollutants from water and soil. Specific areas on which he is working are the microbial technology for rhizoremediation of polluted soil and targeting suitable plants and rhizospheric microorganisms from these areas along with enzymes responsible for remediation of organic and inorganic pollutants. Besides this, he guided six PhD, four MPhil, 31 MTech/MSc and 25 BTech/BSc students, which proves his skills and passion for teaching and research.

Contributors

Norhayati Abdullah
Department of Chemical and Environmental Engineering, Malaysia-Japan International Institute of Technology (MJIIT)
Universiti Teknologi Malaysia
Kuala Lumpur, Malaysia

Marpu Adilakshmi
Department of Biotechnology, School of Bioengineering and Biosciences
Lovely Professional University
Phagwara, India

Mohit Aggarwal
School of Infrastructure
Indian Institute of Technology Bhubaneswar
Odisha, India

Imran Ahmad
Malaysia-Japan International Institute of Technology (MJIIT),
Universiti Teknologi Malaysia
Kuala Lumpur, Malaysia

Myriam A. Amezcua-Allieri
Biomass Conversion Division
Mexican Institute of Petroleum
Mexico City, Mexico

Tesnim Arfi
Department of Bioengineering and Biotechnology
Birla Institute of Technology
Ranchi, India

Sudipti Arora
Dr. B. Lal Institute of Biotechnology
Jaipur, India

Thilagavathi Arumugham
Malaysia-Japan International Institute of Technology (MJIIT)
Universiti Teknologi Malaysia
Kuala Lumpur, Malaysia

Shailendra Kumar Arya
Department of Biotechnology, University Institute of Engineering and Technology
Panjab University
Chandigarh, India

Rima Biswas
Wastewater Technology Division
CSIR-National Environmental Engineering Research Institute
Nagpur, India
Academy of Science and Innovative Research (AcSIR)
Ghaziabad, India

Prasenjit Chakraborty
Department of Chemistry
The University of Burdwan
Burdwan, India

Sankha Chakrabortty
School of Chemical Technology
KIIT Deemed to be University
Bhubaneswar, India

Himani Chandel
EMBL - Environmental Microbiology and Biotechnology Laboratory, EERG - Ecotoxicology and Environmental Remediation Group, School of Biotechnology
Shoolini University of Biotechnology and Management Sciences
Solan, India

Aishwarya Singh Chauhan
Center for Public health Kinetics
New Delhi, India

Preeti Chaurasia
Department of Microbiology, School of Science
Nirwan University
Jaipur, India

Evangeline Christina
Department of Molecular Biology and Genetic Engineering, School of Bioengineering and Biosciences
Lovely Professional University
Phagwara, India

Praveen Dahiya
Amity Institute of Biotechnology
Amity University Uttar Pradesh (AUUP)
Noida, India

Suchismita Das
Department of Life Science and Bioinformatics
Assam University
Silchar, India

Amit Kumar Dutta
Amity Institute of Biotechnology
Amity University Jharkhand
Ranchi, India

Sashita Bindu Ekka
Department of Environmental Science
Indira Gandhi National Tribal University
Amarkantak, India

Aditya Francis
Food Science and Engineering Department
ISARA-Lyon
Lyon, France

Jasiel García
Nanosciences and Nanotechnology Department
CINVESTAV-IPN
Mexico City, Mexico

Divita Gautam
Amity Institute of Biotechnology
Amity University Uttar Pradesh (AUUP)
Noida, India

Sunayana Goswami
Department of Zoology
Biswanath College
Biswanath Chariali, India

Rohan Gupta
Wastewater Technology Division
CSIR-National Environmental Engineering Research Institute
Nagpur, India

Shankramma Kalikeri
Division of Nanoscience and Technology, Department of Water and Health (Faculty of Natural Science)
JSS Academy of Higher Education & Research (Deemed to be University)
Mysore, India

Atya Kapley
Environmental Biotechnology and Genomics Division
CSIR-National Environmental Engineering Research Institute
Nagpur, India
Academy of Scientific and Innovative Research (AcSIR)
Ghaziabad, India

Jyoti Kaushal
Department of Biotechnology, University Institute of Engineering and Technology
Panjab University
Chandigarh, India

Humaira Khanam
Department of Bioengineering and Biotechnology
Birla Institute of Technology
Ranchi, India

Maitri Khanduri
EMBL - Environmental Microbiology and Biotechnology Laboratory, EERG - Ecotoxicology and Environmental Remediation Group, School of Biotechnology
Shoolini University of Biotechnology and Management Sciences
Solan, India

Madhu Khatri
Department of Biotechnology, University Institute of Engineering and Technology
Panjab University
Chandigarh, India

Iwamoto Koji
Department of Chemical and Environmental Engineering, Malaysia-Japan International Institute of Technology (MJIIT)
Universiti Teknologi Malaysia
Kuala Lumpur, Malaysia

Dhirendra Kumar
Department of Botany
Chaudhary Bansi Lal University
Bhiwani, India

Manoj Kumar
School of Environment & Sustainable Development
Central University of Gujarat
Gandhinagar, India

Navneet Kumar
EMBL - Environmental Microbiology and Biotechnology Laboratory, EERG - Ecotoxicology and Environmental Remediation Group, School of Biotechnology
Shoolini University of Biotechnology and Management Sciences
Solan, India

Ramesh Kumar
Department of Earth Resources and Environmental Engineering
Hanyang University
Seoul, Republic of Korea

Sanjeev Kumar
Department of Genetics and Plant Breeding, School of Agriculture
Lovely Professional University
Phagwara, India

Vineet Kumar
Department of Botany, School of Life Sciences
Guru Ghasidas Vishwavidyalaya (A Central University)
Bilaspur, India

Sneh Lata
Department of Environmental Sciences, Central University of Punjab, Bathinda, Punjab -151401, India

Pravin Manekar
Wastewater Technology Division
CSIR-National Environmental Engineering Research Institute
Nagpur, India
Academy of Scientific and Innovative Research (AcSIR)
Ghaziabad, India

Dhanya Mecherivalappil Subramanian
Department of Environmental Sciences, Central University of Punjab, Bathinda, Punjab -151401, India

Sugandha Mishra
Department of Environmental Science
Govt. (PG) College for Women
Rohtak, India

Shaza Eva Mohamad
Department of Chemical and Environmental Engineering, Malaysia-Japan International Institute of Technology (MJIIT)
Universiti Teknologi Malaysia
Kuala Lumpur, Malaysia

Sampurna Nandy
Environmental Biotechnology and Genomics Division
CSIR-National Environmental Engineering Research Institute
Nagpur, India
Academy of Scientific and Innovative Research (AcSIR)
Ghaziabad, India

Jayato Nayak
School of Bio and Chemical Engineering,
Kalasalingam Academy of Research and Education
Srivilliputhur, India

Vinod Kumar Nigam
Department of Bioengineering and Biotechnology
Birla Institute of Technology
Ranchi, India

Sukdeb Pal
Wastewater Technology Division
CSIR-National Environmental Engineering Research Institute
Nagpur, India
Academy of Scientific and Innovative Research (AcSIR)
Ghaziabad, India

Deepak Panchal
Academy of Scientific and Innovative Research (AcSIR)
Ghaziabad, India
Wastewater Technology Division
CSIR-National Environmental Engineering Research Institute
Nagpur, India

Kamini Parmar
Department of Biotechnology
TRS College
Rewa, India

Kalp Bhusan Prajapati
Department of Environmental Studies
Central University of Haryana
Mahendragarh, India

Spriha Raven
Diatom Research Laboratory, Amity Institute of Biotechnology
Amity University
Noida, India

Neelancherry Remya
School of Infrastructure
Indian Institute of Technology Bhubaneswar
Odisha, India

Lipi Rina
Department of Agronomy
SHUATS
Prayagraj, India

Indranil Saha
School of Chemical Technology
KIIT Deemed to be University
Bhubaneswar, India

Bijaylaxmi Sahoo
Wastewater Technology Division
CSIR-National Environmental Engineering Research Institute
Nagpur, India
Academy of Science and Innovative Research (AcSIR)
Ghaziabad, India

Savita Durgappa Sandur
Department of Microbiology, Sahyadri Science College
Kuvempu University
Shivamogga, India

Sakshi Saraswat
Institute of Environmental and Occupational Health Sciences, School of Medicine
National Yang Ming Chiao Tung University
Taipei City, Taiwan

Satyendra
Wastewater Technology Division
CSIR-National Environmental Engineering Research Institute
Nagpur, India
Academy of Scientific and Innovative Research (AcSIR)
Ghaziabad, India

Gaurav Saxena
EMBL - Environmental Microbiology and Biotechnology Laboratory, EERG - Ecotoxicology and Environmental Remediation Group, School of Biotechnology
Shoolini University of Biotechnology and Management Sciences
Solan, India

Samvida Saxena
Dr. B. Lal Institute of Biotechnology
Jaipur, India

Shradhanjali Sethi
Wastewater Technology Division
CSIR-National Environmental Engineering Research Institute
Nagpur, India
Academy of Science and Innovative Research (AcSIR)
Ghaziabad, India

Megha Shah
Department of Microbiology, Parul Institute of Applied Sciences
Parul University
Vadodara, India

Anamika Sharma
Hyper Filteration Private Limited
Sahibabad, India

Asha Sharma
Department of Botany
Maharshi Dayanand University
Rohtak, India

Geetansh Sharma
EMBL - Environmental Microbiology and Biotechnology Laboratory, EERG - Ecotoxicology and Environmental Remediation Group, School of Biotechnology
Shoolini University of Biotechnology and Management Sciences
Solan, India

Rajesh Kumar Sharma
Department of Physics
MNS Govt. College
Bhiwani, India

Shalini Sharma
Department of Biotechnology and Microbiology
Meerut Institute of Engineering and Technology
Meerut, India

Anupama Shrivastav
Department of Microbiology, Parul Institute of Applied Sciences
Parul University
Vadodara, India

Amol Shukla
Wastewater Technology Division
CSIR-National Environmental Engineering Research Institute
Nagpur, India

Kirti Shyam
EMBL - Environmental Microbiology and Biotechnology Laboratory, EERG - Ecotoxicology and Environmental Remediation Group, School of Biotechnology
Shoolini University of Biotechnology and Management Sciences
Solan, India

Bishal Singh
School of Medical Science and Technology
Indian Institute of Technology
Kharagpur, India

Chaudhary Shalu Singh
Central Laboratory Mohan Meakin Limited
Ghaziabad, India

Neeraj Kumar Singh
School of Environment & Sustainable Development
Central University of Gujarat
Gandhinagar, India

R.S. Singh
Department of Chemical Engineering & Technology
Indian Institute of Technology (BHU)
Varanasi, India

Rajesh Singh
School of Environment & Sustainable Development, Central University of Gujarat, Gandhinagar, Gujarat, India

Vikky Sinha
Dr. B. Lal Institute of Biotechnology
Jaipur, India

Heilyn Pérez Soler
Nanosciences and Nanotechnology Department
CINVESTAV-IPN
Mexico City, Mexico

Devanshi Sutaria
Dr. B. Lal Institute of Biotechnology
Jaipur, India

Saurabh Thakur
EMBL - Environmental Microbiology and Biotechnology Laboratory, EERG - Ecotoxicology and Environmental Remediation Group, School of Biotechnology
Shoolini University of Biotechnology and Management Sciences
Solan, India

Abhishek Thakurr
Department of Plant Pathology
Chaudhary Charan Singh Haryana Agricultural University
Hisar, India

Archana Tiwari
Diatom Research Laboratory, Amity Institute of Biotechnology
Amity University
Noida, India

Himanshu Tiwari
Department of Chemical Engineering & Technology
Indian Institute of Technology (BHU)
Varanasi, India

Ritesh Vijay
Wastewater Technology Division
CSIR-National Environmental Engineering Research Institute
Nagpur, India
Academy of Scientific and Innovative Research (AcSIR)
Ghaziabad, India

Ali Yuzir
Department of Chemical and Environmental Engineering, Malaysia-Japan International Institute of Technology (MJIIT)
Universiti Teknologi Malaysia
Kuala Lumpur, Malaysia

Section I

Introduction to Wastewater and Remediation Technologies

1

Wastewater Pollution, Toxicity Profile, and Their Treatment Approaches: A Review

Sugandha Mishra
Govt. (PG) College for Women

Dhirendra Kumar
Chaudhary Bansi Lal University

Rajesh Kumar Sharma
MNS Govt. College

Amit Kumar Dutta
Amity University

Asha Sharma
Maharshi Dayanand University

CONTENTS

1.1 Introduction

Water is essential and one of the most critical resources that nature possesses for all living systems on earth (Pati, 2015: Dutta et al., 2021). For any metabolic reaction to happen, water must be a part of that reaction without which it is impossible to observe any metabolic reactions in the human body, this forms the basis of growth, nourishment, and maintenance of life (Abu-Zeid, 1998; Siegel, 2008). Pure water lacks in colour, odour, or taste at room temperature however the properties may change as per the given conditions, for

DOI: 10.1201/9781003231738-2

instance, in a thin layer it seems to be colourless (Tortajada, 2014). However, the case is not true if the layer is thick which is a common scenario in lakes or oceans where the colour turns to greenish-blue. Water has a higher range of thermal conductivity, specific heat, and surface tension if compared to other compounds (specifically liquids) in the same category (Miller, 2019). Today, approximately 2.2 billion people globally do not have access to safe water at home, and a further about 1.37 billion people globally lack hand washing facilities at home. According to a global study by World Health Organization (WHO) and UNICEF, Joint Monitoring Programme (JMP) these figures are uprising day by day (Demir et al., 2021). Recently we all have faced the COVID-19 pandemic, taking into consideration that one of the most effective ways to stop the spread of COVID-19 and other pathogens is proper hand hygiene (Teixeira and Gomes, 2021; Gift et al., 2020). Most people who fall sick with COVID-19 will experience mild to moderate symptoms and recover without special treatment, whereas already approximately 21.7 crores fell sick and more than 5 billion died, here these numbers are alarming. Almost 2 billion people around the world rely on health care facilities that do not have basic water services (Quinete and Hauser-Davis, 2021). As we know, hand washing with soap has been proven to be one of the best preventative life-saving measures available but the reality is that globally, one out of six (16%±1%) health care facilities has no functional hand hygiene facilities available at either point of care or toilets. In January 2020, World Health Organization (WHO) Director-General Tedros Adhanom Ghebreyesus cited the absence of water, toilets, soap, and waste management in health care facilities among the most urgent global challenges of this decade (Quinete and Hauser-Davis, 2021; Mishra and Rani, 2020). As we have seen in recent months, global health has a direct impact on economies and more importantly, the well-being of people. As we navigate this global health crisis, we need to protect populations all over the world and make sure that our collective recovery is synonymous with positive change (Usman et al., 2021)

Therefore, we all should support the global call to action to prioritize WASH in all healthcare facilities (WHO. *Health challenges for the next decade*. 2020) (Howard et al., 2020). So the question is How? and the answer is by investing in WASH ("water, sanitation and hygiene") infrastructure, and providing technical assistance in behaviour change approaches towards the adoption of key behaviours that protect both care providers and patients (Quinete and Rauser-Davis, 2021; Usman et al., 2021). Access to clean water impacts education is a need of the day, without safe water, kids miss a lot of schools (584 million children lacked a basic drinking water service at their school). Another recent report of United Nations Children's Fund (UNICEF) and World Health Organization (WHO) focused on the progress on drinking water, sanitation and (Howard et al., 2020) cases in New York, 2020 (Schuchat and Team, 2020). Considering these and other facts about water quality and effects of pollution the properties of water can be summarized as in Figure 1.1 as properties of water. Some of the unique properties which water possesses are described as below.

1.2 The Unique Properties of Water Available Globally

1.2.1 Biological Properties and Clinical Benefits of Water

It is an acknowledged fact that provisions for living beings without water are not possible. While water is essential for transporting useful nutrients throughout the human body (Cooke and Kuntz, 1974), it also constitutes a major portion of plant and animal bodies (Tortajada, 2014; Ball, 2017). It also acts as a temperature stabilization product for the human body and clears the toxins from the body in form of human excreta, i.e. Urine, Tears, etc (Abu-Zeid, 1998; Palmer, 2012). Aquatic life is completely dependent on the presence of water in sufficient quantities. It is observable that in the winter season cycle there are numerous layers of ice that float on water. To imagine, these layers of ice act as a thermal insulator and preserves the aquatic life thriving below the sea level. Health experts recommend a bare minimum of nine glasses of water per day to maintain good health (Demir et al., 2021). It is imperative to keep ourselves hydrated because if we do not follow this, we may face Kidney problems towards the later part of life. There are numerous applications of water in the field of medical science. It balances out the fluctuations that happen in the body, helps as a stress buster, refreshes our body and useful in providing therapy (Rani et al., 2018; Montgomery, 2002) to cure one of the worst diseases on the planet.

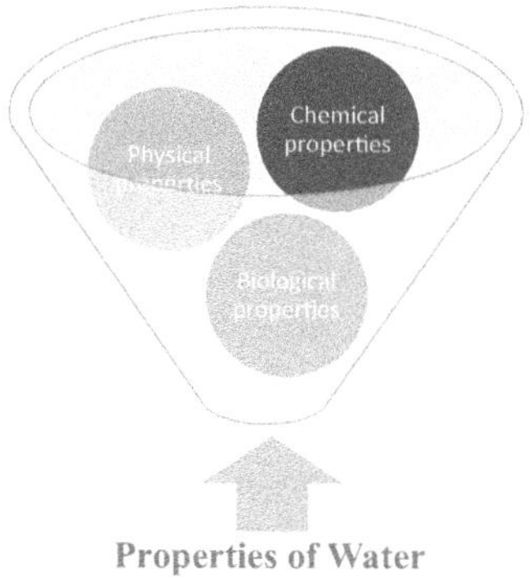

FIGURE 1.1 Illustration of properties of water.

1.2.2 Physical Properties of Water

Water is the only compound that is capable of residing in trio state, namely gas, liquid and solid. The common triplet point of water is described as the point at which all phases, i.e., Liquid, Gas & solid phase co-exist. (T=0.007562°C and P=4.583 mm). Freezing point of water stands at 0° whereas the boiling point stands at 100°C at one atmospheric pressure. These sections describe the physical properties of water in detail (Russell, 2007). The polar nature of water is an interesting area of study, whereas the water molecule which contains oxygen and hydrogen atoms is structured in a way that the former's electro-negativity is higher than that of the latter atom (van Oss and Giese, 2011). Due to this electro-negativity difference, the former occupies a partially negative charge and the latter occupies a partially positive charge. Hence it is easily deducible that water is a polar molecule. The fact is also strengthened by the observed bond moment value which is 1.83 Debye and its dielectric constant value (77.94). The shape of water is V in terms of its structure and has a bond angle of approximately 104°4′. Another significant physical property of water is its hydrogen bonding raised due to electrostatic attraction force (Ludwig, 2001). Here a quick look on the physical properties of water is given to readers as summarized in Table 1.1. Water is known for its high number (value) of melting point, boiling point, high enthalpy of fusion and a high enthalpy of vaporization due to this intermolecular hydrogen bond. The main reason behind water being present in liquid form is the presence of this intermolecular hydrogen bonding. The solid form of water (ice) also comprises the same Hydrogen bonding that is also responsible for the liquid form of water. Solid form of water (ice) also has hydrogen bonding (García-Álvarez et al., 2018). At 98°C, maximum density of water stands at 1.000 g/cm^3. Beyond or below this value of temperature, the water density begins to decrease. While it is evident from the fact that solid state water (ice) floats on the Liquid state water because the solidity/density of ice is lesser when compared to that of water density of water starts decreasing (Niinomi et al., 2020)

TABLE 1.1

A Glance on Physical Properties of Water

S. No	Property	Unit of Measurement	Value
1	Melting-point	K	273.15
2	Boiling-point	K	373.15
3	Density (298 K)	g/cm^3	1
4	Enthalpy of formation	kJ/mol	–285.91
5	Enthalpy of fusion	kJ/mol	6.02
6	Enthalpy of vapourisation	kJ/mol	40.67
7	Molar mass	g/mol	18.049
8	Viscosity	cP	0.9
9	Dielectric constant	C^2/Nm^2	78.4
10	Electric conductivity (293K)	ohm/cm	5.7×10^{-8}

Density of solid water (ice) is less than water. To investigate further, Ice exhibits an open cage-like structure where the water molecules are seen to be packed loosely, and hence the density is on a lower side. At an atmospheric pressure ~1, the measured electrical conductivity of water is 10^{-8}/ohm cm. When water is subjected to extremely high pressure, its dissociation is increased by a factor 10^{12} and it turns out into an excellent conductor of electricity. This is observed by a different group of researchers that the electrical property of water depends largely on the temperature and show direct effects on different life forms (Jeong et al., 2013; Yilmaz and Hao, 2011). The existence of water in its associated form is evident from the fact that it is a polar molecule while the value of its dielectric constant is pretty high at around 77.92 which is a clear indicator of water being a good solvent and for that matter, it is also well-known universal solvent as in liquid form (Zhang and Hoogenboom, 2015). Some of the available ionic and molecular compounds are readily dissolvable in water. Henceforth it is imperative to make sure that due to the dipole interaction of ions the dissolution of ionic compounds takes place within water and because of the genesis of the H-H bond (Shang and Liu, 2011) the dissolution of molecular compounds takes place.

1.2.3 The Chemical Properties of Water

Due to its unique chemical properties, water finds its application in many places, such as Ionic Dissociation and Ionic Product are formed due to the reaction of water disassociation with the reactant. The concentration of hydrogen ion is 10^{-7} mol/L & the same for hydroxide ion is 10^{-7} mol/L in water at a given temperature of 298 K. The product of the concentration of hydrogen ion and hydroxide ion is known as the ionic consequence of H_2O (Kw). The Ionic product of $H_2O$is $1.08 \times 10^{-14} mol^2/L^2$ at 298 K. Another significant property of water is its amphoteric nature (Biswas, 2018). The pH of distilled water which is theoretically determined is 7.0. Its neutrality towards litmus paper is clearly evident. Water is amphoteric in nature as per the concept created by Bronsted-Lowry concept of acid bases Water can give proton (acid) or accept proton (base). The redox nature of water and its associated reactions have been reported widely. In many cases when water reacts with active metals like sodium, lithium, or potassium, it turns out to be a weak oxidizing agent. While in reaction with these metals, hydrogen gas is also released (Al-Anazi et al., 2020; Roy et al., 2016). When water reacts with metals such as Chloride, Fluorine, etc., oxygen gas is released. Salts like halides, metallic hydride, metallic nitride, phosphides and carbides can easily be decomposed by water. On dissolution of salt in water, there are two possibilities of producing either an aqueous solution which is basic or an aqueous solution which is acidic in nature (Orwat et al., 2017). To further deep dive, if the cation of the salt is from a family of strong base, the resultant solution will primarily be basic in nature however if the anion of the salt belongs to a family of strong acids, then the resultant solution is bound to be acidic. Hydrates can be formed by water when it reacts with a metal salt and has primarily three types of hydrates (Gao et al., 2019). Firstly coordination bonding water: coordinate bond with metal ion e.g., $[Cr(H_2O)_6]^{+3}\ 3Cl^-$, secondly hydrogen bonding water: hydrogen bond with certain anion containing oxygen atom for example $CuSO_4.5H_2O$. This is a unique case where a coordinate bond is formed by four water molecules and one more water molecule forms a hydrogen bond. $\left[Cu(H_2O)_4\right] + 2SO_4^{-2}.H_2O$, thirdly, interstitial water: Interstitial site in the crystal lattice is captured by water and a cage-like compound i.e. $BaCl_2.2H_2O$ is formed. A single part of Deuterium oxide is contained in 6,000 parts of normal water. This simple yet impactful observation was established by a well-known American chemist named Urey and such a water sample is heavy water, the formula for which is D_2O. This heavy water can be used to control a nuclear fission reaction and be (Ueda et al., 2015). The purpose of a moderator in a nuclear fission reaction is to tone down the speed of quickly moving neutrons.

1.3 World's Water Resources and Their Distribution

Liquid water exists in large quantities on earth compared to other planets in the universe and has a large proportion of earth is covered by water. An approximate 380 bn gallons of water is present on the earth's surface. A perfectly plane surface and an ocean of around 4 km deep could cover the entire earth (Karato, 2011). The hydrological cycle of water means that it gets continuously recycled and is never in a state of rest. Gradient of energy is the primary driver in a hydrological cycle which governs the movements that happen in water. Solar Electromagnetic radiation is the paramount origin of energy for the evaporation of water (Tian and Ida, 2015). Earth absorbs electromagnetic radiation and increases its temperature at the surface in order to provide energy source and enable evaporation. Although three-quarters of the earth's surface is covered by water alone, every water source is not fit for drinkable or domestic purposes as there is a high amount of salt in this water (Bar-On et al., 2018). More than 95% of water is found in the oceans and it is very costly to cleanse this water in order to prepare its fitment for domestic or personal usage. The remaining 5% of water sources are constituted by surface, ground and rainwater sources. Each of these sources is described in detail here and can be summarized as in Figure 1.2 as major sources of water. This water source is an indirect factor in determining the economic strength of any country. One of the primary sources of getting water is rain where surface water evaporation contributes to its production. In its crudest form rainwater is the purest water source however when it mixes with some of the harmful atmospheric gases like NO_2, SO_2, and CO_2, its level of impurity increases to a great extent (Jeong et al., 2013). Water comes from a prominent source, i.e., Surface water which is again a prominent source. We get this water from water bodies like lakes, ponds, and rivers; however, lot of pollutants get mixed with this and make it not fit for consumption by the living beings is derived from Surface water is another source of water. It is present in form of river, ponds, and lakes. It is important to treat and purify this water before providing it to the community at large. Water from the ground source is critical because the surface pollutants are not able to contaminate it however if present at a lower depth major pollution from industrial activity and human use contaminates it. Numerous ecological, geological, physiological, and climatic conditions are responsible for the availability of this water source while if its distribution is to be considered around the entire earth, it is very varyingly distributed where some parts have a heavier distribution and the other parts have a lighter distribution. The storage is in form of compartments that are placed globally. Many scientists, researchers, and geologists have revealed that global compartments where water is stored is in terms of oceans, Polar Icebergs, ponds, rivers, soil moisture, lakes, atmospheric water vapours, soil moistures, etc. where oceans and polar icebergs emerged to be one of the major containers of water. The distribution of water is summarized in Figures 1.2 and 1.3 while the percentage of total water in different compartments is summarized in Table 1.2.

Ocean contains 96% of all water in the world, but it is usually extra salty for consumption by human beings. Refer to the Figure 1.2. Ocean covers three-quarters of the earth surface. It plays an important role in regulating the earth temperature. Around 2% of water is contained in Polar ice and glaciers from

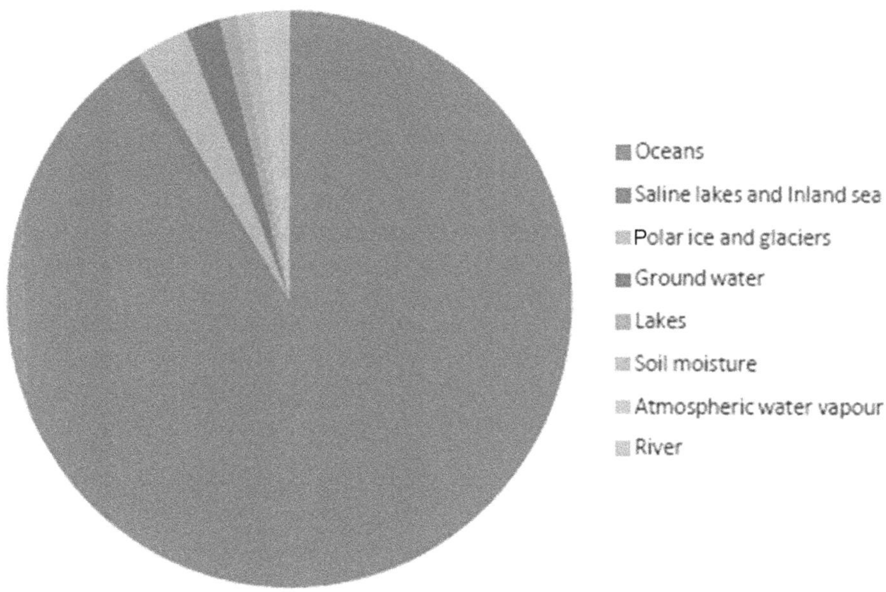

FIGURE 1.2 The distribution of water on the earth.

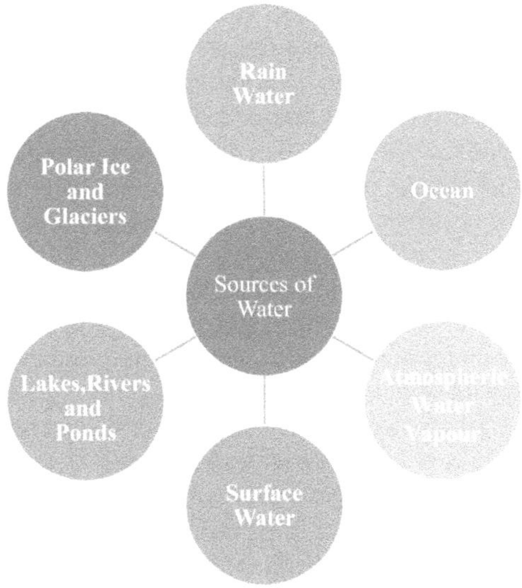

FIGURE 1.3 Sources distribution of water on the earth.

TABLE 1.2

Break-up of Water Availability by Different Resources

S. No.	Source	% Distribution
1	Oceans	96.3
2	Saline lakes and inland sea	0.009
3	Polar ice and glaciers	2.54
4	Ground water	0.62
5	Lakes	0.008
6	Soil moisture	0.006
7	Atmospheric water vapour	0.005
8	Rivers	0.0011

the overall water content in the world and is mostly found in places that have a high altitude and latitudes containing 2.04% of all the water in world. Around 86% of the world's glaciers are contained in Antarctica which is mostly covered by ice. Around 0.62% of water is contained by the groundwater sources from all the water sources available on earth. After oceans, it is the largest water compartment and this water source is mainly used by most of the population for drinkable and other domestic purposes. Around 0.0011% of all water in the world is contained in the rivers. Waters which flow from rains are pulled by gravity inwards towards the sea and the rivulets meet together towards the formation of streams which eventually forms the rivers. About 0.006% of all world's water is found in Soil in form of moisture and it is extremely important for the growth of plants. Approximately, 0.008% of all the water in the world is found in Lakes and ponds almost ten times of what is found in Rivers. These small water bodies form a crucial water source for nearby population whose livelihood depends largely on this source only.

It contains about 0.005% of all water in world is contained in form of vapours, i.e. Atmospheric vapours. It is a unique form in which water is found.

1.4 Water Pollution

About 72% of the surface water available, only 3.5% is in freshwater, and this is found in ice, ice caps forming 75% of the allocation followed by 25% fresh water in stable bodies such as lakes, lakes, rivers, and springs. There are countless activities for people who deploy H_2O for domestic or household purposes including those of consumption, dish-washing, cloth cleansing, maintaining personal hygiene, or other known purposes. Clean water is usually suitable for human consumption but the full pollution thereby has contributed to the pollution of lakes, rivers, lakes, springs, surface, and soil, and this condition is called water pollution. The word pollution is acquired from within the Latin word pollution, meaning pollution. Water is known to be polluted if it is not of high quality and suitable for current or future use. The implication of water pollution is the presence of any toxic component of water that lowers the water level to create a hazard or technically interferes with its usefulness for any human purpose. Although water pollution is a problem of the past, in modern times, difficulties such as high levels of sewage, industrial waste, high levels of radioactive waste have polluted groundwater so much that about 71% of rivers and streams not only in the Indian subcontinent but other countries also contain contaminated water. Many rivers in urban and developing economies are countries with high levels of emissions and, if not treated properly will pollute and deeply pollute our surface and groundwater resources. On the other hand, industrial growth is also the backbone of the growth and development of any country. Industries such as the textile industry, pulp-paper, pharmaceutical giants, fertilizer processing, tannery, electroplating production, alcohol production, detergents, textile production, pesticide oil, and leather industries came and used large amounts of water in various processes (Kumar et al., 2020a; 2021d; Agrawal et al., 2021; Kumar and Chandra, 2020a).

These large industries discharge wastewater into a nearby river, body of water, and open space. This polluted water mixes with river water and pollutes the entire water system (Chandra and Kumar, 2017a; Kumar et al., 2021a, b, c). Many mature industries have greenhouse gas (ETP) plants but these do not perform the physical and chemical treatment process satisfactorily for economic and ROI reasons. Such water treatment technology is very expensive in the market and especially SMEs cannot afford the technology of this level. Industries are in short supply of high-quality engineers and chemists who can brainwash and come up with new technological solutions and have a pool of efficient engineers and pharmacists that do not show good work to treat polluted water industries. Industries do not hire highly skilled talent from abroad and nor do they invest in and appoint low-cost, low-cost chemists that do not perform satisfactory medical procedures. So much of the groundwater is polluted by the wastewater industry. Industries are economically viable but have a detrimental effect on the environment through water pollution (Kumar and Shah, 2021). Water pollution alters the quality of surface water and groundwater to the point where its suitability or human consumption or support of natural human life processes decreases or ceases. Radioactive plants are also key to the nation's growth and development. In India, most of the electricity is also generated with the help of water. The hydropower project is also responsible for water pollution. These projects have used a large amount of water to generate electricity and have subsequently diverted wastewater into the river without adequate treatment. Today groundwater is polluted with various forms and is unsafe for drinking, watering, cooking, and so on. At present human activities also pollute groundwater at risk Groundwater pollution cannot be reversed, if polluted it becomes difficult to restore the original quality of water for drinking purposes. India is a country of villages and a large number of villages depend on river water, lake water, tap water by hand, and well water. This important resource needs to be protected, preserved and used wisely by man but is not visible. Our rivers, lakes, groundwater, and other water sources are polluted by various activities. Water is a basic human need and it is therefore important to maintain water quality.

1.5 The Global Water Pollution

The World Health Organization has recommended that 2,000 cum of fresh water be it is necessary for each person per year for a healthy lifestyle but the top countries of the world have fallen below this level. The Environmental Protection Agency (UNITED STATES) in 1996 announced that about 16.6% of top water in the US and its territories is contaminated with toxic chemicals and other pollutants. About 20% of the world's population has been plagued by alcoholism water. It is noted that in Poland (a country in middle of European Union), more than 95% of available water is potable. In Malaysia, 42 major rivers are reported as natural disasters. In the Philippines, 60%–70% of Manila's total volume. The passing river is polluted by domestic sewage. The maximum number of locations they used the river water not only for the maintenance of personal- hygiene bathing but additionally for drinkable and purposes of food preparation too. In China, 57 of the 78 Rivers are reported to be critically polluted and 41 of the 44 cities use polluted water goods. The EPA (US) estimates that 4.5 trillion (1.2 trillion) of polluted water is available per day land in the United States from municipalities, industrial landfills, septic cesspools tanks, landfills, and agricultural and forestry areas. In Midwest's Corn Belt (farmland), a thousand drinking water sources are closed down in the state due to excessive levels of toxic ethylene dibromide. UAE, as a country has about 2.5 million L of storage underground and the EPA estimates that 42 million petrol a year are leaked into underground fuel tanks. In Europe, nitrate levels in groundwater have been reported to be hazardous in many areas. According to UNWWAP (2003), every day 2 million tons in the wild too agricultural waste are discharged into groundwater equivalent to the total weight of 6.8 billion people. Water emanating from industry sources is another major contributor to contribute in pollution of water sources. According to a 2008 UNICEF and World Health Organization report, 2.45 bn world populations do not have proper access to improvised sanitation and more than 70% of these people live in Asia. According to a 2009 UN water report, 70% of industrial waste in developing countries is dumped into the water without treatment polluting the existing water supply.

1.6 Water Pollution in India

Water pollution is a major and severe problem in India. Most of India's groundwater is polluted by various types and to the level of chemical, physical and physical impurities. About 70% of surface water resources and groundwater resources are polluted by natural, inorganic, and biological pollutants. According to India's 2011 census, the population of India is 1,210,193,422 (Yilmaz and Hao, 2011). It has 29 countries, most of which suffer from water pollution. The river water in India is polluted and the piped water of major cities is also polluted. The Hindustan Times is a well-known Indian newspaper that reported that if all the pipelines in the wild the people of Bombay release once a year are allowed to accumulate then we will soon have another Mount Everest. The National Centre for Oceanography has reported that Goa and Bombay emit more than 2,000 m^3 of seawater a year long enough to fill a 500-m-long, 500-m-wide, 29,000-m-high building. Mercury pollution is reported in two areas of the Thana River in Bombay and the Rushikulya River in Orissa. The waters of many southern rivers were clear a few years ago but now the day of industrial pollution is changing the color of most of the river water. The Damodar in ban gal is black, the Chaliyar River in Kerala is brown and the Ganga River is in Kanpur and most parts of Uttar Pradesh are black or brown. About 500 large and medium-sized industries have polluted India but their contribution to total pollution is less than 10%. All other load pollution is due to the home region. According to the CBPWP, only 13% of the total waste disposal in Bombay factories and 89% of domestic origin waste in Calcutta. The Central Control Pollution Board (CPCB) did a study in the year 1995 and found several areas in 18 major rivers in India. The bulk of this expansion is found in large urban areas and surrounding areas. Geogenic, salt, iron, fluoride, and arsenic pollutants have affected groundwater in more than 200 regions spread across 19 regions.

1.7 Major Water Pollutants

In water about 2,000 pollutants are present. Type of water pollutants, source of water pollutants, and their effect are often correlated. Water pollution sources are of two types: point and non-point pollution sources. Point sources discharge the pollutant from specific locations. Drains pipes, sewerage out falls, coal mines, textile industries, powerpoint, oil wells, and factories are categorized as the origination points (Chandra and Kumar, 2017b; Kumar et al., 2021a, b). Non-point sources emanate the pollutant out of non-categorized locations. Construction sites, lawn gardens, logging areas, and parking lots are characterized as non-specific origins. Specific origination sources of pollutants are predictable and Non-specific sources of pollutants are not predictable. According to the Indian Environment Protection Act 1986 every solid-material, liquid-material or gaseous materials that is available in a

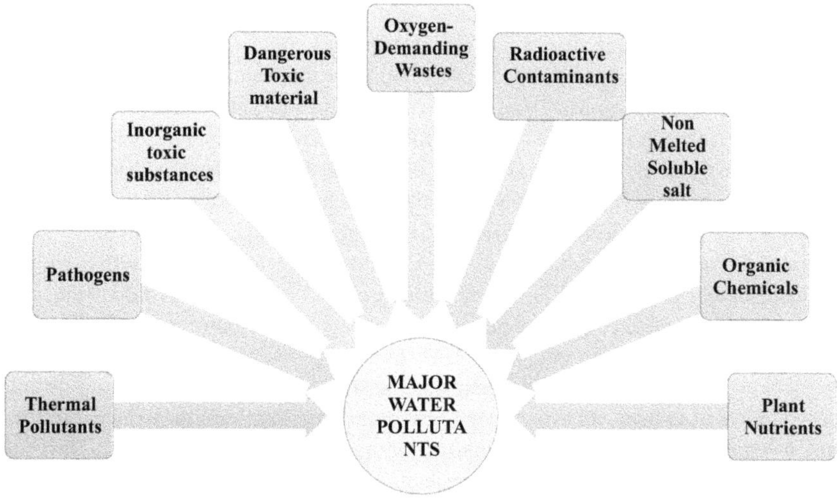

FIGURE 1.4 Major pollutants of water on the earth.

specific quantity that tends to harm environment in any possible manner or tends to be harmful for living beings is termed to be as a pollutant. Something which causes a deviation from the normal composition of an environment is called contaminant. Pollutant is contaminant only if contaminant can adversely affect something that operates values and is present in high concentration (Kumar et al., 2020b, 2021a). Thus, any substance that disallows or prevents the normal use of water is considered as a water pollutant (Kumar and Chandra, 2020b). Water has never been found naturally in a completely pure form. Even are the most unpolluted geographical areas and rainwater may also carry gases, dust, particles, and other particles in suspension form. The water pollutants are unwanted byproducts of our activities that have served intended purposes and tend to impair our economy and life. According to UNCED, 80% of the total disease and 35% of deaths are attributed to water that gets contaminated for consumption of the living beings. There are many pollutants found in water some of them are represented in Figure 1.4 and described here. Bacteria are the main pollutants in human health. Typhoid polio enteritis, cholera, viral diarrhoea, and amoebic, hepatitis is many water-borne diseases. The main source of infection is improperly treated through untreated human waste, food processing factories can be adequately treated plant and animal waste. The organization of World Health (WHO) predicts that in less developed nations, less than fourth/fifth of diseases and illnesses can be caused by water-borne diseases. The mixture of oxygen in water is a true indicator of the objectives of water quality. Water with more than 6 ppm of oxygen is capable of supporting marine life whereas water that has less than 2 ppm of oxygen supports micro-organisms, fungi, bacteria, worms as well as decomposer material. Addition of certain organic matter such as paper pulp, sewage, and food processing waste to irrigate pleasant oxygen consumption by decomposers. The presence of these substances in water has also affected the need for natural oxygen demand for chemical oxygen supply. Water pools are low in organic matter and high in oxygen called oligotrophic (low nutrient). Water has a high biological value, the richest in the body is called eutrophication. High biological production is often reflected in algae blooms. High phosphorus or Nitrogen levels have encouraged the growth of aquatic plants. A large amount of organic also increase the number of viruses

Unusual toxic substances such as lead (Pb), Tin (Sb) and cadmium (Cd) presented in the water due to man-made activity has become one of the top drivers of water pollution sources. Inorganic substances such as acid, base, salt, nitrate, and chlorine discharged in the water due to man-led actions is not toxic at low concentrations but have given negative effects on human societies in high concentration. Many rare and heavy metals including cadmium, nickel, lead including mercury are very highly toxic. Mercury suffers from Minamata disease. A famous case of mercury poisoning was reported in Japan in 1950. Mercury is used to hold gold as well as part of the gold in the field. It is reported that 130 tons of mercury per year is placed in the Amazon River. Lead pipes are used in most drinking water the supply system is a critical source of contaminated drinking water. Mining entry waste and mine extraction are critical sources of heavy metal pollution in drinking water. Desert soil contains toxic non-metallic salts such as selenium and arsenic, i.e., they meet in the water for irrigation and water supply. Serious case of selenium Poising was reported in Kesterson Marsh of California when thousands of migratory birds were killed in 1980. The salt of sodium chloride and calcium chloride popularly known as NaCl and $CaCl_2$, respectively, is used to break the road ice in the Northern provinces. Road salt entry (NaCl, $CaCl_2$) can have a negative impact on the aquatic ecosystem. Many acids base and salts are released as a result of various industries processes such as smelting metal and the application of fuel to the skin and coal mines. They are important sources of water pollution. Coil and Oil heat also leads to the formation of atmospheric nitric acid and sulfuric acid. This acid of nitrate which is popularly known as HNO_3 and sulphuric acid which is denoted by the symbol H_2SO_4 are added to surface water through the long-distance travel process. Many chemical industries used thousands of different organisms and synthetics organic chemicals to prepare pesticides, pharmaceuticals pigment and plastics namely it pollutes our groundwater and surface water. Dioxins or a dozen very dirty persistent natural contamination of the environment. Dioxin exposure even to certain components each quadrillion can cause genetic disorders, birth defects, and cancer. DDT and more chlorinated hydrocarbon have been shown to accumulate to dangerous levels itself the ecosystem to create health problems for people. Other organic sources of toxins chemicals are industrial waste, household waste, and other areas where they are used in abundance. Soil is open sand, clay, and other soil particles that settle in the middle under water. As per the agency that is responsible for the protection of environment, the sediment is the most common pollutant of rivers and lakes. Pollution waste has two sizes, one is physical and the other is chemical. High levels of sewage in penetration of the water limit into the body of the water body and prevent the growth of algae and standing aquatic plants. Soil has also been identified as a major cause of degradation and destruction of coral reefs. Lowering and raising the water temperature from the normal level can be bad affects aquatic life and water quality. Lowering the temperature of the tropical oceans by even 1° can kill coral reefs and corals. Similarly increasing the temperature of water can have a detrimental effect on aquatic life. The rate of oxygen dissipation decreases water at the rise in water temperature so species need a high level of oxygen. Electrical plants, petroleum refineries, paper mills, food processing industries, and molten metals change the temperature of the river and lakes by evacuation warm or cool water. The use of radioactive isotopes in the medical industry is a research and application of radioactive material for nuclear power plants and the use of weapons is the cause radiation. Nuclear power plants produce many types of pollution such as low levels of radioactive waste, product production, and heat fluid and gas debris from fuel fuels. Extractions of radioactive waste from the water bodies create a problem that is harmful to the human body and to aquatic life. All nuclear weapons testing and deployment results in a significant amount of strontium collapse. The Sr_{90} can it is introduced into the ground with dirty water and once inside the human body it lives in the bones and teeth and causes cancer. Oil is also a toxic substance in water and acts as a solvent polluting and affecting aquatic organisms. It has been reported in many cases that oil and diesel act as pollutants in seawater. On September 25, 1974, the American oil tanker Trans Heron crashed into a coral reef in India and oil formed a dense layer on the ocean floor near the Lakshadweep Islands. Seabirds began to die in the area by the thousands and a massive loss of tuna fish. Oil present in water bodies produces various toxin that kills and disrupts the body or directly dysfunction basic activities of aquifers. Oil or a badge tied with tar can serve as an aesthetic.

1.8 An Understanding of Wastewater Treatment Methods

Wastewater treatment processes are categorized as source treatment, pretreatment, primary treatment, secondary treatment, and tertiary or advanced wastewater treatment. Source treatment is used to remove toxics and/or other undesirable contaminants to prevent intermingling with other waste streams (Shingare et al., 2019; Kundu et al., 2014; Gautam et al., 2007). This approach offers opportunities for the reuse of these constituents such as metals, etc. Conventional treatment includes pre- and primary treatment followed by secondary-treatment processes. When necessary, tertiary treatment processes are included in the treatment sequence to remove specific constituents to very low residue levels. Pretreatment is employed to render the raw wastewater compatible and/or amenable for subsequent treatment processes. Consideration is given to those constituents that pass through, interfere with, or accumulate in the sludge or are otherwise incompatible with following treatment processes. Equalization, spill retention, neutralization for pH adjustment, nutrient addition, toxics or inhibitory substance removal, oil and grease removal, and solids removal by flotation, sedimentation, or filtration are typical pretreatment processes. Primary treatment is a subset of pretreatment methods and involves physical separation by screening, grit removal, and sedimentation. Depending on the number of organics contained in the solid material, primary treatment may remove a significant portion of the oxygen-demanding substances (BOD). A well-designed and operated primary plant may remove as much as 35%–40% of the BOD and as much as 60%–65% of the settleable solids for municipal wastewaters. The secondary treatment adds a biological process after primary treatment, which is commonly either activated sludge or trickling filtration for municipal wastewaters. Typically activated sludge or a modification of suspended growth treatment systems is used for industrial wastewater to achieve a high-quality effluent. These biochemical processes are typically aerobic and are the same as previously described as occurring in a river where organics are oxidized to carbon dioxide and water. A well-operated and designed secondary treatment plant can be expected to remove 85%–95% of both BOD and suspended solids. Under existing regulations in the United States, all discharges must be subjected to at least secondary treatment. Conventional treatment is described by Best Conventional Treatment Technology (BCT) and is designed to remove conventional pollutants such as BOD, TSS, etc. common to municipal wastewaters. However, BCT does not effectively remove many constituents of present-day concern, especially those of industrial wastewater origin. These include many VOCs, toxics, non-biodegradable organics, persistent organic pollutants (POPs), nutrients and emerging contaminants such as EDCs and PPCPs. Hence, additional treatment technology is required. Tertiary treatment may be defined as treatment in addition to primary and secondary processes. It may include precipitation, filtration, coagulation and flocculation, air stripping, ion exchange, adsorption, membrane processes, nitrification, and/or denitrification, and other processes. Those processes may be integrated into the secondary treatment plant or added onto the secondary effluent. Tertiary or advanced wastewater treatment can attain virtually any removal efficiency desired. However, as previously noted, as the percentage of contaminant removal increases, so does the cost of attaining it. These technologies are often referred to as Best Available Technology Economically Achievable (BTEA) and are applied to meet stream standards or to comply with TMDL requirements. Tertiary treatment is employed to remove toxics, persistent organics, nonconventional pollutants, nutrients, etc., and is typically considered as BATEA. Tertiary treatment systems following secondary treatment, however, may not be efficient at industrial facilities because of the need for treatment of a large volume of flow with low contaminant concentrations. Many tertiary processes also may not be pollutant-specific. Generally, the most cost-effective approach is to address the problem at the source where flows are low and specific pollutants are present at high concentrations. Applicable technologies for source treatment will depend on the constituents in the process of wastewater targeted for removal. For example, if VOCs and ammonia are to be removed, then air or steam stripping should be evaluated. If heavy metals are of concern, then oxidation/reduction, precipitation, filtration, ion exchange, and membrane processes may be investigated. Organic chemicals may require chemical oxidation, wet air oxidation, anaerobic treatment, granular activated carbon (GAC), polymeric resins, or reverse osmosis for effective removal. Special mention should be made of the solid material removed in any of these processes since residual disposal is a major problem in wastewater treatment. Sludges are particularly troublesome because of their high water content, concomitant large volumes, and concentration of heavy metals, viruses, protozoa, and other constituents capable of causing environmental and public health harm. Costs of treatment may be one-half or more of that of the aqueous waste stream. Many of the processes mentioned in the preceding paragraphs produce solids. Solids from primary and secondary clarifiers must be effectively managed. Residual solids usually undergo a series of treatment steps involving thickening, dewatering, and final disposition or reuse. Organic sludge may also require stabilization prior to final disposal. Stabilization may be achieved by digestion [aerobic or anaerobic (no oxygen present)], lime stabilization, or by other means. Sludge thickening is typically accomplished by gravity, flotation, or centrifugation methods. Centrifuges, belt filters, and filter presses are dewatering options. The selection of unit processes and their sequence will depend primarily on the characteristics and volume of the sludge and on the final disposition or reuse option selected. Ultimate disposal choices include incineration, landfill, land disposal (lagooning or application to land for reuse) or other reuse alternatives. Since this material is a resource, it is important that opportunities for reuse be fully considered. It is the ultimate disposal method selected that may significantly influence the unit operations selected for treatment.

1.9 Waste Water Pollutants and Their Treatment Approaches with Nanotechnology and Nanomaterials

As discussed earlier there are different sources and their remedial measure are available in nature. Nanomaterials are the basis of science related to nanotechnology and its related materials. This branch of science is one of those key areas where relentless efforts are being made in terms of R&D and numerous costs are invested in this topic. The related advancements in this area are growing at a rapid pace because this is one of those rarest areas which have the ability to modify the manner in which materials are constructed additionally with the prime quality of products that are produced (Amith et al., 2021; Singh et al., 2021). While this also brings in additional costs associated which undermines the fact that it will have a major commercial consequence on R&D efforts in the near future. There was an event organized by the American Physical Society meeting where Richard Feynman who is a Nobel Laureate in the subject of Physics spoke confidently on this subject thereby undermining the fact that the concept is not new amongst the gentry. This session took place at California Institute of technology in 1959. The word nano comes from a Greek word meaning little. The word is typically deployed in areas where it acts as an initial point/step for units like metres, seconds, etc. and it translates to a billion units. It is used as the starting point for any unit such as a second or meter, and means a billion units. Hence if a meter is divided by 1 billion, it will be nano-meter as a scale. Or in mathematical terms, it is 10^{-9}. The scale that measures a nanometer is typically indicated as between one and 100 nm however the objects which are a part of nano-science and nano-tech are usually >100 nm. Nanoscale objects are those where at a bare minimum one size has a scale of <100 nm. A typical size comparison of the Nano-meter can be done directly with a human hair strand, where nanometer can be translated as one million mm or 1/300,000 the size of hair. The nanometer is one million millimeters; about 100,000 three times the size of a human hair. Nanomaterials overall are interesting branch to conduct further studies and research because at this scale there are different material and other objects appear. These emerging structures are dynamic with major impacts on all aspects of our lives. If we were to compare a nanometer it can be said that ten hydrogen atoms are placed in a single line or for instance five silicon atoms are put in a straight line.

1.10 Role of Nanoparticles in Water Decontamination

Nanoparticles in recent years have gained increased attention due to their capabilities of removing heavy metal ions from wastewater and decontaminating the drinkable water sources. Several other technological advancements have been made in this area, and nanomaterial technology is one of those areas. Iron-oxide-based nanomaterials are capable of rapidly removing heavy metal ions from water. Our study will include a theoretical review of novel nanomaterial, carbon-based nanomaterial, and zero-valent nanomaterial and their various applications including the removal of heavy metal ions from water. There have also been alternate technologies such as ion exchange, adsorption, membrane filtration, electrochemical treatment, and chemical precipitation which solve the similar purpose of taking out heavy metals from water. Carbon-based nanomaterials have a large surface area and have an ability to remove both organic and inorganic pollutants from water. Some more examples of nanomaterial are summarized as DESs-CNTs Adsorbent and Hg (II) adsorbate; highest capacity of adsorption (mg/g) is 186.97. Amino-functionalized Fe_3O_4/MWCNTS adsorbent and Cu (II) adsorbate; Highest capacity of adsorption (mg/g) is 30.94. Nano-haematite adsorbent and Zinc (II) Lead (II), Cadmium (II), and Copper (II) adsorbate.

1.11 Conclusion and Future Prospects

Everyone knows the cause of water-borne disease, but what matters the most is how this disease affects the world at large. Here are some interesting facts that one needs to know about water-borne diseases. A recent study by the World Health Organization (WHO) states that 80% of most diseases that occur worldwide are water-borne diseases. 3.1% of the deaths worldwide are due to unhygienic water. On average, 1 lakh people every year in India due to water-borne diseases. One-third of the 600 district's groundwater in India is unfit for drinking because it contains high levels of fluoride, iron, and arsenic levels that are beyond tolerance levels. The effects of Water-Borne Diseases range from diarrhoea to meningitis, water-borne diseases affect the human body in numerous ways and can leave the individual with a life-threatening illness. Most of these water borne diseases are raised due to wastewater pollution. Here we have summarized and presented an overlook of the same and how to avoid such ailments. It is recommended that day by day new future technologies are knocking to mankind and we should adopt the newer technologies as per the need of the hour.

Acknowledgement

We would like to thank our Hon'ble Vice Chancellor (s), Prof. Raj Kumar Mittal, Chaudhary Bansi Lal University, Bhiwani, Haryana, India, and Prof. (Dr.) Raman Kumar Jha, Amity University Jharkhand, Ranchi, for giving us permission to carry out this study. We sincerely thank all the faculty and staff members of the Colleges as well as Universities especially of Govt. (PG) College for Women, Rohtak, Haryana; Department of Botany, Maharshi Dayanand University, Rohtak, Haryana; Amity Institute of Biotechnology, Amity University Jharkhand, Ranchi; Department of Botany, Chaudhary Bansi Lal University, Bhiwani, Haryana, India for their immense support and guidance in our work.

REFERENCES

Abu-Zeid, M.A., Water and sustainable development: the vision for world water, life and the environment. *Water Policy*, 1998. **1**(1): pp. 9–19.

Agrawal, N., V. Kumar, and S.K. Shahi, Biodegradation and detoxification of phenanthrene in in-vitro and in-vivo conditions by a newly isolated ligninolytic fungus *Coriolopsis byrsina* strain APC5 and characterization of their metabolites for environmental safety. *Environmental Science and Pollution Research*, 2021. https://doi.org/10.1007/s11356-021-15271-w

Al-Anazi, A., et al., Novel franklinite-like synthetic zinc-ferrite redox nanomaterial: synthesis, and evaluation for degradation of diclofenac in water. *Applied Catalysis B: Environmental*, 2020. **275**: p. 119098.

Amith, A., R. Sukanya, S. Subramanian, V. Kumar, and P. Ramamurthy, Chromium (VI) detection by microbial carbon dots: microwave synthesis and mechanistic study. *Journal of Basic Microbiology*, 2021. pp. 1–10. https://doi.org/10.1002/jobm.202100394

Ball, P., Water is an active matrix of life for cell and molecular biology. *Proceedings of the National Academy of Sciences*, 2017. **114**(51): pp. 13327–13335.

Bar-On, Y.M., R. Phillips, and R. Milo, The biomass distribution on Earth. *Proceedings of the National Academy of Sciences*, 2018. **115**(25): pp. 6506–6511.

Biswas, R., *Amphoteric Nature of Water is a Reflection of Change of Molecular Polarizability–A Theoretical Justification*. 2018. International Journal of Engineering, Science and Mathematics 7, 4(2).https://www.academia.edu/37472413/Amphoteric_Nature_of_Water_isa_reflection_of_change_of_molecular_polarizability_A_Theoretical_Justification

Chandra, R. and V. Kumar, Detection of *Bacillus* and *Stenotrophomonas* species growing in an organic acid and endocrine-disrupting chemicals rich environment of distillery spent wash and its phytotoxicity. *Environmental Monitoring and Assessment*, 2017a. **189**: p. 26. https://doi.org/10.1007/s10661-016-5746-9

Chandra, R. and V. Kumar, Detection of androgenic-mutagenic compounds and potential autochthonous bacterial communities during in-situ bioremediation of post methanated distillery sludge. *Frontiers in Microbiology*, 2017b. **8**: p. 87. https://doi.org/10.3389/fmicb.2017.00887

Cooke, R. and I. Kuntz, The properties of water in biological systems. *Annual Review of Biophysics and Bioengineering*, 1974. **3**(1): pp. 95–126.

Demir, F., M.Y. Ozen, and E.M. Derun, Determination of essential and non-essential element contents of drinking water and baby water for infant's nutrition. *Main Group Metal Chemistry*, 2021. **44**(1): pp. 194–202.

Dutta, D., S. Arya, and S. Kumar, Industrial wastewater treatment: current trends, bottlenecks, and best practices. *Chemosphere* 2021. **285**: p. 131245. https://doi.org/10.1016/j.chemosphere.2021.131245

Gao, J., et al., Effect of ionic liquid/inorganic salt/water pretreatment on the composition, structure and enzymatic hydrolysis of rice straw. *Bioresource Technology Reports*, 2019. **5**: pp. 355–358.

García-Álvarez, J., E. Hevia, and V. Capriati, The future of polar organometallic chemistry written in bio-based solvents and water. *Chemistry-A European Journal*, 2018. **24**(56): pp. 14854–14863.

Gautam, A.K., S. Kumar, and P.C. Sabumon, 2007. Preliminary study of physico-chemical treatment options for hospital wastewater. *Journal of Environmental Management,* **83**: pp. 298–306.

Gift, R.A., R.M. Olalekan, O.E. Owobi, R.M. Oluwakemi, B. Anu, and A.A. Funmilayo, Nigerians crying for availability of electricity and water: a key driver to life coping measures for deepening stay at home inclusion to slow covid-19 spread. *Open Access Journal of Science*, 2020. **4**(3): pp. 69–80.

Howard, G., et al., COVID-19: urgent actions, critical reflections and future relevance of 'WaSH': lessons for the current and future pandemics. *Journal of Water and Health*, 2020. **18**(5): pp. 613–630.

Jeong, C., et al., Analysis of electrical property changes of skin by oil-in-water emulsion components. *International Journal of Cosmetic Science*, 2013. **35**(4): pp. 402–410.

Karato, S.-I., Water distribution across the mantle transition zone and its implications for global material circulation. *Earth and Planetary Science Letters*, 2011. **301**(3–4): pp. 413–423.

Kumar, V. and R. Chandra, Bioremediation of melanoidins containing distillery waste for environmental safety, in Bharagava R. and G. Saxena, (Eds), *Bioremediation of Industrial Waste for Environmental Safety.* 2020a, Springer, Singapore. https://doi.org/10.1007/978-981-13-3426-9_20

Kumar, V. and R. Chandra, Metagenomics analysis of rhizospheric bacterial communities of *Saccharum arundinaceum* growing on organometallic sludge of sugarcane molasses-based distillery. *3 Biotech*, 2020b. **10**(7): p. 316. https://doi.org/10.1007/s13205-020-02310-5

Kumar, V. and M.P. Shah, 2021. Role of fungi and their enzymes in degradation and decolorization of distillery effluent for environmental safety, in Sharma, V.K., M.P. Shah, S. Kumar, and A. Kumar, (Eds.), *Fungi Bio-Prospects in Sustainable Agriculture, Environment and Nano-technology-Volume 2: Extremophilic Fungi and Myco-Mediated Environmental Management.* Academic Press. https://doi.org/10.1016/B978-0-12-821925-6.00013-7

Kumar, V., S.K. Shahi, L.F.R. Ferreira, M. Bilal, J.K. Biswas, and L. Bulgariu, Detection and characterization of refractory organic and inorganic pollutants discharged in biomethanated distillery effluent and their phytotoxicity, cytotoxicity, and genotoxicity assessment using *Phaseolus aureus* L. and *Allium cepa* L. *Environmental Research*, 2021a. **201**: p. 111551. https://doi.org/10.1016/j.envres.2021.111551

Kumar, V., K. Singh, and M.P. Shah, Advanced oxidation processes for complex wastewater treatment, in Shah, M.P., (Ed.), *Advance Oxidation Process for Industrial Effluent Treatment.* 2021b. Elsevier. https://doi.org/10.1016/B978-0-12-821011-6.00001-3

Kumar, V., K. Singh, M.P. Shah, A.K. Singh, A. Kumar, and Y. Kumar, Application of omics technologies for microbial community structure and function analysis in contaminated environment, in Shah, M.P., A. Sarkar, and S. Mandal, (Eds.), *Wastewater Treatment: Cutting Edge Molecular Tools, Techniques & Applied Aspects in Waste Water Treatment.* 2021c, Elsevier. https://doi.org/10.1016/B978-0-12-821925-6.00013-7

Kumar, V., S. Srivastava, and I.S. Thakur, Enhanced recovery of polyhydroxyalkanoates from secondary wastewater sludge of sewage treatment plant: analysis and process parameters optimization. *Bioresource Technology Reports*, 2021d. **15**: p. 100783.

Kumar, V., I.S. Thakur, and M.P. Shah, Bioremediation approaches for pulp and paper industry wastewater treatment: recent advances and challenges, in Shah, M.P. (Ed.), *Microbial Bioremediation & Biodegradation.* 2020a, Springer, Singapore. https://doi.org/10.1007/978-981-15-1812-6_1

Kumar, V., I.S. Thakur, A.K. Singh, and M.P. Shah. Application of metagenomics in remediation of contaminated sites and environmental restoration, in Shah M., S. Rodriguez-Couto, and S.S. Sengor (Eds.), *Emerging Technologies in Environmental Bioremediation.* 2020b, Elsevier. https://doi.org/10.1016/B978-0-12-819860-5.00008-0.

Kundu, P., A. Debsarkar, S.N. Mukherjee, and S. Kumar, Artificial neural network modelling in biological removal of organic carbon and nitrogen for the treatment of slaughterhouse wastewater in a batch reactor. *Environmental Technology*, 2014. **35**: pp. 1296–1306.

Ludwig, R., Water: from clusters to the bulk. *Angewandte Chemie International Edition*, 2001. **40**(10): pp. 1808–1827.

Miller, J., The battle with drought: water, the essential element (1940–49), in *La Niña and the Making of Climate Optimism.* 2019, Springer Nature Switzerland AG. pp. 115–142.

Mishra, R. and R. Rani, Healthy life style practices to combat COVID-19 pandemic–a mini review. *IP Indian Journal of Immunology and Respiratory Medicine*, 2020. **5**(3): pp. 137–40.

Montgomery, K.S., Nutrition column: an update on water needs during pregnancy and beyond. *The Journal of Perinatal Education*, 2002. **11**(3): pp. 40–42.

Niinomi, H., et al., High-density liquid water at a water–ice interface. *The Journal of Physical Chemistry Letters*, 2020. **11**(16): pp. 6779–6784.

Orwat, K., P. Bernard, and A. Migdał-Mikuli, Alternative conceptions of common salt hydrolysis among upper-secondary-school students. *Journal of Baltic Science Education*, 2017. **16**(1): pp. 64–76.

Palmer, N., Water adequacy as an essential element for empowering people. *ILSA Journal of International and Comparative Law*, 2012. **19**: p. 31.

Pati, R., *Water: An Essential Element for Life*, 2015, HeinOnline. https://www.vatican.va/roman_curia/pontifical_councils/justpeace/documents/rc_pc_justpeace_doc_20060322_mexico-water_en.html

Quinete, N. and R.A. Hauser-Davis, Drinking water pollutants may affect the immune system: concerns regarding COVID-19 health effects. *Environmental Science and Pollution Research*, 2021. **28**(1): pp. 1235–1246.

Rani, J., U.K. Sharma, and D. Sharma, Role of adequate water intake in purification of body. *Environment Conservation Journal*, 2018. **19**(1&2): pp. 183–186.

Roy, B., E. Zangrando, and P.S. Mukherjee, Self-assembly of a redox active water soluble Pd 6L 8 'molecular dice'. *Chemical Communications*, 2016. **52**(24): pp. 4489–4492.

Russell, C.A., Hydrotheology: towards a natural theology for water. *Science & Christian Belief*, 2007. **19**(2): pp. 161–184.

Schuchat, A., C. Covid, and R. Team, Public health response to the initiation and spread of pandemic COVID-19 in the United States, February 24–April 21, 2020. *Morbidity and Mortality Weekly Report*, 2020. **69**(18): p. 551.

Shang, C. and Z.-P. Liu, Origin and activity of gold nanoparticles as aerobic oxidation catalysts in aqueous solution. *Journal of the American Chemical Society*, 2011. **133**(25): pp. 9938–9947.

Shingare, R.P., P.R. Thawale, K. Raghunathan, A. Mishra and S. Kumar, Constructed wetland for wastewater reuse: role and efficiency in removing enteric pathogens. *Journal of Environmental Management*, 2019. **246**: pp. 444–461.

Siegel, F.R., Water: an essential, limited, renewable resource, in *Demands of Expanding Populations and Development Planning*, 2008, Springer Berlin, Heidelberg. pp. 83–112.

Singh, S., et al., Sustainable removal of Cr(VI) using graphene oxide-zinc oxide nanohybrid: adsorption kinetics, isotherms, and thermodynamics. *Environmental Research*, 2021. **203**: p. 111891. https://doi.org/10.1016/j.envres.2021.111891

Teixeira, F.J. and C.S. Gomes, Mineral water: essential to life, health, and wellness, in *Minerals Latu Sensu and Human Health.* 2021, Springer, Cham DOI https://doi.org/10.1007/978-3-030-65706-2_1. pp. 517–556.

Tian, F. and S. Ida, Water contents of Earth-mass planets around M dwarfs. *Nature Geoscience*, 2015. **8**(3): pp. 177–180.

Tortajada, C., Water infrastructure as an essential element for human development. *International Journal of Water Resources Development*, 2014. **30**(1): pp. 8–19.

Ueda, H., H. Tanaka, and Y. Sakurai, Reprint of the improvement of the energy resolution in epi-thermal neutron region of Bonner sphere using boric acid water solution moderator. *Applied Radiation and Isotopes*, 2015. **106**: pp. 107–110.

Usman, M., et al., COVID-19: the challenges and opportunities for water, air, agriculture and energy sectors. *International Journal of Agriculture and Biology*, 2021. **25**(5): pp. 1085–1095.

van Oss, C.J. and R.F. Giese. Role of the polar properties of water in separation methods. *Separation & Purification Reviews*, 2011. **40**(3): pp. 163–208.

Yilmaz, T. and Y. Hao. Electrical property characterization of blood glucose for on-body sensors, in *Proceedings of the 5th European Conference on Antennas and Propagation (EUCAP)*. 2011, IEEE.

Zhang, Q. and R. Hoogenboom, Polymers with upper critical solution temperature behavior in alcohol/water solvent mixtures. *Progress in Polymer Science*, 2015. **48**: pp. 122–142.

2

Bioremediation: A Sustainable Approach towards Clean Environment

Humaira Khanam, Tesnim Arfi, and Vinod Kumar Nigam
Birla Institute of Technology

CONTENTS

DOI: 10.1201/9781003231738-3

2.1 Introduction

The population explosion in the world has led to increased pressure on our natural reserves like air, water, and land resources. To meet the demands of the people, there is the rapid expansion of agricultural sector, health care, industries and thus leading to increased effluents which are ultimately contributing to the pollution (Dutta et al., 2021; Paliya et al., 2021). Pollution is the change in the quality of air, water or soil due to the introduction of contaminants that are harmful to the environment and its inhabitants. There are different sources responsible for causing pollution and that include domestic wastes, agricultural wastes, and industrial wastes like cyanide-containing compounds, chlorinated solvents, petroleum hydrocarbons, and other heavy metals, etc. which are toxic and even cause cancer (Ghorani-Azam et al., 2016; Chandra and Kumar, 2017a, b; Wainaina et al., 2020; Kumar and Chandra, 2020a; Rautela et al., 2021). These toxic wastes are potentially harmful to various life processes and pose a severe risk to the environment such as the air we breathe, the soil we harvest our crops, and even the water we drink. Wastes containing plastic, crude oil, polycyclic aromatic hydrocarbons, heavy metals, persistent organic pollutants, etc., raise public as well as environmental health concerns (Kumar et al., 2020a; 2021b; Kumar, 2018). These upon reaching water bodies causes detrimental effects on the entire water system from plants to animals to humans which depend on it as water source. On entering soil surface, these contaminants produce reproductive errors in plants, soil infertility, and interrupt natural restorative processes. Plastics in particular possess a major health risk as they account for almost 60%–80% of marine wastes (Kumar et al., 2021d). Today, when the need for a clean, healthy, disease-free, toxin-free environment is mankind's biggest concern, bioremediation comes to the rescue with its ability to degrade contaminants with almost no side effects. It has been a widely accepted method of destroying potentially harmful substances present in water bodies and soil. Being less complex and eco-friendly, it is better than its counterpart technologies like incineration which requires advanced equipment leading to the release of a substantial amount of toxins and other greenhouse gases on decomposition/degradation of wastes (Azubuike et al., 2016). Depending upon the nature of pollutants, they are categorized as persistent organic pollutants like DDT, 2–4 D, and some byproducts of pharmaceutical and pesticide manufacturing units and inorganic pollutants that include heavy metals such as Cr, Ni, Co, etc.

2.2 Toxicity of the Pollutants to Living Beings and the Environment

Due to the anthropogenic activities, pollutants are mostly accumulated in soil and water and also in the roots of plants; they get accumulated in the plants increasing their biomass. The amount of different organic pollutants absorbed by plants depends on the plant species, size, growth rate, and climatic conditions. These organic pollutants later enter the food chain as it is taken up by plants which are consumed by the herbivores and the successor organisms and thus consequently increase their exposure to humans (Yu et al., 2011). Some plants are capable of hydrolyzing the organic pollutants with their enzymes and by the rhizosphere microflora (Kumar and Chandra, 2018, 2020b; Kumar, 2021). Several plants are well known for the breakdown and absorption of organic pollutants, such as ryegrass, rice, zucchini, maize, cauliflower, poplar, radish, pumpkin, tobacco, nightshade, etc. (Afzal et al., 2014; Zhu et al., 2014; Chandra et al., 2018). However, some organic compounds such as xenobiotics cannot be degraded easily and they remain in the soil and alter its properties of the soil. These are known as persistent organic pollutants (POPs). POPs are toxic and highly stable chemical substances that can last for remarkably extended periods and travel a long distance through soil, water and air. They are either produced as synthetic chemicals in pesticides, dyes, and plastics or as by-products in the manufacturing of chemicals, and burning of fuels in manufacturing industries and vehicles. POPs can cause several diseases like cancer, gastric ailments, dysfunctional immune system, birth defects, and alter the reproduction system, loss of body weight and interrupt the function of nervous and endocrine systems. Therefore, the synthesis and use of POPs such as aldrin, heptachlor, hexachlorocyclohexane, dichlorodiphenyltrichloroethane, 2, 4-D dibenzofurans, etc. are banned worldwide (Pham and Sillanpää, 2020).

The toxicity of heavy metals depends on their bioavailability and the accomplished dose that causes detrimental effects on microorganisms. The harmful effects of heavy metals comprise several mechanisms that change the physiological and biochemical properties of microorganisms such as the formation of DNA as well as protein, enzymatic dysfunctions, reacting as redox catalysts in the production of reactive oxygen species (ROS), and destructing ion regulation, etc. (Gauthier et al., 2014). Heavy metals like Chromium (Cr) and Cadmium (Cd) can stimulate oxidative damage and denaturation of microorganisms as well as diminish the bioremediation capability of microorganisms (Amith et al., 2021; Singh et al., 2021). They may modify the structure and activity of enzymes by reacting with their carboxyl and thiol groups. Furthermore, they also interact electrostatically with negatively charged phosphate groups of DNA that trigger mutations or alterations in replication, transcription, and consequently translation of proteins. Other heavy metals like Copper (Cu (I) and Cu (II)) may involve in the synthesis of ROS via Fenton and Haber-Weis reactions. Likewise, Aluminium (Al) could stabilize superoxide radicals, which are responsible for DNA damage and also cause ion imbalance by adhering to the cell surface and entering the cells through ion channels or transmembrane carriers. Therefore, remediation of these pollutants has always been a concern for environmentalists, and now it is a global issue because of bioaccumulation and their non-biodegradability nature (Leong et al., 2018).

2.3 Bioremediation

In nature, there are various microorganisms including fungi and bacteria that are constantly working to break down organic compounds as well as the xenobiotics into harmless compounds. Bioremediation is a process involved in the hydrolysis of toxic compounds to non-toxic or less toxic components and repairs the damaged environment using biological factors (Kumar et al., 2018; 2020b). Bioremediation can also be defined as any process that uses microorganisms or their enzymes to restore the healthy state of nature sustainably and maintains ecological immunity. Bioremediation methods are employed to degrade specific contaminants, such as cyanide compounds by nitrilases, chlorinated pesticides by specific bacteria, and oil spills that are broken down using multiple techniques to facilitate the decomposition of crude oil by the use of certain microbes (Nigam et al., 2017; Chandra and Kumar, 2015; Safiyanu et al., 2015). Microorganisms when provided with adequate growth and proliferation conditions such as ideal pH, aeration, electron acceptors, moisture and nutrient content, etc. are able to transform target contaminants. Plants absorb pollutants from roots and soil sub-surface to convert those using specific digestive enzymes to yield products that may or may not be used for their growth purposes and this phenomenon is called phytoremediation. The naturally occurring qualities of microorganisms and plants provide an excellent alternative to chemically driven complex and expensive waste treatment techniques.

2.4 Approaches of Bioremediation

Different approaches used in the bioremediation processes are summarized in Figure 2.1. Each section is subsequently elaborated on the importance and limitation of the treatment mechanism.

2.4.1 *In-situ* Bioremediation

In-situ bioremediation refers to the method of treatment at the specific site of contamination. Manipulations such as aeration, moisture content and addition of nutrients lead to enhancement of enzyme activity in microbes present at the site which ultimately leads to the efficient cleanup of pollutants. This technique decreases the risk of public exposure of toxic wastes. *In-situ* bioremediation is mainly employed for remediation of soil and groundwater. It is further divided into two types: (Azubuike et al., 2016)

a. **Intrinsic *In-situ* Bioremediation**: Microorganisms that are natural habitats at the site of pollution are responsible for converting these into their non-toxic forms. This technique has a limitation i.e. the availability of microorganisms and its slow rate of degradation.
b. **Engineered *In-situ* Bioremediation**: In this case, the microorganisms are provided with optimal growth factors such as nutrients, oxygen, electron acceptors which increase the rate of waste degradation and proliferation of more such microorganisms to complete the remediation process.

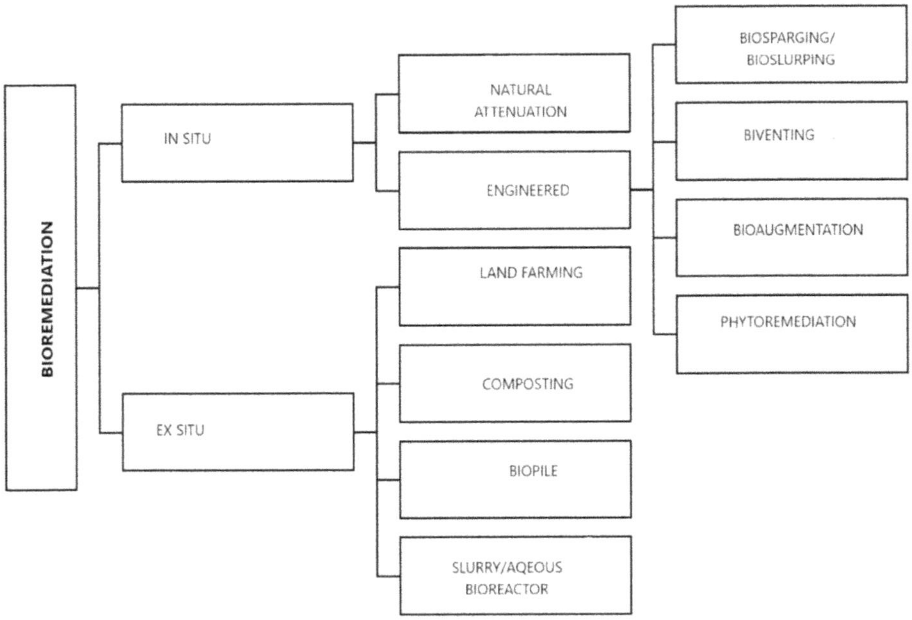

FIGURE 2.1 Schematic representation of different processes used in bioremediation.

In situ bioremediation involves the role of biological systems to clean up toxic wastes present in the environment with the integration of scientific and engineering techniques. Some of the techniques used in the process are mentioned: (Das and Dash, 2014).

2.4.1.1 Biosparging

This technique extensively degrades organic pollutants in groundwater and soil such as mineral oil, toluene, benzene, xylene, etc. It requires the injection of oxygen and essential nutrients into the saturated zone for enhancing the biodegradation process in deep anaerobic layers of soil and groundwater. Biosparging increases the mixing in the saturated zone and thus increases the contact between soil and groundwater. The ease and low cost of installing small-diameter air injection points allow considerable flexibility in the design and construction of the system (Kao et al., 2008).

2.4.1.2 Bioslurping

Bioslurping is a combination of bioventing and vacuum-enhanced free product recovery technique used in the construction process. Vaccum dewatering technique is used to remove light non-aqueous phase liquid contamination and bioventing stimulates aerobic bioremediation of hydrocarbons in unsaturated zones. The system consists of a well with a slurp tube to remove free product during pumping which leads to aeration of the unsaturated zone and thus, increases aerobic biodegradation (Kim et al., 2014). The aim is to regulate the environmental discharge of soil gas and groundwater. The limitation of this technique is that it is less effective in treating residual contamination.

2.4.1.3 Bioventing

Bioventing is a promising new technology that stimulates the natural *in-situ* biodegradation of degradable compounds in presence of oxygen. The indigenous microbial consortium present in the soil is supplied with oxygen at low flow rates to sustain their activity through direct air injections or electric blowers and degrade the wastes gradually. This technique is highly efficient in clearing soils with absorbed fuel and volatile organic compounds (Agarry and Latinwo, 2015).

2.4.1.4 Biostimulation

Biostimulation is known to degrade hydrocarbons and sulphates in the soils assisted by limiting nutrients such as phosphorus, oxygen, nitrogen, and electron donors. The capability of existing microorganisms is enhanced through the addition of supplements such as rate-limiting nutrients, growth factors, etc. Microorganisms that are present at the site perform decomposition of petroleum and light chain hydrocarbons by the introduction of organic nutrients such as cow dung, banana skin, sewage sludge, etc. The addition of nutrients increases the rate of catabolic activity of microbes already present in the soil.

2.4.1.5 Bioaugmentation

Bioaugmentation is the introduction of natural microbial strains or a genetically engineered strain to treat contaminated soil or water, in order to enhance the degradation process. It differs from biostimulation as it involves the addition of microorganisms whereas biostimulation involves the addition of nutritional supplements. It is specially employed for municipal wastewater treatment and at sites where soil and groundwater are contaminated with chlorinated ethane like trichloroethylene, and tetrachloroethylene (Adams et al., 2015).

2.4.1.6 Biopiling

This is a full-scale technology that involves forced aeration and control of pH, heat, and moisture where the contaminants are reduced to carbon dioxide and water. A basic biopile system includes a treatment bed, an aeration system, an irrigation/nutrient system, and a leachate collection system. The irrigation/nutrient system is buried under the soil to pass air and nutrients either by vacuum or positive pressure (Emami et al., 2012).

2.4.2 *Ex-situ* Bioremediation

Ex-situ bioremediation involves the collection of waste and its subsequent treatment away from the original site within a controlled environment. It remediates a wide range of hydrocarbons but leads to disruption of site which is a major drawback. The process involves screening, filtration, mixing, etc. Based on contamination phases, the process of *ex-situ* is divided into:

a. **Solid Phase**: For treatment of solid wastes such as animal excreta, leaves, domestic wastes solid, etc. Composting and land farming are being employed for the treatment of solid wastes
b. **Slurry Phase Treatment**: It is a triphasic system that involves three components; water, suspended particulate matter and air. Solid wastes are made into slurry for treatment in a bioreactor.

   The objectives of slurry phase treatment provide:

   i. Improve microbial growth limiting factors in soil environment such as substrate, nutrients and oxygen availability.
   ii. Promote suitable environmental conditions such as moisture, pH, and temperature for bacterial growth and
   iii. To minimize mass transfer limitations and facilitate desorption of organic material from the soil matrix

2.4.2.1 Composting

It involves the administration of microorganisms in compost for biodegradation of soil and water contaminants. Microorganisms digest the harmful contaminants and turn them into carbon dioxide, water and other harmless inert chemicals. This technique eliminates heavy metals, persistent organic pollutants and dioxins (Figure 2.2). Composting helps in the enrichment of the soil at the end of the process which is an added advantage of this bioremediation technique (Kästner and Miltner, 2016).

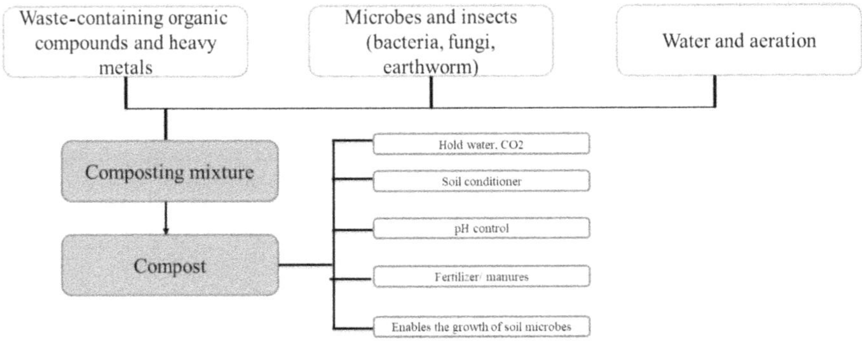

FIGURE 2.2 Composting and its application in agriculture.

2.4.2.2 Land Farming

It is a simple procedure involving the transportation of contaminated soil to a land farming zone where soil conditions are optimized for its pH, moisture content followed by tilling and aeration to stimulate the functions of naturally occurring microflora of soil for degradation of pollutants. The methodology includes spreading contaminated soil in a thin layer at an engineered land farming site followed by intermittent mixing that leads to completion of the process in about 22–24 months. Optimization factors such as oxygen, moisture content, temperature, pH, etc. accelerates the biodegradation of organic contaminants. Thus, land farming has proven to be effective in treatment of soils contaminated with toluene, benzene, polycyclic aromatic hydrocarbons, diesels and crude oil, etc. (Ali et al., 2020). Several factors such as requirement of large space, uncontrolled biodegradation, undegradable inorganic contaminants limit the efficiency of land farming.

2.4.2.3 Biopiles

Biopiles are a synapsis between land farming and composting. Just like land farming, biopiles are also engineered systems that provide oxygen to microbes for the breakdown of contaminants in the soil. The soil is formed into a compost pile and perforated tubes force air into this enclosed pile through a positive pressure air distribution system buried underneath. Biopiles are also known as bioheaps and biocells and are specially designed for the treatment of fuel-contaminated soil.
A comparison between *In-situ* and *Ex-situ* bioremediation is presented in Table 2.1.

2.5 Phytoremediation

It is a plant-based approach to bioremediation that involves the use of plants to extract and solubilize the toxic compounds and thereby lowering their bioavailability in the soils. Plants established rhizosphere in the soil assist in the absorption of ionic compounds present in the soil even at low concentrations by their root system. Plant roots accumulate heavy metals and solubilize them which increases the fertility of soil. Plants are sensitive

TABLE 2.1

Comparison between *In-situ* and *Ex-situ* Bioremediation

***In-situ* Bioremediation**	***Ex-situ* Bioremediation**
Less expensive process but the treatment is slow	Process is quite expensive but fast
Minimal site disruption	Maximum site disruption
No handling expertise required	Excavation, screening, fractioning expertise required
Less effective	More effective
No involvement of bioreactors	Involves the use of bioreactor for microbial growth

TABLE 2.2

Role of Different Plants and Their Applications in Removal of Contaminants

Plant Species	Heavy Metal/Contaminants	References
Pteris vittata, Corrigiola telephiifiola	As, Cr	Kalve et al. (2011)
Phytolacca americana, Hellianthus	Cd, Mn, Pb, radioactive nucleides	Peng et al. (2008); Alaboudi et al. (2018)
Haumaniastrum robertii, Pteris vittata, Achiella millefolium	Co, Cu, As, Hg	Marques et al. (2009); Wang et al. (2012)
Brassica napus	Pb, Se, Zn	Jing et al. (2014)
Alyssum corsicum	Ni, Mn	Lopez et al. (2017)
Thalspi caerulescens	Zn, Cd	Sakakibara et al. (2011)

towards toxic metals and show different responses even at low concentrations (Chandra and Kumar, 2018). The ability of plant to perform stabilization, filtration or uptake mechanisms to complete the remediation process has enabled phytoextraction of accumulated toxic metals from polluted soils and water. It consists of an array of different techniques working in its own natural way in order to immobilize, degrade and even eliminate the contaminants causing no harm to the environment rather providing soil fertility (Chandra et al., 2018; DalCorso et al., 2019). There are several reports which focused on biostimulated degradation by plants like *Aegiceras corniculatum* extracts Brominated diphenyl ethers, *Brassica napus* extracts Selenium (Se) and radionuclides from soil, while *Helianthus* extracts Lead (Pb), Cadmium (Cd) and radionuclides from soil sediments through specialized biochemical interactions and stabilizing their amount of toxicity and purify the soil (Table 2.2).

Moreover, phytoremediation of soil contaminants can be achieved by several techniques which include (Vamerali et al., 2010; Marques et al., 2009)

2.5.1 Phytoextraction/Phytoaccumulation

Phytoextraction involves the absorption of heavy, non-biodegradable metals followed by their storage in the external/superficial parts of plant for detoxification. Plants make it possible by the conversion of heavy metals into water-soluble compounds. When plants are harvested and the dry matter burnt under controlled conditions for disposal of the ash, the contaminants are removed from the soil (Kumar et al., 2021a). It has been found that the recovery of high-price metals such as Nickel (Ni), Thallium (Tl) or Gold (Au) can be harvested from plant material.

2.5.2 Phytostabilization

Phytostabilization involves the conversion of contaminants to less toxic, non-proliferative state, by immobilizing the metal contaminants in their rhizosphere, thus not affecting any other plant part. The toxins are absorbed as energy and their mobility decreases. The growth of roots reduces leaching which adds organic matter in soil and makes the pollutants less or non-available forms. Thus, no secondary waste is generated as fewer metals enter the trophic level in food web.

2.5.2.1 Biochar

The organic material is burned in absence of air to form a charcoal-like substance. It helps in the efficient remediation of multi-metal contaminated soils. Immobilization of heavy metals is considered a positive scope for soil remediation. It is inexpensive and reduces the risk of exposure of pollutants to other organisms.

2.5.3 Phytodegradation/Phytotransformation

Plants using their specialized enzymes degrade the harmful organic compounds around their periphery and later these harmless compounds help in the growth processes of plants. These substances may also act as a limiting factor, which is a disadvantage of this technique.

2.5.4 Phytovolatization

Phytovolatization is the process of transpiration of toxic organic compounds and their release into the atmosphere as growing plants uptake water and organic contaminants from soil. Several species of the genus *Astragalus* accumulate and volatilize Selenium. Uptake and evaporation of Mercury are achieved by some bacteria. The genes of bacteria responsible for such activities have been transferred to *Nicotiana* or *Brassica* species and the developed transgenic plants may be useful in cleaning Mercury contaminated soils.

2.5.5 Phytofiltration/Rhizofiltration

This technique is mainly used for cleaning contaminated surface waters or wastewaters by adsorption or precipitation of metals by roots or other submerged organs of metal-tolerant aquatic plants (Kumar, 2021). Plants that have high adsorption surface and can withstand low oxygen concentration are best for rhizofiltration. Indian mustard (*B. juncea*) and sunflower (*H. annuus*) have long and hairy root system which shows good ability of metal accumulation during rhizofiltration (Dhanwal et al., 2017). While rhizodegradation involves decomposition of organic pollutants by the microorganisms present in their rhizosphere (Javed et al., 2019).

2.5.6 Phytorestauration

Phytorestauration is a method where plants are allowed to grow on barren lands which prevents the migration of contaminated soil particles by wind or water run-off. The fast growing plant species are grown in these areas to prevent the erosion of soil and enable plants to colonize and accumulate the toxic compounds and solubilize them into useful and non-toxic compounds. The methods of removal of toxic compounds in the case of plants are presented in Figure 2.3.

2.5.6.1 Advantages of Using Phytoremediation

i. Phytoremediation is an eco-friendly process that reduces the exposure of pollutants to the environment and ecosystem.
ii. Phytoremediation is economically feasible as plants are autotrophic system which utilizes solar energy, water and carbon dioxide for their growth and thus the process is cost-effective.

Phytoremediation

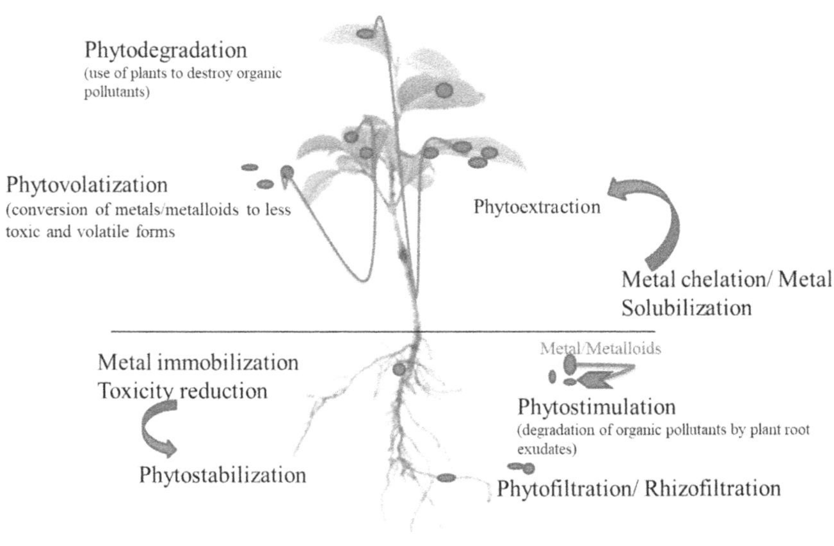

FIGURE 2.3 Phytoremediation techniques.

TABLE 2.3

Application of Microorganisms in Bioremediation Process

Microorganisms	Substrate/Compound	References
Pseudomonas putida	Petrol and diesel polycyclic aromatic hydrocarbons	Safiyanu et al. (2015)
Aspergillus niger, *Monocillium* sp.	Hydrocarbon, crude oil	Odili et al. (2020)
Bacillus cerus, *Pseudomonas aeruginosa*, *Bacillus licheniformis*	Diesel, crude oil	Maliji et al. (2013)
Streptomyces sp.	Nitriles	Agarwal and Nigam (2014)
Saccharomyces cerevisiae	Pb, Hg, Ni	Infante et al. (2014)
Pseudomonas fluorescens, and *Pseudomonas aeruginosa*	Fe, Zn, Pb, Mn and Cu	Paranthaman and Karthikeyan (2015)

iii. It prevents soil erosion and metal leaching by stabilizing heavy metals and reducing the risk of spreading contaminants.
iv. Phytoremediation improves soil fertility by releasing various organic matters to the soil.
v. Phytoremediation can be applied over a large-scale field and even barren areas are utilized and contaminants can be easily disposed.

2.6 Microorganisms in Bioremediation

Various microorganisms are reported that break down the contaminants to less toxic or non-toxic compounds or elemental forms through their different biochemical pathways. Bacteria, archaea and fungi are prime bioremediators, which act as tools to remove the pollutants from soil, water and sediments (Table 2.3). Biotechnological process involving genetically modified microbes, enzymes have enabled the bioremediation process faster and restoring the original natural surroundings and preventing pollution. Many Microorganisms rely on waste materials for food and energy sources as they enjoy specialized enzymes that degrade contaminants into useful substances. Another added benefit is their small size which enables easier contact with contaminants. Capable of genetic manipulation, microorganisms are the best bioremediators with high growth rate and variable metabolic needs making them easy to use and customize according to target pollutants. Reduction of waste load through microbial activity leads to minimal side effects and nearly no environmental damage

2.6.1 Biomining

Biomining is the administration of microbial systems to extract the metals of economic interest from ores and mines. This method is very effective in the case of low-grade ores. Areas particularly polluted by metals are known as action zone where microbes are added with the intention of solubilizing metals into biochemical waste which is then separated by the process of sedimentation. Few of the microorganisms employed in biomining are *Thiobacillus thioxidans* (Zn and Pb), *Thiobacillus organoparpus* (Fe and Cu) and *Sulfilolus acidocalcarius* (Cu, Mo, and Fe), *Bacillus subtilis, Cupriavidus metallidurans* (Cu & Au) (Johnson, 2014; Hong and Valix, 2014). The direct dissolution of metal of interest is called Bioleaching. This process is more cost-effective than smelting processes. Microorganisms break down minerals for their energy needs. It stabilizes sulphate toxins from the mines without causing harm to the environment while also providing the mine with economical profits. However, the process raises certain bio-safety concerns of acid runoff, iron, zinc elimination which may contaminate water and soil. In Germany, a company called BRAIN (Biotechnology Research and Information Network) is using a huge collection of more than 50,000 microbes to identify the most effective microbes for biomining.

2.6.2 Biooxidation

Biooxidation involves microbes that do not solubilize the valuable metals but keeps the metals in solid phase. It is used as a pre-treatment process that degrades mineral sulphides from the ores of valuable metals. *Acidithiobacillus* species and *Sulfolobus* species are mainly employed for extraction of ferrous ion, elemental sulphur and metal sulphides. *A. ferrooxidans* or *L. ferrooxidans* are thermophillic bacteria reported to extract 28.7% of gold using bioxidation (Xu et al., 2020).

2.6.3 Enzyme Mediated Bioremediation

Enzyme-mediated bioremediation makes use of naturally present enzymes in plants and microorganisms for degradation of contamination. It is a quick and easy method of bioremediation, especially in the treatment of xenobiotic compounds such as pesticides and fertilizers, etc. Since it is not a microbial-based process; one does not have to wait for microbial proliferation to conduct this type of remediation. Recent advances in nanobiotechnology and genetic engineering have enabled the direct introduction of enzymes at site of contamination rather than a whole organism. The technique releases fewer or non-toxic by-products making it sustainable and widely accepted. Enzymes such as oxidoreductase, hydrolase, dehalogenases have proven to be specific in their mode of action against toxic pollutants and quickly detoxifies them without releasing disruptive end products (Karigar and Rao, 2011). Enzyme-mediated bioremediation comes especially handy in case of contamination sites in urgent need of revival (Table 2.4). The ligninolytic fungi release the enzyme lignin peroxidase which degrades polycyclic aromatic hydrocarbons as these are structurally similar to lignin minimizing the toxic effect of polycyclic aromatic hydrocarbons. Oyster mushrooms mycelium releases a specific enzyme that breaks down oil, pesticides and heavy metals present in soil and is thus extensively used in oil-spill clean-up sites and thereby turning a toxic substance into food (Akkin, 2021).
A summary of different procedures involved in biological remediation is shown in Figure 2.4.

2.7 Bioreactor-Based Bioremediation

Bioreactors are large vessels in which raw materials are converted to specific product through a series of biological reactions. They work on different modes which include batch, fed-batch, and continuous modes of cultivation. It is designed in a way that it provides the natural optimal conditions (temperature, pH, agitation and aeration rates, substrates and size of inoculums) for the growth of microbial.

TABLE 2.4
Role of Different Enzymes in Bioremediation

Enzyme	Function	Mechanism	References
Hydrolases	Degradation of fats and proteins	Triglyceride is broken down to glycerol and fatty acids and peptide bond is broken down to peptides and amino acids.	Karigar and Rao (2011)
3-Cyanopyridinase, Nitrilase	Hydrolyses nitrile containing compounds	Preforms nucleophillic reactions	Arfi and Nigam (2018); Arfi and Nigam (2020)
Dehalogenase	Removes halogen from carbon and halogen bond	Halogen group is substituted by Hydrogen under aerobic conditions.	Allpress and Gowland (1998)
Laccase	Break down the aromatic ring of organic compounds	Produces free radicals	Shraddha et al. (2011)
Cytochrome 450	Metabolism of xenobiotics, steroids, fatty acids	Performs oxidation and reduction reactions	Guengerich (2018)
Amidase	Catalyzes nitrile amides to carboxylic acid	Performs neucleophillic reactions.	Prabha and Nigam (2020)

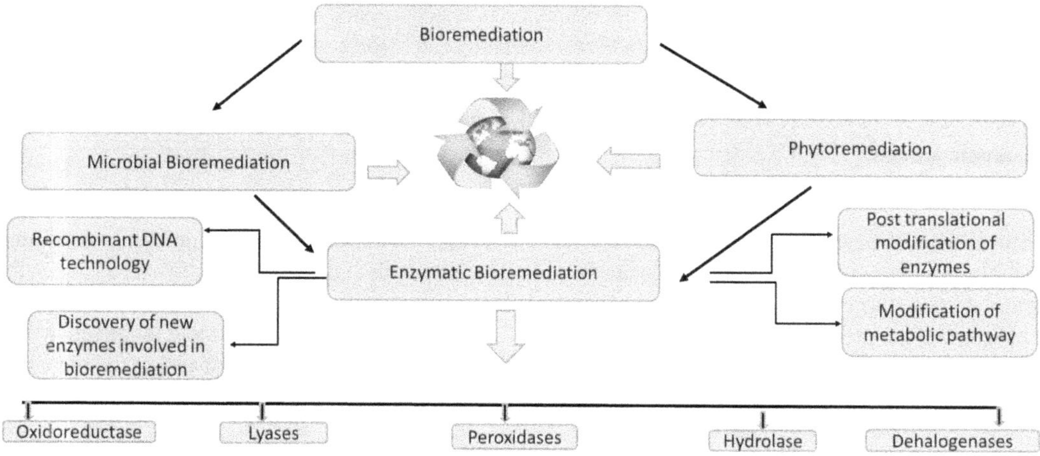

FIGURE 2.4 Types of bioremediation.

The enhanced rate of degradation process and reduced bioremediation time is the advantage in bioreactor-based bioremediation and therefore, it is considered as an efficient and sustainable method of bioremediation. The bioreactor allows maximum biological degradation with minimum losses and also allows the characterisation of bacterial communities involved in bioremediation process (Zangi-Kotler et al., 2015). Biostimulant or bioaugmenting agent is added to the mixture for higher efficiency. Several genetically engineered microorganisms (GEMs) perform bioremediation and work against a wide range of pollutants by increased substrate degradation, with high catalytic ability, low biomass accumulation and speedy recovery of waste polluted sites by neutralizing harmful substances (Abatenh et al., 2017).

2.8 Role of Biosurfactants in Bioremediation

Biosurfactants are surface-active chemicals produced by fungi, bacteria and yeasts. The distinctive characteristics of these biological surfactants are their hydrophobic as well as hydrophilic regions, pH and temperature tolerant, less toxic to surroundings and biodegradability which make them useful in case of enhancing contaminant degradation. The biosurfactants make hydrophobic polycyclic aromatic hydrocarbons (PAH) present in water bodies more soluble and thereby easing the bioremediation of these hydrocarbons (Karlapudi et al., 2018). The biosurfactants also provide the indigenous microorganisms at contamination sites with suitable temperature and pH conditions along with essential nutrients such as oxygen and nitrogen enhancing the growth of microorganisms and subsequent degradation of contaminants. Apart from bioremediation, biosurfactants are extensively used in food industries as it slows stalling, increases shelf life, decreases surface tension, etc. (Akbari et al., 2018).

2.9 Application of Metagenomics in Bioremediation

Bioremediation techniques are forever evolving and growing day by day with advancements in techniques, and one such new technique has emerged as metagenomic bioremediation which works at genetic level by analyzing the biodegrading microbial population (Devarapalli and Kumavath, 2015; Kumar et al., 2020b). This system effectively studies the role of pollutants on microorganisms and how certain microbes develop a specialized defense mechanism to combat contaminants. Metagenomic studies reveal whether a microbial community has functional genes to detoxify a particular characteristic toxin in soil or not (Kumar et al., 2021c). If any species showed the required limit of functional genes, it is administered at site of soil contamination. Reverse transcriptase real-time polymerase chain reaction (RT-qPCR) calculates the expression of toxin/hydrocarbon degradation genes (Mani, 2020). A water body may contain high levels of varied contaminants such as nutrients, slag, toxins, agricultural wastes, sewage, etc. Metagenomic bioremediation of water enables to perform sequencing techniques to categorize the microbial system and the contaminants in order to lower the harmful state of water.

2.10 Constraints in Bioremediation

There are certain limitations in bioremediation. It has been observed that the process is suitable for biodegradable compounds while it is still difficult to degrade non-biodegradable and complex wastes. Additionally, the microbial systems need to be given optimal conditions for effective degradation which is a challenging task. Furthermore, the process of bioremediation is time taking as compared to excavation and incineration, and sometimes degraded products are more toxic than parent compounds and remain persistent in the surrounding. Finally, lack of awareness and trained manpower limit the applications of bioremediation.

2.11 Conclusion

Based on the study, it is inferred that biodegradation of xenobiotics, oil spills and heavy metals, and other pollutants is a major concern due to their toxic effects even at low concentrations. It causes various diseases like intestinal disorder, weight loss, autoimmune disorder, and cancer. Bioremediation is a socially acceptable and sustainable method for cleaning, recovering, and restoring the natural environment and microbial systems are the key agents of bioremediation. It is an advantageous method due to its cost-saving and converting toxic compounds to non-toxic products. In addition, phytoremediation has also played a major role in heavy metal extraction and rejuvenation of contaminated soil along with the prevention of soil erosion and leaching. Similarly, enzyme-mediated bioremediation and bioreactor-based bioremediation have enhanced the efficiency of biodegradation of contaminants. The application of biosurfactants to contaminated sites has also sped up the rate of biodegradation by providing optimum pH, temperature and essential nutrients for the growth of microbes. Likewise, the applications of bioinformatics and metagenomics have provided distinctive futures of microbial community that may be exploited in the process of bioremediation.

2.12 Future Scope

The role of genetic engineering, proteomics, and metagenomics may be further investigated to provide more insight into bioremediation. Also, a combination of these approaches could be beneficial to explore and construct the modified microbes and plants which can specifically and efficiently mineralize the contaminants. Genetically modified plants and microbes are potential agents of bioremediation that should be improved so that genes involved in heavy metal uptake, sequestration, and tolerance can increase the bioavailability of heavy metals, soil health, and promote a cleaner and sustainable environment. In addition, the role of bioreactors must be optimized and characterized in the area of bioremediation.

REFERENCES

Abatenh, E., Gizaw, B., Tsegaye, Z. and Wassie, M., 2017. The role of microorganisms in bioremediation-a review. *Open Journal of Environmental Biology*, *2*(1), pp. 038–046. https://doi.org/10.17352/ojeb.000007

Adams, G.O., Fufeyin, P.T., Okoro, S.E. and Ehinomen, I., 2015. Bioremediation, biostimulation and bioaugmention: a review. *International Journal of Environmental Bioremediation & Biodegradation*, *3*(1), pp. 28–39. https://doi.org/10.12691/ijebb-3-1-5

Afzal, M., Khan, Q.M. and Sessitsch, A., 2014. Endophytic bacteria: prospects and applications for the phytoremediation of organic pollutants. *Chemosphere*, *117*, pp. 232–242. https://doi.org/10.1016/j.chemosphere.2014.06.078

Agarry, S. and Latinwo, G.K., 2015. Biodegradation of diesel oil in soil and its enhancement by application of bioventing and amendment with brewery waste effluents as biostimulation-bioaugmentation agents. *Journal of Ecological Engineering*, *16*(2). https://doi.org/10.12911/22998993/1861

Agarwal, A. and Nigam, V.K., 2014. Nitrilase mediated conversion of indole-3-acetonitrile to indole-3-acetic acid. *Biocatalysis and Agricultural Biotechnology*, *3*(4), pp. 351–357. https://doi.org/10.1016/j.bcab.2014.05.005

Akbari, S., Abdurahman, N.H., Yunus, R.M., Fayaz, F. and Alara, O.R., 2018. Biosurfactants—a new frontier for social and environmental safety: a mini review. *Biotechnology Research and Innovation*, *2*(1), pp. 81–90. https://doi.org/10.1016/j.biori.2018.09.001

Akkin, N., 2021. Mycoremediation by oyster mushroom. *Acta Scientific Agriculture*, *5*(5), pp. 47–48.

Alaboudi, K.A., Ahmed, B. and Brodie, G., 2018. Phytoremediation of Pb and Cd contaminated soils by using sunflower (*Helianthus annuus*) plant. *Annals of Agricultural Sciences*, *63*(1), pp.123–127. https://doi.org/10.1016/j.aoas.2018.05.007

Ali, N., Dashti, N., Khanafer, M., Al-Awadhi, H. and Radwan, S., 2020. Bioremediation of soils saturated with spilled crude oil. *Scientific Reports*, *10*(1), pp. 1–9. doi.org/10.1038/s41598-019-57224-x

Allpress, J.D. and Gowland, P.C., 1998. Dehalogenases: environmental defence mechanism and model of enzyme evolution. *Biochemical Education*, *26*(4), pp. 267–276. https://doi.org/10.1016/S0307-4412(98)00090-9

Amith, A., Sukanya, R., Subramanian, S., Kumar, V., and Ramamurthy, P., 2021. Chromium (VI) detection by microbial carbon dots: microwave synthesis and mechanistic study. *Journal of Basic Microbiology*, pp. 1–10. https://doi.org/10.1002/jobm.202100394

Arfi, T. and Nigam, V. K., 2018. Extraction and quantification of nicotinic acid from 3-cyanopyrdinase catalysed reaction. *Research Journal of Chemistry and Environment*, *22*(9), pp. 1–6.

Arfi, T. and Nigam, V.K., 2020. Studies on a thermostable nitrilase from *Staphylococcus* sp and its *In-silico* characterisation. *Biologia*, *75*(12), pp. 2421–2432. https://doi.org/10.2478/s11756-020-00554-3

Azubuike, C.C., Chikere, C.B. and Okpokwasili, G.C., 2016. Bioremediation techniques–classification based on site of application: principles, advantages, limitations and prospects. *World Journal of Microbiology and Biotechnology*, *32*(11), pp. 1–18. https://doi.org/10.1007/s11274-016-2137-x

Chandra, R. and Kumar, V., 2015. Biotransformation and biodegradation of organophosphates and organohalides. In: Chandra, R. (Ed.) *Environmental Waste Management*. Boca Raton: CRC Press. https://doi.org/10.1201/b19243-17

Chandra, R. and Kumar, V., 2017a. Detection of *Bacillus* and *Stenotrophomonas* species growing in an organic acid and endocrine-disrupting chemicals rich environment of distillery spent wash and its phytotoxicity. *Environmental Monitoring and Assessment*, *189*, p. 26. https://doi.org/10.1007/s10661-016-5746-9

Chandra, R. and Kumar, V., 2017b. Detection of androgenic-mutagenic compounds and potential autochthonous bacterial communities during in-situ bioremediation of post methanated distillery sludge. *Frontiers in Microbiology*, *8*, p. 87. https://doi.org/10.3389/fmicb.2017.00887

Chandra, R. and Kumar, V., 2018. Phytoremediation: a green sustainable technology for industrial waste management. In: Chandra, R., Dubey, N. and Kumar, V., *Phytoremediation of Environmental Pollutants*. Boca Raton: CRC Press, https://doi.org/10.1201/9781315161549-1

Chandra, R., Kumar, V. and Singh, K., 2018. Hyperaccumulator versus nonhyperaccumulator plants for environmental waste management. In: Chandra, R., Dubey, N. and Kumar, V. (Eds.) *Phytoremediation of Environmental Pollutants*. Boca Raton: CRC Press, https://doi.org/10.1201/9781315161549-1

DalCorso, G., Fasani, E., Manara, A., Visioli, G. and Furini, A., 2019. Heavy metal pollutions: state of the art and innovation in phytoremediation. *International Journal of Molecular Sciences*, *20*(14), p. 3412. https://doi.org/10.3390/ijms20143412

Das, S. and Dash, H.R., 2014. Microbial bioremediation: A potential tool for restoration of contaminated areas. In: *Microbial Biodegradation and Bioremediation* (pp. 1–21). https://doi.org/10.1016/B978-0-12-800021-2.00001-7

Devarapalli, P. and Kumavath, R.N., 2015. Metagenomics—a technological drift in bioremediation. In: *Advances in Bioremediation of Wastewater and Polluted Soil* (pp. 73–91). https://doi.org/10.5772/59328

Dhanwal, P., Kumar, A., Dudeja, S., Chhokar, V. and Beniwal, V., 2017. Recent advances in phytoremediation technology. In: *Advances in Environmental Biotechnology* (pp. 227–241). https://doi.org/10.1007/978-981-10-4041-2_14

Dutta, D., Arya, S. and Kumar, S., 2021. Industrial wastewater treatment: Current trends, bottlenecks, and best practices. *Chemosphere*, *285*, p. 131245. https://doi.org/10.1016/j.chemosphere.2021.131245

Emami, S., Pourbabaee, A.A. and Alikhani, H.A., 2012. Bioremediation principles and techniques on petroleum hydrocarbon contaminated soil. *Technical Journal of Engineering and Applied Sciences*, *2*(10), pp. 320–323.

Gauthier, P.T., Norwood, W.P., Prepas, E.E. and Pyle, G.G., 2014. Metal–PAH mixtures in the aquatic environment: a review of co-toxic mechanisms leading to more-than-additive outcomes. *Aquatic Toxicology*, *154*, pp.253–269. https://doi.org/10.1016/j.aquatox.2014.05.026

Ghorani-Azam, A., Riahi-Zanjani, B. and Balali-Mood, M., 2016. Effects of air pollution on human health and practical measures for prevention in Iran. *Journal of Research in Medical Sciences: The Official Journal of Isfahan University of Medical Sciences*, *21*. https://doi.org/10.4103/1735-1995.189646

Guengerich, F.P., 2018. Mechanisms of cytochrome P450-catalyzed oxidations. *ACS Catalysis*, *8*(12), pp. 10964–10976. https://doi.org/10.1021/acscatal.8b03401

Hong, Y. and Valix, M., 2014. Bioleaching of electronic waste using acidophilic sulfur oxidising bacteria. *Journal of Cleaner Production*, *65*, pp. 465–472. https://doi.org/10.1016/j.jclepro.2013.08.043

Infante, J. C., De Arco, R. D. and Angulo, M. E., 2014. Removal of lead, mercury and nickel using the yeast *Saccharomyces cerevisiae*. *Revista MVZ Córdoba*, *19*(2), pp. 4141–4149.

Javed, M.T., Tanwir, K., Akram, M.S., Shahid, M., Niazi, N.K. and Lindberg, S., 2019. Phytoremediation of cadmium-polluted water/sediment by aquatic macrophytes: role of plant-induced pH changes. In: *Cadmium Toxicity and Tolerance in Plants* (pp. 495–529). Academic Press. https://doi.org/10.1016/B978-0-12-814864-8.00020-6

Jing, Y.X., Yan, J.L., He, H.D., Yang, D.J., Xiao, L., Zhong, T., Yuan, M., Cai, X.D. and Li, S.B., 2014. Characterization of bacteria in the rhizosphere soils of *Polygonum pubescens* and their potential in promoting growth and Cd, Pb, Zn uptake by *Brassica napus*. *International Journal of Phytoremediation*, *16*(4), pp. 321–333. https://doi.org/10.1080/15226514.2013.773283

Johnson, D.B., 2014. Biomining—biotechnologies for extracting and recovering metals from ores and waste materials. *Current Opinion in Biotechnology*, *30*, pp. 24–31. https://doi.org/10.1016/j.copbio.2014.04.008

Kalve, S., Sarangi, B.K., Pandey, R.A. and Chakrabarti, T., 2011. Arsenic and chromium hyperaccumulation by an ecotype of *Pteris vittata*–prospective for phytoextraction from contaminated water and soil. *Current Science*, *100*, pp. 888–894.

Kao, C.M., Chen, C.Y., Chen, S.C., Chien, H.Y. and Chen, Y.L., 2008. Application of in situ biosparging to remediate a petroleum-hydrocarbon spill site: Field and microbial evaluation. *Chemosphere*, *70*(8), pp. 1492–1499. https://doi.org/10.1016/j.chemosphere.2007.08.029

Karigar, C.S. and Rao, S.S., 2011. Role of microbial enzymes in the bioremediation of pollutants: a review. *Enzyme Research*, *2011*. https://doi.org/10.4061/2011/805187

Karlapudi, A.P., Venkateswarulu, T.C., Tammineedi, J., Kanumuri, L., Ravuru, B.K., Ramu Dirisala, V. and Kodali, V.P., 2018. Role of biosurfactants in bioremediation of oil pollution-a review. *Petroleum*, *4*(3), pp. 241–249. https://doi.org/10.1016/j.petlm.2018.03.007

Kästner, M. and Miltner, A., 2016. Application of compost for effective bioremediation of organic contaminants and pollutants in soil. *Applied Microbiology and Biotechnology*, *100*(8), pp. 3433–3449. https://doi.org/10.1007/s00253-016-7378-y

Kim, S., Krajmalnik-Brown, R., Kim, J.O. and Chung, J., 2014. Remediation of petroleum hydrocarbon-contaminated sites by DNA diagnosis-based bioslurping technology. *Science of the Total Environment*, *497*, pp. 250–259. https://doi.org/10.1016/j.scitotenv.2014.08.002

Kumar, V., 2018. Mechanism of microbial heavy metal accumulation from polluted environment and bioremediation. In: Sharma, D. and Saharan, B.S., *Microbial Fuel Factories*. Boca Raton: CRC Press.

Kumar, V., 2021. Phytoremediation of distillery effluent: current progress, challenges, and future opportunities. In: Saxena, G., Kumar, V. and Shah, M.P. (Eds.), *Bioremediation for Environmental Sustainability: Toxicity, Mechanisms of Contaminants Degradation, Detoxification and Challenges*. Elsevier. https://doi.org/10.1016/B978-0-12-820524-2.00014-6

Kumar, V. and Chandra, R., 2018. Bacterial assisted phytoremediation of industrial waste pollutants and eco-restoration. In: Chandra, R., Dubey, N.K. and Kumar, V., (Eds.), *Phytoremediation of Environmental Pollutants*. Boca Raton: CRC Press.

Kumar V. and Chandra R., 2020a. Bioremediation of melanoidins containing distillery waste for environmental safety. In: Bharagava R. and Saxena G., (Eds.), *Bioremediation of Industrial Waste for Environmental Safety*. Singapore: Springer. https://doi.org/10.1007/978-981-13-3426-9_20

Kumar, V. and Chandra, R., 2020b. Metagenomics analysis of rhizospheric bacterial communities of *Saccharum arundinaceum* growing on organometallic sludge of sugarcane molasses-based distillery. *3 Biotech*, *10*(7), p. 316. https://doi.org/10.1007/s13205-020-02310-5

Kumar V., Ferreira, L.F.R., Sonkar, M. and Singh, J. 2021a. Phytoextraction of heavy metals and ultra-structural changes of *Ricinus communis* L. grown on complex organometallic sludge discharged from alcohol distillery. *Environmental Technology & Innovation*, *22*, p. 101382. https://doi.org/10.1016/j.eti.2021.101382

Kumar V., Shahi, S.K., Ferreira, L.F.R., Bilal, M., Biswas, J.K. and Bulgariu, L., 2021b. Detection and characterization of refractory organic and inorganic pollutants discharged in biomethanated distillery effluent and their phytotoxicity, cytotoxicity, and genotoxicity assessment using *Phaseolus aureus* L. and *Allium cepa* L. *Environmental Research*, *201*, p. 111551. https://doi.org/10.1016/j.envres.2021.111551

Kumar, V., Shahi, S.K. and Singh, S., 2018. Bioremediation: an eco-sustainable approach for restoration of contaminated sites. In: Singh J., Sharma D., Kumar G. and Sharma N., (Eds.), *Microbial Bioprospecting for Sustainable Development*. Singapore: Springer. https://doi.org/10.1007/978-981-13-0053-0_6

Kumar, V., Singh, K., Shah, M.P., Singh, A.K, Kumar, A. and Kumar, Y., 2021c. Application of omics technologies for microbial community structure and function analysis in contaminated environment. In: Shah, M.P., Sarkar, A. and Mandal, S., (Eds.), *Wastewater Treatment: Cutting Edge Molecular Tools, Techniques & Applied Aspects in Waste Water Treatment*. Elsevier. https://doi.org/10.1016/B978-0-12-821925-6.00013-7

Kumar, V., Srivastava, S. and Thakur, I.S., 2021d. Enhanced recovery of polyhydroxyalkanoates from secondary wastewater sludge of sewage treatment plant: analysis and process parameters optimization. *Bioresource Technology Reports*, *15*, p. 100783.

Kumar, V., Thakur, I.S. and Shah, M.P., 2020a. Bioremediation approaches for pulp and paper industry wastewater treatment: recent advances and challenges. In: Shah, M.P., (Ed.), *Microbial Bioremediation & Biodegradation*. Singapore: Springer. https://doi.org/10.1007/978-981-15-1812-6_1

Kumar, V., Thakur, I.S., Singh, A.K. and Shah, M.P., 2020b. Application of metagenomics in remediation of contaminated sites and environmental restoration. In: Shah, M., Rodriguez-Couto, S. and Sengor, S.S., (Eds.), *Emerging Technologies in Environmental Bioremediation*. Elsevier. https://doi.org/10.1016/B978-0-12-819860-5.00008-0

Leong, W.H., Lim, J.W., Lam, M.K., Uemura, Y., Ho, C.D. and Ho, Y.C., 2018. Co-cultivation of activated sludge and microalgae for the simultaneous enhancements of nitrogen-rich wastewater bioremediation and lipid production. *Journal of the Taiwan Institute of Chemical Engineers*, *87*, pp. 216–224. https://doi.org/10.1016/j.jtice.2018.03.038

Lopez, S., Piutti, S., Vallance, J., Morel, J.L., Echevarria, G. and Benizri, E., 2017. Nickel drives bacterial community diversity in the rhizosphere of the hyperaccumulator *Alyssum murale*. *Soil Biology and Biochemistry*, *114*, pp. 121–130. https://doi.org/10.1016/j.soilbio.2017.07.010

Maliji, D., Olama, Z. and Holail, H., 2013. Environmental studies on the microbial degradation of oil hydrocarbons and its application in Lebanese oil polluted coastal and marine ecosystem. *International Journal of Current Microbiology and Applied Sciences*, *2*(6), pp. 1–18.

Mani, I., 2020. Metagenomics approach for bioremediation: challenges and perspectives. In: *Bioremediation of Pollutants* (pp. 275–285). Elsevier. https://doi.org/10.1016/B978-0-12-819025-8.00012-0

Marques A. P., Rangel, A. O. and Castro, P. M. (2009). Remediation of heavy metal contaminated soils: phytoremediation as a potentially promising clean-up technology. *Critical Reviews in Environmental Science and Technology*, *39*, pp. 622–654. https://doi.org/10.1080/10643380701798272

Nigam, V.K., Arfi, T., Kumar, V. and Shukla, P., 2017. Bioengineering of nitrilases towards its use as green catalyst: applications and perspectives. *Indian Journal of Microbiology*, *57*(2), pp. 131–138. https://doi.org/10.1007/s12088-017-0645-5

Odili, U.C., Ibrahim, F.B., Shaibu-Imodagbe, E.M. and Atta, H.I., 2020. Comparative assessment of crude oil degradation by *Monocillium* sp. and *Aspergillus niger*. *Journal of Applied Sciences and Environmental Management*, *24*(5), pp. 815–820. https://doi.org/10.4314/jasem.v24i5.13

Paliya, S., Mandpe, A., Kumar, M.S. and Kumar, S., 2021. Aerobic degradation of decabrominated diphenyl ether through a novel bacterium isolated from municipal waste dumping site: identification, degradation and metabolic pathway. *Bioresource Technology*, *333*, p. 125208. https://doi.org/10.1016/j.biortech.2021.125208

Paranthaman, S.R. and Karthikeyan, B., 2015. Bioremediation of heavy metal in paper mill effluent using pseudomonas sp. *International Journal of Microbiology*, *1*(4), pp. 1–5.

Peng, K., Luo, C., You, W., Lian, C., Li, X. and Shen, Z., 2008. Manganese uptake and interactions with cadmium in the hyperaccumulator—*Phytolacca Americana* L. *Journal of Hazardous Materials*, *154*(1–3), pp. 674–681. https://doi.org/10.1016/j.jhazmat.2007.10.080

Pham, T.D. and Sillanpää, M., 2020. Ultrasonic and electrokinetic remediation of low permeability soil contaminated with persistent organic pollutants. In: *Advanced Water Treatment* (pp. 227–310). https://doi.org/10.1016/B978-0-12-819227-6.00004-8

Prabha, R. and Nigam, V.K., 2020. Biotransformation of acrylamide to acrylic acid carried through acrylamidase enzyme synthesized from whole cells of *Bacillus tequilensis* (BITNR004). *Biocatalysis and Biotransformation*, *38*(6), pp. 445–456. https://doi.org/10.1080/10242422.2020.1780211

Rautela, R., Arya, S., Vishwakarma, S., Lee, J., Kim, K.H. and Kumar, S., 2021. E-waste management and its effects on the environment and human health. *Science of the Total Environment*, p. 145623. https://doi.org/10.1016/j.scitotenv.2021.145623

Safiyanu, I., Isah, A.A., Abubakar, U.S. and Rita Singh, M., 2015. Review on comparative study on bioremediation for oil spills using microbes. *Research Journal of Pharmaceutical, Biological and Chemical Sciences*, *6*, pp. 783–790.

Sakakibara, M., Ohmori, Y., Ha, N.T.H., Sano, S. and Sera, K., 2011. Phytoremediation of heavy metal-contaminated water and sediment by *Eleocharis acicularis*. *CLEAN–Soil, Air, Water*, *39*(8), pp. 735–741. https://doi.org/10.1002/clen.201000488

Shraddha, R.S., Sehgal, S., Kamthaniya, M., Kumar, A., 2011. Laccase: microbial sources, production, purification, and potential biotechnological applications. *Enzyme Research*, *2011*, pp. 1–11. https://doi.org/10.4061/2011/21761

Singh, S., Anil, A.G., Khasnabis, S., Kumar, V., Nath, B., Sunil Kumar Naik, T.S., Subramanian, S., Kumar, V., Singh, J., Ramamurthy, P.C., 2021. Sustainable removal of Cr(VI) using graphene oxide-zinc oxide nanohybrid: adsorption kinetics, isotherms, and thermodynamics. *Environmental Research*, *203*, p. 111891. https://doi.org/10.1016/j.envres.2021.111891

Vamerali, T., Bandiera, M. and Mosca, G., 2010. Field crops for phytoremediation of metal-contaminated land. A review. *Environmental Chemistry Letters*, *8*(1), pp. 1–17. https://doi.org/10.1007/s10311-009-0268-0

Wainaina, S., Awasthi, M.K., Sarsaiya, S., Chen, H., Singh, E., Kumar, A., Ravindran, B., Awasthi, S.K., Liu, T., Duan, Y. and Kumar, S., 2020. Resource recovery and circular economy from organic solid waste using aerobic and anaerobic digestion technologies. *Bioresource Technology*, *301*, p. 122778. https://doi.org/10.1016/j.biortech.2020.122778

Wang, X., Rathinasabapathi, B., Oliveira, L.M.D., Guilherme, L.R. and Ma, L.Q., 2012. Bacteria-mediated arsenic oxidation and reduction in the growth media of arsenic hyperaccumulator *Pteris vittata*. *Environmental Science & Technology*, *46*(20), pp. 11259–11266. https://doi.org/10.1021/es300454b

Xu, R., Li, Q., Meng, F., Yang, Y., Xu, B., Yin, H. and Jiang, T., 2020. Bio-oxidation of a double refractory gold ore and investigation of preg-robbing of gold from thiourea solution. *Metals*, *10*(9), p. 1216. https://doi.org/10.3390/met10091216

Yu, G.W., Laseter, J. and Mylander, C., 2011. Persistent organic pollutants in serum and several different fat compartments in humans. *Journal of Environmental and Public Health*, *2011*. https://doi.org/10.1155/2011/417980

Zangi-Kotler, M., Ben-Dov, E., Tiehm, A. and Kushmaro, A., 2015. Microbial community structure and dynamics in a membrane bioreactor supplemented with the flame retardant dibromoneopentyl glycol. *Environmental Science and Pollution Research*, *22*(22), pp. 17615–17624. https://doi.org/10.1007/s11356-015-4975-8

Zhu, X., Ni, X., Liu, J. and Gao, Y., 2014. Application of endophytic bacteria to reduce persistent organic pollutants contamination in plants. *CLEAN–Soil, Air, Water*, *42*(3), pp. 306–310. https://doi.org/10.1002/clen.201200314

3

Constructed Wetland-Microbial Fuel Cell Technology during Wastewater Treatment: Progress, Challenges, and Opportunities

Kalp Bhusan Prajapati
Central University of Haryana

Manoj Kumar, Neeraj Kumar Singh, and Rajesh Singh
Central University of Gujarat

CONTENTS

3.1 Introduction

An increase in population, industrialization, and urbanization results in a rise in wastewater discharge volume and a shrinkage of the world's freshwater resources available on earth. In the present era, the demand for the fulfillment of fresh water for the whole world seems impossible. Presently, a large number of wastewater treatment technologies are available, but none of them fulfill the criteria for efficient technology due to their high operational cost and maintenance along with the production of a large quantity of sludge volume (Dutta et al., 2021; Wainaina et al., 2020). There is a requirement for the advancement of a cost-effective, less maintenance requirement and higher energy recovery efficiency eco-friendly technology. Life cycle assessment study reveals constructed wetlands as low-cost engineered systems that rely on nature's technology of wetland for the wastewater treatment. Wastewater treatment through these low-cost engineered treatment processes results in highly efficient wastewater treatment, along with energy recovery.

Constructed wetlands (CWs) are biologically and physically assembled effluent treatment units that are planned and constructed to get the benefit of usual practices along with the interactions between flora present in wetland, soil, and microorganisms that aid in regenerating wastewater (Vymazal,

DOI: 10.1201/9781003231738-4

2010, Yang et al., 2019; Shingare et al., 2019). This technology of wastewater treatment was first developed by Dr. Kathe Seidel in the 1960s and by Reinhold Kickuth in the 1970s (Srivastava et al., 2019). Earlier these developed constructed wetlands were applicable for domestic wastewater treatment and later their application begins in the treatment of a wide variety of wastewaters. The mechanism involved in wastewater treatment through constructed wetlands includes different steps like plant uptake, sedimentation, microbial activity, filtration, volatilization, precipitation, adsorption, etc. (Vymazal, 2007; Kadlec and Wallace, 2009; Faulwetter et al., 2009; US EPA 2000). According to Corbella Vidal (2017) constructed wetland has the potential to treat a wide range of wastewaters i.e., industrial, municipal and agricultural. These are efficient in maintenance and operation and can transform the complex persistent compounds that are present in wastewater into simple by-products. In the last two decades, the use of manmade wetlands for the treatment of wastewater has grown in popularity as a result of their installation, operation, and however, the long treatment procedure is the fundamental downside of this method from a maintenance perspective and to overcome these several efforts like effluent recirculation, artificial aeration, bioaugmentation, electron donor supplementation, etc. have been implemented (Wu et al., 2014). Later Yadav (2010) successfully integrate the concept of MFCs (Microbial Fuel Cells) with constructed wetlands to speed up the treatment process. And this integrated new and emerging technology CW-MFCs gain vast interest over the years due to their additional advantage of electricity production along with contaminant removal. Therefore the present chapter aimed to summarize the mechanism, types, applications, factors affecting the operation mechanism, advancement in the CW-MFCs technology, challenges and future perspectives were systematically discussed.

3.2 Constructed Wetland and Microbial Fuel Cells' Operational Mechanism

The microbial fuel cell and constructed wetlands are hybrid expertise that integrates MFCs with engineered wetlands by the combined action of the substrate, plants, and microorganisms, physically, chemically, and biologically resulting in wastewater remediation and bioenergy production. This integrating hybrid technology associates the benefit of two technologies to be capable of fulfilling the enhanced level of wastewater treatment and bioenergy recovery (Li et al., 2021). CW-MFCs comprise two electrodes, a cathode and an anode the anode electrode is placed in the anaerobic zone of the constructed wetland, while the cathode is placed in the aerobic zone. A fibrous substance or a proton exchange membrane separates these two zones, which are made up of anode and cathode (Figure 3.1). The lower anaerobic

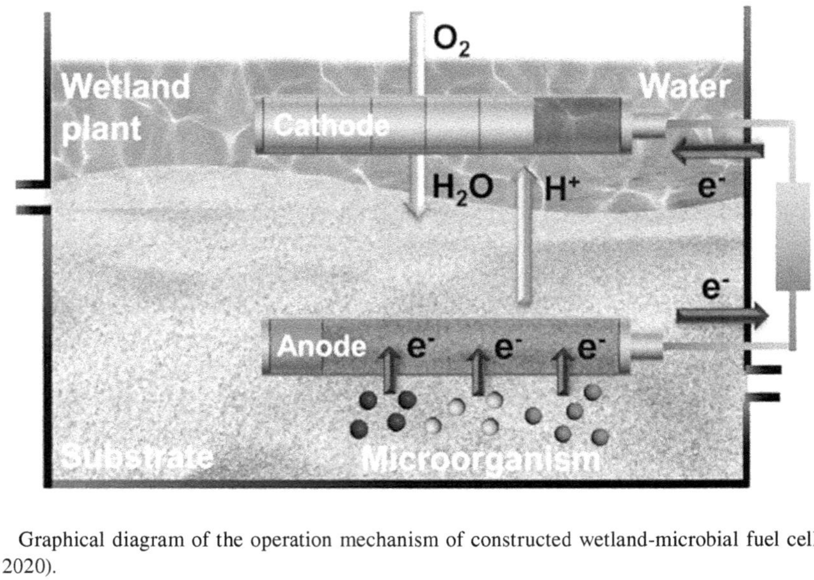

FIGURE 3.1 Graphical diagram of the operation mechanism of constructed wetland-microbial fuel cells (CW-MFCs) (Wang et al., 2020).

zone of the reactor is filled through a stratum of gravels or boulders or a wide variety of media, e.g., granular activated carbon (GAC), alum sludge or zeolite, etc. as well as for the anodic area electrode's support material. This supporting intermediate also helps in creating the anticipated surroundings for the oxidation of contaminants by microbes along with electrons transfer processes (Yadav et al., 2012, 2018; Yang et al., 2018a; Doherty et al., 2015). The generated electrons are shifted towards the anode, from where they move towards the cathode with the help of a conductive substrate and electron acceptor with a high redox potential (i.e., oxygen) (Logan et al., 2006; Rabaey et al., 2007).

The electron transfer phenomenon is one of the greatest significant aspects influencing bioenergy production in CW-MFCs, and it can occur either directly between the electrogens and the biofilm produced on the electrode surface or indirectly via an electron mediator (Logan, 2009) However, the phenomenon of direct transfer of electrons might also be restricted because microbes' that work as a catalytic are extremely affected by deviations in the surrounding environment such as variations in pH, temperature, dissolved oxygen and chemical species present (Cao et al., 2019; Kumar et al., 2016; 2017). At the anode, the compounds mainly oxidized include simple carbohydrates like glucose or acetate that exist naturally in the surrounding medium or result from the microbial breakdown of persistent organic material present in wastewater effluent (Min and Logan, 2004; Reimers et al. 2001; Rabaey and Vertraete, 2005). Several inter-dependent chemicals, biological and operational factors including rate of substrate conversion, internal resistance, chemical oxygen demand (COD) load in the anodic and cathodic region, pH, temperature, type of electrode material, the electrode's total surface area, When measuring the efficiency of CW-MFCs, the distance between the electrodes is also considered (Srivastava et al., 2019). Whereas some of the features like over-potential of anode and cathode, rate of substrate conversion and type of electrode material, adjust the redox conditions and regulate the thermodynamically favorable and unfavorable circumstances for the flow of electrons (Srivastava et al., 2020)

3.3 Various Designs and Operations of Constructed Wetlands

According to the dominant plant vegetation species, the CW-MFCs for effluent treatment and bioenergy recovery have been classified into submerged Macrophytes, floating leaf, free floating, and rooted emergent (Brix and Schierup, 1989). Whereas Srivastava et al. (2019) based on wastewater flow path classified into-free water surface and subsurface slow. In the case of free water surface, the wastewater is drained above the ground, whereas in subsurface the water leach out through the porous media such as gravels or boulders.

3.4 Factors Affecting the Operational Mechanisms

The level and circumstances of numerous aspects before wastewater pre-treatment and post-treatment operation of technology affect the performance and output efficiency (Du et al., 2007). Bacteria are used in MFCs to transform organic substrate into electricity via electron transfer (Bond et al., 2002; Logan and Regan, 2006a, b). Microorganisms contribute electrons to the anode in three ways (Logan et al., 2006). Ocean sediments, fresh-water sediments, soil, wastewater, and activated sludge are also abundant sources of these microorganisms (Du et al., 2007). The use of bacteria had previously been studied by screening and detecting bacteria, then creating a chromosomal library of these bacteria that may generate electric power through organic matter degradation (Logan et al., 2006).

3.4.1 Electrode Material and Separators

In terms of e-allocation rate, electrochemical effectiveness, and microbial adhesion, MFC performance is usually determined by the electrode materials. MFCs should provide the best power density at the lowest cost if they are used properly. Also, they must have catalytic capabilities in cathodic materials for O_2 reduction, and anodic and cathodic materials must share the following characteristics, the porosity of electrical conductivity; stability and durability of surface area; availability, and price (Mustakeem, 2015; Aelterman, 2009; Aghababaie et al., 2015; Allen and Bennetto, 1993).

3.4.2 Anode Materials

The anode compartment for anaerobic bacteria, which has a substrate inside, is the most important aspect of MFCs, the microbes, non-meditators, and mediators, and the electron acceptor role performed by the electrode anode. The MFC performance is prominently affected by the type of material used in the development of electrodes and the set-up of the equipment (Logan et al., 2006; Oh et al., 2004). The electrode material should ideally have a large surface area; excellent conductivity; low resistance; excellent compatibility with microbes; durability; chemical stability along with anti-corrosion characteristics. As the anode, graphite rods, carbon cloths, and fiber brushes often comprise carbon material that contains graphite (Logan et al., 2006).

3.4.3 Cathode Materials

The power capacity of MFCs is largely determined by the cathode, the electrode materials should be easy to catch protons and have a significant oxidation state. Presently, Carbon cloths are employed in this technology, whereas graphite electrodes and carbon paper are generally used as a cathode (Zhou et al., 2011). The use of platinum as a cathode is extremely novel, which enhances the effectiveness of microbial fuel cells by boosting the rate of the reactions and lowering the stimulation energy of biochemical responses that happens by the cathode (Zhou et al., 2011; Moon et al., 2006). The researchers discovered that combining graphite and platinum triples the power density of pure graphite (Logan et al., 2006) (Table 3.1).

3.4.4 Separator

In microbial fuel cells, the separator is crucial. Although separators have advanced rapidly in recent years, there are still certain obstacles to overcome, such as oxygen leaks and proton transfer limits, the internal resistance rises, lowering MFC performance, as a result, MFCs' practical application is limited.

TABLE 3.1

Efficiency of Different Types of Materials Used in Electrode Preparation

Elect(rode Material	Maximum Power Density	References
Graphite plate	15.7[a]	Yadav et al. (2012)
Graphite plate	9.4[a]	Zhao et al. (2013)
GAC	0.302[c]	Fang et al. (2013)
Graphite plate	0.15[a]	Villasenor et al. (2013)
GAC	43.63[a]	Srivastava et al. (2015)
Anode-GAC Cathode- GAC	0.852[c]	Fang et al. (2015)
Granular graphite	0.268[c]	Doherty et al. (2015)
Carbon felt	6.12[a]	Oon et al. (2015)
Cylindrical graphite rod	36[a]	Corbella et al. (2015)
Activated carbon	93[b]	Oon et al. 2015)
Cylindrical graphite rod	131[a]	Xu et al. (2016b)
Graphite plate	112[a]	Wu et al. (2017)
Anode-graphite felt Cathode-carbon cloth	31.04[b]	Srivastava et al. (2017)
GAC	0.20[c]	Song et al. (2017)

Source: Adapted and modified table from previous studiesYadav et al. (2018).

[a] Maximum Power Density (mW/m2)

[b] Maximum Power Density (mW/m3)

[c] Maximum Power Density (W/m3)

TABLE 3.2

Major Drawbacks and Advantages of Different Types of Separators

Drawbacks	Advantages	Separator
Significant internal resistance due to excessive electrode spacing, oxygen permeation, major loss of substrate, fouling of cathode and deactivation	High rate of Proton transfer, power density is high, costs are low, and the configuration is simple	None
Proton transport restrictions, pH splitting, membrane fouling, and high costs	Low oxygen permeability and substrate loss, effective separation of anode and cathode	Ion exchange membranes
Internal resistance and oxygen permeability are both high	Proton transport is high, pH accumulation is minimal, and expenses are mild	Microporous membranes
Biofilm on the filter causes a lot of substrate loss, and the filter isn't very durable	Cost-efficient, high proton transport efficiency, and low pH	Coarse-pore structures
Internal resistance is high	Configuration is simple, and the costs are modest	Salt bridge

Source: Adapted from previous study from Li et al. (2011).

When it comes to MFCs, the electrons generated in the anodic chamber are sent to the anodic electrode and then to the cathodic zone through an external connection, when protons travel from the anode region towards the cathode they have to pass through the separator (Logan et al., 2006). Although it has been claimed that a single-chamber MFC without a separator produces a better power density (Liu and Logan, 2004; Yang et al., 2009). Lack of separator enhances the diffusion of substrate and oxygen and thereby reducing the population of anodic microorganisms, bio electrocatalytic activity, and coulombic efficiency (Liu et al., 2005). To mitigate the detrimental effects of O_2 dissemination, a sufficient distance between electrodes must be maintained, consequently, there will be a great deal of internal opposition (Liu et al., 2005; Cheng et al., 2006) (Table 3.2).

3.4.5 Types of Substrates

An important component of CW-MFCs is the substrate, also called media, which serves as a support matrix/material or filler material. Substrates play an important function in the physical, eliminating chemical and more significant biological functions (filtration, trap, adsorption, and biodegradation) (Yang et al. 2018b). The hydraulic conductivity and the potential for pollutant absorption are considered while selecting a substrate. In the case of a major obstruction, there would be a lack of hydraulic conductivity, reducing the capacity of the process, and low substrate biosorption could potentially affect long period elimination efficiencies of CWs (Wang et al., 2010). The selection of CWs substrates has been the subject of countless experiments, in particular, to effectively eliminate phosphorus from wastewater, and the most commonly utilized substrates are mostly naturally existing materials, artificially synthesized media, and by-products obtained from industries (Saeed et al., 2012, Yan et al., 2014, Chong et al., 2013, Rahman et al., 2020).

3.4.6 Constructed Wetland Plants

In developed wetlands, just a few plant kinds were commonly used (Vymazal, 2013). The focus of the ongoing study on green CW designs should be on the selection of plants used in CWs (Vymazal, 2011). As seen in Table 3.3 comparison of COD removal performance between different macrophytes based on earlier studies. Apart from adapting to extreme weather conditions, resilience to both waterlogged-anoxic and hyper-eutrophic conditions, for plant selection, factors such as pollution absorption capabilities are recommended. Macrophytes commonly used in CW management include growing p and waterlogged plants, including free-floating plants. While over 150 different species of macrophyte have already been applied in CWs around the world, but in reality, only a few plant species are generally cultivated in engineered constructed wetlands (Vymazal, 2013).

TABLE 3.3
Comparison of COD Removal Performance between Different Macrophytes Based on Earlier Studies

S. No.	Macrophytes	COD Removal (%)	Reference
1	*Canna indica*	75.00%	Yadav et al. (2012)
2	*Ipomoea aquatic*	85.70%	Fang et al. (2013)
3	*Phragmites australis*	90.0%–95.0%	Villaseñor et al. (2013)
4	*Ipomoea aquatic*	85.70%	Liu et al. (2013)
5	*Iris pseudacorus*	85.70%	Wu et al. (2017)
6	*Phragmites australis*	64.00%	Doherty et al. (2015a, b)
7	*Typha latifolia*	91.20%	Oon et al. (2016)
8	*Elodea nuttallii*	97%–98%	Oon et al. (2017)
9	*Eichhornia crassipes*	85.14%, 74.70% & 83.9%	Kumar and Singh., (2020a, b, 2021)

Source: Adapted from previous study Jingyu et al. (2020).

3.5 Application of Constructed Wetland-Microbial Fuel Cells

Constructed wetlands integrated microbial fuel cells are renewable energy sources for both bioelectricity and the treatment of wastewater. At the air-water contact, two very different technologies feature an aerobic area and an anaerobic area at the bottom, where another anode and cathode are positioned properly (Guadarrama-Pérez et al., 2019). Furthermore, the MFCs tool for wastewater treatment, which recovers bioenergy from the waste to offset the impact of climate change, will be an appropriate contender. This technique has a lot of potential uses, including bioelectricity generation, biogas generation, treatment of wastewater, as well as the elimination of hazardous chemicals (Giorgi and Leccese, 2013). At this time, treatments using bioelectrochemical systems (BESs) have emerged as a possible method for producing bioelectricity from organic substrates acquired during the wastewater treatment process (Mohanakrishna et al., 2015).

3.6 Advancement in Constructed Wetlands by Integration of Microbial Fuel Cell

The treatment of wastewater technology based on MFC has been investigated for decades. The concept combination of MFC with CWs was firstly reported by (Yadav et al., 2012). Integration of a CW system with MFCs is a technology that seems promising for bio-electricity production and wastewater treatment that is both sustainable and beneficial (Doherty et al., 2015a; Oon et al., 2015). This combined technology CW-MFC generates electricity by degradation of an organic compound by bacteria, through an oxidation process in the soil of CWs (Oon et al., 2015). Nowadays this integrated technology CW-MFC has widely used in dye degradation in wastewater (Fang et al., 2015), Sewage treatment (Yadav et al., 2012), Swine slurry treatment (Doherty et al., 2015b). Kumar and Singh (2020a) investigated the CWs by combining them with BES. Their integrated laboratory BES design was of a working volume of 8.5 L (Figure 3.2). It can efficiently remove nearly 83.98% of SO_4^{2-}, 91.04% of PO_4^{3-} -P, 83.09%, 83.09% of NO_3^- -N, and 96.66% of NH_4^+-N. Their MFCs design was able to produce an average current density of 5.423 and 1.723 mW/m^2 of power density interestingly they noticed that an applied voltage of 15 mV does not favour the accumulation of Nitrate. Whereas applied voltage of 30 mV does not favour the removal of ammonia although removal of other pollutants at this voltage was reported by them. Their design was very beneficial in further designing of Microbial electrolysis cells (MECs) integrated CWs treatment systems and can be a substitute solution for the water-energy during the treatment of wastewater effluent.

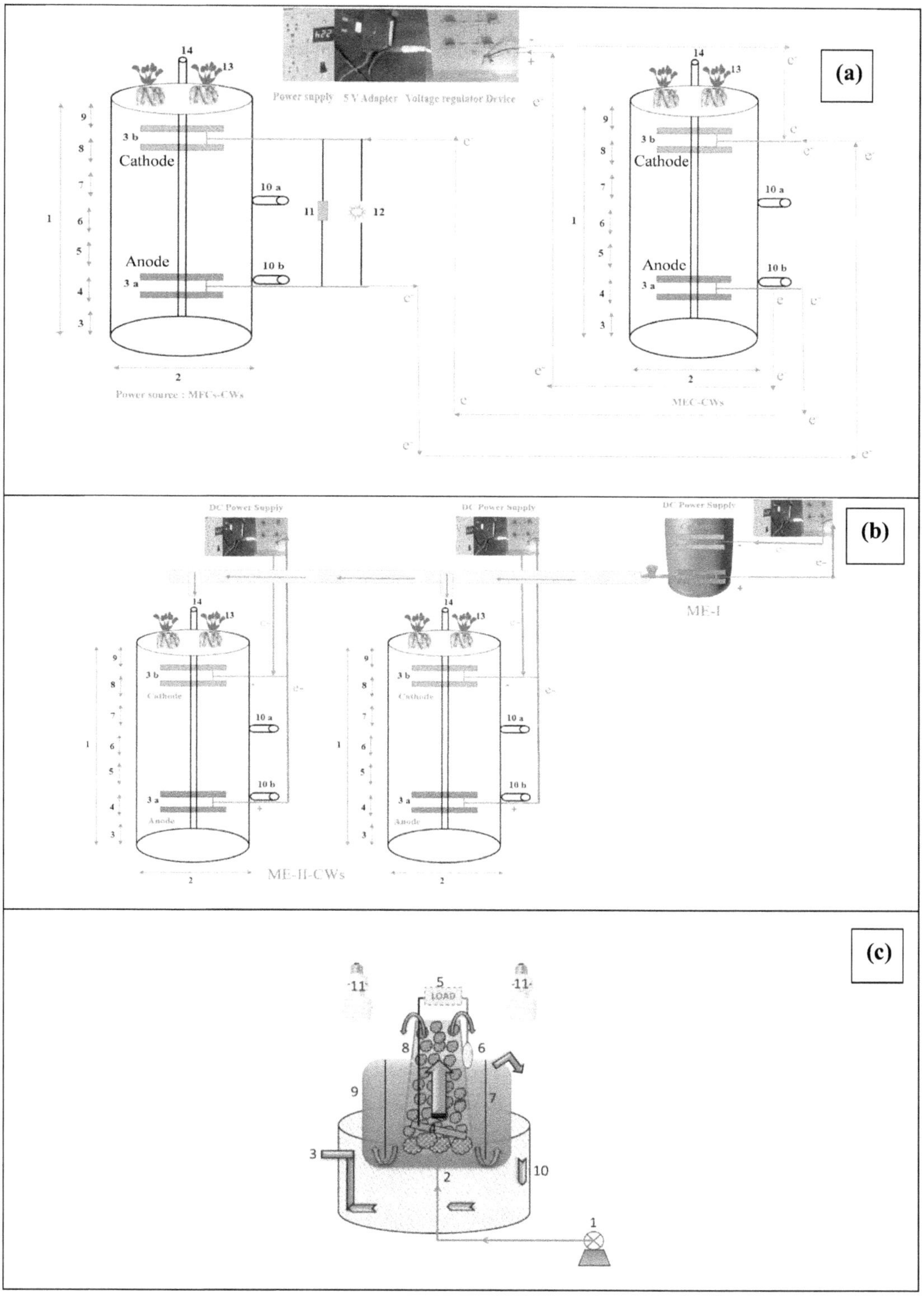

FIGURE 3.2 Microbial fuel cells (MFCs) integrated constructed wetlands (CWs). (Adapted from previous study (a) Kumar and Singh (2020a); (b) Kumar and Singh (2020b); (c) Gupta et al. (2021).)

TABLE 3.4

Detail Performance of CWs-MFCs with their COD Removal Efficiency, Electrical Performance and Type of Wastewater Used

S. No.	Type of Waste Water (Synthetic Wastewater[a])	COD Removal	Volume of CWs/Flow Rate	Electrical Performance	Reference
1	SWW	75%	5L	15.73 mW/m^3	Yadav et al. (2012)
2	SWW	95%	150L	1161 mV	Villasenor et al. (2013)
3	Domestic wastewater	65%	30 L/d	250 mV	Camacho et al. (2014)
4	SWW	92.1%	83 mL/h	15.1 mW/m^3	Yakar et al. (2018)
5	SWW	67.92%–78.3%	—	100–250 mV	Xie et al. (2018)
6	SWW domestic	91.7%	30L	592–623 mV	Tang et al. (2019)
7	SWW	95.4%–98.9%	65.0L	426–674 mV	Srivastava et al. (2020)
8	Sewage wastewater	72.3% and 75.3%	8.5L	1.723 mW	Kumar et al. (2020a)

SWW=Synthetic Wastewater

Kumar and Singh (2020b) designed two-phase straight down flow constructed wetlands integrated with a BES (Bioelectrochemical System) operated in continuous mode for pretreatment of higher pollutants in continuous mode. Their pre- cleaner design consists of a 25 L barrel-shaped drum installed with two graphite electrodes having a length of 6 cm and diameter of 0.6 cm at depth of 17 cm from the bottom (Figure 3.2a). Pre-cleaner BESs system removes nearly 25.55%–35.18% of NO_3^--N, 60.41%–85.78% of NH_4^+-N, 57.80–84.65% of total nitrogen, 37.24%–70.08% of TKN, 38.89%–63.40% of PO_4^{3-}-P, 49.53%–76.06% of SO_4^{2-} and 25.83%–74.70% of COD from municipal sewage water. This specific installed BES system in CWs could be applicable for wastewater treatment.

Tamta et al. (2020) designed 12 vertically-arranged CWs of stack type. The overall working volume of the stack was 12.0 L. These CWs were coupled with MFCs. CWs show NH_4^+-N (90.4%) at 15 kΩ, NO_3^--N (86.9%) at 0.08 kΩ, and COD degradation efficiency 98.5% at 0.08 kΩ. Whereas at 15 kΩ the extreme Power density of 30.85 mW/m^3 was produced.

Gupta et al. (2021) designed algal-assisted CW-MFCs for effluent treatment followed by electricity production (Figure 3.2c). The algal biomass was used to aerate the cathode region. Maximum current densities and power density produced by these CWs-MFCs were reported 235.0 and 33.14 mW/m^3 respectively. Various study on the integration of CWs-MFCs with their p erformance and energy output has been shown in Table 3.4.

3.7 Challenges and Future Perspectives

The major constraints associated with the wide applicability of CWs in the field of wastewater treatment are slow rate treatment. Intensive research is required to increase the efficiency and faster the process of wastewater treatment in CWs. The CWs operational condition has been modified by integrating it with effluent recirculation process, artificial aeration process, drop aeration process, bioaugmentation process, and integrating of CWs with vermicomposting technology could fasten the biodegradation process. CW also faces major issues during its applicability in geographical areas of cold climate conditions in treating low C/N ratio wastewater. Sustainable innovative research in the area of CWs by collaborating engineers and scientists has to be done to overcome these challenges. Lab-scale designed scale when operated for their field applicability sometimes failed to treat wastewater due to differences in environmental operational conditions has to be carefully addressed with the reason for the failure of CWs operations and its field suitability.

Recent studies on laboratory designs of CWs coupled with MFC and MEC technology have been found to remove the pollutants in an eco-friendly and cost-effective way. These integrated designs are applied at the industrial level for bioelectricity along with wastewater treatment (Clauwaert et al., 2007). However, despite extraordinary advancements in research in this field, it faces major obstacles like low power density during wastewater treatment (Xu et al., 2016a). Effective research should be carried out in this field of design so that it could treat wastewater in large volumes with high electricity yields. The cost of CW depends on the cost of electrodes used in the constructed wetlands; therefore cheap electrodes should be explored. The potential area of exploring wastewater characteristics like high organics, organic pollutants, heavy metal removal, nitrification, high denitrification, pesticides, etc. for high energy yield. The performance of MFCs in CWs is hampered by methanogens (Arends et al., 2014; Fang et al., 2013). Therefore successful mechanism for the suppression of methanogens in the CWs should be explored.

3.8 Conclusion

The laboratory scale-up design CW-MFCs has shown interesting results. The gap between reality and practical scalability of CW-MFCs technology lies in the design and operational condition. There are several drawbacks associated with design and operation that need to be overcome for making this technology more proficient. This book chapter provides a summary of the CWs applications, their design and operations, their mechanism and factor affecting the performance CW-MFCs and its challenges. Additionally, this chapter compiles various obligatory studies with their concise result studied in the field of constructed wetlands.

REFERENCES

Aelterman, P. 2009. *Microbial Fuel Cells for the Treatment of Waste Streams with Energy Recovery.* Ghent University, Ghent.

Aghababaie, M., Farhadian, M., Jeihanipour, A., Biria, D. 2015. Effective factors on the performance of microbial fuel cells in wastewater treatment a review. *Environmental Technology Reviews*, *4*(1), 71–89.

Allen, R.M., Bennetto, H.P. 1993. Microbial fuel-cells. *Applied Biochemistry and Biotechnology*, *39*(1), 27–40.

Arends, J.B., Speeckaert, J., Blondeel, E., De Vrieze, J., Boeckx, P., Verstraete, W., Rabaey, K., Boon, N., 2014. Greenhouse gas emissions from rice microcosms amended with a plant microbial fuel cell. *Applied Microbiology and Biotechnology*, *98*(7), 3205–3217.

Bond, D.R., Holmes, D.E., Tender, L.M., Lovley, D.R. 2002. Electrode-reducing microorganisms that harvest energy from marine sediments. *Science*, *295*(5554), 483–485.

Brix, H.; Schierup, H.H. 1989. The use of aquatic macrophytes in water-pollution control. *Ambio*, 28, 100–107

Camacho, J.V., Vico, M.D.C.M., Andrés, M., Rodrigo, R., Morales, F.J.F., Cañizares, P.C. 2014. Energy production from wastewater using horizontal and vertical subsurface flow constructed wetlands. *Environmental Engineering and Management Journal*, *13*(10), 2517–2523.

Cao, Y., Mu, H., Liu, W., Zhang, R., Guo, J., Xian, M., Liu, H. 2019. Electricigens in the anode of microbial fuel cells: pure cultures versus mixed communities. *Microbial Cell Factories*, *18*(1), 1–14.

Cheng, S., Liu, H., Logan, B.E. 2006. Increased power generation in a continuous flow MFC with advective flow through the porous anode and reduced electrode spacing. *Environmental Science & Technology*, *40*, 2426–2432.

Chong, H.L., Chia, P.S., Ahmad, M.N. 2013. The adsorption of heavy metal by Bornean oil palm shell and its potential application as constructed wetland media. *Bioresource Technology*, *130*, 181–186.

Clauwaert, P., Van der Ha, D., Boon, N., Verbeken, K., Verhaege, M., Rabaey, K., Verstraete, W. 2007. Open air biocathode enables effective electricity generation with microbial fuel cells. *Environmental Science & Technology*, *41*(21), 7564–7569.

Corbella Vidal, C. 2017. Constructed wetland microbial fuel cells: electricity generation, treatment efficiency improvement, COD bioindication and clogging assessment.

Corbella, C., Guivernau, M., Viñas, M., Puigagut, J., 2015. Operational, design and microbial aspects related to power production with microbial fuel cells implemented in constructed wetlands. *Water Research*, *84*, 232–242.

Doherty, L., Zhao, Y., Zhao, X., Hu, Y., Hao, X., Xu, L., Liu, R., 2015. A review of a recently emerged technology: constructed wetland-microbial fuel cells. *Water Research*, *85*, 38–45.

Doherty, L., Zhao, Y., Zhao, X., Wang, W. 2015a Nutrient and organics removal from swine slurry with simultaneous electricity generation in an alum sludge-based constructed wetland incorporating microbial fuel cell technology. *Chemical Engineering Journal, 266*, 74–81

Doherty, L., Zhao, X., Zhao, Y., Wang, W. 2015b. The effects of electrode spacing and flow direction on the performance of microbial fuel cell-constructed wetland. *Ecological Engineering, 79*, 8–14.

Du, Z., Li, H., Gu, T., 2007. A state-of-the-art review on microbial fuel cells: a promising technology for wastewater treatment and bioenergy. *Biotechnology Advances, 25*(5), 464–482

Dutta, D., Arya, S., Kumar, S. 2021. Industrial wastewater treatment: current trends, bottlenecks, and best practices. *Chemosphere, 285*, 131245. https://doi.org/10.1016/j.chemosphere.2021.131245

Fang, Z., Song, H., Cang, N., Li, X. 2015. Electricity production from Azo dye wastewater using a microbial fuel cell coupled constructed wetland operating under different operating conditions, *Biosensors and Bioelectronics, 68*, 135–141.

Fang, Z., Song, H.-L., Cang, N., Li, X.-N. 2013. Performance of microbial fuel cell coupled constructed wetland system for decolorization of azo dye and bioelectricity generation. *Bioresource Technology, 144*, 165–171.

Faulwetter, J. L., Gagnon, V., Sundberg, C., Chazarenc, F., Burr, M. D., Brisson, J., Stein, O. R. 2009. Microbial processes influencing performance of treatment wetlands: a review. *Ecological Engineering, 35*(6), 987–1004.

Giorgi L, Leccese F. 2013. Fuel cells: technologies and applications. *The Open Fuel Cells Journal, 6*(1), 1–20.

Guadarrama-Pérez, O., Gutiérrez-Macías, T., García-Sánchez, L., Guadarrama-Pérez, V.H., Estrada-Arriaga, E.B., 2019. Recent advances in constructed wetland-microbial fuel cells for simultaneous bioelectricity production and wastewater treatment: a review. *International Journal of Energy Research, 43*(10), 5106–5127.

Gupta, S., Nayak, A., Roy, C., Yadav, A.K. 2021. An algal assisted constructed wetland-microbial fuel cell integrated with sand filter for efficient wastewater treatment and electricity production. *Chemosphere, 263*, 128132.

Jingyu, H., Miwornunyuie, N., Ewusi-Mensah, D., Koomson, D.A. 2020. Assessing the factors influencing the performance of constructed wetland–microbial fuel cell integration. *Water Science and Technology, 81*(4), 631–643.

Kadlec, R. H., Wallace, S. 2009. *Treatment Wetlands*. CRC Press. Boca Raton, FL.

Kumar, A., Hsu, L. H. H., Kavanagh, P., Barrière, F., Lens, P. N., Lapinsonnière, L., Leech, D. 2017. The ins and outs of microorganism–electrode electron transfer reactions. *Nature Reviews Chemistry, 1*(3), 1–13.

Kumar, M., Singh, R. 2020a. Sewage water treatment with energy recovery using constructed wetlands integrated with a bioelectrochemical system. *Environmental Science: Water Research & Technology, 6*(3), 795–808.

Kumar, M., Singh, R. 2020b. Installation of a bioelectrochemical system as a pre-cleaner in a constructed wetland with higher pollutant loading under continuous mode. *Environmental Science: Water Research & Technology, 6*(12), 3326–3340.

Kumar, M., Singh, R. 2021. Area-based speciation kinetic analysis of multipollutant removal in constructed wetlands to enhance the treatment efficiency. *Environmental Science: Water Research & Technology, 7*(6), 1090–1102.

Kumar, R., Singh, L., Zularisam, A.W. 2016. Exoelectrogens: recent advances in molecular drivers involved in extracellular electron transfer and strategies used to improve it for microbial fuel cell applications. *Renewable and Sustainable Energy Reviews, 56*, 1322–1336.

Li, K., Qi, J., Zhang, F., Miwornunyuie, N., Amaniampong, P. S., Koomson, D. A., Samwini, A.M.N. 2021. The role of wetland plants on wastewater treatment and electricity generation in constructed wetland coupled with microbial fuel cell. *Applied Sciences, 11*(16), 7454.

Li, W.W., Sheng, G.P., Liu, X.W., Yu, H.Q. 2011. Recent advances in the separators for microbial fuel cells. *Bioresource Technology, 102*(1), 244–252.

Liu, H., Cheng, S., Logan, B.E., 2005. Power generation in fed-batch microbial fuel cells as a function of ionic strength, temperature, and reactor configuration. *Environmental Science & Technology, 39*, 5488–5493.

Liu, H., Logan, B.E. 2004. Electricity generation using an air-cathode single chamber microbial fuel cell in the presence and absence of a proton exchange membrane. *Environmental Science & Technology, 38*, 4040–4046.

Logan, B.E. 2009. Exoelectrogenic bacteria that power microbial fuel cells. *Nature Reviews Microbiology*, *7*(5), 375–381.

Logan, B.E., Hamelers, B., Rozendal, R., Schröder, U., Keller, J., Freguia, S., Aelterman, P., Verstraete, W., Rabaey, K., 2006a. Microbial fuel cells: methodology and technology. *Environmental Science & Technology*, *40*(17), 5181–5192.

Logan, B. E., Hamelers, B., Rozendal, R., Schröder, U., Keller, J., Freguia, S., Rabaey, K. 2006b. Microbial fuel cells: methodology and technology. *Environmental Science & Technology*, *40*(17), 5181–5192.

Logan, B.E., Regan, J.M., 2006a. Electricity-producing bacterial communities in microbial fuel cells. *Trends Microbiology*, *14*(12), 512–518.

Logan, B.E., Regan, J.M., 2006b. Microbial fuel cells—challenges and applications. *Environmental Science & Technology*, *40*(17), 5172–5180.

Min, B., Logan, B. E. 2004. Continuous electricity generation from domestic wastewater and organic substrates in a flat plate microbial fuel cell. *Environmental Science & Technology*, *38*(21), 5809–5814.

Mohanakrishna G, Srikanth S, Pant D. 2015. Bioelectrochemical systems (BES) for microbial electroremediation: an advanced wastewater treatment technology. In: *Applied Environmental Biotechnology: Present Scenario and Future Trends* (G. Koushik, ed.). Salmon Tower Building, New York City, pp. 145–167.

Moon, H., Chang, I.S., Kim, B.H. 2006. Continuous electricity production from artificial wastewater using a mediator-less microbial fuel cell. *Bioresource Technology*, 97(4):621–627.

Mustakeem, M. 2015. Electrodematerials formicrobial fuel cells: nanomaterial approach. *Materials for Renewable and Sustainable Energy*, *4*(4), 1–11.

Oh, S., Min, B., Logan, B.E. 2004. Cathode performance as a factor in electricity generation in microbial fuel cells. *Environmental Science & Technology*, *38*(18), 4900–4904.

Oon, Y.L., Ong, S.A., Ho, L.N., Wong, Y.S., Dahalan, F.A., Oon, Y.S., Lehl, H.K., Thung, W.E. 2016. Synergistic effect of up-flow constructed wetland and microbial fuel cell for simultaneous wastewater treatment and energy recovery. *Bioresource Technology*, *203*, 190–197.

Oon, Y. L., Ong, S. A., Ho, L. N., Wong, Y. S., Dahalan, F. A., Oon, Y.S., Lehl, H.K., Thung, W.E., Nordin, N. 2017. Role of macrophyte and effect of supplementary aeration in up-flow constructed wetland microbial fuel cell for simultaneous wastewater treatment and energy recovery. *Bioresource Technology*, *224*, 265–275.

Oon, Y.L., Ong, S.A., Ho, L.N., Wong, Y.S., Oon, Y.S., Lehl, H.K., Thung, W.E. 2015. Hybrid system up-flow constructed wetland integrated with microbial fuel cell for simultaneous wastewater treatment and electricity generation. *Bioresource Technology*, *186*, 270–275.

Rabaey, K., Rodríguez, J., Blackall, L.L., Keller, J., Gross, P., Batstone, D., Nealson, K.H. 2007. Microbial ecology meets electrochemistry: electricity-driven and driving communities. *The ISME Journal*, *1*(1), 9–18.

Rabaey, K., Verstraete, W. 2005. Microbial fuel cells: novel biotechnology for energy generation. *Trends in Biotechnology*, *23*(6), 291–298.

Rahman, M.E., Bin Halmi, M.I.E., Bin Abd Samad, M.Y., Uddin, M.K., Mahmud, K., Abd Shukor, M.Y., Sheikh Abdullah, S.R., Shamsuzzaman, S.M. 2020. Design, operation and optimization of constructed wetland for removal of pollutant. *International Journal of Environmental Research and Public Health*, *17*(22), 8339.

Reimers, C. E., Tender, L. M., Fertig, S., Wang, W. 2001. Harvesting energy from the marine sediment-water interface. *Environmental Science & Technology*, *35*(1), 192–195.

Saeed, T.; Afrin, R.; Muyeed, A.A., Sun, G. 2012. Treatment of tannery wastewater in a pilot-scale hybrid constructed wetland system in Bangladesh. *Chemosphere*, *88*, 1065–1073.

Shingare, R.P., Thawale, P.R., Raghunathan, K., Mishra, A., Kumar, S. 2019. Constructed wetland for wastewater reuse: role and efficiency in removing enteric pathogens. *Journal of Environmental Management*, *246*, 444–461.

Song, H., Zhang, S., Long, X., Yang, X., Li, H., Xiang, W. 2017. Optimization of bioelectricity generation in constructed wetland-coupled microbial fuel cell systems. *Water*, *9*(3), 185.

Srivastava, P., Abbassi, R., Garaniya, V., Lewis, T., Yadav, A.K. 2020. Performance of pilot-scale horizontal subsurface flow constructed wetland coupled with a microbial fuel cell for treating wastewater. *Journal of Water Process Engineering*, *33*, 100994.

Srivastava, P., Dwivedi, S., Kumar, N., Abbassi, R., Garaniya, V., Yadav, A.K. 2017. Performance assessment of aeration and radial oxygen loss assisted cathode based integrated constructed wetland-microbial fuel cell systems, *Bioresource Technology*, *244*, 1178–1182.

Srivastava, P., Yadav, A.K., Garaniya, V., Abbassi, R. 2019. Constructed wetland coupled microbial fuel cell technology: development and potential applications. In: *Microbial Electrochemical Technology* (S. Venkata Mohan, S. Varjani, A. Pandey, eds.). Elsevier, Amsterdam, pp. 1021–1036.

Srivastava, P., Yadav, A.K., Mishra, B.K. 2015. The effects of microbial fuel cell integration into constructed wetland on the performance of constructed wetland. *Bioresource Technology, 195*, 223–230.

Tamta, P., Rani, N., Yadav, A.K. 2020. Enhanced wastewater treatment and electricity generation using stacked constructed wetland–microbial fuel cells. *Environmental Chemistry Letters, 18*(3), 871–879.

Tang, C., Zhao, Y., Kang, C., Yang, Y., Morgan, D., Xu, L., 2019. Towards concurrent pollutants removal and high energy harvesting in a pilot-scale CW-MFC: insight into the cathode conditions and electrodes connection. *Chemical Engineering Journal, 373*, 150–160.

US EPA, 2000. *Manual Constructed Wetlands Treatment of Municipal Wastewaters*, US Environmental Protection Agency, Cincinnati, OH.

Villasenor, J., Capilla, P., Rodrigo, M. A., Cañizares, P., Fernández, F.J. 2013. Operation of a horizontal subsurface flow constructed wetland – microbial fuel cell treating wastewater under different organic loading rates. *Water Research, 47*(17), 6731–6738.

Vymazal, J. 2007. Removal of nutrients in various types of constructed wetlands. *Science of the Total Environment, 380*(1–3), 48–65.

Vymazal, J. 2010. Constructed wetlands for wastewater treatment. *Water, 2*, 530–549

Vymazal, J. 2011. Constructed wetlands for wastewater treatment: five decades of experience. *Environmental Science & Technology, 45*, 61–69.

Vymazal, J. 2013. Emergent plants used in free water surface constructed wetlands: a review. *Ecological Engineering, 61*, 582–592.

Wainaina, S., Awasthi, M.K., Sarsaiya, S., Chen, H., Singh, E., Kumar, A., Ravindran, B., Awasthi, S.K., Liu, T., Duan, Y., Kumar, S. 2020. Resource recovery and circular economy from organic solid waste using aerobic and anaerobic digestion technologies. *Bioresource Technology, 301*, 122778. https://doi.org/10.1016/j.biortech.2020.122778

Wang, R., Korboulewsky, N., Prudent, P., Domeizel, M., Rolando, C., Bonin, G. 2010. Feasibility of using an organic substrate in a wetland system treating sewage sludge: impact of plant species. *Bioresource Technology, 101*, 51–57.

Wang, W., Zhang, Y., Li, M., Wei, X., Wang, Y., Liu, L., Shen, S. 2020. Operation mechanism of constructed wetland-microbial fuel cells for wastewater treatment and electricity generation: a review. *Bioresource Technology, 314*, 123808.

Wu, S., Kuschk, P., Brix, H., Vymazal, J., Dong, R. 2014. Development of constructed wetlands in performance intensifications for wastewater treatment: a nitrogen and organic matter targeted review. *Water Research, 57*, 40–55.

Wu, S., Lv, T., Lu, Q., Ajmal, Z., Dong, R. 2017. Treatment of anaerobic digestate supernatant in microbial fuel cell coupled constructed wetlands: evaluation of nitrogen removal, electricity generation, and bacterial community response. *Science of the Total Environment, 580*, 339–346.

Xie, T., Jing, Z., Hu, J., Yuan, P., Liu, Y., Cao, S., 2018. Degradation of nitrobenzene-containing wastewater by a microbial-fuel-cell-coupled constructed wetland. *Ecological Engineering, 112*, 65–71.

Xu, L., Zhao, Y., Doherty, L., Hu, Y., Hao, X., 2016a. Promoting the bio-cathode formation of a constructed wetland microbial fuel cell by using powder activated carbon modified alum sludge in anode chamber. *Science Report, 6*, 26514.

Xu, L., Zhao, Y., Doherty, L., Hu, Y., Hao, X., 2016b. The integrated processes for wastewater treatment based on the principle of microbial fuel cells: a review. *Critical Reviews in Environmental Science and Technology, 46*(1), 60–91.

Yadav, A.K., Dash, P., Mohanty, A., Abbassi, R., Mishra, B. K. 2012. Performance assessment of innovative constructed wetland-microbial fuel cell for electricity production and dye removal. *Ecological Engineering, 47*(June), 126–131.

Yadav, A.K., Srivastava, P., Kumar, N., Abbassi, R., Mishra, B. K. 2018. Constructed wetland–microbial fuel cell: an emerging integrated technology for potential industrial wastewater treatment and bio-electricity generation. In: *Constructed Wetlands for Industrial Wastewater Treatment* (A.I. Stefanakis, ed.). Wiley, Chichester, pp. 493–510.

Yadav A.K. 2010. Design and development of novel constructed wetland cum microbial fuel cell for electricity production and wastewater treatment. In: *Proceedings of the 12th IWA International Conference on Wetland Systems for Water Pollution Control*, Venice, Italy, pp. 4–9.

Yakar, A., Türe, C., Türker, O.C., Vymazal, J., Saz, Ç. 2018. Impacts of various filtration media on wastewater treatment and bioelectric production in up-flow constructed wetland combined with microbial fuel cell (UCW-MFC). *Ecological Engineering*, *117*, 120–132.

Yan, Y.; Xu, J. 2014 Improving winter performance of constructed wetlands for wastewater treatment in Northern China: a review. *Wetlands*, *34*, 243–253.

Yang, E., Chae, K.-J., Choi, M.-J., He, Z., Kim, I.S. 2019 Critical review of bioelectrochemical systems integrated with membrane-based technologies for desalination, energy self-sufficiency, and high-efficiency water and wastewater treatment. Desalination, *452*, 40–67

Yang, S.Q., Jia, B.Y., Liu, H., 2009. Effects of the Pt loading side and cathode-biofilm on the performance of a membrane-less and single-chamber microbial fuel cell. *Bioresource Technology*, *100*, 1197–1202.

Yang, Z., Yang, L., Wei, C., Wu, W., Zhao, X., Lu, T. 2018a. Enhanced nitrogen removal using solid carbon source in constructed wetland with limited aeration. *Bioresource Technology*, *248*, 98–103.

Yang, Y., Zhao, Y., Liu, R., Morgan, D. 2018b. Global development of various emerged substrates utilized in constructed wetlands. *Bioresource Technology*, 261, 441–452

Zhao, Y., Collum, S., Phelan, M., Goodbody, T., Doherty, L., Hu, Y. 2013. Preliminary investigation of constructed wetland incorporating microbial fuel cell: batch and continuous flow trials. *Chemical Engineering Journal*, *229*, 364–370.

Zhou, M., Chi, M., Luo, J., He, H., Jin, T., 2011. An overview of electrode materials in microbial fuel cells. *Journal of Power Sources*, *196*(10), 4427–4435.

4

Genetically Engineered Microorganisms (GEMs) for a Sustainable Environment: A Promising Biotechnological Tool

Savita Durgappa Sandur
Kuvempu University

Shankramma Kalikeri
JSS Academy of Higher Education & Research (Deemed to be University)

CONTENTS

4.1 Introduction

Anthropogenic activities such as contemporary farming techniques, industrialisation, overcrowding, and unhealthy rivalry for supremacy are wreaking havoc on the planet. These activities produce an unprecedented number of phenols, azo dyes, polyaromatic hydrocarbons (PAHs), polychlorinated chemicals,

DOI: 10.1201/9781003231738-5

pesticides, heavy metals, and other refractory contaminants (Kumar et al., 2021a, b; Agrawal et al., 2021; Mishra et al., 2021). These chemicals are resistant to biodegradation and can persist for a long time in the environment, causing acute and chronic injury to ecosystem biotic components (Bhandari et al., 2021; Elekwachi, 2014; Singh et al., 2021). In people and animals, these pollutants are teratogenic, carcinogenic, mutagenic, and poisonous, and have had a global impact (Bhandari et al., 2021). As a result, issues about the disposal of organic waste from the environment are crucial (Paliya et al., 2021; Wainaina et al., 2020; Kumar et al., 2019. these conventional methods for removing xenobiotic pollutants include throwing rubbish into a pit, incineration at a high temperature, and organic and ultraviolet (UV) breakdown. Though, owing to a shortage of space, expensive costs, complex procedures, stringent regulatory restrictions imposed on decontamination by various countries, and widespread public opposition, these physical and chemical techniques are losing ground. These methods come with several drawbacks, including the production of significant sums of slush, which necessitates benign disposal, and the generation of harmful derivative impurities (Bhandari et al., 2021; Karigar and Rao, 2011). As a result, an instant and a swift transition from these environmentally unfavourable physical and chemical procedures to *eco-friendly* methods are required.

The principle of bioremediation underpins current techniques, which is the sole viable option for dealing with contamination (Kumar et al., 2018). Biodegradation is the organic/biological process through which a material is biologically destroyed. It is a natural procedure that takes place without the involvement of humans. The engineering approach of using biological means to decompose a substance (including algae fungi, bacteria and others) to destroy a material is known as bioremediation (Pant et al., 2021). Biodegradation is a time-consuming process, but bioremediation is a faster one. Temperature modulation, food or nutrient availability, and other forms of human interaction are utilised to alter the rate of bioremediation (Chandra and Kumar, 2015). It continuously converts, modifies, and utilises hazardous pollutants to get energy and create biomass, based on the capabilities of particular microbes (Tang et al., 2007). The collection of harmful and persistent substances, notably xenobiotics (chemicals detected within an organism that are neither typically formed nor expected to be present), cannot effectively decay in the environment due to the natural metabolic diversity of local microbes. Due to the lack of relevant metabolic pathways, the new properties of these pollutants include unusual chemical bonds or additional functional groups that allow them to resist the natural degradation of microorganisms (Kumar et al., 2020a, b).

Bioremediation is an important process used to reduce pollution (Kumar and Chandra, 2020). Today, we are seeing increasing attention to xenobiotic compounds, such as sucralose (artificial sweetener), poly fluoroalkanes and alkenoates (PFAS), diclofenac or carbamazepine (drugs). The lack of biodegradation process and its widespread use led to the widespread and persistent existence of pollutants in soil, groundwater and external water. At the exit of the chemical treatment plant, even if the effluent undergoes different stages of biological treatment, compounds such as dimethoxymethane or dimethyl urethane can also be detected. Bioremediation has been used to restore surface water, groundwater, soil, and ecosystems with great success.

Timmis and Pieper (1999) recently defined many opportunities to use genetic engineering strategies to advance the performance of degradation. For example, rate-limiting steps in existing metabolic pathways can be genetically changed to accelerate breakdown, or a completely new metabolic pathway might be added into bacterial strains to degrade hitherto refractory substances. Bioremediation function is based on the concept of the biogeochemical cycle. In situ bioremediation occurs when the process takes place in the same location. This technology modifies two types of pollutants, namely organic pollutants and inorganic pollutants. In contrast, Ex-situ bioremediation is the process of transporting polluted materials (soil and water) to various areas in order to speed up biocatalysts (Dua et al., 2002).

Genetically Engineered Microorganisms (GEMs) have increased degradability and have been shown to successfully degrade a variety of contaminants under specific conditions. However, environmental and ecological issues and regulatory restrictions are the main obstacles to testing GEMs in this area. To overcome these problems, GEMs can propose an efficient cleaning process at a reasonable cost. Pesticides and hydrocarbons are frequently degraded by aerobic bacteria such as *Mycobacterium, Sphingomonas, Rhodococcus, Alcaligenes,* and *Pseudomonas.* Anaerobic bacterial enzymes, on the other hand, have been employed to bioremediate polychlorinated biphenyls (PCBs) and dechlorinate trichloroethylene (TCE) and chloroform. Microorganisms' most common enzymes and GEMs are responsible for digesting various forms of pollutants used in the bioremediation process including protease, cytochrome P450, hydrolase, laccase, dehydrogenase, dehalogenase and lipases.

4.2 Microbial Bioremediation Is Influenced by a Variety of Factors

The use of bacteria, fungi, and plants to degrade, fixate, modify, remove, or detoxify various chemical and physical toxins in the environment is known as bioremediation. Microorganisms recycle a variety of substances and compounds, and they can only act on pollutants if they aid in the production of energy and nutrition, which allows them to make more cells. The physical and chemical qualities, chemical properties, contaminant groups, and accessibility to microorganisms are all well-organised bioremediation variables (El Fantroussi and Agathos, 2005).

4.2.1 Biological Factors

Bacterial competition for limited carbon sources, hostile interactions amongst microbes, and so forth, or predation on microbes by protozoa and pages are all abiotic processes that influence the breakdown of organic molecules. The rate at which pollutants degrade is usually determined by their concentration and the amount of "catalyst" present. The rate of deterioration of pollutants can be amplified or reduced by the expression of certain enzymes in the cell (Kumar and Chandra, 2018). In addition, specific enzymes must participate in the degree of pollutants metabolism and their "affinity" to pollutants, and the availability of pollutants is also mainly required. Mutation, horizontal gene transfer, enzyme activity, interaction (competition, inheritance, and predation), their own growth to critical biomass, population size, and composition are some of the primary biological aspects described here (Naik and Duraphe, 2012; Boopathy, 2000).

4.2.1.1 Environmental Factors

Humidity, pH, temperature, water solubility, nutrients, soil structure, redox potential, place features, oxygen concentration, in this field, there is a scarcity of well-educated personnel, and contaminants all influence microorganism growth and activity (pollutants). Type, solubility, concentration, chemical structure, and toxicity are all affected by physical and chemical bioavailability (Omokhagbor Adams et al., 2020). Biodegradation can happen over a wide variety of pH levels; however, in most marine and terrestrial systems, pH levels between 6.5 and 8.5 are best for biodegradation. The following are the most important environmental factors: Accessibility to nutrients Appropriate nutrition, particularly the availability of vital nutrients like N and P, can boost biodegradation efficacy by raising bacteria's C: N: P ratio. Several nutrients, such as carbon, nitrogen, and phosphorus, are compulsory for microbes to survive and continue their microbial activity. Microorganisms' metabolic activity is being increased by adding sufficient nutrition as a removal method. Oily microorganisms, like other organisms, and nutrients are required for proper growth and development.

4.2.1.1.1 Temperature

Temperature is the most important physical component that influences the existence of microbes. In cold climates, sub-zero temperatures cause the transit stations in bacterial cells to close or even freeze, rendering most lipophilic microorganisms metabolically inactive. (Yang et al., 2009; Macaulay, 2015). Some compounds require a specific temperature to break down. Temperature has a big impact on microorganisms' physiological characteristics; therefore, it can speed up or slow down the bioremediation process. As the temperature rises, the rate of microbial activity increases, peaking at the optimal temperature. The temperature drops sharply as it climbs or declines, eventually coming to a halt when it reaches a certain point.

4.2.1.1.2 Oxygen Concentration

Different creatures require different amounts of oxygen, whereas others do not. It promotes biodegradation at the appropriate rate for your needs. Because oxygen is the gas that most organisms require, both aerobic and anaerobic settings are conducive to biodegradation. The presence of oxygen can boost hydrocarbon metabolism in certain circumstances. The effects of dissolved oxygen concentration on microbial growth and respiration rate have been the focus of most of the research. Bacterial respiration appears to be unaffected above essential dissolved oxygen concentrations in general (Longmuir, 1954).

4.2.1.1.3 Moisture Content

Microorganisms require a sufficient amount of water to grow. The amount of moisture in the soil has a detrimental impact on bio degradants.

4.2.1.1.4 pH

pH of compounds that are the acidity, basic and alkaline properties of the compound, which influence itself the metabolic activity of the microorganism and also increases and decreases the disposal process. The pH measurement on the ground can indicate the possibility of microbial growth (Asira, 2013). The highest or lower pH values reveal lower results. The metabolic process is extremely sensitive to a slight difference in pH. Waste sludge hydrolysis is greatly improved at alkaline pH compared to other pHs, and thermophilic anaerobic fermentation can be severely impeded under strong alkaline or strong acid circumstances (Meng et al., 2018).

4.2.1.1.5 Metallic Ions

Metals are extremely important for bacteria and fungi, but in their presence, cellular metabolic activity is limited. The metallic compounds have a direct and indirect effect on the degradation rates. Mortal compounds with large amounts of toxic pollutants may have toxic poisoning and prevent decontamination. The mechanisms and extension of toxicity may differ from certain poisons, their exposed concentrations and microorganisms. Toxic compounds can be found in both organic and inorganic compounds, their objective organiations. Several inorganic metals, such as magnesium (Mg), nickel (Ni), chromium, copper, calcium, manganese, and sodium, as well as zinc, are essential for metabolic and redox processes in modest quantities. Aluminum, lead, cadmium, gold, mercury, and silver are heavy metals that have no biological function and are hazardous to living creatures (Turpeinen et al., 2004).

4.3 Types of Bio-Remediation

There are different types of technologies or treatment methods in the bioremediation process. The main types of bio-remediation are as follows:

4.3.1 Bio-Stimulation

This strategy is to stimulate the activities of local microorganisms by inoculating specific nutrients in places (soil/groundwater). It is different in stimulating indigenous or inherent bacterial and fungal communities. The presence of trace contaminants can also stimulate the conversion of operons into bioremediation enzymes. These nutrients are the fundamental components of life, allowing microorganisms to produce fundamental necessities including energy, cellular biomass, and pollution-degrading enzymes. Carbon, nitrogen, and phosphorus are required for all of them. (Naik and Duraphe, 2012). Biological deterioration Biological or natural deterioration refers to the elimination of concentrations of pollutants from the immediate surroundings. It can take part in biological processes, physical phenomena, and chemical reactions.

4.3.2 Bioaugmentation

The inclusion of microorganisms is known as bioaugmentation, which is designed to act as bioremediation agents in a system in order to remove complicated toxins fast and thoroughly. Furthermore, genetically modified microbes have been found to improve the degrading efficiency of a variety of contaminants. Because natural species cannot break down some substances quickly enough, they must be genetically engineered by modifying DNA; genetically modified microbes break down contaminants quicker than natural species and are extremely competitive with native species, predators, and abiotic influences. Because of the lowering conditions in the subterranean environment, the majority of bioaugmentation attempts for chlorinated ethane-contaminated groundwater have concentrated on anaerobic processes rather than aerobic/co-metabolic dichlorination processes (Moo-Young, 2019).

4.3.3 Biovent

The biovent is a procedure that restores the natural presence of biodegradable pollutants in the soil by delivering air or oxygen to soil microorganisms, as shown in Figure 4.1. Bioventes use a modest airflow to supply just enough oxygen to keep microbial activity going. Many researchers have investigated Biovent's efficacy of bioremediating oil-contaminated soil (Agarry and Latinwo, 2015; Lee et al., 2006).

4.3.4 Biopiles

A Biopile (also known as a Biosel, Bio cell, or Biomass) is used to reduce the concentration of petroleum pollutants in perforated soil during biodegradation. Aeration, irrigation, nutrient and leachate collecting systems, and a treatment bed are all part of this process. Ex-situ technique is gaining popularity due to its advantages, which include cost-effectiveness and the capacity to accomplish great biodegradation when nutrition, aeration, and temperature are all appropriately handled (Whelan et al., 2015). The use of biopile in contaminated places can help to reduce the volatilisation of Low Molecular Weight (LMW) toxins, and it can also be used to effectively repair polluted severe settings like very cold locales (Dias et al., 2015; Gomez and Sartaj, 2014; Whelan et al., 2015) (Figure 4.2).

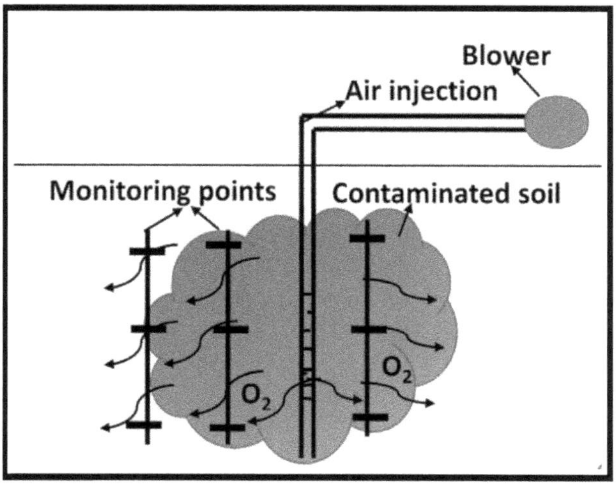

FIGURE 4.1 A well and a blower pump air into the soil through the well making up a basic bioventing system. (https://frtr.gov/matrix/Bioventing/.)

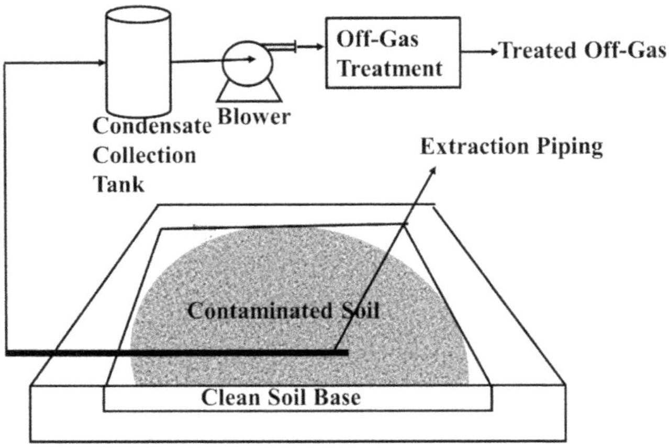

FIGURE 4.2 Schematic diagram of biopile: The air is delivered to a biominated system by a pump and pump system that circulates air through a battery or pipe under negative pressure in a positive pressure system. (https://frtr.gov/matrix/Biopiles/.)

4.4 Microbial Enzymes for Bioremediation

4.4.1 Cytochrome P450

Cytochrome P450 is a superfamily of global heme enzymes present in eukaryotes, bacteria, and archaebacteria that perform a variety of tasks ranging from the production of complex natural products to the development of biological systems. From drug metabolism to biotransformation, there is a long way to go (Bak et al., 2011; Li et al., 2020). Deadly compounds in our ecosystem P450 have the inherent ability to destroy xenobiotics through chemical transformations (such as dealkylation, aliphatic hydroxylation, epoxidation, dehalogenation, and various deactivation-based mechanisms). CYP101, CYP102, CYP1A1, CYP1A2, and CYP1B1 are known to absorb polycyclic aromatic hydrocarbons, with p-dibenzodioxin (DD) and mono-, di-, and trichloro DD exhibiting high activity in CYP1A1, and CYP1A1 mutant F240A exhibiting high activity against 2, 3, 7, and 8 Tetrachloride. To produce carbon substrates and oxidation products, they utilise molecular oxygen as well as NADH or NADPH as cofactors. Ferredoxin and ferredoxin reductase were also employed as electron sources for catalytic activity. Das et al. (2016) reported that numerous microbes such *as Rhodococcus, Mycobacterium, Gordonia* and *Pseudomonas* carry plasmids articulating P450, catabolic genes and genomes, which are used to degrade and eliminate persistent organic pollutants in the environment. Awad and Garnier (2021) discovered that the *Bacillus megaterium* P450 BM3 (CYP102A1) designed by *E. coli* BL21 can deactivate hollow Pt/TiO2Cu nanoparticles under solar radiation to degrade a large amount of polluting organic gases. The continuous supply of electrons through nanoparticles was shown to be much higher at a pH of 7.0, room temperature, and a concentration of 20 mg/L (95%). The YCJY1 strain obtained from *Sphinobium* sp. degrades Bisphenol A (BPA), a plasticizer.

Table 4.1 exhibits isolated enzymes from infectious sources that are involved in contaminant bioremediation (Cytochrome P450) whereas Table 4.2 demonstrates the function of the microbial enzymes involved in bioremediation.

4.4.2 Laccase

Laccase is an extracellular enzyme that contains more copper, with monomer, dimer, and tetramer glycoproteins specified by plants, fungi, and bacteria. Microbial laccases from a variety of microorganisms, including Streptomyces laccases from actinomycetes, have mostly been found, categorised and investigated. *S. coelicolor, S. bikiniensis, S. cyaneus,* and *S. ipomoea* are some of these species, with S. coelicolor being the most widespread. Agricultural wastes including banana peels, sawdust, and rice bran contain lignin and phenolic chemicals that can boost laccase synthesis. (Muthukumarasamy et al., 2015).

TABLE 4.1

Microbial Enzymes from Various Microorganisms: Properties and Uses

Microorganisms and Enzymes	Substrate	Optimum temperature (°C)	Optimum pH	Applications	Reference
Pseudomonas putida F6 (cell-free extract)	Syringaldazine (SGZ)	30	7.0	Synthetic dye degradation	Bhandari et al. (2021)
Bacillus subtilis (cell-free extracts lipase)	Olive oil	45	7.0	Wastewater bioremediation	Chouhan et al. (2020)
Bacillus megaterium (purified P450)	Dichlorobenzo-p-dioxin (2,3-DCCD), 2,3,7-trichloro-dibenzo-p-dioxin (2,3,7-triCDD)	30	7.5	PCCDS hydroxylation	Gustafsson et al. (2004)
Mycobacterium sp. (cell-free extracts P450)	–	30	6.6	Morpholine biodegradation	Sielaff et al. (2001)

TABLE 4.2

Microbial Enzymes Involved in Bioremediation and Their Function

Enzymes	Mechanism	Function
Cytochrome P450	By reducing or oxidising heme iron, performs electron transfer processes and catalyses.	Steroids, fatty acids, and xenobiotics are oxidised through the synthesis and metabolism of numerous compounds and substances within cells.
Laccase	Reduction of the O_2 molecule, which includes one-electron oxidation with a variety of aromatic chemicals.	In aromatic compounds, ring cleavage reduces one molecule of oxygen in the water, resulting in free radicles.
Dehalogenase	Reductive mechanism	Removes the halogens by breaking the carbon-halogen bond.
Dehydrogenase	The reversible conversion of alcohol to aldehyde or ketone is catalyzed by the microbial alcohol dehydrogenase.	Creating energy by oxidising organic molecules
Hydrolase	In triglyceride hydrolysis, one mole triglyceride (T) is combined with three moles of water to produce one mole glycerol (G) and three moles fatty acids. The peptide link in proteins is also broken away by hydrolysis.	Fats and proteins are degraded.
Protease	Catalyze the dissolution of protein peptide linkages.	Degradation of proteins such as keratin and casein, as well as dehairing of leather and wastewater treatment
Lipase	Transfer of a proton	This enzyme catalyses the hydrolysis of mono, di, and triglycerides into fatty acids and glycerol.

Laccases' hardness at different temperatures, hardness in organic solvents, pH, and salt concentration are the most perceptible biochemical features. It's a high-stability enzyme with a 75°C thermal stability that's suitable for industrial use. More information of the molecular properties of water bodies is needed to find and create new laccases that can function under difficult conditions, as well as novel tactics and processes for their effective usage in treating contaminated water.

4.4.3 Dehydrogenase

Dehydrogenase belongs to the family of oxidoreductase, and is found in the range of yeast, bacteria and plants. Aldehyde dehydrogenase $(NADP)^+$. Aldehyde oxidation-dependent to carboxylic acids. The activity of polyethylene glycol dehydrogenase is attracting attention in the free extract of cells of bacterial degradation, and polyethylene glycol of industrially dissimilar molecular weight is produced. Some *Sphingomonas* sp. when oxidising the terminal alcohol group of the polymer chain, polyethylene glycol is used as a foundation of carbon and energy. The aldehyde is oxidised with raw enzymes and purified enzymes, but the process is slow. Spingobium sp. free extract the presence of PPG dehydrogenase activity in a PW1 strain cultured in 0.5% PPG verifies PPG's oxidative metabolic route. Furthermore, it was discovered that the polyvinyl alcohol of recombinant dehydrogenase oxidised glycols such as polypropylene glycol and 1,3 butane/cyclohexanodiol and 2.4 pentanodiol, as well as primary and secondary alcohols, and that the Primary secondary alcohol was also oxidised.

The 3,4 dehydroadipyl-CoA semialdehyde dehydrogenase gene (orf9) in *Azoarcus evansii* is responsible for the production of 3,4 dehydroadipyl-CoA semialdehyde dehydrogenase, which is employed to oxidise 3,4 dehydroadipyl-CoA semialdehyde and has been efficiently expressed. *Rhodococcus* NCIMB12038 uses naphthalene as its only carbon source, and heterologous expression of NCIMB12038 cisnaphthalene dihydrodiol dehydrogenase was shown to have 39% amino acid content when compared to *Pseudomonas putida* G7 identity cisnaphthalene dihydrodiol dehydrogenase

4.4.4 Hydrolase

Hydrolases use water to break chemical bonds and minimise the toxicity of pollutants, bigger molecules are converted to smaller molecules. They promote the cleavage of C-C, C-O, C-N, SS, SN, SP, CP and other bonds via water, and catalyze a number of related processes, including condensation reactions and alcohol reactions. The main advantages of this type of enzyme are easy availability, economy, environmental protection, lack of stereoselectivity in cofactors, and resistance to the addition of water-soluble solvents.

Microbial hydrolases (lipase, esterase, amylase, protease, cellulase, nitrilase, peroxidase, cutinase, etc.) can be used to manage the degradation of residues, plastics and pesticides generated during food processing, biofilm sediment treatment and contamination by hydrocarbons. Microbial hydrolases from *Achromobacter, Pseudomonas, Flavobacterium, Nocardia,* and *Bacillus cereus* have been utilised to convert carbofuran, carbaryl or parathion, diazinon, and coumaphos, as well as other pollutants, in the hydrolysis process.

Organophosphate (OP) chemicals are neurotoxins with a high lethality (Chandra and Kumar, 2015). Pesticides are extensively used in agriculture, posing harm to the biological environment. Pyrethroids and malathion are two of them that are harmful to the environment. Hydrolysis of phosphodiester links can detoxify OP pesticides, while hydrolysis of carboxylic ester bonds can detoxify pyrethroids and malathion. Many bacteria, including Bacillus licheniformis, use malathion as a carbon source, so the hydrolase from Bacillus licheniformis aids in the bioremediation of malathion-contaminated soil. As a result, the microbial enzyme OP hydrolase, OP acid hydrolase, or methyl parathion hydrolase (MPH) has attracted attention as a non-toxic bioremediation approach. *Pseudomonas* diminuta's organophosphate hydrolase gene (GenBank accession number M20392) was also expressed. The detoxifying ability of *Escherichia coli* is demonstrated by methyl concentrations of 10%–80% and 3.6%–45%.

The breadth of the cleanup problem is significant, as the harmful consequences of metal and radionuclide exposure and ingestion on human health, the environment, and the food supply are well established (Kasprzak, 2002). Furthermore, the presence of these contaminants has an adverse effect on ecosystem resilience and contributes to biodiversity loss (Philp et al., 2014).

Phthalates are common contaminants found in the environment. Inside plastic polymers, phthalates can serve as important building elements. In these circumstances, phthalates (most commonly TPA) are attached to bivalent alcohols such as ethylene glycol, 1,4-butanediol, 1,6-hexanediol, and so on via ester bonds. Needless to say, these covalently bonded TPA residues, like manmade polymers in nature, are problematic to break from their polymeric backbones (Aftring et al., 1981). Commonly used as plasticizers, phthalates (PE) and short-chain alkyl phthalates (C1-C4) are a class of deadly organic compounds, extensively used as plasticizer additives (softeners) or made of plastic. Due to its estrogen, carcinogenic and endocrine-disrupting properties, these have become major environmental problems worldwide. A recent study showed that, compared with other hydrolases such as lipases, polyester hydrolases have the potential to enzymatically degrade synthetic polyesters obtained from *Actinomyces thermophilus* (ermobifida fusca KW3).

4.5 Examples of Genetic Engineering of Microorganisms and Biodegradation

The use of GEMs in bioremediation has gotten a lot of press, but it's mostly restricted to the lab. This is owing to challenges with regulatory risk assessment, as well as their actual impact on field conditions and delivery uncertainty. Table 4.3 depicts the efficacy of genetic engineering in biodegradation processes, while Table 4.4 depicts genetic engineering for contaminant biodegradation. There are at least four main methods for developing GEM for bioremediation applications.

1. Changes in the specificity and affinity of enzymes
2. Development and control of pathways
3. Development, monitoring, and control of biological processes
4. End point analysis and reduction using bio-affinity sensors in chemical detection and toxicity

There have been numerous cases of bacteria degrading environmental contaminants. Only conceivable bioremediation linked to GEM will be examined, as well as the building and deployment of GEM. The purpose of environmental biotechnology is to produce efficient and lucrative bioremediation techniques Successful in situ bioremediation with GEM requires a mixture of microbiology and environmental expertise, biochemical mechanisms, and on-site engineering design. Degradation (Catabolism) Gene: Gives a thorough analysis of biochemical pathways, operon structure, and molecular biology of biodegradation pathways that are significant in bioremediation. The majority of the data is important for developing a good GEM; however, it is limited to aerobic catabolism and co-metabolic pathways (Kuiper et al., 2004).

TABLE 4.3

Genetic Engineering for Biodegradation Process Efficacy

Contaminants	Microorganism	Application
PCB	*A. eutrophus* H850Lr	Process monitoring
TCE, BTEX	*Pseudomonas putida* TVA8	
Naphthalene, antrhacene, phenanthrene	*Pseudomonas fluorescence* HK44	
2,4-D	*Bacillus cepacia* BRI600L	Strain monitoring
BTEX	*Pseudomonas flourescence* 0586s/Pucd607	Stress response
Cd, 2,4-dinitrophenol, hydroquinone	*Pseudomonas strain* Shk1	Toxicity assessment
Nonpolar narcotics	*A. eutrophus* 2050	End point analysis

TABLE 4.4

Genetic Engineering for Biodegradation of Contaminants

Contaminants	Microorganism	Modification
Mono/dichlorobenzoates	*Pseudomonas* sp. B13	Pathway
4-Ethylbenzoate	*Pseudomonas putida*	
Toluene/benzoate	*Pseudomonas putida* KT2442	
Chloro-, methylbenzoates	Pseudomonas sp. FR1	
o-, p-Monochlorobiphenyls	*C.testosteroni* VP44	Substrate specificity
PCB	*Pseudomonas sp.* LB400	
PCB, benzene, toluene	*Escherichia coli* JM109(pSHF1003)	
TCE, toluene, benzene	*Pseudomonas pseudoalkaligenes* KF707-D2	

4.5.1 Branched-Chain Aromatics

4.5.1.1 *Pseudomonas putida* Plasmid TOL Pathway

The plasmid TOL route of *Pseudomonas putida* has been extensively manipulated, according to Ramos et al. (1986), to boost its catabolism competence into numerous branching compounds. The substrate for TOL's metabolic pathway has been altered to 4-ethylbenzoate. The XylS pathway regulator or the enzyme's substrate selectivity (catechol 2,3 dioxygenase) has been altered in these recombinant strains (Ramos et al., 1986).

4.5.2 Chlorinated Compounds

4.5.2.1 Chlorobenzoate

4.5.2.1.1 *Pseudomonas B13-Hybridisation Pathway*

A plasmid containing 3-chlorobenzoate and 4-chlorophenol degradation genes has been significantly changed in strain B 13 pWR1, resulting in a hybrid pathway that extends the vertical degradation of halogenated aromatic compounds. The TOL plasmid pWW0 from *Pseudomonas putida* mt2 was presented into strain B13 for the first time by Latorre et al. (1984). It was discovered that transconjugants can use monochlorobenzoate and dichlorobenzoate as growth substrates. The inactivation of pathway (due to the lack of activity of catechol 2,3 dioxygenase) necessitates the use of 4-chlorobenzoic acid as the sole carbon source by the hybrid strain *Pseudomonas* sp. WR211, as expressed by the plasmid TOL.

Alcaligenes, a benzoate and phenol degrading agent, was reported by Schwien and Schmid (1982). Conjugation was performed to obtain A7, which is strain B13's chlorcatechol degradation capacity. Strains of Alcaligenes that have been recombinantised A72 show how three isomeric chlorophenols can be employed as the sole source of carbon and energy.

4.5.2.1.2 *Pseudomonas aeruginosa AC869 (pAC31), Pseudomonas aeruginosa AC869 (pAC31), Pseudomonas aeruginosa AC869*

Malodor includes plasmid pAC25, which has been demonstrated to be structurally similar to pWR1 and encodes the chlorobenzoic acid degradation gene 3. A genomic rearrangement between plasmid pAC25 and TOL occurred during chemostat maintenance, resulting in pAC31. The xylDGEF gene is found on the chromosome in the new plasmid. The strain AC869 (pAC31) was shown to be capable of degrading 3,5 dichlorobenzoate as well as 3 and 4 chlorobenzoate (Chatterjee and Chakrabarty, 1983).

4.5.2.1.3 *Former Monochlorobenzoate*

Former monochlorobenzoate from *Pseudomonas* genusUS1 and 2,4D as a degrading agent, strain US1 ex. Conjugation was used to create the molecule containing pJP4. The donor was ME. *Escherichia coli* JMP397 harbouring plasmid pJP4 (without expression), while the acceptor was the 4-chlorobenzoate degradation agent US1. When cultivated in the respective chloroaromatic hydrocarbons as a carbon source, the new strains release stoichiometric amounts of chloride (Sahasrabudhe et al., 1991). In freshwater ecosystems and activated sludge units, horizontal transfer of the 3-chlorobenzoate catabolic plasmid (pBRC60) has been observed (pD10) (McClure and colleagues, 1989).

4.5.2.2 Polychlorinated Biphenyls (PCBs) and Chlorinated Biphenyls

4.5.2.2.1 *Alcal igenes eutrophus (Formerly Known as Alcaligenes eutrophus) AE707/AE1216*

The chromosome contains *R. eutropha* A5 (SHIELDS et al., 19PETchroma catabolic gene). Natural combination was used to transfer the catabolic PCB genes and Alcaligenes dentrificans sp. JB1 to the heavy metal resistant *R. eutropha* CH34 strain. Biphenyl and monochlorobiphenyl are consistently degraded to benzoate and chlorobenzoate by all contributing strains. Benzoate is further metabolised by the lysis mechanism in strains A5 and LBS1C1, while strain JB1 utilises the lysis pathway. In strain A5, the catabolic transposon Tn4317 is identified, which contains biphenyl and

4-chlorobiphenyl degradation genes. The dehalogenase activity of strain A5 is mediated by the plasmid pSS50 (Layton et al., 1992).

4.5.2.2.2 Pseudomonas Strain Mixed with Pseudomonas

The UCR2 strain was created by combining the chlorobenzoate degrader Pseudomonas aeruginosa JB2 with the 2-chlorobiphenyl user Arthrobacter (Adams et al., 1992). B1Barc is a bacterial strain. The UCR2 strain displayed 2-chloro and 2,5-dichlorobiphenyl mineralisation. The UCR2 and JB2 strains exhibit more phenotypic and genomic DNA homology than the other strains. There was no hybridisation when the parent strains were tested against each other. *Pseudomonas putida* BN10 eats biphenyl and produces the metabolites 2,4-chlorobenzoate and 3-chlorocatechol from the monochlorobiphenyl it eats. *Pseudomonas aeruginosa* produces a chlorinated biphenyl degradation agent. The BN210 (obtained chlorcatechol degradation gene) and B131 (obtained biphenyl degradation gene) Pseudomonas cultures that resulted may thrive on 3-chlorobiphenyl.

4.5.2.2.3 Pseudomonas cepacia JHR22

Havel and Reineke (1991) described Pseudomonas cepacia JHR2, a hybrid bacterium that grows by consuming biphenyl donor and chlorobenzoic acid (*P. cepacia*). JH230 is made up of mating filters. Strain 230 is a hybrid strain that was created when chlorcatechol degradation genes from strain B13 were transferred to *Pseudomonas aeruginosa* (Hartman, 1989). On 3 and 4 chlorobiphenyls, the JHR2 strain can develop. JHR22, a 2-chlorobiphenyl degradation agent, was created by inoculating JHR2 with 1-mM 4-chlorobenzoate in the presence of 2-chlorobiphenyl (3 mM). The ability of the new hybrid strain *P. cepacia* JHR22 to utilise 2-chloro, 3-chloro, 4-chloro, 2,4-dichloro, and 3,5-dichlorobiphenyl as the sole carbon and energy source has been demonstrated. When tested in soil, the JHR22 strain was able to decompose all monochlorobiphenyl in Aroclor 1221T (Havel and Reineke, 1993).

4.5.2.2 4 Pseudomonas acidovorans M3GY

Pseudomonas acidovorans M3GY by mixing *P. acidovorans* CC1 (a chloroacetate and biphenyl degrading agent) and *Pseudomonas* sp. to produce a recombinant bacterial strain M3GY in a multi-chemostat culture on ceramic beads covered with 3,3-dichlorobiphenyl, strain CB15 (hybrid strain) (McCullar et al., 1994). The catabolic ability of the M3GY strain to utilise 3,4-dichlorobiphenyl as the sole source of carbon and energy is demonstrated. The recipient strain was chosen based on phenotypic and genetic similarities between the M3GY and CC1 strains.

4.5.2.2.5 Pseudomonas putida (pDA261)

Al-Ahmad et al., (1997) reported that the degradability of several biphenyl degradation genes (bphABCD) from *Trichomonas testicularis* B356 was investigated after they were subcloned into *Pseudomonas putida* and *Escherichia coli*. Sylvestre, (1995) reported that the hphC and bphD genes are well expressed in both cells; however, even though they are situated downstream of the tac promoter, the bphA and bphB genes in *E. coli* are not well expressed.

4.5.2.2.6 Trichomonas testicularis VP44 (pE43)/VP44 (pPC3)

The plasmid pE43 harbouring the ohbAB gene encodes the iron-sulfur protein (ISPOHB) of halobenzoate1,2 dioxygenase in *P. aeruginosa* 142. The plasmid pE43 was used to produce strain VP44. The recombinant bacteria VP44 mineralised 2-chlorobiphenyl and 2-chlorobenzoate (pE43). In another recombinant strain, VP44, the fcbABC gene (from Arthrobacter globosa KZT1) catalyses the hydrolysis and dechlorination of 4-chlorobenzoic acid (pPC3). This strain has been found to flourish on 4-chlorobiphenyl and 4-chlorobenzoate.

4.5.2.2.7 E. Coli JM109 (PSHF1003)/(PSHF1007)

Pseudomonas alkaline pseudo Kf707 and *Burkholderia cepacia* LB400 BPHA1 are a hybrid bipenil dioxygenase. DNA Whuffling generates a random number. (Kumamaru et al., 1998). The BFHA1 SHAVED

GEN was cloned upstream of BPHA2A3A4BC A and replaced the BPHA1 gene that was deleted in PJHF18 (Muli). (Hirose et al., 1994). Erickson and Mondello, (1993). Studied the Electroporation was used to transform the plasmids harbouring the BPHA1 (SHUFFLE) and BPHA2A3A4BC obtained (KF707) into *E. coli* JM109. Previously, sytedireted mutagenesis on Gen LB400 BPHA was described. The relevant KF707 sequences were converted from four amino acids in LB400. With PCB, the modified Bifegenil dioxens demonstrated a higher substrate specificity. *P. putida* OU 83's BHB gene, which encodes cisvifenyldhydrozodohydrogenase, was cloned into a pqe31 vector and expressed in *E. coli* Iptginded recombinant cells.

4.5.2.2.8 Pseudomonas putida IPL5 (Field Application Vector)

The Field Application Vector (FAV) is a combination of host (non-native) microorganisms, a simple selective substrate, and a cloning vector that offers a temporary niche for the host in hostile conditions. Bacteria are microorganisms that live in the (Lajoie and Shapiro, 1992). AVF can aid to stabilise and speeding up the expression of foreign genes in contaminated areas. PCB catabolic genes are encoded on the Pseudomonas chromosome (bphABC). The ENV307 strain was cloned onto the plasmid pRK293, which can be used in a variety of hosts. Using the non-ionic surfactant Igepal CO720T (IGP) as a selective substrate, the plasmid was transferred to the host *Sphingomonas pauginata* 1IGP. The recombinant strain 1IGP4 (pCL3) was able to break down a single PCB homolog in Aroclor 1242T without using biphenyl as an inducer (Lajoie et al., 1993). The transposon-encoded PCB degradation gene (bphABC) is more stable in the presence of surfactants than the plasmid expressed by *P. putida* IPL5. 1992 (Lajoie and Shapiro).

4.5.2.3 *Trichlorethylene (TCE)*

4.5.2.3.1 HB101/pMY402 and FM5/pKY287 Escherichia coli Strains

Piece of XhoI (4.7 kb) Containing toluene monooxygenase (TMO) from *Pseudomonas mendocera* KR1 in *Escherichia coli* HB101/pMY402 and FM5/pKY287 in Escherichia coli to form plasmid pMY402, the gene was subcloned onto a number of host vectors, including pMMB66EH, which contains an E. The identical section was inserted into another E. *coli* expression vector, pCFM1146, which contains the temperature-inducible coliphage PL promoter 1, to create plasmid pKY287. The recombinant *E. coli* strains HB101/pMY402 and FM5/pKY287 may oxidise toluene and TCE (Winter et al., 1989).

4.5.2.3.2 Pseudomonas pseudoalcaliges KF707D2 and Pseodomones putida KF715D5

E. coli JM109 Tod and Operon BPH have a hybrid dioxygenase genetic cluster. By deleting 1.3 KB of PPMI segments from PJHF10 to the BHB and BPHC genes, the plasmid PJHF101 carrying Todcl: BFHA2ORF3A3A4 was created. PJHF10 was built with the vector PUC118. This recombinant strain, E. TCE, degraded MLP1 HP1 at an initial rate of 1.8 g/g, which is significantly faster than *E. coli* cells with a Toluxioxigenase gene (*TDC1C2BA*) or a bifenyl dioxygenase gene (*BFA1A2A3A4*, Furukawa et al., 1994). *TDC1: PHA2ORF3A3A4,* Gene Cluster Hybrid,

4.5.2.3.3 E. coli JM109 (PDTG601)

P. toruhaneoxigenase. putida F1 has been identified as an enzyme involved in TCE breakdown (Wackett and Gibson, 1988). Toluxigenase structural genes (*TDC1C2BA*) were cloned from the plasmid PKK2233 and inserted into the JM109 strain. *E. coli* JM109 (PDTG601) is the resulting strain. TCE degraded at a slower beginning rate than *P. putida* F39/D or the mutant strain F1 of the F1 strain that does not include *Cystoluene dihydroidiol* dehydrogenase (Zylstra and Gibson, 1989).

4.5.2.4 2,4-Dichlorophenoxyacetic acid (2.4D)

4.5.2.4.1 Pseodomones putida PPO300 (PRO 101) and PPO 301 (PRO103) PJP4 (the chlorobenzate degradation gene) plasmid in *Ralstonia eutropha* JMP134 in 2.4D and 3 Code (Don and Pemberton, 1985). Gems with the catabolic gene for the breakdown of 2.4D to 2) chlorinel acetate. The Pro 101 plasmid (PJP 4: TN1721) was created by inserting the TN1721 transposon into PJP4 and then conjugating it to other Pseudomonas strains. In the presence of a 2.4D track inductor, *P. putida* PPO 300 (PRO 101) employs phenol hydroxylase encoded on chromosomes to break down phenoxyacetic acid storage (2,4D or 3 chlorobenzoate).

4.5.2.4.2 Pseudomonas CEPACIA RHJ1 Burkholderia CEPACIA (previously P. CEPACIA) RH1 is a recombinant strain resulting from the cross of *R. eutropha* JMP134 (2.4D) and B. Cepacia AC1100 (2,4,5 degrading trichlorophenoxyacetate drug). *Alcaligenes eutrophus* JMP134, PJP4 2,4,5 TRICHROPENOCINE ACEGIC ACID (2,4,5T) Degrading Device B. CEP1100. This novel strain, known as RHJ1, was able to break down 2, 4D and 2, 4, 5t combinations at the same time.

4.6 Recombinant DNA (r DNA) Technology for Microorganisms in Bioremediation

r DNA technology is the practice of modifying genetic material outside of an organism to get superior and desirable features in living organisms or as their products. Using this procedure, DNA fragments from various sources are inserted into a vector encoding the required gene sequence (Harvey et al., 2000).

The method utilised to develop recombinant strains for the overproduction of proteins is very different from the strategy for controlling any microbial cells that can break down certain substances and keep them alive for numerous generations. Various research groups have determined the number of distinct expression systems available that incorporate varied promoters and their regulators in order to create an effective microorganism (Endo and Silver, 1995); de Lorenzo et al., 1993). A technique to improve the stability of mRNA has been devised by inserting a DNA, untranslated fragment of the appropriate gene (Carrier and Keasling, 1997).

Using recombinant DNA technology, rational design entails creating a single bacterium with a chosen biodegradation pathway or enzymes from multiple species to accomplish a given reaction (Ang et al., 2005). Because the fine-folding dynamics of any protein can readily destroy its confirmation and lead to loss, prior information on its mechanical, dynamic, and structural qualities is required to successfully adapt it to meet a specified purpose through rational design catalytic activity. However, the ability to rationally design technologies has significantly improved because of the use of more effective functional bioinformatics and X-ray crystallographic technology. Kellner (1997) improved the substrate selectivity and catalytic efficiency of the P450 monooxygenase system. The sensible design of P45O to begin with requires an understanding of the structural and functional connections of distinct P45O cytochromes catalyse the biodegradation reactions you want. It was discovered that even a single amino acid change to the substrate-binding site can improve the enzyme's binding affinity flexibility.

De La Pea Mattozzi et al. (2006) provide an encouraging example of a catabolic route that has been logically constructed. After mutation, a strain of *P. putida* was created that can break down the organophosphorus chemical paraoxon and utilise it as the only carbon and nitrogen source. The *Pseudomonas* pnp operon. The strain ENV2030 was injected into the organism and simultaneously expressed the synthetic operon encoding the phosphodiesterase from Delftia acidovorans (encoded by pde), a Flavobacterium species The hydrolase of organophosphorus (encoded by opd). ATCC 27551 and alkaline phosphatase (encoded by phoA) from *P. aeruginosa* HN854 (De La Pea Mattozzi et al., 2006). A constitutive promoter controls the entire assembly process. De La Pea Mattozzi et al. (2006) found that the recombinant strain can use 275 g/L paraoxon as an energy source.

4.7 Bioremediation Potential

Assessing the overall success of an on-site bioremediation program is often a difficult process, whether using genetically modified microorganisms or intrinsic microorganisms. These difficulties lie mainly in inferring the real contribution of microorganisms to the degradation process and recognising that volatilisation and chemical transformations and other factors occur simultaneously in the system. Specifically, for

GEM use, due to the local microbial community, there may also be problems distinguishing GEM-specific degradation and biodegradation. Another obstacle is that due to the highly heterogeneous distribution of contaminants, the effectiveness of bioremediation cannot be measured statistically.

The glutamate decarboxylase (GAD)-dependent system, the F1-F0-ATPase proton pump, and macromolecule protection or repair are among the acid-resistance systems that have evolved in bacteria (Singhvi and colleagues, 2018). One or two glutamate decarboxylases (GadA/B) and one antiporter, GadC, make up the GAD-dependent system in bacteria (Kanjee and Houry, 2013). GadA/B transforms protonated glutamate to -aminobutyric acid (GABA), while GadC consumes intracellular protons and exports GABA in exchange for a new extracellular glutamate molecule. In *Deinococcus radiodurans*, IrrE, a global regulator, plays an important role in the bacterium's radiation resistance. Various ways of resistance to *Escherichia coli* to ethanol osmotic pressure, butanol, acetate, oxidative stress and inorganic acid are also involved in this IrrE regulator.

4.8 Bioremediation Impact on Human and Environmental Health

Increased industrialisation and resource exploitation have resulted in widespread environmental contamination. A vast amount of toxic waste has been strewn throughout our country in millions of contaminated sites. The threats to human and environmental health are growing by the day, and research suggests that this combination of contaminants has played a crucial role in the global epidemic of cancer along with other fatal diseases.

Bioremediation is the use of living assessment on biodiversity to reduce (and to the extent possible completely eliminate) the harmful effects of environmental pollutants in specific locations. Its operating principle is the biogeochemical cycle. In situ bioremediation occurs when the process takes place in the same location. Natural impedance (little or no human action), biological stimulation or enhancement, and the intentional insertion of innate or manipulated microorganisms to speed up the requisite catalytic capacity are all considered to be part of bioremediation. As a result, bioremediation, phytoremediation, and rhizome remediation have a substantial impact on hazardous waste fate and can be employed to remove undesired substances from the biosphere.

Pollutants and their negative consequences Heavy metals, metalloids, fluorides, cyanides, phenol, pesticides (DDT), and benzene are examples of inorganic and organic pollutants that damage land and water bodies. The following are a few of them: Arsenic: Arsenic (As) is a very poisonous element that can be found in soil and water. Arsenic contamination alters biological activity and has a negative impact on the immune system as a teratogen, carcinogen, and mutagen. Pesticide fate and degradation in agricultural soils have been studied for more than 50 years. Bacterial enzymes are involved in comprehensive catabolic processes. Bacteria and fungi possess the ability in order to break down xenobiotics.

4.9 Conclusion

GEMs have been used in several studies on various contaminants universally. Biodegradation of metabolic pathways, genes, and enzymes has all been found. These studies have revealed that GE bacteria have a strong potential for bioremediation of contaminants lot more flourishing than traditional techniques. Furthermore, it has been recognised that pollution prevention and remediation of contaminated places are both necessary for a safer environment. As a result, the only way to meet the demand for waste bioremediation is to focus on the development of more GE bacteria, which can then generate the management of impurities. Although, due to different ecological and environmental conditions, the usage of GE bacteria in the field is limited. As a result, knowing about assimilation coaction is the only basis that needs to be considered further in order to improve the regulation and developmental techniques of GE bacteria. This can only be accomplished through a variety of living cycle analyses with field experiments that balance field risks with microbiological advantages.

REFERENCES

Adams R.H., Huang, C.M., Higson, F.K., Brenner, V., Focht, D.D., 1992. Construction of a 3-chlorobiphenyl-utilizing recombinant from an intergeneric mating. Appl. Environ. Microbiol. doi:10.1128/aem.58.2.647-654.1992

Aftring R.P., Chalker, B.E., Taylor, B.F., 1981. Degradation of phthalic acids by denitrifying, mixed cultures of bacteria. *Appl. Environ. Microbiol.* doi:10.1128/aem.41.5.1177-1183.1981

Agarry, S., Latinwo, G.K., 2015. Biodegradation of diesel oil in soil and its enhancement by application of bioventing and amendment with brewery waste effluents as biostimulation-bioaugmentation agents. *J. Ecol. Eng.* doi:10.12911/22998993/1861

Agrawal, N., Kumar, V., Shahi, S.K., 2021. Biodegradation and detoxification of phenanthrene in in-vitro and in-vivo conditions by a newly isolated ligninolytic fungus *Coriolopsis byrsina* strain APC5 and characterization of their metabolites for environmental safety. *Environ. Sci. Pollut. Res.* doi:10.1007/s11356-021-15271-w

Al-Ahmad, A., Kümmerer, K., Schön, G., 1997. Biodegradation and toxicity of the antineoplastics mitoxantron hydrochloride and treosulfane in the closed bottle test (OECD 301 D). *Bull. Environ. Contam. Toxicol.* doi:10.1007/s001289900390

Ang, K.H., Chong, G., Li, Y., 2005. PID control system analysis, design, and technology. *IEEE Trans. Control Syst. Technol.* doi:10.1109/TCST.2005.847331

Asira, E.E., 2013. Factors that determine bioremediation of organic compounds in the soil. *Acad. J. Interdiscip. Stud.* doi:10.5901/ajis.2013.v2n13p125

Awad, G., Garnier, A., 2021. Promising optimization of bacterial cytochrome P450BM3 enzyme production by engineered *Escherichia coli* BL21. *Biocatal. Agric. Biotechnol.* doi:10.1016/j.bcab.2021.101917

Bak, S., Beisson, F., Bishop, G., Hamberger, B., Höfer, R., Paquette, S., Werck-Reichhart, D., 2011. Cytochromes P450. *Arabidopsis Book 9*, e0144.

Bhandari, S., Poudel, D.K., Marahatha, R., Dawadi, S., Khadayat, K., Phuyal, S., Shrestha, S., Gaire, S., Basnet, K., Khadka, U., Parajuli, N., 2021. Microbial enzymes used in bioremediation. *J. Chem. 2021*, 8849512. doi:10.1155/2021/8849512

Boopathy, R., 2000. Factors limiting bioremediation technologies. Bioresour. Technol. doi:10.1016/S0960-8524(99)00144-3

Carrier, T.A., Keasling, J.D., 1997. Controlling messenger RNA stability in bacteria: Strategies for engineering gene expression. Biotechnol. Prog. doi:10.1021/bp970095h

Chandra, R., Kumar, V., 2015. Biotransformation and biodegradation of organophosphates and organohalides. In Chandra, R. (Ed.) *Environmental Waste Management*. Boca Raton: CRC Press. doi:10.1201/b19243-17

Chatterjee, D.K., Chakrabarty, A.M., 1983. Genetic homology between independently isolated chlorobenzoate-degradative plasmids. *J. Bacteriol.* doi:10.1128/jb.153.1.532-534.1983

Chouhan, R., Lapaka, S., Alpula, N., Podeti, S., 2020. Optimization and molecular characterization of lipase producing *Bacillus subtilis* strain Rcps3 and *Bacillus fumarioli* strain Rcps4 from oil-contaminated soils of Warangal. *J. Pure Appl. Microbiol.* doi:10.22207/JPAM.14.3.47

Das, S., Dash, H.R., Chakraborty, J., 2016. Genetic basis and importance of metal resistant genes in bacteria for bioremediation of contaminated environments with toxic metal pollutants. *Appl. Microbiol. Biotechnol.* doi:10.1007/s00253-016-7364-4

De La Peña Mattozzi, M., Tehara, S.K., Hong, T., Keasling, J.D., 2006. Mineralization of paraoxon and its use as a sole C and P source by a rationally designed catabolic pathway in Pseudomonas putida. Appl. Environ. Microbiol. doi:10.1128/AEM.00907-06

de Lorenzo, V., Eltis, L., Kessler, B., Timmis, K.N., 1993. Analysis of Pseudomonas gene products using lacIq/Ptrp-lac plasmids and transposons that confer conditional phenotypes. Gene. doi:10.1016/0378-1119(93)90533-9

Dias, R.L., Ruberto, L., Calabró, A., Balbo, A. Lo, Del Panno, M.T., Mac Cormack, W.P., 2015. Hydrocarbon removal and bacterial community structure in on-site biostimulated biopile systems designed for bioremediation of diesel-contaminated Antarctic soil. Polar Biol. doi:10.1007/s00300-014-1630-7

Don, R.H., Pemberton, J.M., 1985. Genetic and physical map of the 2,4-dichlorophenoxyacetic acid-degradative plasmid pJP4. J. Bacteriol. doi:10.1128/jb.161.1.466-468.1985

Dua, M., Singh, A., Sethunathan, N., Johri, A., 2002. Biotechnology and bioremediation: Successes and limitations. *Appl. Microbiol. Biotechnol.* doi:10.1007/s00253-002-1024-6

El Fantroussi, S., Agathos, S.N., 2005. Is bioaugmentation a feasible strategy for pollutant removal and site remediation? *Curr. Opin. Microbiol.* doi:10.1016/j.mib.2005.04.011

Endo, G., Ji, G., Silver, S., 1995. Heavy Metal Resistance Plasmids and Use in Bioremediation, in: Environmental Biotechnology. doi:10.1007/978-94-017-1435-8_5

Erickson, B.D., Mondello, F.J., 1993. Enhanced biodegradation of polychlorinated biphenyls after site-directed mutagenesis of a biphenyl dioxygenase gene. *Appl. Environ. Microbiol.* doi:10.1128/aem.59.11.3858-3862.1993

Furukawa, K., Inagaki, H., Hotta, Y., 1994. Identification and cloning of an mRNA coding for a germ cell-specific A-type lamin in mice. Exp. Cell Res. doi:10.1006/excr.1994.1164

Gilbert, E.S., Crowley, D.E., 1997. Plant compounds that induce polychlorinated biphenyl biodegradation by Arthrobacter sp. Strain B1B. Appl. Environ. Microbiol. doi:10.1128/aem.63.5.1933-1938.1997

Gomez, F., Sartaj, M., 2014. Optimization of field scale biopiles for bioremediation of petroleum hydrocarbon contaminated soil at low temperature conditions by response surface methodology (RSM). Int. Biodeterior. Biodegrad. doi:10.1016/j.ibiod.2014.01.010

Gustafsson, M.C.U., Roitel, O., Marshall, K.R., Noble, M.A., Chapman, S.K., Pessegueiro, A., Fulco, A.J., Cheesman, M.R., Von Wachenfeldt, C., Munro, A.W., 2004. Expression, purification, and characterization of *Bacillus subtilis* cytochromes P450 CYP102A2 and CYP102A3: Flavocytochrome homologues of P450 BM3 from *Bacillus megaterium*. *Biochemistry*. doi:10.1021/bi035904m

Hartman, J., 1989. Degradation of 2-chlorobenzoate by in vivo constructed hybrid pseudomonads. FEMS Microbiol. Lett. doi:10.1016/0378-1097(89)90162-6

Harvey, L., Arnold, B., S Lawrence, Z., Pau, l M., David, B., James, D., 2000. Molecular Cell Biology. 4th edition, Journal of the American Society for Mass Spectrometry. doi:10.1016/j.jasms.2009.08.001

Havel, J., Reineke, W., 1993. Degradation of Aroclor 1221 in soil by a hybrid pseudomonad. FEMS Microbiol. Lett. doi:10.1111/j.1574-6968.1993.tb06101.x

Havel, J., Reineke, W., 1991. Total degradation of various chlorobiphenyls by cocultures and in vivo constructed hybrid pseudomonads. FEMS Microbiol. Lett. doi:10.1111/j.1574-6968.1991.tb04437.x

Hirose, T., Smith, R.J., Jetten, A.M., 1994. ROR-γ: The Third Member of ROR/RZR Orphan Receptor Subfamily That Is Highly Expressed in Skeletal Muscle. Biochem. Biophys. Res. Commun. doi:10.1006/bbrc.1994.2902

Huang, W.C., Tang, I.C., 2007. Bacterial and Yeast Cultures-Process Characteristics, Products, and Applications, in: Bioprocessing for Value-Added Products from Renewable Resources. doi:10.1016/B978-044452114-9/50009-8

Karigar, C.S., Rao, S.S., 2011. Role of microbial enzymes in the bioremediation of pollutants: A review. *Enzyme Res.* doi:10.4061/2011/805187

Kanjee, U., Houry, W.A., 2013. Mechanisms of acid resistance in escherichia coli. Annu. Rev. Microbiol. doi:10.1146/annurev-micro-092412-155708

Kasprzak, K.S., 2002. Oxidative DNA and protein damage in metal-induced toxicity and carcinogenesis. *Free Radic. Biol. Med.* doi:10.1016/S0891-5849(02)00809-2

Kellner, D.G., Maves, S.A., Sligar, S.G., 1997. Engineering cytochrome P450s for bioremediation. Curr. Opin. Biotechnol. doi:10.1016/S0958-1669(97)80003-1

Kuiper, I., Lagendijk, E.L., Bloemberg, G. V., Lugtenberg, B.J.J., 2004. Rhizoremediation: A beneficial plant-microbe interaction. *Mol. Plant-Microbe Interact.* doi:10.1094/MPMI.2004.17.1.6

Kumamaru, T., Suenaga, H., Mitsuoka, M., Watanabe, T., Furukawa, K., 1998. Enhanced degradation of polychlorinated biphenyls by directed evolution of biphenyl dioxygenase. *Nat. Biotechnol.* doi:10.1038/nbt0798-663

Kumar, V., Chandra, R., 2018. Characterisation of manganese peroxidase and laccase producing bacteria capable for degradation of sucrose glutamic acid-Maillard reaction products at different nutritional and environmental conditions. *World J. Microbiol. Biotechnol. 34*, 32.

Kumar V., Chandra R., 2020. Bioremediation of melanoidins containing distillery waste for environmental safety. In: Bharagava R., Saxena G. (eds) *Bioremediation of Industrial Waste for Environmental Safety*. Springer, Singapore. doi:10.1007/978-981-13-3426-9_20

Kumar, S., Dhar, H., Nair, V.V., Govani, J., Arya, S., Bhattacharya, J.K., Vaidya, A.N., Akolkar, A.B., 2019. Environmental quality monitoring and impact assessment of solid waste dumpsites in high altitude subtropical regions. *J. Environ. Manage. 252*, 109681. doi:10.1016/j.jenvman.2019.109681

Kumar, V., Ferreira, L.F.R., Sonkar, M., Singh, J. 2021a. Phytoextraction of heavy metals and ultrastructural changes of *Ricinus communis* L. grown on complex organometallic sludge discharged from alcohol distillery. *Environ. Technol. Innov. 22*, 101382. doi:10.1016/j.eti.2021.101382

Kumar, V., Shahi, S.K., Ferreira, L.F.R., Bilal, M., Biswas, J.K., Bulgariu, L., 2021b. Detection and characterization of refractory organic and inorganic pollutants discharged in biomethanated distillery effluent and their phytotoxicity, cytotoxicity, and genotoxicity assessment using *Phaseolus aureus* L. and *Allium cepa* L. *Environ. Res. 201*, 111551. doi:10.1016/j.envres.2021.111551

Kumar, V., Shahi, S.K., Singh, S., 2018. Bioremediation: An eco-sustainable approach for restoration of contaminated sites. In: Singh J., Sharma D., Kumar G., Sharma N. (eds) *Microbial Bioprospecting for Sustainable Development*. Springer, Singapore. doi:10.1007/978-981-13-0053-0_6

Kumar, V., Thakur, I.S., Shah, M.P., 2020a. Bioremediation approaches for pulp and paper industry wastewater treatment: recent advances and challenges. In: Shah, M.P. (ed) *Microbial Bioremediation & Biodegradation*. Springer, Singapore. doi:10.1007/978-981-15-1812-6_1

Kumar, V., Thakur, I.S., Singh, A.K., Shah, M.P., 2020b. Application of metagenomics in remediation of contaminated sites and environmental restoration. In: Shah, M., Rodriguez-Couto, S., Sengor, S.S. (eds) *Emerging Technologies in Environmental Bioremediation*. Elsevier. doi:10.1016/B978-0-12-819860-5.00008-0

Latorre, J., Reineke, W., Knackmuss, H.J., 1984. Microbial metabolism of chloroanilines: enhanced evolution by natural genetic exchange. Arch. Microbiol. doi:10.1007/BF00454919

Layton, A.C., Sanseverino, J., Wallace, W., Corcoran, C., Sayler, G.S., 1992. Evidence for 4-chlorobenzoic acid dehalogenation mediated by plasmids related to pSS50. Appl. Environ. Microbiol. doi:10.1128/aem.58.1.399-402.1992

Lee, T.H., Byun, I.G., Kim, Y.O., Hwang, I.S., Park, T.J., 2006. Monitoring biodegradation of diesel fuel in bioventing processes using in situ respiration rate. *Water Sci. Technol.* doi:10.2166/wst.2006.131

Li, Z., Jiang, Y., Guengerich, X.F.P., Ma, L., Li, S., Zhang, W., 2020. Engineering cytochrome P450 enzyme systems for biomedical and biotechnological applications. *J. Biol. Chem.* doi:10.1074/jbc.REV119.008758

Longmuir, I.S., 1954. Respiration rate of bacteria as a function of oxygen concentration. *Biochem. J.* doi:10.1042/bj0570081

Macaulay, B.M., 2015. Understanding the behaviour of oil-degrading micro-organisms to enhance the microbial remediation of spilled petroleum. Appl. Ecol. Environ. Res. doi:10.15666/aeer/1301_247262

McClure, N.C., Fry, J.C., Weightman, A.J., 1991. Survival and catabolic activity of natural and genetically engineered bacteria in a laboratory-scale activated-sludge unit. Appl. Environ. Microbiol. doi:10.1128/aem.57.2.366-373.1991

McCullar, M. V., Brenner, V., Adams, R.H., Focht, D.D., 1994. Construction of a novel polychlorinated biphenyl-degrading bacterium: Utilization of 3,4'-dichlorobiphenyl by Pseudomonas acidovorans M3GY. *Appl. Environ. Microbiol.* doi:10.1128/aem.60.10.3833-3839.1994

Meng, X., Yu, D., Wei, Y., Zhang, Y., Zhang, Q., Wang, Z., Liu, J., Wang, Y., 2018. Endogenous ternary pH buffer system with ammonia-carbonates-VFAs in high solid anaerobic digestion of swine manure: An alternative for alleviating ammonia inhibition? *Process Biochem.* doi:10.1016/j.procbio.2018.03.015

Mishra, D., Kumari, S., Jaiswal, A., Arya, S., Yadav, B.R., Thul, S.T., Kumar, S., Pandey, R., Kumar, R., Pandey, A., 2021. Evaluation of distillery sludge as a soil amendment for improving soil quality and sugarcane (CO-265) yield. *Environ. Technol. Innov. 23*, 101624. doi:10.1016/j.eti.2021.101624

Moo-Young, M., 2019. Comprehensive biotechnology. *Ann. Inst. Pasteur Virol.* doi:10.1016/s0769-2617(86)80233-7

Muthukumarasamy, N.P., Jackson, B., Joseph Raj, A., Sevanan, M., 2015. Production of extracellular laccase from *Bacillus subtilis* MTCC 2414 using agroresidues as a potential substrate. *Biochem. Res. Int.* doi:10.1155/2015/765190

Naik, M.G., Duraphe, M.D., 2012. Review paper on-parameters affecting bioremediation. *Int. J. Life Sci. Pharma Res.*, *2*, L-77–L-80.

O. Elekwachi, C., 2014. Global Use of Bioremediation Technologies for Decontamination of Ecosystems. J. Bioremediation Biodegrad. doi:10.4172/2155-6199.1000225

Omokhagbor Adams, G., Tawari Fufeyin, P., Eruke Okoro, S., Ehinomen, I., 2020. Bioremediation, biostimulation and bioaugmention: A review. *Int. J. Environ. Bioremediat. Biodegrad.* doi:10.12691/ijebb-3-1-5

Paliya, S., Mandpe, A., Kumar, M.S. and Kumar, S., 2021. Aerobic degradation of decabrominated diphenyl ether through a novel bacterium isolated from municipal waste dumping site: Identification, degradation and metabolic pathway. *Bioresour. Technol.*, *333*, 125208. doi:10.1016/j.biortech.2021.125208

Pant, G., Garlapati, D., Agrawal, U., Prasuna, R.G., Mathimani, T., Pugazhendhi, A., 2021. Biological approaches practised using genetically engineered microbes for a sustainable environment: A review. *J. Hazard. Mater.* doi:10.1016/j.jhazmat.2020.124631

Philp, J.C., Bamforth, S.M., Singleton, I., Atlas, R.M., 2014. Environmental pollution and restoration: A role for bioremediation. In: *Bioremediation*. doi:10.1128/9781555817596.ch1

Ramos, J.L., Stolz, A., Reineke, W., Timmis, K.N., 1986. Altered effector specificities in regulators of gene expression: TOL plasmid xylS mutants and their use to engineer expansion of the range of aromatics degraded by bacteria. Proc. Natl. Acad. Sci. U. S. A. doi:10.1073/pnas.83.22.8467

Sahasrabudhe, A., Pande, A., Modi, V., 1991. Dehalogenation of a mixture of chloroaromatics by immobilized *Pseudomonas* sp. US1 ex cells. *Appl. Microbiol. Biotechnol.* doi:10.1007/BF00169904

Schwien, U., Schmidt, E., 1982. Improved degradation of monochlorophenols by a constructed strain. Appl. Environ. Microbiol. doi:10.1128/aem.44.1.33-39.1982

Sielaff, B., Andreesen, J.R., Schräder, T., 2001. A cytochrome p450 and a ferredoxin isolated from *Mycobacterium* sp. strain HE_5 after growth on morpholine. *Appl. Microbiol. Biotechnol.* doi:10.1007/s002530100634

Singh, E., Kumar, A., Mishra, R., You, S., Singh, L., Kumar, S., Kumar, R., 2021. Pyrolysis of waste biomass and plastics for production of biochar and its use for removal of heavy metals from aqueous solution. *Bioresour. Technol.*, *320*, 124278. doi.org/10.1016/j.biortech.2020.124278

Singhvi, M., Zendo, T., Sonomoto, K., 2018. Free lactic acid production under acidic conditions by lactic acid bacteria strains: challenges and future prospects. Appl. Microbiol. Biotechnol. doi:10.1007/s00253-018-9092-4

Sylvestre, M., 1995. Biphenyl/chlorobiphenyls catabolic pathway of *Comamonas testosteroni* B-356: Prospect for use in bioremediation. *Int. Biodeterior. Biodegrad.* doi:10.1016/0964-8305(95)00034-3

Timmis, K.N., Pieper, D.H., 1999. Bacteria designed for bioremediation. *Trends Biotechnol.* doi:10.1016/S0167-7799(98)01295-5

Turpeinen, R., Kairesalo, T., Häggblom, M.M., 2004. Microbial community structure and activity in arsenic-, chromium- and copper-contaminated soils. *FEMS Microbiol. Ecol.* doi:10.1016/S0168-6496(03)00232-0

Wackett, L.P., Gibson, D.T., 1988. Degradation of trichloroethylene by toluene dioxygenase in whole-cell studies with Pseudomonas putida F1. Appl. Environ. Microbiol. doi:10.1128/aem.54.7.1703-1708.1988

Wainaina, S., Awasthi, M.K., Sarsaiya, S., Chen, H., Singh, E., Kumar, A., Ravindran, B., Awasthi, S.K., Liu, T., Duan, Y., Kumar, S., 2020. Resource recovery and circular economy from organic solid waste using aerobic and anaerobic digestion technologies. *Bioresource Technol.*, *301*, 122778. doi:10.1016/j.biortech.2020.122778

Whelan, M.J., Coulon, F., Hince, G., Rayner, J., McWatters, R., Spedding, T., Snape, I., 2015. Fate and transport of petroleum hydrocarbons in engineered biopiles in polar regions. Chemosphere. doi:10.1016/j.chemosphere.2014.10.088

Winter, R.B., Yen, K.M., Ensley, B.D., 1989. Efficient degradation of trichloroethylene by a recombinant escherichia coli. Bio/Technology. doi:10.1038/nbt0389-282

YANG, S.Z., JIN, H.J., WEI, Z., HE, R.X., JI, Y.J., LI, X.M., YU, S.P., 2009. Bioremediation of Oil Spills in Cold Environments: A Review. Pedosphere. doi:10.1016/S1002-0160(09)60128-4

Zylstra, G.J., Gibson, D.T., 1989. Toluene degradation by Pseudomonas putida F1. Nucleotide sequence of the todC1C2BADE genes and their expression in Escherichia coli. J. Biol. Chem. doi:10.1016/S0021-9258(18)63793-7

5

Performance of Anammox in Industrial Wastewater Treatment: Recent Advances and Future Prospects

Thilagavathi Arumugham, Imran Ahmad, Norhayati Abdullah, Ali Yuzir, Iwamoto Koji, and Shaza Eva Mohamad
Universiti Teknologi Malaysia

CONTENTS

5.1 Introduction of Anammox

Biological wastewater treatment systems can be divided into two branches that are aerobic and anaerobic (Muralikrishna & Manickam, 2017; Dutta et al., 2021). The biological treatment method depends on the availability of oxygen for the processes that will be carried out by the bacteria community in the wastewater. The application of anaerobic treatment is more preferred compared to aerobic due to unique characteristics like reduction of sludge production, methane generation as the energy source, and the efficient removal of pollutants (Luo et al., 2020; Ahmad et al., 2020). Anaerobic digestion technology is applied generally in wastewater treatment to treat the organic waste and sewage sludge due to an eco-friendly and economical process (Hanum et al., 2019). In anaerobic digestion, a high concentration of ammonium will be produced through the decomposition of the nitrogen components like the protein (Qiao et al., 2016; Yang et al., 2018). Hence, the biological nitrogen removal technology has been explored to treat the nitrogen in wastewater in order to prevent pollution occurrence in the environment (Wang et al., 2015).

Recently, wastewater treatment technology has focused on completely autotrophic nitrogen removal (ANR) particularly in anaerobic ammonium oxidation (anammox) (Connan et al., 2018). This is because anammox was established as the shortcut technology in the biogeochemical nitrogen cycle for sustainable nitrogen removal from wastewater (Lawson et al., 2021). Anammox was discovered in 1995 and many intensive studies were conducted (Mulder et al., 1995; Tomaszewski et al., 2017). The anammox process converts the ammonium into nitrogen gas by utilizing the nitrite as an electron acceptor and saves energy resources (Sri Shalini & Joseph, 2018; Güven et al., 2004; Henze et al., 2008). The anammox bacteria which is an anaerobic bacteria are classified under the order Planctomycetales. The five genera branching under

DOI: 10.1201/9781003231738-6

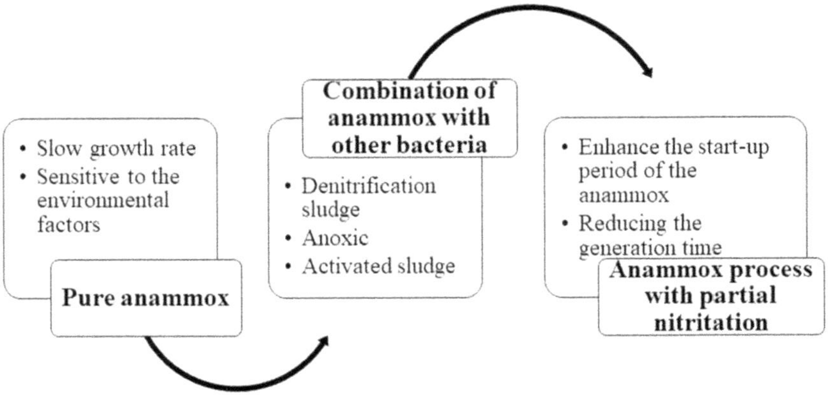

FIGURE 5.1 The revolution of the anammox process.

this order are: *Candidatus Brocadia, Candidatus Kuenenia, Candidatus Jettenia, Candidatus Scalindua* and *Candidatus Anammoxoglobus* (Arumugham et al., 2020). The anammoxosome is the membrane layer of the anammox which is dense and has a low permeability capacity in order to regulate the concentration gradients for the catabolism process (Ma et al., 2016).

However, the growth rate of the anammox bacteria is slow, which causes the application in a wide range of conditions to pose difficulty. This is because the anammox bacteria are sensitive to environmental factors like temperature, pH, the concentration of nitrogen and dissolved oxygen (DO) (Kallistova et al., 2016; Gonzalez-Martinez et al., 2018). Additionally, the constraint on obtaining pure cultures of the anammox community has limited the practical application of anammox (Ni and Zhang, 2013). Therefore, exclusive studies were conducted upon the inoculation of anammox seed sludge mixed with other sources like denitrification sludge, anoxic and activated sludge (Ding et al., 2017). The purpose was to enhance the start-up period of the anammox process besides reducing the generation time (Lin et al., 2018). In addition, extensive research has been conducted on combining the anammox process with partial nitritation as an alternative method to treat wastewater with a high ammonium content (Zhang et al., 2020; Kampschreur et al., 2009). Figure 5.1 shows the revolution of the anammox process.

This chapter summarises the synergy between partial nitritation and the anammox (PN/A) process for nitrogen removal in industrial wastewater. The chapter provides an insight into the biological nitrogen removal technologies applied to the wastewater treatment system. Furthermore, the chapter also incorporated a few potential techniques to overcome the obstacles hindering the anammox applications in the treatment system. The chapter is expected to contribute to the area of industrial wastewater treatment by providing knowledge on the PN/A application for the wastewater treatment system, which is both eco-friendly and cost-effective.

5.2 Contribution of Anammox in the Wastewater Treatment Plants (WWTPs)

The anammox application in the wastewater treatment sector has reduced the high operational cost and time consumption for the nitrogen removal process (Azari et al., 2017). According to previous studies, the engineering application of the anammox process in wastewater treatment has been implemented and is in operation at more than 100 full-scale plants (Mao et al., 2017; Lackner et al., 2014). The application of anammox for nitrogen-rich wastewater has shown excellent results in terms of economics and efficient nitrogen removal (Kawagoshi et al., 2019; van der Star et al., 2007; Jetten et al., 2005). The nitrogen removal efficiency through the anammox process is interconnected with the nitrogen loading rate in the wastewater. This is because ammonium is recognised as the primary source of nitrogen for the anammox process, besides simultaneously interpreting the nitrogen removal rate, the development of anammox bacteria and efficient anammox activity (Cho et al, 2020). Therefore, the correlation between the ammonium concentration and the density of the anammox bacteria should be balanced in order to achieve high anammox activity (Strous et al., 1999).

TABLE 5.1
Application of Anammox Process

Wastewater Type	Nitrogen Loading Rate	Nitrogen Removal (%)	References
Municipal wastewater treatment plant	13.32±2.35 mg N/L.d	54±14	Vázquez-Padín et al. (2018)
Slaughterhouse wastewater	1.3–6.3 g N/L.d	44.8±19.7–87.7±5.6	Silveira et al. (2021)
Synthetic wastewater	0.58–0.66 g N/L.d	81–89	Zhang et al. (2021a)
Synthetic wastewater	2.02 kg N/m^3/d	84.8±3.5	Huang et al. (2020)
Mined waste	1.2 kg N/m^3/d	84.0	Sri Shalini and Joseph (2018)
Liquid-ammonia mercerization wastewater	0.82 kg N/m^3/d	70.0	Zheng et al. (2020)

The inappropriate nitrogen loading rate via ammonium and nitrite may lead to the inhibition of the anammox process due to the stimulation of unwanted bacteria growth, high rate of nitrate production, high levels of salt, heavy metals and inorganic carbon (Cho et al., 2020). Thus, the nitrogen concentration can be calibrated by calculating the stoichiometric of the nitrite to ammonium molar ratio to obtain maximum specific anammox activity (Martins et al., 2017). The granular sludge form is preferable compared to the suspended anammox biofilm for the long term of the anammox process because of granular anammox adaption towards different concentrations of nitrogen (Gilbert et al., 2015). Nevertheless, the anammox biomass is selected for the bioreactors to prevent the anammox process interruption and to promote the adaption towards various levels of nitrogen concentration and loading, which is consequently able to optimise the process (Wang et al., 2016a). Table 5.1 summarises the application of the anammox process for different wastewater with various concentrations of nitrogen loading rate.

Table 5.1 shows the wastewater sourced from different sources composed with various values of nitrogen loading due to inconsistent ammonium and nitrite concentration. Nitrogen pollution in wastewater from various sectors necessitates treatment prior to discharge to the environment or reuse (van der Hoek et al., 2018). The anammox process application at different rates of nitrogen loading rate is able to achieve excellent nitrogen removal up to 89%. Table 5.1 shows the higher nitrogen loading rate (2.02 kg N/m^3/d) able to perform better (84.8%±3.5%) (Huang et al., 2020) as compared to the 1.3-6.3 g N/L.d with the removal efficiency of 44.8±19.7–87.7±5.6 (Silveira et al., 2021). This may be due to the interruption by factors like the temperature and pH, which are the crucial parameters that may significantly impact the wastewater treatment and anammox process (Tomaszewski et al., 2017).

Therefore, this chapter aims to discuss the anammox application for industrial wastewater containing a variety of contaminants and the challenges posed by anammox in industrial application. Furthermore, the proposed solutions may aid the anammox application in the industrial wastewater to be improved and yield a high quality of effluent that can be discharged directly into the streamline or reused. Hence, the following sections of the chapter will elaborate on the industrial wastewater characterization from various industries and the biological nitrogen removal methods applied in the wastewater system to reduce the water pollution due to nitrogen. The PN/A which is a sustainable technology in the industrial wastewater treatment system was facing obstacles in achieving a stable process for the long term. Therefore, a few solutions were proposed in this chapter to overcome the challenges as the key opportunity to expand the application widely in the wastewater system.

5.3 Industrial Wastewater Characteristic

The wastewater sourced from industry generally contains a high number of heavy metals and organic compounds, which consequently lead to water pollution phenomena (Verma et al., 2021; Kumar et al. 2021a, b; Chandra and Kumar 2017a; Moustafa, 2014). The high nitrogen concentration in the wastewater may lead to the eutrophication phenomenon, which consequently can cause an interruption in the aquatic ecosystem chain and a deterioration in the water quality (Rodriguez-Sanchez et al., 2016; Soudejani et al., 2019; Weralupitiya et al., 2021). Therefore, industries are concerned with treating the wastewater in order

TABLE 5.2

Industrial Wastewater with Different Rates of Ammonia Concentrations

Wastewater Type	Ammonia Concentration (mg NH_4^+-N/L)	References
Optoelectronic industry	3,712±120	Daverey et al. (2013)
Textile industry	1,257±72	Yaseen & Scholz (2019)
Palm oil effluents	750 mg/L of nitrogen	Khadaroo et al. (2021)
Brewery industry	22.78±3.91	Gebeyehu et al. (2018)
Dairy industry	1462 mg/L of Total Kjeldahl nitrogen (TKN)	Gil-Pulido et al. (2018)
Rubber industry	55	Nasir et al. (2019)
Petroleum industry	10–300	Ghimire and Wang (2019)
Paper industry	30.8±2.5 of total nitrogen	Van Den Hende et al. (2017)
Agriculture industry	200–250	Zhang et al. (2021b)
Pharmaceutical industry	800–1,000 ppm of Total Kjeldahl nitrogen (TKN)	Pal (2017)

to comply with government regulations prior to discharging it to the environment (Nancharaiah and Reddy, 2018). The industries are focusing on developing sustainable and cost-effective nitrogen removal wastewater treatment technologies (Weralupitiya et al., 2021). However, the nitrogen pollutant, which is ammonia, in the wastewater consists of variant concentrations because it depends on the type of industrial wastewater. Table 5.2 summarises the ammonia concentrations in the different types of industrial wastewater bodies.

Table 5.2 indicates the ammonia concentration in the wastewater sourced from various backgrounds of industries is high. Therefore, industries are required to treat the wastewater in order to prevent water pollution. For instance, the nitrate-contaminated wastewater treatment via adsorption or co-precipitation was not attainable due to the stability and solubility of the nitrate, leading to extreme energy consumption and expense (Rezvani et al., 2019). Hence, the most general and economical technology used for nitrogen removal in the wastewater system is the biological nitrogen removal method (Rahimi et al., 2020).

5.3.1 Biological Nitrogen Removal Treatments in Industrial Wastewater

Industrial wastewater commonly contains various contaminants like hazardous compounds and toxic materials (Huy Quoc Anh et al., 2015; Chandra and Kumar, 2017b). Hence, it was a prerequisite for each industry to treat the wastewater prior to discharge into the environment, besides complying with the regulations regulated by the governments (Manipura et al., 2005; Kumar and Shah, 2021). The effluents from the industrial sectors, agriculture sectors and human activities have contributed to the nitrogen element in the wastewater, which has concurrently resulted in the occurrence of eutrophication in the water (Islam et al., 2009; Nourmohammadi et al., 2013; van der Hoek et al., 2018). The most general contributors to the total effluent nitrogen are ammonium, nitrite, nitrate and soluble organic nitrogen compounds. Due to the variability of the nitrogen concentration in various industries, hence, many biological nitrogen removal methods are being implemented to treat the wastewater. Figure 5.2 shows the biological nitrogen removal processes in wastewater treatment for nitrogen removal.

5.3.1.1 Conventional Nitrification-Denitrification Process

The most common technology used in biological nitrogen removal is nitrification-denitrification as shown in Figure 5.1 (Winkler and Straka, 2019). Nitrification is the biological conversion of ammonia to nitrate that occurs under oxidative conditions by ammonium oxidizing bacteria (AOB) and nitrite oxidizing bacteria (NOB). Meanwhile, denitrification is the conversion of nitrate to nitrogen gas that occurs under anoxic conditions with the association of organic carbon (Farazaki and Gikas, 2019). The nitrification-denitrification processes require sufficient organic carbon, which is a 3:1 ratio of soluble carbonaceous biological oxygen demand (BOD) to nitrite or nitrate ion. However, the influent with insufficient of organic content, obsolete treatment facilities and short hydraulic retention time have resulted in the bottleneck of the nitrification-denitrification processes in the wastewater treatment plants

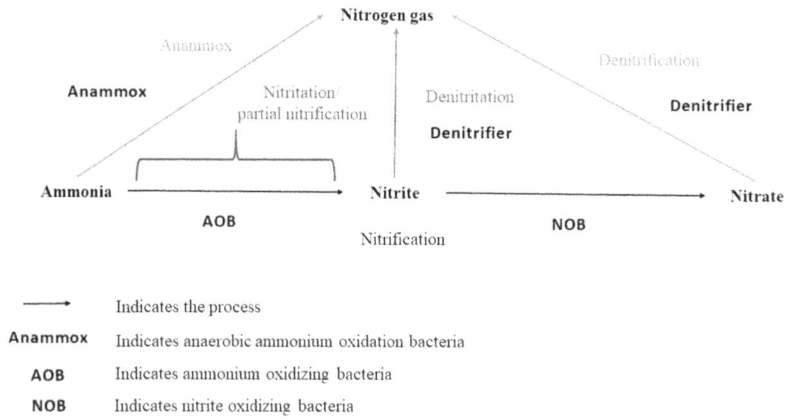

FIGURE 5.2 Biological nitrogen removal processes and the bacteria responsible for each process.

(WWTPs) (Wang et al., 2016b). Therefore, in order to enhance the carbon sources for denitrification, the WWTPs applied the traditional method by substituting alternative organic substrates like methanol, acetic acid, propionic acid, glucose and benzoic acid (Elefsiniotis & Li, 2006). Consequently, this has caused the nitrification-denitrification processes to require energy intensive and led to the consideration of alternative technology.

5.3.1.2 Nitritation-Denitritation Process

Nitritation is the process of converting the ammonia into nitrite or also known as partial nitrification as shown in Figure 5.1 (Sun et al., 2015). Meanwhile, denitritation is the conversion of nitrite to nitrogen gas under the 3.43 mgO_2/mgN condition compared to full nitrification which needs 4.57 mgO_2/mgN (Winkler and Straka, 2019; Jenicek et al., 2004). The combination of nitritation and denitritation is recognised as the short-cut biological nitrogen removal (BNR) technology that can enhance the AOB and inhibit the NOB. Therefore, intermittent aeration was introduced in order to enrich AOB and control NOB development (Gu et al., 2018; Sun et al., 2017). The drawback of the process is that at low levels of dissolved oxygen (DO), the nitrous oxide production is triggered and results in the emission of greenhouse gases (Massara et al., 2017).

5.3.1.3 Partial Nitritation/Anammox (PN/A) Process

The alternative technology called the partial nitritation coupled with the anammox process as shown in Figures 5.2 and 5.3 was implemented in the WWTPs for the nitrogen removal. Firstly, in the aerobic zone, the ammonium is partially oxidised to nitrite by the AOB which is known as partial nitritation. Next, in the anammox process, the anammox bacteria oxidises the ammonium into nitrogen gas in the anaerobic zone by utilizing the nitrite as shown in Figure 5.3 (Cao et al., 2017 Chen et al., 2019). The PN/A has distinct advantages as an autotrophic nitrogen removal process: (i) reduces the oxygen requirement up to 60% for partial nitritation, (ii) practically terminates the carbon usage by 100% for denitrification, and (iii) reduces the sludge generation to 80% (Jetten et al., 1997; Wett, 2007; Daigger, 2014).

5.4 Anammox in Industrial Wastewater

The high concentration of ammonia in industrial wastewater has attracted the anammox process. Furthermore, the anammox bacteria tolerance towards the organic waste, different temperatures and other contaminants such as fluorine and phenol has made the application in the wastewater system a successful process (Liu et al., 2019). Nowadays, the PN/A process has been implemented in the wastewater

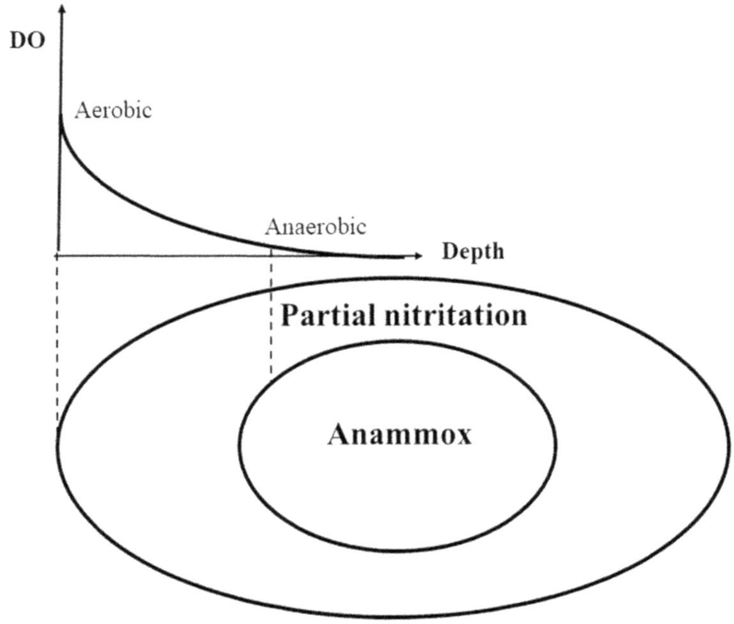

FIGURE 5.3 The PN/A process at different level of DO concentrations in the granules.

treatment system for warm and concentrated effluents (Weralupitiya et al., 2021). The nitrogen removal rate from the industrial wastewater using the anammox process is summarised in Table 5.3.

Table 5.3 indicates the application of the anammox process in many countries for industrial wastewater has contributed to good success in the nitrogen removal process. This co-currently shows the effluent discharged from the industries was treated using the most economical process, besides enabling them to prevent the environmental constraints (Gonzalez-Martinez et al., 2018).

5.5 Challenges

The challenges of implementing the PN/A process widely in the WWTPs are the high carbon to nitrogen (C/N) ratio, which consequently leads to the heterotrophic bacteria (HB) community being dominant in the sludge. Hence, this phenomenon may cause inhibition of the growth and activities of anammox and AOB followed by resulting low nitrogen removal rate (NRR) in the WWTPs (Metcalf and Eddy, 2003; Cao, van Loosdrecht, and Daigger, 2017). According to (Metcalf & Eddy, 2003), the lower ammonium concentration in the wastewater (30–100 mg NH_4-N/L) may be challenging to suppress the NOB and may result in major constraints on the mainstream PN/A process. The anammox activity was low at temperatures below 15°C, compared to the seasonal temperature range of 10°C–25°C (Vlaeminck et al., 2012).

TABLE 5.3

Anammox Technology Implemented at Different Industrial Wastewater

Country	Influent	Nitrogen Removal Rate	References
Italy	Refinery wastewater produced by the integrated gasification combined cycle (IGCC)	71%±3%	Milia et al. (2017)
China	Monosodium glutamate industry	90%	Muñoz-Palazon et al. (2017)
Japan	Slurry	48%–98%	Suto et al. (2017)
Germany	Landfill leachate	94%	Azari et al. (2017)
Taiwan	Optoelectronic	89%	Daverey et al. (2013)

This is because the sharper specific activity of the anammox was observed at 15°C while the activity dropped approximately ten times when the temperature decreased from approximately 30°C–10°C (Lotti et al., 2015; Cao et al., 2017).

5.6 The Opportunity and Future Prospects

According to (García-Ruiz et al., 2018), the wastewater containing organic matter with the concentration of 100 mg COD/L able to achieve an excellent nitrogen removal efficiency through the combination of anammox bacteria with other microbes like AOB and denitrifiers. The collaboration of several microbial metabolisms resulted in a total nitrogen removal efficiency of 92% at DO levels of less than 1.5 mg O_2/L and 0 mg NO_2^--NL. Currently, extensive studies are being conducted in order to support the hypothesis that anammox is able to oxidise the multitude of organic compounds like the acetate, propionate and methylamines in the PN/A process (Gonzalez-Martinez et al., 2018; Cho et al., 2020). Interestingly, many studies have been carried out to treat the pharmaceutical effluent, particularly hospital wastewater with various antibiotics using the anammox approach. Based on (Sguanci et al., 2017), the increase of biomass concentration in the anammox granular reactor is able to enhance the resistance of anammox bacteria to antibiotics due to a combination of culture metabolisms. The collaboration of a variety of microbes like the AOB and denitrifiers is able to aid the anammox process, besides tackling the inhibition of the anammox activity.

Many investigations are being carried out in order to enhance the anammox process at low temperatures by inhibiting the NOB. The continuous stirring of the reactor is suggested as an option to suppress the NOB and maintain the high reaction rate (Cao et al., 2017). Furthermore, the low competition and high anammox retention at 10°C have proven to have stable specific ammonium removal rates. This is because there are many anammox communities in the wastewater and the suitable genus at the low temperature was Candidatus *Kuenenia* (De Cocker et al., 2018). According to Li et al. (2019), the NOB activity can be inhibited by controlling the DO to reduce the NOB oxidation rate of nitrite to 0.04 kg/(m^3·d). Therefore, the stable nitritation process can be achieved in the system.

5.7 Conclusion

The combination of the microbes like the AOB and denitrifiers with anammox is able to overcome many obstacles faced by the anammox process in the full-scale application. For instance, like enrichment of anammox bacteria, which is known as the slow-growing bacteria, inhibition by other microbes like NOB and stable performance of the nitrogen removal process. The collaboration between the variant microbes via mixed inoculation of anammox sludge and conventional sludge has attracted many researchers to explore further to achieve stable nitrogen removal performance in order to apply widely in full-scale plants. Furthermore, many research studies are being conducted in order to apply PN/A widely in the open wastewater treatment system by considering potential solutions to the obligations that may affect the anammox process.

This chapter elaborates on the synergy between partial nitritation and the anammox process for nitrogen removal in industrial wastewater. This is because the wastewater contains a high level of ammonia concentration and requires an extensive process in order to prevent pollution to the environment. The various industries contribute to various concentrations of ammonia in the wastewater. Hence, this chapter has provided the biological nitrogen removal technologies applied to the wastewater treatment system. The benefits of anammox have enticed industrial stakeholders to incorporate them into their wastewater treatment systems. However, the application has been challenging for the industrial stakeholders due to high carbon to nitrogen (C/N) ratio, inhibition by NOB and low activity at low temperature. Hence, this chapter has suggested a few potential techniques to overcome the obstacles hindering the anammox application in the treatment system. In conclusion, this chapter was able to contribute to industrial wastewater by providing knowledge on the PN/A application for the wastewater treatment system, which is both eco-friendly and cost-effective.

Acknowledgements

The authors would like to thank the Malaysia-Japan International Institute of Technology (MJIIT) for providing great support to conduct the study and to the Ministry of Higher Education under the Fundamental Research Grant Scheme (FRGS/1/2021/STG04/UTM/01/1) for funding this research.

REFERENCES

Ahmad, I., Abdullah, N., Chelliapan, S., Yuzir, A., Iwamoto, K., Al-Dailami, A., & Arumugham, T. (2020). *Effectiveness of Anaerobic Technologies in the Treatment of Landfill Leachate*. IntechOpen, London. https://doi.org/10.5772/intechopen.94741

Arumugham, T., Husaini, N. A., Yuzir, A., Mohamad, S. E., Iwamoto, K., & Abdullah, N. (2020). Anammox reactor treating low strength domestic wastewater: A review. *IOP Conference Series: Earth and Environmental Science*, *479*(1). https://doi.org/10.1088/1755-1315/479/1/012021

Azari, M., Walter, U., Rekers, V., Gu, J. D., & Denecke, M. (2017). More than a decade of experience of landfill leachate treatment with a full-scale anammox plant combining activated sludge and activated carbon biofilm. *Chemosphere*, *174*, 117–126. https://doi.org/10.1016/j.chemosphere.2017.01.123

Cao, Y., van Loosdrecht, M. C. M., & Daigger, G. T. (2017). Mainstream partial nitritation–anammox in municipal wastewater treatment: Status, bottlenecks, and further studies. *Applied Microbiology and Biotechnology*, *101*(4), 1365–1383. https://doi.org/10.1007/s00253-016-8058-7

Chandra, R., & Kumar, V. (2017a). Detection of *Bacillus* and *Stenotrophomonas* species growing in an organic acid and endocrine-disrupting chemicals rich environment of distillery spent wash and its phytotoxicity. *Environmental Monitoring and Assessment*, *189*, 26. https://doi.org/10.1007/s10661-016-5746-9

Chandra, R., & Kumar, V. (2017b). Detection of androgenic-mutagenic compounds and potential autochthonous bacterial communities during in-situ bioremediation of post methanated distillery sludge. *Frontiers in Microbiology*, *8*, 87. https://doi.org/10.3389/fmicb.2017.00887

Chen, R., Ji, J., Chen, Y., Takemura, Y., Liu, Y., Kubota, K., & Li, Y.-Y. (2019). Successful operation performance and syntrophic micro-granule in partial nitritation and anammox reactor treating low-strength ammonia wastewater. *Water Research*, *155*, 288–299. https://doi.org/10.1016/j.watres.2019.02.041

Cho, S., Kambey, C., & Nguyen, V. K. (2020). Performance of anammox processes for wastewater treatment: A critical review on effects of operational conditions and environmental stresses. *Water*, *12*(1). https://doi.org/10.3390/w12010020

Connan, R., Dabert, P., Moya-Espinosa, M., Bridoux, G., Béline, F., & Magrí, A. (2018). Coupling of partial nitritation and anammox in two- and one-stage systems: Process operation, N_2O emission and microbial community. *Journal of Cleaner Production*, *203*, 559–573. https://doi.org/10.1016/j.jclepro.2018.08.258

Daigger, G. T. (2014). Oxygen and carbon requirements for biological nitrogen removal processes accomplishing nitrification, nitritation, and anammox. *Water Environment Research*, *86*(3), 204–209. https://doi.org/10.2175/106143013x13807328849459

Daverey, A., Su, S. H., Huang, Y. T., Chen, S. S., Sung, S., & Lin, J. G. (2013). Partial nitrification and anammox process: A method for high strength optoelectronic industrial wastewater treatment. *Water Research*, *47*(9), 2929–2937. https://doi.org/10.1016/j.watres.2013.01.028

De Cocker, P., Bessiere, Y., Hernandez-Raquet, G., Dubos, S., Mozo, I., Gaval, G., & Sperandio, M. (2018). Enrichment and adaptation yield high anammox conversion rates under low temperatures. *Bioresource Technology*, *250*(November), 505–512. https://doi.org/10.1016/j.biortech.2017.11.079

Ding, Z., Ventorino, V., Panico, A., Pepe, O., van Hullebusch, E. D., Pirozzi, F., & Esposito, G. (2017). Enrichment of anammox biomass from different seeding sludge: Process strategy and microbial diversity. *Water, Air, and Soil Pollution*, *228*(1), 1–13. https://doi.org/10.1007/s11270-016-3181-8

Dutta, D., Arya, S. & Kumar, S. (2021). Industrial wastewater treatment: Current trends, bottlenecks, and best practices. *Chemosphere*, *285*, 131245. https://doi.org/10.1016/j.chemosphere.2021.131245

Elefsiniotis, P., & Li, D. (2006). The effect of temperature and carbon source on denitrification using volatile fatty acids. *Biochemical Engineering Journal*, *28*(2), 148–155. https://doi.org/10.1016/j.bej.2005.10.004

Farazaki, M., & Gikas, P. (2019). Nitrification-denitrification of municipal wastewater without recirculation, using encapsulated microorganisms. *Journal of Environmental Management*, *242*(April), 258–265. https://doi.org/10.1016/j.jenvman.2019.04.054

García-Ruiz, M. J., Maza-Márquez, P., González-Martínez, A., Campos, E., González-López, J., & Osorio, F. (2018). Performance and bacterial community structure in three autotrophic submerged biofilters operated under different conditions. *Journal of Chemical Technology and Biotechnology, 93*(8), 2429–2439. https://doi.org/10.1002/jctb.5598

Gebeyehu, A., Shebeshe, N., Kloos, H., & Belay, S. (2018). Suitability of nutrients removal from brewery wastewater using a hydroponic technology with *Typha latifolia*. *BMC Biotechnology, 18*(1), 1–13. https://doi.org/10.1186/s12896-018-0484-4

Ghimire, N., & Wang, S. (2019). Biological treatment of petrochemical wastewater. *Petroleum Chemicals - Recent Insight.* https://doi.org/10.5772/intechopen.79655

Gilbert, E. M., Agrawal, S., Schwartz, T., Horn, H., & Lackner, S. (2015). Comparing different reactor configurations for Partial Nitritation/Anammox at low temperatures. *Water Research, 81*, 92–100. https://doi.org/10.1016/j.watres.2015.05.022

Gil-Pulido, B., Tarpey, E., Almeida, E. L., Finnegan, W., Zhan, X., Dobson, A. D. W., & O'Leary, N. (2018). Evaluation of dairy processing wastewater biotreatment in an IASBR system: Aeration rate impacts on performance and microbial ecology. *Biotechnology Reports, 19*, e00263. https://doi.org/10.1016/j.btre.2018.e00263

Gonzalez-Martinez, A., Muñoz-Palazon, B., Rodriguez-Sanchez, A., & Gonzalez-Lopez, J. (2018). New concepts in anammox processes for wastewater nitrogen removal: Recent advances and future prospects. *FEMS Microbiology Letters, 365*(6), 1–10. https://doi.org/10.1093/femsle/fny031

Gu, J., Yang, Q., & Liu, Y. (2018). A novel strategy towards sustainable and stable nitritation-denitritation in an A-B process for mainstream municipal wastewater treatment. *Chemosphere, 193*, 921–927. https://doi.org/10.1016/j.chemosphere.2017.11.038

Güven, D., Van De Pas-Schoonen, K., Schmid, M. C., Strous, M., Jetten, M. S. M., Sözen, S., & Schmidt, I. (2004). Implementation of the anammox process for improved nitrogen removal. *Journal of Environmental Science and Health - Part A Toxic/Hazardous Substances and Environmental Engineering, 39*(7), 1729–1738. https://doi.org/10.1081/ESE-120037873

Hanum, F., Yuan, L. C., Kamahara, H., Aziz, H. A., Atsuta, Y., Yamada, T., & Daimon, H. (2019). Treatment of sewage sludge using anaerobic digestion in Malaysia: Current state and challenges. *Frontiers in Energy Research, 7*(March), 1–7. https://doi.org/10.3389/fenrg.2019.00019

Henze, M., Loosdrecht, M.C.M.V., Ekama, G., & Bridjanovic, D. (2008). *Biological Wastewater Treatment, Principles, Modelling and Design*. IWA Publishing, London, pp. 139–151.

Huang, X., Mi, W., Hong, N., Ito, H., & Kawagoshi, Y. (2020). Efficient transition from partial nitritation to partial nitritation/Anammox in a membrane bioreactor with activated sludge as the sole seed source. *Chemosphere, 253*, 126719. https://doi.org/10.1016/j.chemosphere.2020.126719

Huy Quoc Anh, D., Tantayotai, P., Cheenkachorn, K., & Sriariyanun, M. (2015). Anammox process: The principle, the technological development and recent industrial applications. *KMUTNB International Journal of Applied Science and Technology, 8*(4), 1–8. https://doi.org/10.14416/j.ijast.2015.08.003

Islam, M., George, N., Zhu, J., & Chowdhury, N. (2009). Impact of carbon to nitrogen ratio on nutrient removal in a liquid-solid circulating fluidized bed bioreactor (LSCFB). *Process Biochemistry, 44*(5), 578–583. https://doi.org/10.1016/j.procbio.2009.02.003

Jenicek, P., Svehia, P., Zabranska, J., & Dohanyos, M. (2004). Factors affecting nitrogen removal by nitritation/denitritation. *Water Science and Technology, 49*(5–6), 73–79. https://doi.org/10.2166/wst.2004.0739

Jetten, M. S. M., Cirpus, I., Kartal, B., Van Niftrik, L., Van De Pas-Schoonen, K. T., Sliekers, O., & Strous, M. (2005). 1994–2004: 10 Years of research on the anaerobic oxidation of ammonium. *Biochemical Society Transactions, 33*(1), 119–123. https://doi.org/10.1042/BST0330119

Jetten, M. S. M., Horn, S. J., & Van Loosdrecht, M. C. M. (1997). Towards a more sustainable municipal wastewater treatment system. *Water Science and Technology, 35*(9), 171–180. https://doi.org/10.1016/S0273-1223(97)00195-9

Kallistova, A. Y., Dorofeev, A. G., Nikolaev, Y. A., Kozlov, M. N., Kevbrina, M. V., & Pimenov, N. V. (2016). Role of anammox bacteria in removal of nitrogen compounds from wastewater. *Microbiology, 85*(2), 140–156. https://doi.org/10.1134/S0026261716020089

Kampschreur, M. J., Poldermans, R., Kleerebezem, R., Van Der Star, W. R. L., Haarhuis, R., Abma, W. R., & Van Loosdrecht, M. C. M. (2009). Emission of nitrous oxide and nitric oxide from a full-scale single-stage nitritation-anammox reactor. *Water Science and Technology, 60*(12), 3211–3217. https://doi.org/10.2166/wst.2009.608

Kawagoshi, Y., Yamashita, Y., Van Duc, L., Hama, T., & Ito, H. (2019). Changes of nitrogen-removal performance and that of the bacterial community in a mixed culture comprising freshwater and marine anammox bacteria under averaged environmental condition. *Chemosphere*, 609–617. https://doi.org/10.1016/j.chemosphere.2018.11.047

Khadaroo, S. N. B. A., Grassia, P., Gouwanda, D., He, J., & Poh, P. E. (2021). Enhancing the biogas production and the treated effluent quality via an alternative Palm Oil Mill Effluent (POME) treatment process: Integration of thermal pretreatment and dewatering. *Biomass and Bioenergy*, *151*(May), 106167. https://doi.org/10.1016/j.biombioe.2021.106167

Kumar, V. & Shah, M.P. (2021). Role of fungi and their enzymes in degradation and decolorization of distillery effluent for environmental safety. In: Sharma, V.K., Shah, M.P., Kumar, S., Kumar, A., (Eds). *Fungi Bio-prospects in Sustainable Agriculture, Environment and Nano-technology-Volume 2: Extremophilic Fungi and Myco-mediated Environmental Management*. Academic Press. https://doi.org/10.1016/B978-0-12-821925-6.00013-7

Kumar, V., Shahi, S.K., Ferreira, L.F.R., Bilal, M., Biswas, J.K., & Bulgariu, L. (2021a). Detection and characterization of refractory organic and inorganic pollutants discharged in biomethanated distillery effluent and their phytotoxicity, cytotoxicity, and genotoxicity assessment using *Phaseolus aureus* L. and *Allium cepa* L. *Environmental Research*, *201*, 111551. https://doi.org/10.1016/j.envres.2021.111551

Kumar, V., Singh, K., & Shah, M.P. (2021b). Advanced oxidation processes for complex wastewater treatment. In: Shah, M.P. (Eds), *Advance Oxidation Process for Industrial Effluent Treatment*. Elsevier. https://doi.org/10.1016/B978-0-12-821011-6.00001-3

Lackner, S., Gilbert, E. M., Vlaeminck, S. E., Joss, A., Horn, H., & van Loosdrecht, M.C. (2014). Full-scale partial nitritation/anammox experiences: An application survey. *Water Research*, *55*, 292–303. https://doi.org/10.1016/j.watres.2014.02.032

Lawson, C. E., de Graaf, R. M., Nuijten, G., Jacobson, T. B., Pabst, M., Stevenson, D. M., & Lucker, S. (2021). Autotrophic and mixotrophic metabolic network fluxes suggest versatile lifestyle for the anammox bacterium *Candidatus* Kuenenia stuttgartiensis. *The ISME Journal*, *15*, 673–687.

Li, X., Yuan, Y., Huang, Y., Bi, Z., & Lin, X. (2019). Inhibition of nitrite oxidizing bacterial activity based on low nitrite concentration exposure in an auto-recycling PN-Anammox process under mainstream conditions. *Bioresource Technology*, *281*, 303–308. https://doi.org/10.1016/j.biortech.2019.114

Lin, J., Wang, Y., Wang, W., Wang, Q., Hu, Z.-H., Tang, R., & Zhan, X. (2018). Selection of seeding strategy for fast start-up of Anammox process with low concentration of Anammox sludge inoculum. *Bioresource Technology*, *268*(June), 638–647. https://doi.org/10.1016/j.biortech.2018.08.056

Liu, Y., Ngo, H. H., Guo, W., Peng, L., Wang, D., & Ni, B. (2019). The roles of free ammonia (FA) in biological wastewater treatment processes: A review. *Environment International*, *123*(October), 10–19. https://doi.org/10.1016/j.envint.2018.11.039

Lotti, T., Kleerebezem, R., Abelleira-Pereira, J. M., Abbas, B., & van Loosdrecht, M. C. M. (2015). Faster through training: The anammox case. *Water Research*, *81*, 261–268. https://doi.org/10.1016/j.watres.2015.06.001

Luo, H., Zeng, Y., Cheng, Y., He, D., & Pan, X. (2020). Recent advances in municipal landfill leachate: A review focusing on its characteristics, treatment, and toxicity assessment. *Science of the Total Environment*, *703*, 135468. https://doi.org/10.1016/j.scitotenv.2019.135468

Ma, B., Wang, S., Cao, S., Miao, Y., Jia, F., Du, R., & Peng, Y. (2016). Biological nitrogen removal from sewage via anammox: Recent advances. *Bioresource Technology*, *200*, 981–990. https://doi.org/10.1016/j.biortech.2015.10.074

Manipura, A., Duncan, J. R., Roman, H. J., & Burgess, J. E. (2005). Potential biological processes available for removal of nitrogenous compounds from metal industry wastewater. *Process Safety and Environmental Protection*, *83*(5 B), 472–480. https://doi.org/10.1205/psep.04271

Mao, N., Ren, H., Geng, J., Ding, L., & Xu, K. (2017). Engineering application of anaerobic ammonium oxidation process in wastewater treatment. *World Journal of Microbiology and Biotechnology*, *33*(8), 153. https://doi.org/10.1007/s11274-017-2313-7

Martins, T. H., Souza, T. S. O., Bernadete, M., & Varesche, A. (2017). The influence of stirring speed, temperature and initial nitrogen concentration on specific anammox activity. *Biological and Applied Sciences*, *60*(2), 1–11. http://dx.doi.org/10.1590/1678-4324-2017160421

Massara, T. M., Malamis, S., Guisasola, A., Baeza, J. A., Noutsopoulos, C., & Katsou, E. (2017). A review on nitrous oxide (N_2O) emissions during biological nutrient removal from municipal wastewater and sludge reject water. *Science of the Total Environment, 596–597,* 106–123. https://doi.org/10.1016/j.scitotenv.2017.03.191

Metcalf, E. & Eddy, I. (2003). *Wastewater Engineering Treatment and Reuse.* McGraw-Hill Companies, Inc, NewYork.

Milia, S., Perra, M., Tocco, G., & Carucci, A. (2017). The start-up of an anammox reactor as the second step for the treatment of ammonium rich refinery (IGCC) wastewater with high Corg/N ratio. *Ecological Engineering, 106*(2), 358–368. https://doi.org/10.1016/j.ecoleng.2017.04.068

Moustafa, T. (2014). *Aerobic Granular Sludge–Study of Applications for Industrial and Domestic Wastewater* (Master's thesis).

Mulder, A., van de Graaf, A. A., Robertson, L. A., & Kuenen, J. G. (1995). Anaerobic ammonium oxidation discovered in a denitrifying fluidized bed reactor. *FEMS Microbiology Ecology, 16*(3), 177–183. https://doi.org/10.1016/0168-6496(94)00081-7

Muñoz-Palazon, B., Rodriguez-Sanchez, A., Castellano-Hinojosa, A., Gonzalez-Lopez, J., C.M. van Loosdrecth, M., Vahala, R., & Gonzalez-Martinez, A. (2017). Quantitative and qualitative studies of microorganisms involved in full-scale autotrophic nitrogen removal performance. *AIChE Journal, 64(2).* https://doi.org/10.1002/aic

Muralikrishna, I. V., & Manickam, V. (2017). Wastewater treatment technologies. In: *Environmental Management.* https://doi.org/10.1016/b978-0-12-811989-1.00012-9

Nancharaiah, Y. V., & Reddy, G.K.K. (2018). Aerobic granular sludge technology: Mechanisms of granulation and biotechnological applications. *Bioresource Technology, 247,* 1128–1143. https://doi.org/10.1016/j.biortech.2017.09.131

Nasir, N., Daud, Z., Abd, A., Aziz, A., Latiff, A., Ahmad, B. & Abdul, A. (2019). Removal of ammonia nitrogen from rubber industry wastewater using zeolite as adsorbent. *Malaysian Journal of Fundamental and Applied Sciences, 15*(6), 862–866.

Ni, S. Q., & Zhang, J. (2013). Anaerobic ammonium oxidation: From laboratory to full-scale application. *BioMed Research International,* 2013. https://doi.org/10.1155/2013/469360

Nourmohammadi, D., Esmaeeli, M. B., Akbarian, H., & Ghasemian, M. (2013). Nitrogen removal in a full-scale domestic wastewater treatment plant with activated sludge and trickling filter. *Journal of Environmental and Public Health,* 2013. https://doi.org/10.1155/2013/504705

Pal, P. (2017). Treatment and disposal of pharmaceutical wastewater: Toward the sustainable strategy. *Separation and Purification Reviews, 47*(3), 179–198. https://doi.org/10.1080/15422119.2017.1354888

Qiao, S., Yin, X., Tian, T., Jin, R., & Zhou, J. (2016). Hydrazine production by anammox biomass with NO reversible inhibition effects. *Green Chemistry, 18*(18), 4908–4915. https://doi.org/10.1039/c6gc01261b

Rahimi, S., Modin, O., & Mijakovic, I. (2020). Technologies for biological removal and recovery of nitrogen from wastewater. *Biotechnology Advances, 43,* 107570. https://doi.org/10.1016/j.biotechadv.2020.107570

Rezvani, F., Sarrafzadeh, M. H., Ebrahimi, S., & Oh, H. M. (2019). Nitrate removal from drinking water with a focus on biological methods: A review. *Environmental Science and Pollution Research, 26*(2), 1124–1141. https://doi.org/10.1007/s11356-017-9185-0

Rodriguez-Sanchez, A., Purswani, J., Lotti, T., Maza-Marquez, P., van Loosdrecht, M. C. M., Vahala, R., & Gonzalez-Martinez, A. (2016). Distribution and microbial community structure analysis of a single-stage partial nitritation/anammox granular sludge bioreactor operating at low temperature. *Environmental Technology, 37*(18), 2281–2291. https://doi.org/10.1080/09593330.2016.1147613

Sguanci, S., Lotti, T., Caretti, C., Caffaz, S., Dockhorn, T., & Lubello, C. (2017). Inhibitory effects of veterinary antibiotics on anammox activity: Short- and long-term tests. *Environmental Technology, 38*(21), 2661–2667. https://doi.org/10.1080/09593330.2016.1272640

Silveira, N. C., Oliveira, G. H. D., Damianovic, M. H. R. Z., & Foresti, E. (2021). Two-stage partial nitrification-anammox process for nitrogen removal from slaughterhouse wastewater: Evaluation of the nitrogen loading rate and microbial community analysis. *Journal of Environmental Management, 296*(April), 1–10. https://doi.org/10.1016/j.jenvman.2021.113214

Soudejani, H. T., Kazemian, H., Inglezakis, V. J., & Zorpas, A. A. (2019). Application of zeolites in organic waste composting: A review. *Biocatalysis and Agricultural Biotechnology, 22,* 101396. https://doi.org/10.1016/j.bcab.2019.101396

Sri Shalini, S., & Joseph, K. (2018). Combined SHARON and ANAMMOX processes for ammoniacal nitrogen stabilisation in landfill bioreactors. *Bioresource Technology*, *250*, 723–732. https://doi.org/10.1016/j.biortech.2017.10.077

Strous, M., Fuerst, J. A., Kramer, E. H. M., Logemann, S., Muyzer, G., Van De Pas-Schoonen, K.T., Webb, R., & Jetten, M. S. M. (1999). Missing lithotroph identified as new planctomycete. *Nature*, *400*(6743), 446–449. https://doi.org/10.1038/22749

Sun, Y., Guan, Y., Pan, M., Zhan, X., Hu, Z., & Wu, G. (2017). Enhanced biological nitrogen removal and N_2O emission characteristics of the intermittent aeration activated sludge process. *Reviews in Environmental Science and Biotechnology*, *16*(4), 761–780. https://doi.org/10.1007/s11157-017-9444-z

Sun, H., Peng, Y., & Shi, X. (2015). Advanced treatment of landfill leachate using anaerobic-aerobic process: Organic removal by simultaneous denitritation and methanogenesis and nitrogen removal via nitrite. *Bioresource Technology*, *177*, 337–345. https://doi.org/10.1016/j.biortech.2014.10.152

Suto, R., Ishimoto, C., Chikyu, M., Aihara, Y., Matsumoto, T., Uenishi, H., & Waki, M. (2017). Anammox biofilm in activated sludge swine wastewater treatment plants. *Chemosphere*, *167*, 300–307. https://doi.org/10.1016/j.chemosphere.2016.09.121

Tomaszewski, M., Cema, G., & Ziembińska-Buczyńska, A. (2017). Influence of temperature and pH on the anammox process: A review and meta-analysis. *Chemosphere*, *182*, 203–214. https://doi.org/10.1016/j.chemosphere.2017.05.003

Van Den Hende, S., Rodrigues, A., Hamaekers, H., Sonnenholzner, S., Vervaeren, H., & Boon, N. (2017). Microalgal bacterial flocs treating paper mill effluent: A sunlight-based approach for removing carbon, nitrogen, phosphorus, and calcium. *New Biotechnology*, *39*, 1–10. https://doi.org/10.1016/j.nbt.2017.03.004

van der Hoek, J. P., Duijff, R., & Reinstra, O. (2018). Nitrogen recovery from wastewater: Possibilities, competition with other resources, and adaptation pathways. *Sustainability*, *10*(12), 4605. https://doi.org/10.3390/su10124605

van der Star, W. R. L., Abma, W. R., Blommers, D., Mulder, J. W., Tokutomi, T., Strous, M., & van Loosdrecht, M. C. M. (2007). Startup of reactors for anoxic ammonium oxidation: Experiences from the first full-scale anammox reactor in Rotterdam. *Water Research*, *41*(18), 4149–4163. https://doi.org/10.1016/j.watres.2007.03.044

Vázquez-Padín, J. R., Morales, N., Aiartza, I., Pedrouso, A., Campos, J. L., Val del Rio, A., & Mosquera-Corral, A. (2018). Pilot-scale ELAN ® process applied to treat primary settled urban wastewater at low temperature via partial nitritation-anammox processes. *Separation and Purification Technology*, *200*(December), 94–101. https://doi.org/10.1016/j.seppur.2018.02.017

Verma, R. K., Sankhla, M. S., Rathod, N., & Sonone, S. S. (2021). Eradication of fatal textile industrial dyes by wastewater treatment. *Biointerface Research in Applied Chemistry*, *12*(1), 567–587. https://doi.org/10.33263/briac121.567587

Vlaeminck, S. E., De Clippeleir, H., & Verstraete, W. (2012). Microbial resource management of one-stage partial nitritation/anammox. *Microbial Biotechnology*, *5*(3), 433–448. https://doi.org/10.1111/j.1751-7915.2012.00341.x

Wang, B., Peng, Y., Guo, Y., Yuan, Y., Zhao, M., & Wang, S. (2016a). Impact of partial nitritation degree and C/N ratio on simultaneous sludge fermentation, denitrification and anammox process. *Bioresource Technology*, *219*, 411–419. https://doi.org/10.1016/j.biortech.2016.07.114

Wang, D., Wang, Q., Laloo, A., Xu, Y., Bond, P. L., & Yuan, Z. (2016b). Achieving stable nitritation for mainstream deammonification by combining free nitrous acid-based sludge treatment and oxygen limitation. *Scientific Reports*, *6*(December), 25547. https://doi.org/10.1038/srep25547

Wang, H., Guan, Y., Li, L., & Wu, G. (2015). Characteristics of biological nitrogen removal in a multiple anoxic and aerobic biological nutrient removal process. *BioMed Research International*, *2015*, 531015. https://doi.org/10.1155/2015/531015

Weralupitiya, C., Wanigatunge, R., Joseph, S., Athapattu, B. C. L., Lee, T. H., Kumar Biswas, J., & Vithanage, M. (2021). Anammox bacteria in treating ammonium rich wastewater: Recent perspective and appraisal. *Bioresource Technology*, *334*(March), 125240. https://doi.org/10.1016/j.biortech.2021.125240

Wett, B. (2007). Development and implementation of a robust deammonification process. *Water Science and Technology*, *56*(7), 81–88. https://doi.org/10.2166/wst.2007.611

Winkler, M. K., & Straka, L. (2019). New directions in biological nitrogen removal and recovery from wastewater. *Current Opinion in Biotechnology*, *57*, 50–55. https://doi.org/10.1016/j.copbio.2018.12.007

Yang, Y., Zhang, Y., Li, Y., Zhao, H., & Peng, H. (2018). Nitrogen removal during anaerobic digestion of wasted activated sludge under supplementing Fe(III) compounds. *Chemical Engineering Journal, 332*(July), 711–716. https://doi.org/10.1016/j.cej.2017.09.133

Yaseen, D. A., & Scholz, M. (2019). Textile dye wastewater characteristics and constituents of synthetic effluents: A critical review. *International Journal of Environmental Science and Technology, 16*, 1193–1226. https://doi.org/10.1007/s13762-018-2130-z

Zhang, D., Wang, G., & Dai, X. (2020). Operation of pilot-scale nitrification-anammox reactors for the treatment of reject-water produced from the anaerobic digestion of thermal hydrolysis-treated sludge. *Bioresource Technology, 314*(April), 123717. https://doi.org/10.1016/j.biortech.2020.123717

Zhang, J., Miao, Y., Sun, Y., Zhang, Q., Dai, J., & Peng, Y. (2021a). An effective strategy for in situ start-up of mainstream anammox process treating domestic sewage. *Bioresource Technology, 339*(July), 125525. https://doi.org/10.1016/j.biortech.2021.125525

Zheng, L., Chen, Y., Zhou, S., Chen, Y., Wang, X., Wang, X., & Chen, Z. (2020). Nitrogen removal for liquid-ammonia mercerization wastewater via partial nitritation/anammox based on zeolite sequencing batch reactor. *Water, 12*(8). https://doi.org/10.3390/w12082234

Zhang, X., Zhang, H., Chen, Z., Wei, D., Song, Y., Ma, Y., & Zhang, H. (2021b). Achieving biogas production and efficient pollutants removal from nitrogenous fertilizer wastewater using combined anaerobic digestion and autotrophic nitrogen removal process. *Bioresource Technology, 339*(July), 125659. https://doi.org/10.1016/j.biortech.2021.125659

6

Vermifiltration Technology: Earthworm Assisted Green Technology for Wastewater Treatment

Sudipti Arora, Samvida Saxena, Vikky Sinha, and Devanshi Sutaria
Dr. B. Lal Institute of Biotechnology

Sakshi Saraswat
National Yang Ming Chiao Tung University

CONTENTS

6.1 Introduction

Fresh water is a quintessential source for the survival of life on earth but it is running out fast due to active anthropogenic activities such as urbanization and industrialization which in turn result in polluting this precious source. The United Nations World Water Development Report 2018 also provides an update on the proper availability of water and its future requirement. This shortage of clean water is a major problem for the current 7.7 billion people worldwide. Additionally, future demand for water from all sectors will require up to 25%–40% of the water to be diverted from productivity below higher productivity and employment, especially in water-prone areas. Wastewater that is generated contains hazardous chemicals had an extremely high load of organic matter which is harmful to all living beings (Dutta et al., 2021). In the next two decades,

DOI: 10.1201/9781003231738-7

global demand for water will increase in three components - industry, domestic, and agriculture. It requires proper management of how the water will be managed in the agriculture sector as well as other sectors too in the future. Hence, an alarming situation has been created for years for the treatment of wastewater. When it comes to solutions, the major purpose of the treatment plant is to make the effluent contaminant-free before it is released into the environment. Now the major challenge faced is the gap between the wastewater produced and its treatment (Kumar and Chandra, 2018; Kumar et al., 2021a). Wastewater is dispensed in the environment but when it comes to treatment, the impacts of centralized and decentralized systems need to be checked. A rise in the CO_2 level and the harmful emissions generated from these treatment plants are leading to global warming and disruption of healthy lifestyles. Due to insufficient treatment infrastructure, wastewater generation and treatment have become a major health concern in developing countries (Singh et al., 2017). The most significant source of contamination of water resources is the discharge of untreated sewage into surface and subsurface water streams. It has been observed that in developing countries, the rural and urban populations are mostly dependent on onsite wastewater treatment systems since they are not capital intensive (Sharma and Kazmi, 2015). Sewage discharge is the primary source of pollution in waterways. Untreated sewage contains a huge amount of biodegradable organic matter as well as a variety of disease-causing pathogenic organisms, all of which are harmful to human health. Various wastewater treatment technologies have been thoroughly researched (Agrawal et al., 2021; Kumar et al., 2020a). Septic tanks, aerobic biological treatment units, fixed activated sludge treatment, membrane bioreactors, recirculating sand filters, trickling filters, and engineered wetlands are among the on-site treatment technologies that are operational and successful in eliminating contaminants. Wastewater treatment plants work on the principle of filtration and decomposition of organic matter by the community of microorganisms in combination with light and oxygen. The aim of wastewater treatment is the conversion of waste material into a stable oxidized product that can be disposed properly or suitable for reuse, which will solve the problems of water scarcity as well as water pollution.

However, in the past decade, it has been gradually realized that these technologies come with their own limitations due to the extensive financial and labour dependency which often lead to their failure in underdeveloped nations (Muga and Mihelcic, 2008). As a result, an alternative onsite wastewater treatment procedure that is cost-effective, environmentally sustainable, and socially acceptable is required (Wu et al., 2013). Sustainable management of water resources is therefore essential for the smooth working of the sustainable development goals (SDGs). Novel innovative solutions such as constructed wetlands, soil infiltration trenches, vegetation-based wastewater treatment, and vermifiltration are among the many options used to treat residential wastewater, particularly in rural areas and developing economy (Sinha et al., 2008). Vermifiltration might be a propitious technology of the future since it incorporates the traditional microbial metabolic activity with the degradation activity of earthworms to work on a green and sustainable mechanism. The earthworms along with the decomposed community of microbes actively degrade the organic matter and enhance the physicochemical composition of water and the filter media layer in which they dwell.

Such actions are meant to protect sustainable management and usually restore natural or modified ecosystems, also the societal challenges are being addressed effectively and adaptively come under nature-based solutions (NBS) (Cohen-Shacham et al., 2016). NBS involves wastewater management technologies that mimic the natural processes in urban landscapes. Different NBS includes constructed wetlands, vermifiltration, zoo plankton-based treatments, hydroponics, macrophytes-assisted remediation, etc. The effectiveness of these systems in wastewater processing is due to a strong interplay between plants, biofilms, substrate, atmosphere, and nutrients from waste. The associated contact between roots, substrate, and biofilm promotes various basic mechanisms for removing pollutants and pathogens, such as sedimentation and filtration as physical processes, precipitation and adsorption as chemical processes, and microbiological degradation and plant uptake as biological processes. A large amount of mechanisms involved implies considerable variability in the elimination efficiency which is discussed in the further sections.

Vermifiltration technology is a green nature-based solution, which has gained attention in the last decade due to its high efficacy in the removal of contaminants. Vermifilters involve the use of earthworms and associated microorganisms, to treat wastewater through the enzymatic activity of microorganisms. A recent study done by Arora et al. (2020a, b) suggested a symbiotic interrelationship between earthworms and microorganisms for elucidating the mechanisms of organic matter biodegradation and removal of pathogens from wastewater for clinical & domestic wastewater. Vermifiltration technology for wastewater treatment is implemented in many developing countries owing to its simplicity and advancements in natural treatment methodologies. However, with the few advancements in integration of this technology with other NBS, such as plants, macrophytes, and zooplanktons, this technology can gain better perspectives, due to its increased efficiency. The chapter discussed various case studies on vermifiltration technology and its implications/enhancements for water treatment & reuse.

6.2 Vermifiltration Technology

In the year 1992, University of Chile researchers invented vermifiltration for the first time (Arora et al., 2014). Vermifiltration is a technique for degrading organic matter in wastewater that combines the use of microorganisms with the synergistic action of earthworms. Traditional vermicomposting processes are adapted for wastewater treatment in this technology. A vermifilter is a typical wastewater treatment filtration system that is loaded with worms and a combination of gravel and sand media. It's also known as a lumbrifilter or a vermi-digestor. The wastewater trickles down the filter media layer where the organic and inorganic matters undergo physicochemical and microbiological degradation. This process results in vermicompost, vermiwash, and a variety of other value-added end products. Bacteria and other aerobic microorganisms participate in the metabolic breakdown of organic materials found in wastewater. The burrowing activity of the earthworm prevents anaerobic breakdown thus, it is also an odor-free process. Earthworms feed on microbes, which helps to coordinate microbial biomass and activity. Organic matter and biofilm which are consumed by the earthworms are then digested into the organic matter which turns into humus which is called vermicompost. Vermicompost made by the earthworms are rich in minerals and nutrients as the earthworms consume the organic matter into their gut and we get degraded humus. It can be applied to the soil for enhancing fertility of the soil. The technology has an added advantage for its zero formation of sludge in the process which otherwise requires an accessory expenditure on landfill disposal. The treated wastewater can be reused for agricultural and recreation purposes.

6.2.1 Earthworms: Earth's Trash Managers

Earthworms are classified mainly based on their nutrient consumption and their dwelling environments. Some commonly used species in vermifiltration are described in Table 6.1 along with their niche and feeding habits.

Eisenia fetida is an epigeic earthworm species, which survives in organic wastes and requires proper high moisture content, enough organic matter, suitable bedding material for proper growth and development. They may be found in a variety of environments, but are most common in dark, damp places. They can endure temperatures ranging from 20°C to 25°C. Furthermore, their intestines symbiotically occupy a large diversity of microflora with biodegradable properties that diminish the starch, cellulose, lipids, proteins, heavy metals, and so on from wastewater. Furthermore, they play a vital role in carbon turnover, humus formation, nitrification, bio-accumulation, and heavy metal and pesticide biodegradation. The ideal stocking density for these worms is 10,000 worms/m^3 (Arora et al., 2014; Kumar et al., 2015). Vermifiltration for wastewater treatment results in significant organic matter removal and nutrient fixation up to 95% in the effluent as compared to the influent. One of the most fundamental reports in this field stated the treatment efficiencies of biochemical oxygen demand (BOC), chemical oxygen demand (COD), total dissolved solid (TDS), and total soluble solid

TABLE 6.1

Some Common Species of Earthworms Used in Vermifiltration Technology

Earthworm Species	Niche	Feeding Habit
Eisenia fetida	Epigeic	Detrivores
Lumbricus terrestris	Anecic	Detritivores
Eisenia andrei	Epigeic	Detritivores
Perionyx excavatus	Epigeic	Geophages
Eudrilus eugeniae	Epigeic	Geophages
Eisenia veneta	Epigeic	Detritivores

(TSS) of the order greater than 90%, 80%–90%, 90%–92%, and 90%–95%, respectively (Sinha et al., 2008). In a recent study by K. Ispolnov in 2021, it was observed that the earthworms also alter some microbial communities, by favouring betaproteobacteria and denitrifiers while opposing pathogenic species.

6.2.2 Design of the Vermifilter

Industrial, municipal, and commercial activities release toxic and recalcitrant compounds in effluent which leads to environmental deterioration, and hence, needs to be treated before disposal (Mandpe et al., 2019; 2021a, b). The self-purification ability to receive water is highly limited and they cannot withstand increasing loads of organic and inorganic pollutants (Murugesan and Rajkumari, 2019). Wastewater treatment plants work on the principle of filtration and decomposition of organic matter by the community of microorganisms in combination with light and oxygen (Kumar et al., 2020b; 2021b; Kumar and Chandra, 2020). On a similar idea, a technology that can prove to be the most promising approach to solve this wastewater generation problem is vermifiltration technology. Vermifiltration- an earthworm-assisted remediation system- is a new, clean, eco-friendly, and innovative technology. Naming this technique, which employs earthworms and their harmonious interaction with microorganisms to clean and filter excess sludge and contaminated water. This technique is being provided as an alternative approach to treating sewage effluent in poor nations where operating and replacement costs for wastewater treatment equipment are high. It functions similarly to conventional wastewater treatment plants, but with the addition of earthworms to maximize their potential. A multi-purpose filtration technique because it can adjust the BOD, COD, turbidity, pH, and other parameters of the sample in addition to filtering it. The appropriate selection and design of the vermifilter are very essential in creating an artificial environment in municipal wastewater treatment. Vermifilter design optimization not only defines the best commercial procedure but also improves in terms of investment. A schematic diagram of vermifilter depicting the outline of the unit consisting of plants and many different layers of gravels assisted with this new integrated vermifiltration technology is shown in Figure 6.1. A field

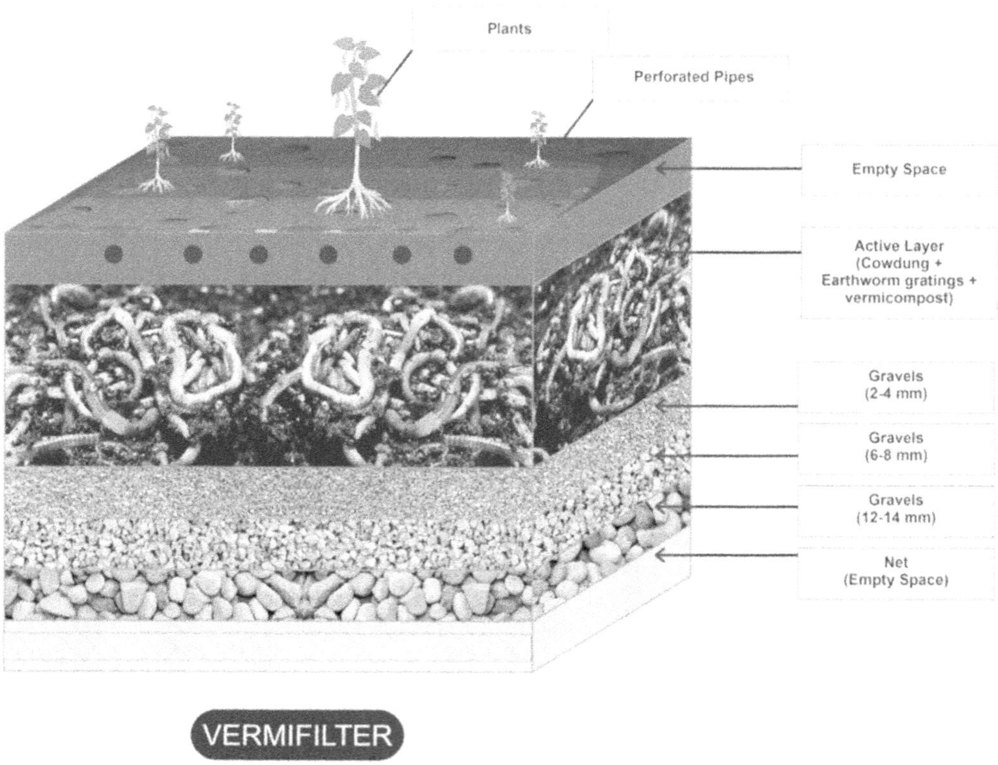

FIGURE 6.1 A schematic diagram depicting the outline plan of integrated vermifiltration.

scale vermifilter is an enclosed reactor that can be of cuboid or cylindrical shape with a carrying capacity of generally 1,200–1,500 L. The Assembly of the reactor is composed of durable materials, mainly plastic or concrete, that can eliminate the entry of vermin and protect it from environmental fluctuations. This model can process up to 400–1,000 L of water per day. The main filtration unit is connected to a large collection and storage sub reservoirs where sewage gets collected. Here, the entire primary waste removal from wastewater has been completed. Another tank with continuous agitation is installed which provides the proper mixing of wastewater containing organic and inorganic substances and avoids the settling of organic waste at the bottom of the tank. The depth of the main tank has been further divided into four-five zones consisting of different layers of gravel of different sizes (filled from bottom to top) as shown in Figure 6.1 and the vermicompost rich with earthworms. At the top of the unit, an empty space is kept for aeration purposes. The first two layers comprise the most active layer which is filled with plants and a mixture of vermigratings, cow dung, vermicompost, and gravels of a fine size where the earthworm species *Eisenia fetida* is inoculated at 10,000 worms/m^3 stocking density. *E. fetida*, commonly known as red wigglers, appears to be the most common earthworm species across the world. The acclimatization period of worms is around ~1–2 weeks for them to habituate to the filter media layer. A protective mesh is also installed at the bottom of the vermifilter unit which allows only water to pass through. This mesh is followed by an empty space where the treated effluent gets collected.

The inlet system is equipped with a polypropylene/Polyvinyl chloride (PVC) pipe with holes (which acts as a sprinkler) which ensure uniform distribution of the influent (wastewater/sewage) over the vermifilter bed. The gravity flow of wastewater is arranged in a manner that lets the uniform distribution of influent over the active layer. Based upon the flow rate and reactor configurations, the hydraulic retention time (HRT) and hydraulic loading rate (HLR) of the vermifiltration are decided. The treated effluent was collected from the bottom in a separate effluent collection tank. Table 6.2 describes design parameters of VF and gives the description of the filter bed layer.

6.2.3 Types of Earthworms

In addition to the previously mentioned earthworm species (Table 6.1), some of the most commonly reported species in India are *E. fetida, Perionyx excavatus, Eudrillus eugeniae, Eisenia hortensis, E. andrei,* and *Pheretima peguana* (Kumar and Ghosh, 2019). Because of their strong resilience to environmental fluctuations and better capacity for burrowing, *Eisenia fetida* is the most targeted species for wastewater treatment by vermifiltration. These species consume garbage equivalent to their body weight each day in order to grow their population by twofold every 2–2.5 months. Furthermore, their intestines symbiotically occupy a large diversity of microflora with biodegradable properties that diminish the starch, cellulose, lipids, proteins, heavy metals, and so on from wastewater. Furthermore, they play a vital role in carbon turnover, humus formation, nitrification, bio-accumulation, and heavy metal bio-degradation.

6.2.4 Hydraulic Retention Time (HRT) and Hydraulic Loading Rate (HLR)

The HLR and HRT are two important parameters that considerably influence the process of vermifiltration. The effectiveness of VF is heavily influenced by the rate of flow of wastewater and hence, it is especially important to optimize these two parameters. HLR is the rate of flow of wastewater through

TABLE 6.2
Design Parameters of VF

Filter type	Cuboid, cylindrical
Material used	Concrete, plastic
Filter bed material	Gravels, ceram site, quartz, charcoal
Earthworm type	*Eisenia fetida, Eudrilus eugeniae, Perionyx excavates*
HLR (hydraulic loading rate)	1 m^3/m^2/day
HRT (hydraulic retention time)	8 hours
Stocking density	5,000–10,000 worms/m^3

the unit area of the vermifilter reactor for unit time. HRT is the total time that the wastewater interacts with the vermifilter bed in which earthworms reside. Both characteristics are crucial in maintaining the efficacy of the vermifilter for waste removal, and they have an impact on the symbiotic connection between earthworms and microorganisms. HRT is an essential metric because it specifies the average amount of time wastewater spends in the bed reactor and in touch with the earthworms. During this contact period, earthworms manage to remove the organics and inorganics from the wastewater by carrying out the physical, chemical, and biochemical processes. This reduces BOD, COD, total dissolved solids (TDS), and pathogen burden. The longer the HRT, the more effective vermifiltration. HRT can thus be enhanced by lowering the HLR, (i) resulting in a slower rate of infiltration into the bed, and (ii) increasing the depth of the vermibed profile.

6.3 Mechanism of Vermifiltration Technology for Wastewater Treatment

After studying the design parameters and feasibility of vermifiltration technology, this section emphasizes the working mechanism of how the wastewater gets treated by this system. Also, it reports the effect of filter bed material composition on the performance of the vermifilter. The influent in the form of concentrated sewage gets uniformly sprinkled over the vermifilter bed with different layers comprising gravels/soil/cow dung/and earthworms. The filter media serves multiple purposes. First and foremost, it acts as a supporting matrix and feeding layer for the earthworms. Thus, it helps in the active removal of pollutants and provides nourishment to the worms. It also acts as a sorption medium for the microorganisms to flourish along with the earthworm and work in a symbiotic way. Microbe heterogeneity aids in waste reduction by enzyme secretion. It has a large surface area, which allows for the required wastewater retention period and therefore extends the time of contact between wastewater and earthworms.

Moreover, earthworms play a crucial role in this technology. These worms are dark brown in color, cylindrical, bilaterally symmetrical, and segmented animals with a delicate cuticle over the body. They are found in a variety of habitats, particularly in dark, damp areas. They can withstand 20°–25° of temperature. 50%–60% of the water content is optimum for their survival. These burrowing creatures serve many additional benefits to the filtration unit by reinforcing the water property, humus formation, and waste decomposition. In addition to this, by tilting of soil and vermicast they can create an aerobic atmosphere inside the filtration chamber which avoids the production of bad odour as produced in conventional wastewater treatment systems. Few researchers reported the major changes due to earthworms, as they ensure the dissolving of more oxygen by offering maximum contact due to the newly formed burrows and channels (Singh et al., 2018, 2019b). They eat up the organic waste and nutrients present in the wastewater and show tunneling actions which intensify the aerobic degradation activities. It also diversifies the microbial population by providing favorable conditions (Singh et al., 2018, Samal et al., 2020). Furthermore, the burrowing action can assist in improving the hydraulic conductivity of the vermifilter bed. Earthworms promote the growth of beneficial decomposers bacteria that helps in degrading organic matter. Some bacteria and fungi are encouraged by earthworms which produces 'antibiotics' that kill the harmful pathogens present in wastewater and make the medium virtually sterile and odorless. The examples included Fungus named *Penicillium* spp. which produces antibiotic 'penicillin' has been reported from the intestine of earthworm (Sinha et al., 2008).

The microbial and vermiprocess simultaneously work in vermifiltration system to treat wastewater due to the joint activity of earthworms and microorganisms. Various researchers reported that the earthworm gut is rich in microflora which helps in multiplying the action of earthworm waste degradation (Sinha et al., 2008; Arora et al., 2014, 2016, 2020). These microbial diversity acts synergistically with earthworms to treat sewage. Also, symbiotic relationships get established between the earthworm and microorganisms. Earthworms get food and a suitable environment to dwell in and in return they provide a suitable environment to survive. Vermifilters with various layers such as gravels, glass beads, soil, ceramsite, quartz, and so on are built. Vermigratings are put to the top of the vermifilter medium. Vermigratings include vermicompost and earthworms. Wastewater is carefully sprayed over the vermifilters. By secreting enzymes, the microflora present in vermifilter bed medium and earthworm coelomic fluid promotes biodegradation. Wastewater is absorbed via the earthworm's body walls, crushed inside the gizzard, and given to microorganisms for further degradation via the action of aerobic microgranism present in earthworm gut. Earthworms fed on a large number of microbes (decomposers/biodegrader) present inside and excrete them in soil along with nutrients Nitrogen (N), Phosphorus (P) in their vermicast (vermiprocess). Also, vermicast contains N and P which are further used by microbes for multiplication. Earthworms improve their number by improving aeration by burrowing action. Lower layers of the vermifilter unit also help in the thriving of aerobic microbes and the castings secreted by the worms help in building better hydraulic conductivity.

The Action mechanism of earthworms increases the dissolved oxygen content due to the burrowing action up to 58%–63% total organic carbon (TOC), 98%–99% BOD, and 70%–100% COD. Earthworm treatments have also been shown to remove ammonia up to 67% by vermicomposting and up to 97% by vermifiltration and increase nitrate up to 75% by vermicomposting. The C:N ratio has been reported to decrease in solid and liquid wastes treated by earthworms due to the loss of carbon as CO_2.

Earthworms work in a symbiotic relationship with different kinds of microbes:

1. **Earthworm Gut-Associated Microbes: Microorganisms Which Reside in the Alimentary Canal of the Earthworm**: Earthworms ingest solids, organics, and pathogens from wastewater absorbed in the active layer and the gut-associated microbes help in the breakdown of organics and solids through enzymatic activity (they release enzymes like amylase, cellulase, and pectinase) so here the earthworms provide a residing place for microbes and in turn, the microbes help in the digestion of their food. This is the endosymbiotic association.
2. **Earthworm Cast Associated Microbes: Microorganisms That Are Excreted Along with the Cast of the Earthworm**: Earthworms release certain beneficial microbes along with cast into the active layer which helps in mineralization and humification of the organics present in the layer. This is why the active layer needs to be removed periodically when the vermifilter is run continuously for a very long time as it gets converted into nutrient-enriched vermicompost.
3. **Indigenous Microbes of the Earthworm Residing Layer: Microorganisms Originally Present in the Active Layer**: Again the microbes present in the soil show antibacterial and enzymatic activity against pathogens present in wastewater. These microbes also help in the formation of a biofilm which further helps in the degradation process. The coelomic fluid released by the earthworms helps in further trapping and degradation of the remaining pathogens. Additionally, aeration is also provided to the microbes by the earthworms through their burrowing activity. Here both the earthworms and microbes work in synergy against the pathogens.

   Thus, we say that earthworms and microorganisms work symbiotically and synergistically against pathogens and for the treatment of wastewater.

6.4 Integration of Earthworms in Other Nature-Based Solutions

There are many integrated systems where vermifilters can be equipped or integrated with other NBS to enhance their efficacy. Figure 6.2 shows the different integrations. As compared to the traditional technologies, these NBS come with economic and labour extensive operations. One of the most widely studied integration is constructed wetlands along with vermifilters. It can be a successful technology, especially for rural areas or agricultural lands and might even prove as a circular, green technology if planned carefully. In this study, hog farm wastewater was treated using a system that combined a vermifilter with a downstream hydroponic unit. Hydroponic cultures can be utilized to achieve two major goals: eliminating eutrophicating pollutants from vermifilter effluent and producing nutritious crops for human and/or animal use. Hydroponic plant cultures used in conjunction with a vermifilter can extract inorganic phosphate and nitrate from previously treated wastewater, ensuring optimal plant health and growth rate while also contributing to improved wastewater treatment. Plant degradation includes artificial wetlands and rhizofiltration. However, it should be noted that the plant species should be chosen precisely, suitable for the physical environment of constructed wetlands rich in organic content.

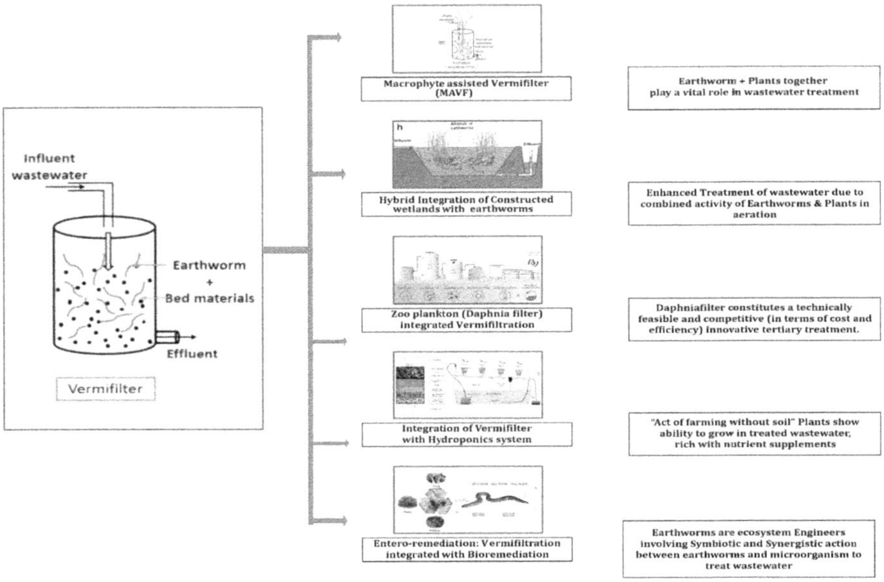

FIGURE 6.2 Integration of vermifilters with other nature based solutions.

6.5 Case Studies

6.5.1 Integration of Constructed Wetlands with Vermifilter to Treat Feedlot Runoff Wastewater – A Case Study in the US

Recently, in 2021 Singh et al., integrated constructed wetlands with vermifilter to treat feedlot runoff. As expected, the study revealed enhancement in the removal potential after the incorporation of macrophytes in the vermifilter or incorporation of macrophytes and earthworms together in a biofilter. The study design was particularly designed to achieve the desired treatment conditions. This group has been recurrently reporting on the advantages of macrophyte assisted vermifilters (MAVF) such as high nutrients and organic removing ability and increment in the life span of the vermifilter. In this study, they were successfully able to treat feedlot runoff and highly recommend this technology for future use (Figure 6.3).

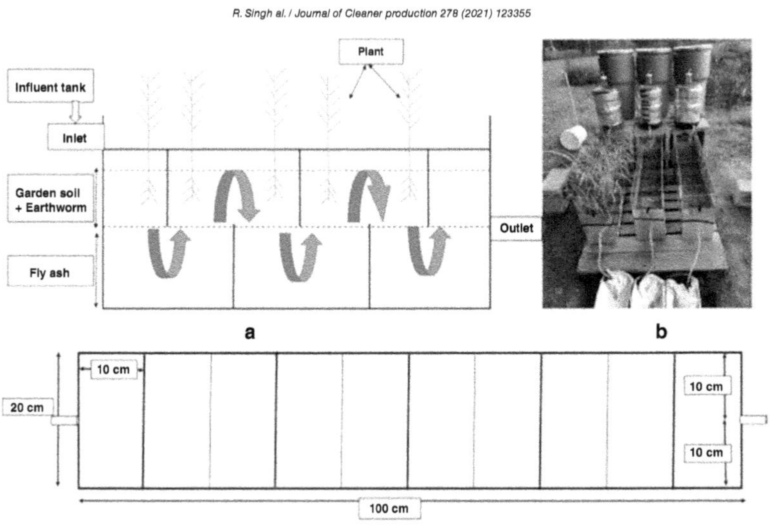

FIGURE 6.3 Case study of University of Nebraska-Lincoln, NE, the USA on feedlot runoff.

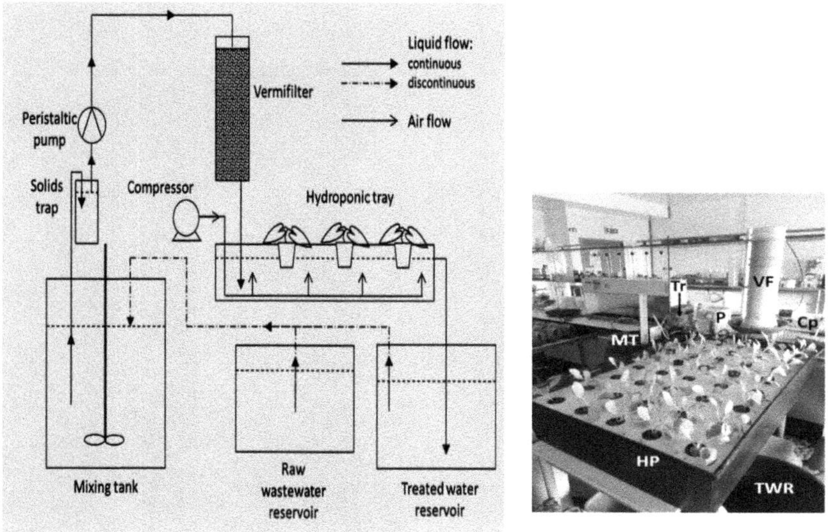

FIGURE 6.4 Case study on combined vermifiltration-hydroponic system (Ispolnovet al., 2021.)

6.5.2 Integration of Vermifiltration and Hydroponic System for Swine Wastewater Treatment – A Case Study in Portugal (Ispolnov et al., 2021)

In 2021, Ispolnov et al., worked treatment of swine wastewater by the integration of vermifiltration along with the hydroponic crop cultivation. They further used the treated effluent as feed dilution water which resulted in a decrease in BOD and stabilization of nitrites. This novel technique was also applicable for the removal of heavy metals and the treated swine wastewater was successfully applied for the plantation of pointed cabbage (*Brassica oleracea* L., var. capitata) with a 10 fold dilution (Figure 6.4).

6.5.3 Indian Institute of Technology (IIT) Bhubaneshwar on Macrophyte Assisted Vermifilter for Dairy Wastewater (Samal et al., 2018)

The important findings to this research revealed that a combination of horizontal subsurface flow vermifilter system and vertical downflow vermifilter system planted with *Canna indica* were capable of effective substantial reduction of BOD, COD, TSS, TDS, NH_4, TN, and TP from dairy wastewater tolerating the fluctuations of pollutant loads. Their vertical flow system removed more pollutants compared to the horizontal flow system. Under tropical climatic conditions, the area-based removal rate constants seem to be remarkably higher than previously reported for macrophyte filter systems in temperate areas, and the higher water temperatures are the most credible explanation. The present study reports that MAVF systems do perform very well and the promising results will stimulate to a broader use of MAVF systems as strong, naive, cost-effective, and structured wastewater management systems (Figure 6.5).

6.5.4 Earthworms Help in Dealing with the Clogging Issue of HSFCWs – A Case Study of Australia

Horizontal Subsurface Flow Constructed Wetland systems (HSFCWs) are commonly applied in Australia for in situ treatment of wastewater characteristically in the rural region and industries. However, due to their mechanistic design, they often face the clogging problem which sometimes also leads to the presence of unpleasant odour. The earthworms have the habit of active burrowing in their dwellings and thus help in the prevention of clogging. This phenomenon was reported in Australia as it was observed that many non-clogged HSFCWs which were treating Grey water and solid-liquid wastewater were hosting earthworms. This study focused on the potential of animal wastewater treatment as a promising solution to the clogging and sludge production problem through the introduction of earthworms into the wastewater systems.

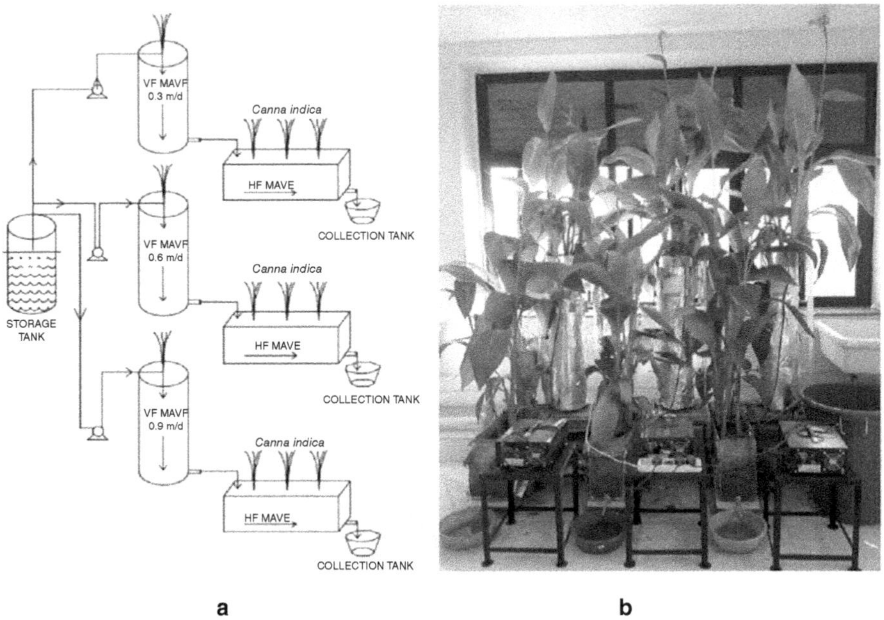

FIGURE 6.5 Case study of Indian Institute of Technology (IIT)- Bhubaneshwar, microphyte-assisted vermifilter (Samal et al., 2018.)

6.5.5 Potential Effects of Applying Earthworms into Constructed Wetlands Ecosystem – A Case Study in Thailand

In 2004, Edwards et al. investigated the potential effects of applying earthworms to a constructed wetland ecosystem. This can be considered as one of the inception studies working upon the integration of two nature-based technologies. The wetland received swine wastewater and was treated with the earthworm species *Pheretima peguana* and *E. fetida*. The tolerance power of each species varied under different temperatures but there was no significant alteration observed in moisture content, C:N ratio, or ammonia. After this study, a surge in the import of *E. fetida* was observed in Thailand, specifically for the purpose of vermicomposting (Figure 6.6).

FIGURE 6.6 Illustration of the pilot-scale swine wastewater treatment system.

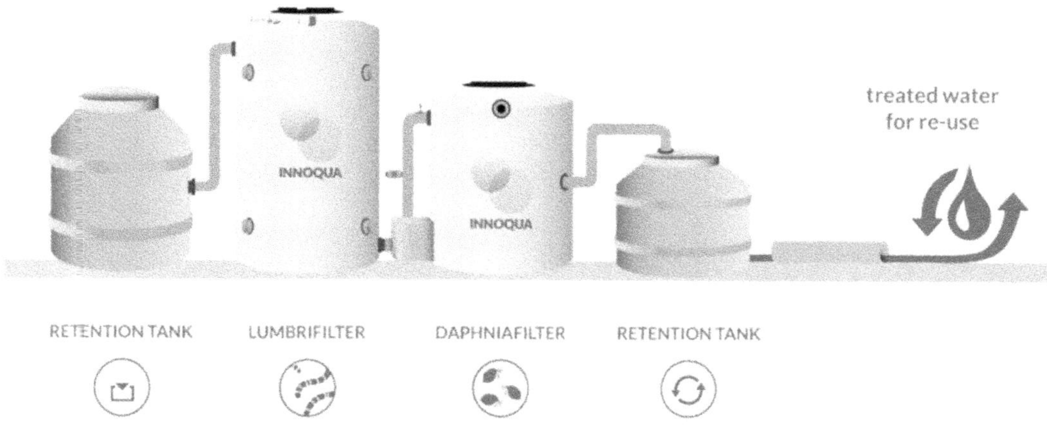

FIGURE 6.7 Schematic diagram of the pilot-scale integration of vermifilter and Daphnia filter by Scottish Water and INNOQUA (https://innoqua-project.eu/.)

6.5.6 INNOQUA Project (2020)

INNOQUA is an EU-funded project through the Horizon 2020 research and innovation programme launched in June 2016 which aims to promote sustainable water sanitation technologies capable of performing a whole water treatment cycle and available in multiple modular configurations adapted to local contexts and markets. This project has 20 partners have along with them they have developed and demonstrated the potential novel, modular nature-based, decentralized wastewater treatment solutions in India and ten other countries. These solutions combine the different treatment capabilities of microalgae, bacteria, biofilms, Daphnia ("water fleas"), and UV light – but most of the treatment is delivered by interactions between earthworms and bacteria in vermifilters. These units have delivered robust performance under a range of environmental conditions in each installation. Most notable technologies include the Tiger Biofilter technology in India where vermifilters were used to process human waste products effectively in a reallife scenario and a pilot-scale integration of vermifilter (which utilize earthworms) and Daphnia filter (which utilize water fleas) installed by Scottish Water at the Littlemill Wastewater Treatment Works as part of the INNOQUA project. This is used to treat a side stream from an existing wastewater treatment plant (Figure 6.7).

6.6 Advantages of the Integration

From all the previously reported studies, it can be concluded that the incorporation of plants in the VF significantly improves the physicochemical properties of the effluent as the degradation activity enhances due to the mutual effect of plants and earthworms. Plants result in extension in the working span of the VF by active recycling of nutrients which is reflected in the hydraulic conductivity. Earthworms are naturally capable of cleaning the substrate and depositing their cast on or near the surface, which aids in the relief of clogging issues. Microorganisms present in the soil, also actively degrade the organic matter and thus help in resolving the clogging issue. The microbial community, in turn is also associated with plants and certain plants provide them a niche and help in their propagation. Thus, it can be rightfully said that nature-based solutions such as vermitechnology are more helpful and sustainable than any other traditional approaches. Such solutions are more beneficial than any other conventional techniques and sustainable towards green technology (Figure 6.8).

> Nobody and nothing can be compared with earthworms and their positive influence on the whole living Nature. They create soil & improve soil's fertility and provide critical biosphere's functions: disinfecting, neutralizing, protective and productive.
>
> **"Dr. Anatoly Igonin" (Kumar and Ghosh, 2019)**

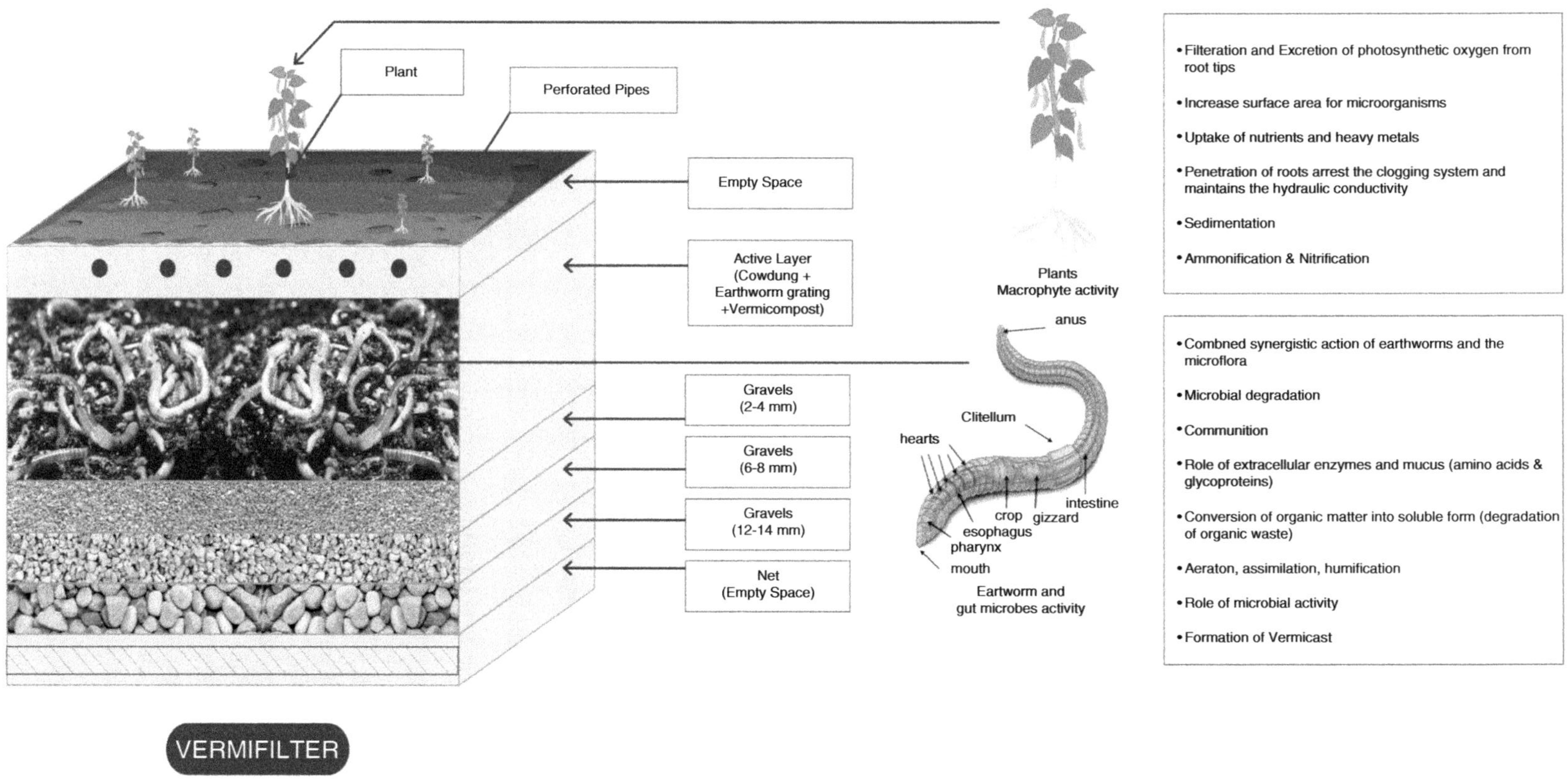

FIGURE 6.8 Mechanisms of integration of vermifiltration with nature-based solutions.

REFERENCES

Agrawal, N., Kumar, V. and Shahi, S.K., 2021. Biodegradation and detoxification of phenanthrene in in-vitro and in-vivo conditions by a newly isolated ligninolytic fungus *Coriolopsis byrsina* strain APC5 and characterization of their metabolites for environmental safety. *Environmental Science and Pollution Research*. https://doi.org/10.1007/s11356-021-15271-w

Arora, S., Rajpal, A., Kumar, T., Bhargava, R. and Kazmi, A.A., 2014. A comparative study for pathogen removal using different filter media during vermifiltration. *Water Science and Technology*, *70*(6), pp. 996–1003.

Arora, S., Rajpal, A., and Kazmi, A.A. (2016). Antimicrobial activity of bacterial community for removal of pathogens during vermifiltration. *Journal of Environmental Engineering*, *142*(5), 04016012.

Arora, S., Saraswat, S., Mishra, R., Rajvanshi, J., Sethi, J., Verma, A., Nag, A. and Saxena, S., 2020a. Design, performance evaluation and investigation of the dynamic mechanisms of earthworm-microorganisms interactions for wastewater treatment through vermifiltration technology. *Bioresource Technology Reports*, *12*, p. 100603.

Arora, S., Saraswat, S., Rajpal, A., Shringi, H., Mishra, R., Sethi, J., Rajvanshi, J., Nag, A., Saxena, S. and Kazmi, A.A., 2020b. Effect of earthworms in removal and fate of Antibiotic resistant bacteria (ARB) and antibiotic resistant genes (ARG) during clinical laboratory wastewater treatment by vermifiltration. *bioRxiv*.

Cohen-Shacham, E., Walters, G., Janzen, C. and Maginnis, S., 2016, *Nature-Based Solutions to Address Global Societal Challenges*. IUCN: Gland, Switzerland. https://innoqua-project.eu/

Dutta, D., Arya, S. and Kumar, S., 2021. Industrial wastewater treatment: current trends, bottlenecks, and best practices. *Chemosphere*, *285*, 131245. https://doi.org/10.1016/j.chemosphere.2021.131245

Ispolnov, K., Luis M.I.A., Lourenço, N.D. and Vieira, J.S., 2021. A combined vermifiltration-hydroponic system for swine wastewater treatment. *Applied Sciences*, *11*(11), p. 5064.

Kumar, C. and Ghosh, A.K., 2019, Fabrication of a vermifiltration unit for wastewater recycling and performance of vermifiltered water (vermiaqua) on onion (*Allium cepa*). *International Journal of Recycling of Organic Waste in Agriculture*, *8*(4), pp. 405–415.

Kumar, V. and Chandra, R., 2018. Characterisation of manganese peroxidase and laccase producing bacteria capable for degradation of sucrose glutamic acid-Maillard reaction products at different nutritional and environmental conditions. *World Journal of Microbiology and Biotechnology*, *34*, p. 32.

Kumar, V. and Chandra, R., 2020. Metagenomics analysis of rhizospheric bacterial communities of *Saccharum arundinaceum* growing on organometallic sludge of sugarcane molasses-based distillery. *3 Biotech*, *10*(7), p. 316. https://doi.org/10.1007/s13205-020-02310-5

Kumar, T., Bhargava, R., Prasad, K.H. and Pruthi, V., 2015. Evaluation of vermifiltration process using natural ingredients for effective wastewater treatment. *Ecological Engineering*, *75*, pp. 370–377.

Kumar, V., Singh, K., Shah, M.P., 2021a. Advanced oxidation processes for complex wastewater treatment. In: Shah, M.P. (Ed), *Advance Oxidation Process for Industrial Effluent Treatment*. Elsevier. https://doi.org/10.1016/B978-0-12-821011-6.00001-3

Kumar, V., Singh, K., Shah, M.P., Singh, A.K, Kumar, A., Kumar, Y., 2021b. Application of omics technologies for microbial community structure and function analysis in contaminated environment. In Shah, M.P., Sarkar, A., and Mandal, S., (Eds), *Wastewater Treatment: Cutting Edge Molecular Tools, Techniques & Applied Aspects in Waste Water Treatment*. Elsevier. https://doi.org/10.1016/B978-0-12-821925-6.00013-7

Kumar, V., Thakur, I.S., Singh, A.K., Shah, M.P., 2020a. Application of metagenomics in remediation of contaminated sites and environmental restoration. In: Shah, M., Rodriguez-Couto, S., and Sengor, S.S. (Eds), *Emerging Technologies in Environmental Bioremediation*. Elsevier. https://doi.org/10.1016/B978-0-12-819860-5.00008-0

Kumar, V., Thakur, I.S., Shah, M.P., 2020b. Bioremediation approaches for pulp and paper industry wastewater treatment: recent advances and challenges. In: Shah, M.P. (Ed) *Microbial Bioremediation & Biodegradation*. Springer: Singapore. https://doi.org/10.1007/978-981-15-1812-6_1

Mandpe, A., Paliya, S., Kumar, S. and Kumar, R., 2019. Fly ash as an additive for enhancing microbial and enzymatic activities in in-vessel composting of organic wastes. *Bioresource Technology*, *293*, p. 122047. https://doi.org/10.1016/j.biortech.2019.122047

Mandpe, A., Tyagi, L., Paliya, S., Chaudhry, S., Motghare, A. and Kumar, S., 2021a. Rapid-in-house composting of organic solid wastes with fly ash supplementation: performance evaluation at thermophilic exposures. *Bioresource Technology*, *337*, p. 125386. https://doi.org/10.1016/j.biortech.2021.125386

Mandpe, A., Yadav, N., Paliya, S., Tyagi, L., Yadav, B.R., Singh, L., Kumar, S. and Kumar, R., 2021b. Exploring the synergic effect of fly ash and garbage enzymes on biotransformation of organic wastes in in-vessel composting system. *Bioresource Technology*, *322*, p. 124557. https://doi.org/10.1016/j.biortech.2020.124557

Muga, H.E. and Mihelcic, J.R., 2008. Sustainability of wastewater treatment technologies. *Journal of Environmental Management*, *88*(3), pp. 437–447.

Murugesan, A. G., and Rajakumari, C., 2019. *Environmental Science and Biotechnology: Theory and Techniques*. MJP Publisher, Chennai.

Samal, K., Dash, R.R. and Bhunia, P., 2018. Effect of hydraulic loading rate and pollutants degradation kinetics in two stage hybrid macrophyte assisted vermifiltration system. *Biochemical Engineering Journal*, *132*, pp. 47–59.

Samal, K., Singh, R., Dash, R.R., and Bhunia, P. (2020). In: Ajay S. Kalamdhad (Ed.), Investigation on the effect of planting Canna indica in two-stage vermifilter for synthetic dairy wastewater treatment. In Recent developments in waste management (pp. 289–297). Springer, Singapore.

Sharma, M.K., and Kazmi, A.A. (2015). Effect of physical property of supporting media and variable hydraulic loading on hydraulic characteristics of advanced onsite wastewater treatment system. *Environmental Technology*, *36*(11), 1414–1422.

Singh, R., Bhunia, P., and Dash, R.R. (2017). A mechanistic review on vermifiltration of wastewater: Design, operation and performance. *Journal of Environmental Management*, 197, 656–672.

Singh, R., Bhunia, P. and Dash, R.R., 2018. Understanding intricacies of clogging and its alleviation by introducing earthworms in soil biofilters. *Science of the Total Environment*, *633*, pp. 145–156.

Singh, R., D'Alessio, M., Meneses, Y., Bartelt-Hunt, S.L., Woodbury, B. and Ray, C., 2021. Development and performance assessment of an integrated vermifiltration based treatment system for the treatment of feedlot runoff. *Journal of Cleaner Production*, *278*, p. 123355.

Singh, R., Samal, K., Dash, R.R. and Bhunia, P., 2019b. Vermifiltration as a sustainable natural treatment technology for the treatment and reuse of wastewater: a review. *Journal of Environmental Management*, 247, pp. 140–151.

Sinha, R.K., Bharambe, G. and Chaudhari, U., 2008. Sewage treatment by vermifiltration with synchronous treatment of sludge by earthworms: a low-cost sustainable technology over conventional systems with potential for decentralization. *The Environmentalist*, *28*(4), pp. 409–420.

Wu, Y., Zhu, W. and Lu, X., 2013. Identifying key parameters in a novel multistep bio-ecological wastewater treatment process for rural areas. *Ecological Engineering*, *61*, pp. 166–173.

7

Amalgamation of Constructed Wetland and Microbial Fuel Cell Systems as a Sustainable Approach towards Wastewater Treatment and Energy Recovery

Amol Shukla
CSIR-National Environmental Engineering Research Institute

Satyendra, Pravin Manekar, Rima Biswas, Ritesh Vijay*, and Sukdeb Pal*
CSIR-National Environmental Engineering Research Institute
Academy of Scientific and Innovative Research (AcSIR)

CONTENTS

7.1 Introduction

Population explosion and scarce investment in wastewater treatment infrastructures, particularly in low-income countries have resulted in an illicit discharge of pollutants in the environment thus causing negative impacts on health and the environment (Wu et al., 2014). Direct disposal of wastewater from industries, commercial buildings and municipality lead to various environmental problems such as eutrophication. In addition to this, wastewater treatment plants generate a considerable amount of greenhouse gases. Therefore, it is necessary to develop a sustainable and energy-positive approach to wastewater treatment.

* Corresponding authors.

DOI: 10.1201/9781003231738-8

Further, the sewage treatment plants suffer a huge work-energy imbalance to adhere to strict disposal values (Yang et al., 2018). Furthermore, it is observed that sewage treatment plants require about 0.5 kWh/m^3 of wastewater treated but interestingly wastewater is a renewable source of energy which could be 3–10 times the energy needed for its treatment (Pandey et al., 2016; Gude, 2015). Hence, the development of low-cost and low-energy intensive wastewater treatment technologies is gaining popularity to make the overall treatment as energy neutral process.

But as long as humans have disposed sewage into the environment, wetlands have been more or less involved in the cleaning of the wastewater through various removal mechanisms like biological degradation, chemical precipitation, adsorption on media surface, volatilization of Ammonia (NH_3), plants uptake, filtration, sedimentation, etc. (Carvalho et al., 2017; Liu et al., 2016). To imitate these processes in a more controlled fashion, Constructed Wetlands (CWs) are used. CWs have been widely used because of their lower carbon footprints, low operation and maintenance cost, and ease of operation (Vymazal, 2006; Zhang et al., 2011). Constructed wetlands are efficient for both municipal and industrial wastewater. It is been successfully used to treat polluted river water, municipal wastewater, molasses wastewater, swine wastewater aquaculture wastewater by Wu et al. (2010); Butterworth et al. (2013); Sohsalam and Sirianuntapiboon (2008); Stone et al. (2004) and Lin et al. (2002) respectively. However, the CWs are sustainable wastewater treatment systems but they have certain drawbacks like large area requirements and frequent clogging issues. CWs are intensively researched and several approaches were used to improve their efficiency by manipulating flow pattern, media, vegetation, intermittent aeration, effluent recirculation, baffles in CW, adding external carbon sources. But the efficiency is always moderate for organic and nutrient removal due to a lack of electron acceptors such as oxygen in CW (Ji et al., 2020). Hence it is imperative to integrate CWs with other low-cost technologies.

It is a fortune that CWs provide a natural base to integrate it with Microbial Fuel Cell (MFC) because CWs have an aerobic area at the top portion and anaerobic area at the lower portion (Yadav et al., 2012). MFCs consist of cathode, anode, separator, and external circuits distributed in different redox conditions. Further MFCs use organic matter for electron transfer through the electrode and generate electricity by utilizing bacteria (Cheng et al., 2006). The anode of the MFC is kept in the anaerobic zone of CW where it acts electron acceptor during organic matter degradation. During the process of oxidation of organic matter, protons are released at the anode which travels to the cathode through wastewater, and electrons travel from the external circuit to the cathode which is finally accepted by oxygen at the aerobic zone of CWs (Logan, 2008).

Various studies have been done to use coupled CW-MFC for wastewater treatment like swine wastewater treatment (Ren et al., 2021), RO rejects treatment (Das et al., 2019), Nitrogen (N) removal improvements (Srivastava et al., 2020), synthetic wastewater (Oon et al., 2015), Bio-energy generation (Wang et al., 2017a), antibiotic treatment (Wen et al., 2020), boron removal (Can and Yakar, 2017; Türker and Yakar, 2017), bisphenol A and ibuprofen degradation (Li et al., 2019), acid orange removal with leaves as a carbon source for energy generation (Fang et al., 2017). This reflects that the CW-MFC is a booming technology for wastewater treatment and bio-energy generation. Hence there is a need for extensive research in this field to make wastewater treatment sustainable with energy recovery options. This book chapter explores the various design considerations that are taken in CWs and MFCs. Further book chapter discusses the resource recovery options from CW and MFCs, life cycle cost of various wastewater technologies, and future research that has to be considered to make CW-MFCs a sustainable wastewater treatment technology.

7.2 Constructed Wetland and Microbial Fuel Cell

7.2.1 Constructed Wetlands and Its Removal Mechanism

CWs are the mimicked structure having plants and media as an exclusive feature. CWs require low construction cost and less energy requirement (i.e., flow takes place under gravity. Based on the hydraulics CWs are categorized into two groups (i) Free Surface Flow (FSF) CW and (ii) Sub-surface Flow (SSF) CW. Further, the SSF CW are classified into two groups (i) Horizontal Flow (HF) CW and (ii) Vertical Flow (VF) CW (Paul et al., 2018). In FSF CW wastewater generally flows over the media surface while SSF CW wastewater flows 10–15 cm (from top level) below the media level. FSF CWs are not preferred

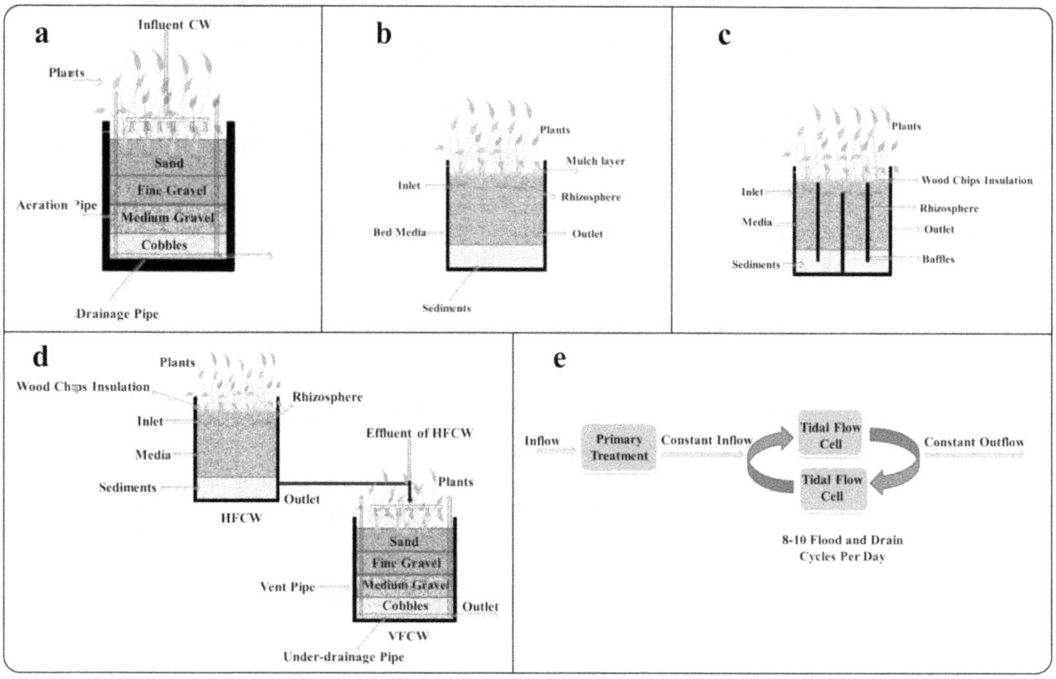

FIGURE 7.1 Schematics of different types of constructed wetlands.

because it creates a nuisance in the environment hence SSF CW are widely preferred (Liu et al., 2009). To further improve the efficiency of CWs innovative fill and drain type tidal flow CWs have been developed to enhance oxic and anoxic treatment processes (Behrends and Chopereno, 2012). Similarly, baffled flow CW has been introduced in which the baffles present in CW guide the wastewater to flow in sinusoidal form for attaining maximum contact between wastewater and microorganisms. Furthermore, hybrid CWs are developed to avail the advantages of both HFCW and VFCW together as nitrification efficiently takes place in VFCW while denitrification efficiently takes place in HFCW. Further VFCWs have higher air circulation and less area requirement compared to HFCWs (Vymazal, 2006). Schematics of VFCW, HFCW, Tidal flow CW, hybrid flow CW and baffled flow CW is shown in Figure 7.1.

7.2.1.1 Removal Mechanism in CW

CWs use physical, chemical and biological techniques to remove contaminants. The removal of suspended solids is mainly a result of abiotic factors like filtration and sedimentation. Further N in wastewater is reduced due to root adsorption, substrate adsorption NH_3 volatilization and biochemical N fixation (Badejo et al., 2018).

Bio-chemical N fixation involves converting Ammoniacal Nitrogen (NH_4-N) into Nitrite (NO_2^- - N) and Nitrate (NO_3^- - N) by nitrification with oxygen, while NO^{2-}-N and NO^{3-}-N are eliminated by denitrification with electronic donors to form N_2/N_2O. In addition to biological N removal, plants are also responsible for N uptake in CWs (Zhao et al., 2018). During nitrification, N acts as electron donor and oxygen acts as electron acceptor while in denitrification carbon acts as electron donor. It is observed that only 25% N is taken up by plants while 50% N fixation occurs due to biochemical transformation via nitrification-denitrification (Uusheimo et al., 2018). It is believed that Biochemical Oxygen Demand (BOD) and Chemical Oxygen Demand (COD) are removed rapidly due to the settling and entrapment of particulates in the void spaces within gravel and rock (EPA, 1993).

Biological growth on the media surfaces and on roots and rhizomes of plants remove soluble organic compounds. According to DeBusk (1999), 50% of the dissolved organic matter is consumed by aerobic microbes as an energy source. For this purpose, oxygen is supplied by plants rhizosphere to convert carbon into carbon dioxide (EPA, 2000). The decomposers mainly bacteria and fungi are responsible for organic removal via mineralization and gasification. Further in anaerobic zone of CW organic matter decompose into carbon dioxide and methane via the fermentation process (Qin and Chen, 2016). In addition to this photo-volatilization is also responsible for BOD removal since wetland plants take up the organic matter and transpire it into atmosphere through evapo-transpiration process (Hong et al., 2001). Organic volatile compounds like phenol are directly removed through volatilization. P is found in wastewater in both organic and inorganic form. Numerous studies show that P removal in CW is mainly achieved by adsorption, micro-organisms accumulation, leaching, mineralization and precipitation (Vymazal, 2007) Dissolved organic P and insoluble form of organic and inorganic P are not biologically available until transformed into soluble inorganics form. Further these soluble inorganic phosphorous are synthesized ATP, DNA, RNA, etc., by uptake and assimilation of plants and it can be removed from the system by harvesting plants (Vymazal, 1995; Qin and Chen, 2016). According to Shen et al. (2011), adsorption by plants will be significant if P is present in low concentration while plant adsorption is insignificant if its concentration is more. Further heavy metal removal in constructed wetland is achieved by chemical precipitation, filtration, plants and media adsorption, mineralization, etc. There is only 1%–2% heavy metal removal by plants and it mainly occurs in aerobic zone (Qin and Chen, 2016).

7.2.2 Microbial Fuel Cell and Its Removal Mechanism

MFCs are the low-cost wastewater treatment unit that converts chemical energy into electrical energy using biological processes. A typical sketch of MFC is shown in Figure 7.2.

In CW-MFCs microbes release electrons and protons due to anaerobic oxidation in an anodic zone by digesting organic matter attached to roots (Kabutey et al., 2019). The protons released during anaerobic oxidation get transferred to cathode through the separator while the electrons are transferred through the current collector and external circuit. This reduces the oxygen in cathodic chamber generating electricity and water as by-product (Santoro et al., 2017). Further electrochemical processes are exergonic thus it does not produce any intermediate products and proceed rapidly with negative free reaction energy. This free energy can be easily converted to standard cell voltage (E). Simple reactions taking place at cathode and anode are given equations (7.1)–(7.3)

$$\text{Anode}: CH_3COO^- + 4H_2O \rightarrow 2HCO_3^- + 9H^+ + 8e^- \ (E = -0.289 \text{ V}) \tag{7.1}$$

$$\text{Cathode}: 2O_2 + 8H^+ + 8e^- \rightarrow 4H_2O \ (E = 0.805 \text{ V}) \tag{7.2}$$

$$\text{Total}: CH_3COO^- + 2O_2 \rightarrow 2HCO_3^- + H^+ \ (E = 1.094 \text{ V}) \tag{7.3}$$

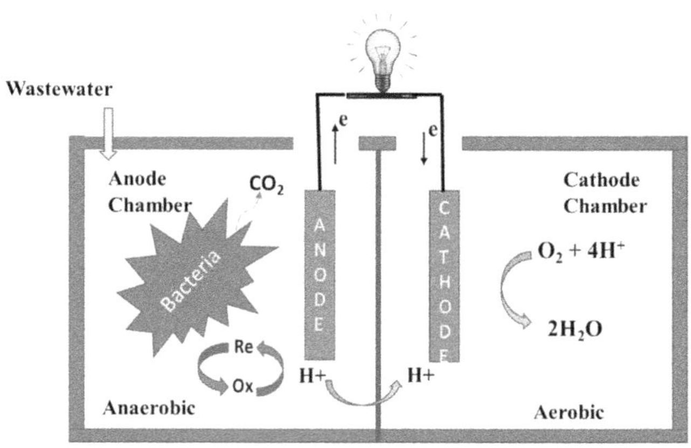

FIGURE 7.2 Schematic of a typical microbial fuel cell.

7.2.3 Integration of CW-MFC

Amalgamation of CWs with MFCs is an innovative approach to treating wastewater with bio-energy generation. Combing these two technologies discards the drawbacks of each other and work efficiently together (Wen et al., 2020, Zhao et al., 2013). It is often seen in CWs that electron acceptors (O_2) gradually exhausted leading to limit the organic removal in anaerobic zone of CW (Yadav et al., 2018). Hence, the integration of MFC in CW act as limitless electron acceptor (anode) and electron donor (Cathode) in deficient zones of CWs (Wang et al., 2016b). Further CW-MFCs expedite the electron movement generated by redox reactions during contaminant removal in CW. Further, the MFCs in CW stimulate the microbial activity in the system thus enhancing the microbial degradation. The plants in CWs exudate the organic matter which acts as the electron donor for electrochemically active bacteria that are naturally formed at anode (kept in anaerobic condition) of MFC. These electrochemically active bacteria oxidize the organic matter and thus, generating electrons, protons and carbon dioxide (Rana et al., 2017; Sophia & Sreeja, 2017). Further, these generated electrons and protons generated by electrochemically active bacteria during organic matter decomposition are transported to anode and cathode respectively. These generated electrons and protons combine with electron acceptor (O_2) generating water and bio-electricity in the system (Doherty et al., 2015; Zhao et al., 2013).

7.3 Design Considerations in Constructed Wetland and Microbial Fuel Cell

7.3.1 Vegetation in CWs

The vegetation (plants) in CWs is an exclusive aspect of constructed wetland which differentiates it from lagoons. These plants in CW accomplish the physical processes (like filtration, adsorption), chemical processes, provide oxygen to aerobic microbes and surface area for microbial growth. Further plants decrease the velocity of flow thus allowing solids to settle. Furthermore, vegetation offers external carbon to maintain C/N ratio and act as an insulator to the CW system (Shelef et al., 2013; Vymazal, 2011). In addition to this plant in CW increases the green scenery and visual appearance of the sewage treatment plants. Characteristics of plants directly affect the hydraulic efficiency of CWs hence, selection of plants is important for CWs design.

According to Ebrahimi et al. (2013), planted CW performed better than unplanted CWs. Further, *Cyperus alternifolius* showed very fast growth and it is capable of removing only organic matter, not P (Ebrahimi et al., 2013). Therefore, multiple plant species should be used along with *cyperus alternifolius* to efficiently remove P from wastewater. According to Mesquita et al. (2013), organic removal in CW is much easier than NH_3 removal since NH_3 removal is carried out by autotrophs who have less respiration rate therefore the N in wastewater is taken up by plants in CW. Carballeira et al. (2016), reported that at a low surface loading rate (2.5 g BOD per m^2 per day) all wetland species perform almost the same. Carballeira et al. (2016), observed 87%, 69%, 94%, and 86% biochemical oxygen demand (BOD) removal in CW planted with *Juncus effusus, Typha latifolia, Phragmites australis and Iris pseudacarus* respectively. Further, it was observed that the evapotranspiration losses of *T. latifolia* were only 21% while the losses in *Phragmites australis* were 27%. Biomass production in *T. latifolia* was maximum which helped more microbial contact with wastewater and hence showed 30% more N removal compared to *Iris pseudacarus. Juncus effuses* showed the highest N removal (70%) with the highest biomass production amongst *T. latifolia, Iris pseudacarus, and T. australis* (Carballeira et al., 2016). Further Dong et al. (2012) used *Acorus calamus L* in VFCW for treating polluted river water observed 48% and 35% COD and P removal respectively. Fan et al. (2013), observed 97% and 95% COD and N removal respectively using *P. australis* which is the most widely used plant species in CW. Saeed and Sun (2013), observed 80% BOD removal and 67% N removal when they used *P. australis* and *D. Scnderina* in VFCW for the treatment of textile wastewater. But the BOD removal (>95%) and N removal (>80%) increased significantly when they used *P. australis, D. sanderina* and *A. platyneuron* together in hybrid CWs. Further, Yazdania and Golestanib (2019), observed the 93.62% COD removal for dairy wastewater. *P. australis* shows better pollutant removal when it is used in VFCW but shows less removal in HFCW (Calheiros et al., 2007). To explore CWs for saline water treatment a halophyte called *Bassia Indica* was used for phytoremediation. Shelef et al. (2013) observed that this halophyte can gain huge biomass (i.e., 9 kg of its dry weight) in a short span and can accrue Sodium (Na) up to 10% of its dry weight.

7.3.2 Substrate

Substrate is also an important part of CWs because it is responsible to perform physical processes like absorption, filtration, adsorption, etc. Further, these substrates provide the surface area for microbial growth (Paul et al., 2018; Gorgoglione and Torretta, 2018). Substrate physical properties like shape, size, structure, permeability, porosity, chemical composition, etc., directly influence the efficiency of CWs (Gorgoglione and Torretta, 2018). Clogging is the most common problem in CWs and it is the function of porosity and time (Bai et al., 2016). The commonly used media in CW are sand, crushed rock, gravel, soil, etc., but these materials are only efficient in BOD removal not for heavy metals removal (Qomariyah et al., 2016 and Haynes, 2015). Shiwei et al. (2019) and El-Khateeb et al. (2019), observed that the porosity of coarse gravel, rubber, and plastic media was reduced by 19%, 12%, and 9% respectively but the plastics media gave better results compared to rubber and gravels because of biological enzyme property of plastic media. Further, Saraiva et al. (2018), observed that plastic substrate is indolent in nature and hence does not affect the effluent characteristics when compared to gravel that releases cations which reacts with hydrogen ion thus increasing the pH of effluent. Size of media impacts the hydraulic performance as well as removal performance of CWs, therefore Pucher and Langergraber (2019) concluded that it is best to load media with fine grain sizes (0.8–4 mm) after longer intervals to ensure sufficient aeration, while coarser materials can be loaded for shorter intervals. Phosphorous (P) removal in a CWs is a challenging act since P removal by wetland species is not compelling and it is mostly removed by surface phenomena like adsorption, filtration, absorption (Pucher and Langergraber, 2019). Nguyen et al. (2020), used white hard clam shells in HFCW to remove P from anaerobically treated swine effluent and observed 89.9% P removal due to the presence of calcium oxide in white hard clam shells which increases pH thus P removal. They observed a sorption range between 61.2 and 95.5 ($g/m^2/year$).

In recent years, there has been a lot of research on the use of industrial waste, agriculture waste, demolished building waste, etc., as substrate in the CWs. Fly ash, ceramic materials, crushed plastic bottles, limestone, etc. are the common industrial wastes that are used as media in CW. Mateus et al. (2012), used limestone in CWs for heavy metal removal and advanced nitrification because it increases the pH which leaches the heavy metals and balances the pH during the nitrification process. Further, Lizama-Allende et al. (2021), used Zeolite and Limestone as supporting media for CW to remove arsenic and heavy metals from highly acidic wastewater. They observed more than 96% arsenic, iron, and lead removal. Limestone can raise pH up to 7.6 which helps in metal precipitation. Furthermore, Lu et al. (2016), compared five different substrates (Charcoal, maifanite, Bamboo, Steel slag and Limestone) for COD removal and concluded that maifanite is softer and more able to sorb carbon dioxide than steel slag and this may account for its superior COD removal. Among different substrates, maifanite was the best at sorption followed by steel slag, bamboo, charcoal, and limestone. Biochar is obtained through the pyrolysis of organic material. Further, Fu et al. (2020), used different combinations of media (ceramsite + activated carbon + sand) in CW A, (sand + activated carbon + ceramsite) in CW B, and sand in CW C. They found that CW A had higher dissolved oxygen indication and higher porosity in (ceramsite + activated carbon + sand) combination. Further, it was observed that N removal (98%) was higher CW B. Though CWs are efficient in pollutant removal but during removal gaseous compounds are released causing Green House Gas (GHG) emissions. Thus, preventive measures are to be taken to curtail GHGs emission. To reduce GHG emissions Guo et al. (2020), used biochar for GHGs reduction. Biochar enhanced the COD and NH_4-N removal but it may result in secondary pollution if not disposed properly since very less information is known for biochar recycling and reuse (Sanjrani et al., 2019). Different plant species and media used by scientists in CWs are shown in Table 7.1.

7.3.3 Materials of Electrode

Electrodes are the heart of MFCs since they are involved in redox reactions, electron transfer and biofilm generation for the removal of pollutants and electricity production. Materials of electrode have been intensively researched. Yadav et al. (2012); Fang et al. (2013); Oon et al. (2015); Srivastav et al. (2015) and Xu et al., 2018 Xu et al. (2018); have respectively used graphite electrodes, granular activated carbon electrode, carbon felt electrode, platinum-coated carbon cloth and titanium mesh and activated carbon. Carbon-based electrodes are mostly used in MFCs due to their bio-compatibility, good conductivity, and low cost with higher durability (Li et al., 2017). Further, these carbon-based electrodes promote the growth of electrochemically active bacteria which in turn improves electron transfer thus felicitating high COD removal and electricity generation. Kalathil and Pant (2016) used graphite-based electrodes and found that graphite has a hybrid structure which makes a crystal form of carbon that causes higher electrical conductivity. In addition to this, anode of MFCs made of graphite fabricated from felts, brushes, and rods but it was found that graphite felt electrodes produce much higher power density compared to graphite rods. Furthermore, inoculated-based graphite electrodes promote bio-enzymatic action which increases the power density and pollutant removal (Gao et al., 2013). Further in this line Yasri and Nakhla (2016) found that doping of electrodes with calcium sulphide, ferric oxide and ferrous sulphide could enhance the bacterial affinity towards the electrode surface. Stainless steel electrodes have also been used in MFCs but these electrodes do not produce thick and robust biofilm which prevents high power density generation (Ketep et al., 2014). With the development of 3D printing technology, the use of 3D printed electrodes has become popular. These 3D-printed electrodes have specific surface area and have a spongy honeycomb structure that inhabits bacterial growth (Tang et al., 2020). The current density production of 3D carbonaceous electrodes was 40.2 A/m² which was 19% higher than Kenaf electrodes due to high diffusion of substrates and microbe in 3D electrodes (Logan et al., 2007). But 3D electrodes have poor strength which limits their use of it in MFCs but their mechanical strengths can be improved using copper-melamine foams (Baudler et al., 2017). Ceramic-based electrodes are popular for wastewater treatment in harsh conditions. Ceramic electrode showed the highest current density when coated with titanium due to their highly porous structures (Massazza et al., 2015). Similar to anodes, cathodes are an integral part of MFCs since they act as an electron acceptor to complete the redox reactions. Cathode electrodes can be classified into two groups: (i) abiotic and (ii) biotic. Cathode which reduces oxygen is highly favourable thermodynamically when the catalyst is present. Platinum-coated carbon electrodes are commonly used oxygen-reducing cathodes in MFCs but platinum-coated carbon electrodes have high cost which limits their use. Hence non-precious electrodes like Fe, MnO, ZnO, Fe_3O_4, etc., are used with doped carbon. An activated-carbon-coated cathode with 0.5 mg coating exhibited a higher power density than platinum-carbon-cloth catalysts (55 mW/m²) (S. Cheng et al., 2006). Abiotic electrodes are expensive hence biotic electrodes are introduced in the research. The oxygen concentration required for platinized abiotic cathodes is 10 times higher than the concentration required for aerobic bacteria which increases the cost. Formation of biofilm on abiotic bacteria helps to get oxygen from water (Milner et al., 2016). Further biocatalyst cathode is self-sustaining and highly stable when subjected to harsh wastewater treatment conditions.

7.3.4 Impacts of Vegetation and Media on Electricity Production

Fuel sources in CW-MFCs are dependent upon the concentration of organic pollutants and rhizosphere deposits in wetland systems (Ong et al., 2010). Further, the bacteria in the rhizosphere play an important role in breaking down the organic pollutants thus releasing electrons in the CW-MFC system. The type of vegetation is directly related with the specific area provided by these plants for microbial growth and bio-electricity generation (Corbella et al. 2015). Saz et al. (2018) observed that different plant species showed different power density due to their rhizosphere microbial growth capability. It was observed that *Typha latifolia, Typha angustifolia, J. gerardii, C. divisa* and unplanted CW-MFC system generated a power density of 13.4, 18.1, 8.1, 8.8 and 5.39 mW/m² respectively. The highest power density was seen in *Typha angustifolia* because it produced the highest biomass which provided a large surface area for

TABLE 7.1

Different Plants Species and Media Used in CW

Wastewater Type, CW Type and Design Parameters	Substrate	Vegetation	Pollutant Removal (%)	Conclusion	Reference
Domestic wastewater, HFCW	Gravel	-	BOD = 44, COD = 43 and TSS = 48	• The specific area of rubber, garvel and plastic is 130, 177 and 283 m^2/m^3 respectively. • Plastic media is non- reactive in nature and possess anti-clogging properties.	Zidan et al. (2015)
	Shredded plastic	-	BOD = 50, COD = 49 and TSS = 60		
	Rubber	-	BOD = 38, COD = 38 and TSS = 45		
Synthetic wastewater, HFCW HLR (Gravel) = 4 cm/d HRT = 5.1 days HLR (Filterlite NR) = 3.5 cm/d HRT = 5.7 days	Gravel	*Phragmites australis*	BOD = 80, NH_4-N = 80	• Filterlite NR (specific surface area 1,250 m^2/m^3 with bed porosity 0.45) performed well compare to gravel but did not show significant removal difference.	Albuquerque et al. (2009)
	Filterlite NR	*Phragmites australis*	BOD = 90, NH_4-N = 85		
Synthetic wastewater, VFCW Aeration rate = 0.8–1.2 L/min Aeration Time = 2–6 h/d HRT = 72 hours	Biochar	*Iris pseudacorus*	COD = 84, NH_4-N = 95	• Biochar as well as intermittent aeration improved nitrogen removal due presence of carbon that helped in denitrification	Feng et al. (2020)
Groundwater treatment, FWS and SSF CW HLR for FWS = 0.12 m/d HLR for SSF CW = 0.07 m/d	River gravel (10–20 mm)	*Phragmites australis*	More than 85% NO^{3-}-N removal was obtained for both FSF and SSF CW at HLR not exceeding 0.04 m/d	• Hybrid CW is recommended to remove nitrogen and reduce area requirement.	Lin et al. (2008)
Domestic wastewater, VFCW HLR= 0.15 m/d	Sand	*Heliconia psittacorum*	BOD = 96, NH_4N = 75, TSS = 85	• HLR should not exceed 0.15 m/d if gravel is used as media in CW	Bohórquez et al. (2017)
	Gravel	*Heliconia psittacorum*	BOD = 47, NH_4N = 36, TSS = 0		
Domestic wastewater, Baffled CW HRT = 2, 3 and 5 days	Gravels and rice husk	*Typha latifolia*	NH_4-N removal efficiency in baffled flow CW was 74%, 84%, 99% for 2, 3 and 5 days respectively while for conventional CW it was 55, 70, 96 for 2, 3, and 5days respectively	• Baffled CWs performed much better than conventional CWs • Oxygen reduction potential at 0.6m depth was low • Rice husk as media increased BOD in effluents	Tee et al. (2012)
Septic tank waste, VFCW, HLR = 13 m/year Solid application rate = 81 kg TS/ m^2/year	Gravel	*Cynodon dectylon pers and Tifton 85 grass*	BOD = 77, NH_4-N = 71, COD = 83	• VFCWs are efficient for septage treatment	Andrade et al. (2017)
Municipal wastewater, VFCW HLR = 156 kg/m^2/d	Cobbles, fly ash, soil and sand	*Pennisetum perpureum schum*	P = 97, NH_4-N = 93, COD = 89	• Maximum removal observed at Fly ash to soil ratio (3:7) by weight and hydraulic load of 156 kg/m^2/d is best suited for VFCW	Guo et al. (2014)

(*Continued*)

TABLE 7.1 (*Continued*)
Different Plants Species and Media Used in CW

Wastewater Type, CW Type and Design Parameters	Substrate	Vegetation	Pollutant Removal (%)	Conclusion	Reference
Domestic wastewater hybrid CW (VFCW unplanted + HFCW planted) HRT = 1–5 days	Zeolite	*Canna indica*	BOD = 93, COD = 94, N = 92	• External carbon addition in HFCW decreases DO level in the system hence increased denitrification.	Xinshan et al., 2010
Domestic wastewater, Hybrid CW HRT = 4, 8, 12, 16, 20, 21 and 28 days	Soil, gravel and sand	*B. reptans* and *T. protulacastrum*	For *B. reptans* TSS = 58, TDS = 63, P=58, Nitrate = 70 For *T. protulacastrum* TSS = 70, BOD = 74, P = 76, Nitrate = 80	• *T. protulacastrum* showed good performance • Nutrient removal was maximum at 20 days HRT but maximum organic removal was achieved in 8 days HRT	Sehar et al., 2015
Municipal wastewater, phytorid baffled CW	Different sizes of gravels	*Phragmites spp, Typha spp, Canna spp*	TSS = 75–95, BOD =70–80, COD = 60–75, TN = 60–70 and P = 50–60	• Baffles increased the retention of wastewater in the system hence more contact time was available for wastewater and microbes	Balpande and Mhaske (2017)
Swine wastewater, VFCW Swine wastewater diluted with tap water (v/v) = 1:3 HLR = 4 cm/d HRT = 5 days	Oyster shell, brick and red soil	*Phragmites australis*	TN = 63.3, NH_4–N = 65.1, COD = 70.5, TP = 92.29	• Vertical up-flow constructed wetlands showed more than 90% nutrient removal.	(Huang et al., 2014)
Domestic wastewater, Hybrid (HFCW+HFCW) Organic loading Rate	Light expanded clay aggregates (LECA)	*Phragmites australis*	COD = 73.90, NH_4-N = 59.3	• A has higher porosity, fast adhesion, surface area for fast growth of plant	Mesquita et al. (2013)
Petrochemical wastewater, HFCW, HRT = 90 days	Gravel and sand	*Phragmites spp. And Typha spp.*	BOD = 90.41, COD = 93.83, Turbidity = 58.97, Phenolic compound = 65-90	• Typha spp. showed better results than Phragmites spp. For industrial wastewater treatment.	Sudarsan et al. (2016)
Run-off wastewater, FWS Low Temperature study = 8°C	Peat	-	NH_4-N = 82 P = 47 TSS = 71	• Temperature had only impacted the COD and NH_4-N removal while other parameters like P, SS	Postila et al., 2015
Domestic wastewater, HFCW Low temperature study = 2°C–5°C	Loess, gravel slag, humus	*Hemerocallis Lilioasphodelus, L. Iris tectorum*	COD= 85, NH_4-N = 70, P= 37	• Green house shed maintained the temperature in the system thus improved the efficiency. The treated water was used for floriculture	Gao and Hu, 2012
Phenolic and petroleum derivative in groundwater, HFCW HRT = 10 days	Fine and coarse gravels	*Phragmites australis*	Phenol-99.8, m-cresol-97.7	• Phenolic compounds did not cause toxicity to wetland plants • Complete removal of phenol and m-cresol observed at 620 and 80 mg/m^2/d loadings respectively	Stefanakis et al. (2016)
Petroleum contaminated groundwater, Hybrid CW (HFCW+VFCW) HRT = 37–43 cm/d DT = 0.61–0.78 day	Soil, sand and gravel	*willows (Salix), reed (Phragmites), bulrush (Scirpus), rush (Juncus), and dogwood*	Benzene-80, total BTEX-88	• Cascade aerator as primary treatment for volatile compound removal using hybrid CW is a good option	Bedessem et al. (2007)

microbes (Saz et al., 2018). Similar conclusions were made by Jingyu et al. (2020), when they used *Typha angustifolia* in the CW-MFC system and found power density of 18.1 mW/m^2 with 88% COD removal. The power densities observed by researchers for various plant species like Phragmites autralis, Canna indica, Ipomoea aquatic, Iris pseudacorus and Elodea nuttallii are 20.76, 15.73, 5.62, 9.6 and 184.55 mW/m^2 respectively (Jingyu et al., 2020). On the other-hand filter media used in CWs influence the electricity production as observed by Jingyu et al. (2020) because these Medias provide the surface are for microbes more surface area means more biological degradation hence the more bio-electricity production. Yadav et al. (2012) used gravel in the CW-MFC system and found power density of 15.7 mW/m^2 while (Xu et al., 2018) found power density of 3714 mW/m^2 when they used sand and ceramsite due to the high specific surface area for these materials compared to gravel. Further, Mu et al., 2021 used four different filter media (Bio-ceramic, Zeolite, Calcite and, Volcanic rock in the CW-MFC system. The output voltage was highest in calcite (0.675 ± 0.021 V) based CW-MFC system followed by volcanic rock (0.595 ± 0.019 V) based CW-MFC systems due to higher organic degradation in the anaerobic zone of CW-MFC filled with calcite (Mu et al., 2021).

7.4 Current Scenario and Advancement in CW-MFCs

The concept of CW-MFC was first proposed by Yadav et al. (2012) for the removal of COD and dye from synthetic wastewater using VFCW and MFC. The technology was later used by various researchers across the globe. Zhao et al. (2013) and Camacho et al. (2014) used it for swine wastewater treatment and synthetic wastewater treatment. Further, with time, many papers have been published to improve the efficiency of purification, electricity production, materials of electrode by Doherty et al. (2015a); Yakar et al. (2018); Tang et al. (2019); Ge et al. (2020); Yang et al. (2020). Further, Mu et al. (2021) studied the effect of filter media on removal efficiency and electricity production. Furthermore, Saeed and Miah (2021) studied the impacts of filter media on the removal of pollutants in tidal flow-based CW integrated with MFC.

7.4.1 Use of Integrated CW-MFC for Various Wastewater Treatment

There are multiple papers being published using integrated CW-MFC approach. Table 7.2 shows the usage of coupled CW-MFC for various wastewater.

7.4.2 Resource Recovery Options from Integrated CW-MFCs

Wastewater treatment and its reuse in irrigation are being adopted worldwide. Thus, integrating CW-MFCs in water, food and energy nexus could maximize the resource use and give both economic and ecological benefits. According to Avellán and Gremillion (2019), 200 PE community could provide the treated effluent of about 11–52 m^3/d which can irrigate 0.2–1.1 ha of maize crops with a yield of 750–3,500 kg of grain per season. Further Abusam et al. (2015), used treated effluent of CW in agriculture which were meeting the irrigation standards laid by Kuwait Environmental Public Authority (KEPA). Furthermore, Ghrair and Al-mashaqbeh (2016) and Nelson et al. (2008) used the treated wastewater for construction and gardening purposes. In addition to this, the harvested parts of the plants can be a good source of energy. Nelson et al. (2018) found that CW of capacity 1–1.2 m^3/d yielded 1,200 kg of biomass which was used for feeding of goats, pigs and chickens. The biomass yielded from *C. papyrus, Phragmites spp., and Typha spp.* are 4297, 2788 and 1755 g/m^2 respectively. The energy produced from *A. donax, C. papyrus, Phragmites spp., and Typha spp. are* 17, 16.8, 18.3 and 18.9 MJ/kg respectively. Further the fly ash produces by burning this dry harvested plant can be used in construction works. Furthermore, biomass can be used with other solid organic matter to generate biogas in anaerobic digestors. Iiyama et al. (2014), (taken from allevien) per-capita wood consumption rate is 0.69 m^3/year while timber production rate is approximately 6.5 m^3/ha/year and for the same 140 ha/year forests are cleared for the community of 200 people. The wetland biomass can reduce this demand by 0.9–12 ha/year which

TABLE 7.2

Usage of Coupled CW-MFC for Various Wastewater

Wastewater Type	CW Type	Plant species	Substrate	Material of Electrode	Removal (%)	Maximum Power Generation (V)	Reference
Synthetic wastewater	VFCW	*Ipomoea aquatic*	Gravels	GAC/stainless steel	COD = 95	0.740	Liu et al. (2013)
Swine wastewater	VFCW	*Phragmites australis*	Gravels	Graphite plate	COD = 76	0.495	Zhao et al. (2012)
RO Reject	VFCW	*Canna indica*	Gravels	Graphite	COD = 70 TDS = 52	0.86	Das et al. (2019)
Dye wastewater treatment	VFCW	*Ipomoea aquatic*	Activated carbon and gravels	Graphite	-	0.56	Fang et al. (2017)
Synthetic wastewater	HFCW HRT = 4 days	*Aglaonema Commutatum, Epipremnum aureum, Dranacaena braunni, Philodendron cordatum*	Red volcanic rock	Carbon felt and brass mesh	-	0.199 for *Philodendron cordatum*	Oscar et al. (2020)
Synthetic wastewater	VFCW	*Phragmites australis*	Gravel	Activated carbon	COD = 94	0.74 1 for 20 cm electrode spacing	Song et al. (2017)
Swine wastewater	Up-flow-downflow CW	*Phragmites australis*	Alum sludge	Graphite	COD = 64	0.434	Doherty et al. (2015)
Synthetic wastewater	Up-flow CW	*Typha latifolia*	Gravels	Carbon felt	COD = 100	0.421	Oon et al. (2015)
Synthetic wastewater	HFCW	*Phragmites australis*	Gravels	Graphite	COD = 90	1.161	Villaseñor et al. (2013)

(*Continued*)

TABLE 7.2 (*Continued*)
Usage of Coupled CW-MFC for Various Wastewater

Wastewater Type	CW Type	Plant species	Substrate	Material of Electrode	Removal (%)	Maximum Power Generation (V)	Reference
Chromium Solution	Up-flow CW	*Phragmites australis*	Bio-ceramic, zeolite, calcite, and volcanic rock	Activated carbon/ steel mesh	COD = 97 and Bio-ceramic removed maximum (24%) chromium through adsorption	0.675 for calcite	Mu et al. (2021)
Synthetic wastewater	HFCW	*Canna indica*	Stone and gravel	Graphite	COD = 98	0.68	Srivastava et al. (2020)
Municipal Wastewater	Tidal flow CW	*Vetiver and Phragmites*	Brick, concrete blocks, steel slag, coal, jute fibre, biochar	Steel plates	COD = 75–100	0.318	Saeed and Miah (2021)
Antibiotic wastewater treatment	Up-flow CW	*Oenanthe javanica*	Soil and gravel	Granular activated carbon	Antibiotic removal = 97–98	0.68 and it decreased with increase in antibiotic concentration	Zhang et al. (2016)
Synthetic wastewater	-	*Canna indica*	Stone gravel	Graphite and platinum coated carbon cloth	-	0.76	Srivastava et al. (2015)
Synthetics wastewater	Hybrid CW(HFCW+VFCW)	*Common sedges*	Pebbles and graphite granules	Graphite plates	NH_4-N = 90–94 COD = 95–99	0.095	Srivastava et al. (2020)
Synthetic wastewater	Up-flow CW	-	Gravels and alum sludge	Steel plates and carbon fibre felt	COD = 97 with aeration and recirculation	0.639 after aeration and recirculation	Tang et al. (2019)

would not only save the forest ecosystem but also reduce the global warming impacts. Further, the wastewater contains high amount of nutrients (P and N) which can be recovered and used as soil conditioners. Furthermore, the use of treated effluent is been researched by many authors. Gharair and Masaqbeh (2016) used treated water in concrete production since chlorides and sulphate were in limit. Similarly, Nelson et al. (2008) used treated wastewater of CW for gardening purposes. Resource recovery option from wastewater is shown in Table 7.3.

7.4.3 Comparison of CW-MFC with Other Treatment Technologies

CW-MFCs are low-cost sustainable technology for wastewater treatment. The presence of microbial fuel cell in CW reduces GHGs emission. Further, the presence of macrophytes in CWs increases the aesthetic view of the wastewater treatment plant. The efficiency of CW-MFCs is comparable with other treatment technologies like ASP, SBR, MBR, MBBR, etc. Furthermore, the incorporation of MFCs reduces the land requirement which was the major drawback of CWs. Comparative study of various wastewater treatment technology is shown in Table 7.4.

7.5 Future Scope and Challenges

CW is found to be more effective for domestic wastewater treatment as compared to industrial wastewater treatment. Hence, its amalgamation with MFCs is a hot topic of research. Till now, CW-MFCs have shown an improvement in efficiency for various wastewater but studies on its design and role of microbes in combined CW-MFCs are the area of study. It is found from the study that CWs lack in P removal therefore denitrifying phosphate accumulating organisms could play an important role in P removal. In addition to this, the use of industrial waste, construction waste, agriculture waste for P removal enhancement should be explored. Further, the use carbonized metal-organic materials, porous covalent organic materials 3D printed substrates as electrode material and CW media need research since these materials have high removal efficiency for heavy metals (Liu et al., 2021; Biswas et al., 2020). Furthermore, these materials have high specific area suitable for microbial growth that must be explored in future. Electron transfer is a key phenomenon in the biological degradation of waste hence the electron transfer efficiency of electrochemically active bacteria is of utmost importance. There is a significant gap in micro-organism electron transfer reactions, pollutant removal and bio-energy generation therefore these gaps must be filled in future studies. Another important phenomenon that takes place in CW is de-nitrification. Many times, it is observed that de-nitrification is incomplete in CW due to the deficiency of electron donor species in wastewater. Hence, the electrode of MFCs could play an important role as an electron donor to complete the de-nitrification in CW. It is observed that the anode of MFC influence the anaerobic oxidation of contaminant and cathode hampers the reduction reactions. Many studies revealed that MFCs have coulombic efficiencies and power densities but CW-MFC combinations have not performed equivalent to MFCs in respect of electrical performance due to the application of a high volume of wastewater to exposed electrode (Wang et al., 2017b). Therefore, sizing of electrode, spacing of electrode, and usage of multiple material electrodes to enhance electron transfer and bio-energy generation is to be researched in the future for broad application of CW-MFC in the area of wastewater treatment. Further, the use of nanoparticle-based electrodes is a good topic of research which must be explored in the future. In addition to this, the improvement of rhizosphere in the anodic chamber with exogenous substrates will improve electricity production. Rhizosphere enhancement is dependent on plants used in CWs therefore detailed plant study is required. Furthermore, CW-MFCs face voltage reversal causes difficulty in power storage which should be addressed in future studies (Wang et al., 2017b). Abiotic factors like temperature and solar radiation affect the function of both plants and micro-organisms and thus need to consider in future studies. Green House Gases (GHGs) emission from wastewater treatment plants is a burning topic. Thus, GHGs emissions from integrated CW-MFCs should be considered in the future. Till now various studies show that integrated CW-MFCs reduce GHGs emission (Wang et al., 2019, Zhang et al., 2020). Coupled CW-MFCs can be a promising technology in the field of wastewater treatment.

TABLE 7.3
Resource Recovery and Bioenergy Recovery Options from Wastewater

Technology	Wastewater Type	COD Removal (%)	Resource/Energy Recovery	Remarks	Reference
MFCs	Paper recycling wastewater	96 for 500 hours batch cycle	0.672 W/m^2	Addition of phosphate buffer in wastewater improved the energy production	Huang and Logan (2008)
MFCs	Starch processing wastewater	98	0.239 W/m^2	High chemical oxygen demand (COD) removal observed for starch processing wastewater using MFCs with energy production	Lu et al. (2009)
MFCs	Tetrachloroaurate wastewater	--	99% Gold recovery (from 200 ppm concentration) 6.58 W/m^2	Anodic concentration separation supported the mechanism of electron transfer	Choi and Hu (2013)
Dual chamber MFCs	Leachate	89	0.04866 kWh/kg COD)	Removal efficiency of N and COD is more than 90% when COD loading was 3 kg COD/m^3/d	Zhang et al., (2015)
MFCs	Wine wastewater	89–91	90% recovery of P and N as struvite at cathode	Suspended P gets easily precipitated than dissolved P at cathode.	Ichihashi and Hirooka (2012)
CWs	Swine wastewater	--	87% nitrogen recovery by plant harvesting (*M. aquaticum*)	*M. aquaticum* as vegetation CWs is a good option for nutrient recovery from wastewater	Luo et al. (2018)
MFCs	Urine	--	94% P recovery as struvite		Zang et al. (2012)
CWs	Dairy wastewater	81	N and P recovery 22.4 g N/m^2/year and 7.4 P/m^2/year, respectively	Duckweed based wetlands provide good biomass rich in nutrients	Adhikari et al. (2015)
MFCs	Mill wastewater	20–70	1.771 W/m^2	MFCs found inefficient in cyanide removal	Kaewkannetra et al. (2011)
MFCs	Dairy wastewater	63	0.20 kW h/kg COD removed	Maximum voltage was obtained in continuous operation	Faria et al. (2017)

TABLE 7.4

Life Cycle Cost of various Treatment Technologies

Technology and HRT	Treatment Capacity	Efficiency	Area Requirement	Energy Consumption	GHGs emission	Capital Cost	Operation and maintenance cost
ASP (4–8 hours)	2,150 m^3/d	COD=80–95, Nitrogen= 65–80, BOD= 85–92	2.04 ha	0.682 (MJ/m^3)	2.5/m^3 of wastewater	US $ 3,99,900	US $ 1,01,050
WSP (8–15 days)	30,000 m^3/d	BOD= 65–80, COD = 65–75, Nitrogen = 65–80	45.9 ha	-	12.8 mmol m^2/d as CO_2 equivalent	US $ 50,10,000	US $ 50,100
UASB (4–10 hours)	36,000 m^3/d	BOD = 75–78, COD = 70–80, Nitrogen = 65–80	14 m^2/MLD	-	0.24–0.90 kg CO_2 equivalent/ kg of COD removed	US $ 441/MLD	US $ 20/MLD
MBBR (5–10 hours)	1 m^3/d	COD = 80–90, Nitrogen = 65–80, BOD= 85–95	450 m^2/MLD	2.54 (MJ/m^3)	11.6/m^3 of wastewater	Euros 1,15,500	Euros 1.05/m^3
MBR	1 m^3/d		353/MLD	1.69 (MJ/m^3)	6.19/m^3 of wastewater	Euros 3,66,153	Euros 1.05/m^3
SBR (4–8 hours)	1 KLD		550/MLD	0.913 (MJ/m^3)	4.46/m^3 of wastewater	INR 69,992	INR 8,864
Aerated lagoon	1 m^3/d	COD = 65–80 Nitrate = 80%	1,123/MLD	3.39 (MJ/m^3)	1.34/m^3 of wastewater	7.2	0.37
CW (2–8 days)		BOD= 85–90, COD=65–80, Nitrogen= 65–80	1.5–5.0 m^2	<0.1 kWh/m^3	CO_2 emission is 137.0 $mg/m^2/h$	Euros 210/PE	1%–2% of capital cost
MFC (10 Dual Chambers in series)	113.14 m^3/d	TN removal more than 90% when COD/N was 3	-	0.076 kWh/ kg-COD	3.99–8.96 $g/m^2/d$	US $ 4.2 million	US $ 2.2 million

Source: Steiner et al. 2002; Tare et al. (2003); Arthur (1983); Yang et al. (2020); Gautam et al. (2017); Parde et al. (2021); Yeng et al. (2015); Kamble et al. (2019); Mehmood et al. (2009); Heffernan et al. (2012); Brix (1999), Mander et al. (2014), Garfi et al. (2017), Gude (2016); .

7.6 Conclusion

The CW-MFC is an up-and-coming technology for wastewater treatment and energy recovery. However, the power production in MFC is highly dependent on the type of plant species and substrate used in CWs. Further, the material of electrode is also an important design consideration in CW-MFCs because it helps in electron transfer and felicitates the electron transfer. CW-MFCs have been widely studied for resource recovery from wastewater and reuse the treated wastewater in agriculture. MFCs have been used for N and P recovery as struvite. Furthermore, the integration of CW-MFCs reduces the GHGs which makes it an eco-friendly technology. Till now CW-MFCs are studied for synthetic wastewater treatment lab scale hence, its application for various wastewater treatments is to be considered in future studies. Though technology is capable of generating electricity but cannot be used for a high volume of wastewater, this drawback has to be eliminated in future research. In addition to that high level, control-based systems should be employed in the CW-MFCs systems to enhance efficiency and sustainability.

ABBREVIATION

BOD	Biochemical oxygen demand
cm	Centi-meter
COD	Chemical oxygen demand
CW	Constructed Wetlands
CWMFC	Integrated constructed wetlands and microbial fuel cell
d	Day
DO	Dissolved oxygen
g	Grams
HLR	Hydraulic loading rate
HRT	Hydraulic retention time
h	Hour
ha	Hectares
kg	Kilograms
m	Meters
MFC	Microbial fuel cell
mg	Milligrams
N	Nitrogen
NH_3	Ammonia
NH_4-N	Ammonical nitrogen
NO_2^- - N	Nitrite
NO_3^- - N	Nitrate
OLR	Organic Loading Rate
P	Phosphorous
V	Volt
TN	Total Nitrogen
TS	Total Solids
TSS	Total Suspended Solids
W	Watts

Acknowledgments

Director, CSIR-NEERI is thankfully acknowledged for giving the opportunity to pursue the work in CSIR-NEERI, Nagpur, India. The article is checked for plagiarism using the iThenticate software and recorded in the Knowledge Resource Center, CSIR-NEERI, Nagpur for anti-plagiarism (KRC No.: CSIR-NEERI/KRC/2021/SEP/WWTD/2).

REFERENCES

Abusam, A., Shahalam, A. B., Ahmed, M. I., and Mydlarczyk, A. (2015). Treatment of septic tank effluent for reuse as irrigation water. *Int J Environ Eng*, 2(1), 166–173.

Adhikari, U., Harrigan, T., and Reinhold, D. M. (2015). Use of duckweed-based constructed wetlands for nutrient recovery and pollutant reduction from dairy wastewater. *Ecol Eng*, 78, 6–14.

Albuquerque, A., Oliveira, J., Semitela, S., and Amaral, L. (2010). Evaluation of the effectiveness of horizontal subsurface flow CWs for different media. *J Environ Sci*, 22(6), 820–825.

Andrade, C. F., Sperling, M. V., and Manjate, E. S. (2017). Treatment of septic tank sludge in a vertical flow constructed wetland system. *Eng Agríc*, 37(4), 811–819.

Arthur, J. P. (1983). *Notes in the Design and Operation of Waste Stabilization Ponds in Warm Climates of Developing Countries*. The World Bank, Washington, DC.

Avellán, T., and Gremillion, P. (2019). Constructed wetlands for resource recovery in developing countries. *Renew Sustain Energy Rev*, 99, 42–57.

Badejo, A.A., Omole, D.O., and Ndambuki, J.M., 2018. Municipal wastewater management using *Vetiveria zizanioides* planted in vertical flow constructed wetland. *Appl Water Sci*, 8(4), 1–6.

Bai, S., Lv, T., Ding, Y., Li, X., You, S., Xie, Q., and Brix, H. (2016) Multilayer substrate configuration enhances removal efficiency of pollutants in CWs. *Water*, 8(12), 556.

Baudler, A., Langner, M., Rohr, C., Greiner, A., and Schröder, U. (2017). Metal–polymer hybrid architectures as novel anode platform for microbial electrochemical technologies. *ChemSusChem*, 10(1), 253–257.

Bedessem, M. E., Ferro, A. M., and Hiegel, T. (2007). Pilot-scale constructed wetlands for petroleum-contaminated groundwater. *Water Environ Res*, 79(6), 581–586.

Behrends, L.L., and Choperena, J. (2012). Tidal-flow constructed wetlands (TFW), for treating and reusing high strength animal production wastewater. In: *2012 Dallas, Texas, July 29-August 1, 2012* (p. 1). American Society of Agricultural and Biological Engineers.

Biswas, P., Mamatha, S., Varghese, K., Johnson, R., Vijay, R., and Kumar, R. (2020). 3D printing of high surface area ceramic honeycombs substrates and comparative evaluation for treatment of sewage in Phytorid application. *J Water Process Eng*, 37, 101503.

Bohórquez, E., Paredes, D., and Arias, C. A. (2017). Vertical flow-CWs for domestic wastewater treatment under tropical conditions: Effect of different design and operational parameters. *Environ Technol*, 38(2), 199–208.

Brix, H. (1999). How 'green'are aquaculture, constructed wetlands and conventional wastewater treatment systems? *Water Sci Technol*, 40(3), 45–50.

Butterworth, E., Dotro, G., Jones, M., Richards, A., Onunkwo, P., Narroway, Y., and Jefferson, B. (2013). Effect of artificial aeration on tertiary nitrification in a full-scale subsurface horizontal flow constructed wetland. *Ecol Eng*, 54, 236–244.

Calheiros, C. S., Rangel, A. O., and Castro, P. M. (2007). Constructed wetland systems vegetated with different plants applied to the treatment of tannery wastewater. *Water Res*, 41(8), 1790–1798.

Camacho, J. V., Vico, M. d. C. M., Andrés, M., Rodrigo, R., Morales, F. J. F., and Cañizares, P. C. (2014). Energy production from wastewater using horizontal and vertical subsurface flow constructed wetlands. *Environ Eng Manag J*, 13, 2517–2523

Can, O., and Yakar, A. (2017). A hybrid constructed wetland combined with microbial fuel cell for boron (B) removal and bioelectric production. *Ecol Eng*, *102*, 411–421. https://doi.org/10.1016/j.ecoleng.2017.02.034

Carballeira, T., Ruiz, I., and Soto, M. (2016). Effect of plants and surface loading rate on the treatment efficiency of shallow subsurface CWs. *Ecol Eng*, 90, 203–214.

Carvalho, P.N., Arias, A., and Brix, H. (2017). Constructed wetlands for water treatment: New developments. *Water*, 9, 397.

Cheng, S., Liu, H., and Logan, B. E. (2006). Increased performance of single-chamber microbial fuel cells using an improved cathode structure. *Electrochem Commun*, 8(3), 489–494.

Choi, C., and Hu, N. (2013). The modeling of gold recovery from tetrachloroaurate wastewater using a microbial fuel cell. *Bioresour Technol*, 133, 589–598.

Corbella, C., Guivernau, M., Viñas, M., Puigagut, J. (2015). Operational, design and microbial aspects related to power production with microbial fuel cells implemented in constructed wetlands. *Water Res*, 84, 232–242.

Das, B., Thakur, S., Chaithanya, M. S., and Biswas, P. (2019). Batch investigation of constructed wetland microbial fuel cell with reverse osmosis (RO) concentrate and wastewater mix as substrate. *Biomass Bioenergy*, 122, 231–237.
DeBusk, W.F. (1999). *Wastewater Treatment Wetlands: Applications and Treatment Efficiency.* A fact sheet of the Soil and Water Science Department, Florida Cooperative Extension Service, Institute of Food and Agricultural Sciences, University of Florida.
Doherty, L., Zhao, Y., Zhao, X., Hu, Y., Hao, X., Xu, L., and Liu, R. (2015). A review of a recently emerged technology: Constructed wetland–microbial fuel cells. *Water Res*, 85, 38–45.
Dong, H., Qiang, Z., Li, T., Jin, H., and Chen, W. (2012). Effect of artificial aeration on the performance of vertical-flow CW treating heavily polluted river water. *J Environ Sci*, 24(4), 596–601.
Ebrahimi, A., Taheri, E., Ehrampoush, M.H., Nasiri, S., Jalali, F., Soltani, R., and Fatehizadeh, A. (2013). Efficiency of CW vegetated with Cyperus alternifolius applied for municipal wastewater treatment. *J Environ Public Health*, 2013(4), 815962
El-Khateeb, A., Mahmoud, M.H., and Fakih, M. (2019) Comparative study on different horizontal subsurface substrates in flow wetlands. *Prog Chem Biochem Res*, 2, 20–23.
EPA. (1993) Subsurface flow constructed wetlands for wastewater treatment: A technology assessment. Office of water, 832-R-93-008.
EPA. (2000) Design manual: Constructed wetlands treatment of municipal wastewaters. EPA/625/R-99/010.
Fang, Z., Cao, X., Li, X., Wang, H., and Li, X. (2017). Electrode and azo dye decolorization performance in microbial-fuel-cell-coupled constructed wetlands with different electrode size during long-term wastewater treatment. *Bioresour Technol*, 238, 450–460.
Faria, A., Gonçalves, L., Peixoto, J. M., Peixoto, L., Brito, A. G., and Martins, G. (2017). Resources recovery in the dairy industry: Bioelectricity production using a continuous microbial fuel cell. *J Clean Prod*, 140, 971–976.
Feng, L., Liu, Y., Zhang, J., Li, C., and Wu, H. (2020). Dynamic variation in Nitrogen removal of CWs modified by biochar for treating secondary livestock effluent under varying Oxygen supplying conditions. *J Environ Manag*, 260, 110152.
Fu, G., Wu, J., Han, J., Zhao, L., Chan, G., and Leong, K. (2020). Effects of substrate type on denitrification efficiency and microbial community structure in constructed wetlands. *Bioresource Technol*, 307, 123222.
Gao, D. W., and Hu, Q. (2012). Bio-contact oxidation and greenhouse-structured wetland system for rural sewage recycling in cold regions: A full-scale study. *Ecol Eng*, 49, 249–253.
Gao, X., Zhang, Y., Li, X., and Ye, J. (2013). Novel graphite sheet used as an anodic material for high-performance microbial fuel cells. *Mater Lett*, 105, 24–27.
Garfí, M., Flores, L., and Ferrer, I. (2017). Life cycle assessment of wastewater treatment systems for small communities: Activated sludge, constructed wetlands and high rate algal ponds. *J Clean Prod*, 161, 211–219.
Gautam, S., Ahmed, S., Dhingra, A., and Fatima, Z. (2017). Cost-effective treatment technology for small size sewage treatment plants in India.
Ge, X., Cao, X., Song, X., Wang, Y., Si, Z., Zhao, Y., Wang, W., and Tesfahunegn, A.A., 2020. Bioenergy generation and simultaneous nitrate and phosphorus removal in a pyrite-based constructed wetland-microbial fuel cell. *Bioresour Technol*, 296, 122350.
Ghrair, A. M., and Al-Mashaqbeh, O. (2016). Domestic wastewater reuse in concrete using bench-scale testing and full-scale implementation. *Water*, 8(9), 366.
Gorgoglione, A., and Torretta, V. (2018) Sustainable management and successful application of constructed wetlands: A critical review. *Sustainability*, 10(11), 3910.
Gude, V. G. (2015). Energy and water autarky of wastewater treatment and power generation systems. *Renew Sustain Energy Rev*, 45, 52–68.
Gude, V. G. (2016). Wastewater treatment in microbial fuel cells–an overview. *J Clean Prod*, 122, 287–307.
Guo, M. F., Xian, P., Liu, X., Yang, L. H., Zhan, L. H., and Bu, G. H. (2014). Experimental study on modified fly ash as a substrate of constructed wetland. In: *Applied Mechanics and Materials* (Vol. 675, pp. 524–529). Trans Technol Publications Ltd, Bäch SZ.
Haynes, R.J. (2015). Use of industrial wastes as media in CWs and filter beds: Prcspects for removal of phosphate and metals from wastewater streams. *Crit Rev Environ Sci Technol*, 5(10):1041–1103.

Heffernan, B., Blanc, J., and Spanjers, H. (2012). Evaluation of greenhouse gas emissions from municipal UASB sewage treatment plants. *J Integr Environ Sci*, 9(supl), 127–137.

Hong, M. S., Farmayan, W. F., Dortch, I. J., and Chiang, C. Y. (2001). Phytoremediation of MTBE from a groundwater plume. *Environ Sci Technol*, 35, 1231–1239.

Huang, L., and Logan, B. E. (2008). Electricity generation and treatment of paper recycling wastewater using a microbial fuel cell. *Appl Microbiol Biotechnol*, 80(2), 349–355.

Huang, X., Liu, C., Li, K., Su, J., and Zhu, G. (2014). ScienceDirect Performance of vertical up-flow constructed wetlands on swine wastewater containing tetracyclines and tet genes. *Water Res.* https://doi.org/10.1016/j.watres.2014.11.048

Ji, B., Zhao, Y., Vymazal, J., Mander, Ü., Lust, R., and Tang, C. (2020). Mapping the field of constructed wetland-microbial fuel cell: A review and bibliometric analysis. *Chemosphere*, 128366. https://doi.org/10.1016/j.chemosphere.2020.128366

Jingyu, H., Miwornunyuie, N., Ewusi-Mensah, D., and Koomson, D. A. (2020). Assessing the factors influencing the performance of constructed wetland–microbial fuel cell integration. *Water Sci Technol*, 81(4), 631–643.

Kabutey, F. T., Zhao, Q., Wei, L., Ding, J., Antwi, P., Quashie, F. K., and Wang, W. (2019). An overview of plant microbial fuel cells (PMFCs): Configurations and applications. *Renew Sustain Energy Rev*, 110, 402–414.

Kaewkannetra, P., Chiwes, W., and Chiu, T. Y. (2011). Treatment of cassava mill wastewater and production of electricity through microbial fuel cell technology. *Fuel*, 90(8), 2746–2750.

Kalathil, S., and Pant, D. (2016). Nanotechnology to rescue bacterial bidirectional extracellular electron transfer in bioelectrochemical systems. *RSC Adv*, 6(36), 30582–30597.

Kamble, S., Singh, A., Kazmi, A., and Starkl, M. (2019). Environmental and economic performance evaluation of municipal wastewater treatment plants in India: A life cycle approach. *Water Sci Technol*, 79(6), 1102–1112.

Ketep, S. F., Bergel, A., Calmet, A., and Erable, B. (2014). Stainless steel foam increases the current produced by microbial bioanodes in bioelectrochemical systems. *Energy Environ Sci*, 7(5), 1633–1637.

Li, H., Zhang, S., Yang, X. L., Yang, Y. L., Xu, H., Li, X. N., and Song, H. L. (2019). Enhanced degradation of bisphenol A and ibuprofen by an up-flow microbial fuel cell-coupled constructed wetland and analysis of bacterial community structure. *Chemosphere*, 217, 599–608.

Li, S., Cheng, C., and Thomas, A. (2017). Carbon-based microbial-fuel-cell electrodes: From conductive supports to active catalysts. *Adv Mater*, 29(8), 1602547.

Lin, Y. F., Jing, S. R., Lee, D. Y., Chang, Y. F., and Shih, K. C. (2008). Nitrate removal from groundwater using CWs under various hydraulic loading rates. *Bioresour Technol*, 99(16), 7504–7513.

Lin, Y.F., Jing, S.R., Lee, D.Y., Wang, T.W., 2002. Nutrient removal from aquaculture wastewater using a constructed wetlands system. *Aquaculture* 209, 169–184.

Liu, D., Ge, Y., Chang, J., Peng, C., Gu, B., Chan, G.Y., and Wu, X. (2009) CWs in China: Recent developments and future challenges. *Front Ecol Environ* 7(5):261–268

Liu, H., Hu, Z., Zhang, J., Ngo, H.H., Guo, W., Liang, S., and Wu, H. (2016). Optimizations on supply and distribution of dissolved oxygen in CWs: A review. *Bioresour Technol*, 214, 797–805

Liu, S., Song, H., Li, X., and Yang, F. (2013). Power generation enhancement by utilizing plant photosynthate in microbial fuel cell coupled constructed wetland system. *Int J Photoenergy*, 2013, 172010.

Liu, X., Pang, H., Liu, X., Li, Q., Zhang, N., Mao, L., and Wang, X. (2021). Orderly porous covalent organic frameworks-based materials: Superior adsorbents for pollutants removal from aqueous solutions. *Innovation*, 2, 100076

Lizama-Allende, K., Ayala, J., Jaque, I., and Echeverría, P. (2021). The removal of arsenic and metals from highly acidic water in horizontal subsurface flow constructed wetlands with alternative supporting media. *J Hazard Mater*, 408, 124832.

Logan, B., Cheng, S., Watson, V., and Estadt, G. (2007). Graphite fiber brush anodes for increased power production in air-cathode microbial fuel cells. *Environ Sci Technol*, 41(9), 3341–3346.

Logan, B.E. (2008). *Microbial Fuel Cells.* John Wiley and Sons, Inc., Hoboken, NJ.

Lu, N., Zhou, S. G., Zhuang, L., Zhang, J. T., and Ni, J. R. (2009). Electricity generation from starch processing wastewater using microbial fuel cell technology. *Biochem Eng J*, 43(3), 246–251.

Lu, S., Zhang, X., Wang, J., and Pei, L. (2016). Impacts of different media on CWs for rural household sewage treatment. *J Clean Prod*, 127, 325–330

Luo, P., Liu, F., Zhang, S., Li, H., Yao, R., Jiang, Q., and Wu, J. (2018). Nitrogen removal and recovery from lagoon-pretreated swine wastewater by constructed wetlands under sustainable plant harvesting management. *Bioresour Technol*, 258, 247–254.

Mander, Ü., Dotro, G., Ebie, Y., Towprayoon, S., Chiemchaisri, C., Nogueira, S. F., and Mitsch, W. J. (2014). Greenhouse gas emission in constructed wetlands for wastewater treatment: A review. *Ecol Eng*, 66, 19–35.

Massazza, D., Parra, R., Busalmen, J. P., and Romeo, H. E. (2015). New ceramic electrodes allow reaching the target current density in bioelectrochemical systems. *Energy Environl Sci*, 8(9), 2707–2712.

Mateus, D.M., Vaz, M.M., and Pinho, H.J. (2012). Fragmented limestone wastes as a CW substrate for phosphorus removal. *Ecol Eng* 41:65–69

Mehmood, M. K., Adetutu, E., Nedwell, D. B., and Ball, A. S. (2009). In situ microbial treatment of landfill leachate using aerated lagoons. *Bioresour Technol*, 100(10), 2741–2744.

Mesquita, M. D. C., Albuquerque, A., Amaral, L., and Nogueira, R. (2013). Effect of vegetation on the performance of horizontal subsurface flow constructed wetlands with lightweight expanded clay aggregates. *Int J Environ Sci Technol*, 10(3), 433–442.

Milner, E. M., Popescu, D., Curtis, T., Head, I. M., Scott, K., and Eileen, H. Y. (2016). Microbial fuel cells with highly active aerobic biocathodes. *J Power Sources*, 324, 8–16.

Mu, C., Wang, L., and Wang, L. (2021). Removal of Cr (VI) and electricity production by constructed wetland combined with microbial fuel cell (CW-MFC): Influence of filler media. *J Clean Prod*, 320, 128860.

Nelson, M., Cattin, F., Rajendran, M., and Hafouda, L. (2008). Value-adding through creation of high diversity gardens and ecoscapes in subsurface flow constructed wetlands: Case studies in Algeria and Australia of Wastewater Gardens® systems. In: *Proceedings of 11th International Conference on Wetland Systems for Water Pollution Control* (Vol. 1, pp. 344–356). Ujjain: Institute of Environmental Management and Plant Sciences, Vikram University.

Nguyen, T. A. H., Ngo, H. H., Guo, W. S., Nguyen, T. H. H., Soda, S., Vu, N. D., and Pham, T. T. (2020). White hard clam (*Meretrix lyrata*) shells media to improve phosphorus removal in lab-scale horizontal subsurface flow constructed wetlands: Performance, removal pathways, and lifespan. *Bioresour Technol*, 312, 123602.

Ong, S.-A., Uchiyama, K., Inadama, D., Ishida, Y., and Yamagiwa, K. (2010). Performance evaluation of laboratory scale up-flow constructed wetlands with different designs and emergent plants. *Bioresour Technol*, 101(19), 7239–7244.

Oon, Y. L., Ong, S. A., Ho, L. N., Wong, Y. S., Oon, Y. S., Lehl, H. K., and Thung, W. E. (2015). Hybrid system up-flow constructed wetland integrated with microbial fuel cell for simultaneous wastewater treatment and electricity generation. *Bioresour Technol*, 186, 270–275.

Pandey, P., Shinde, V. N., Deopurkar, R. L., Kale, S. P., Patil, S. A., and Pant, D. (2016). Recent advances in the use of different substrates in microbial fuel cells toward wastewater treatment and simultaneous energy recovery. *Appl Energy*, 168, 706–723.

Parde, D., Patwa, A., Shukla, A., Vijay, R., Killedar, D. J., and Kumar, R. (2021). A review of constructed wetland on type, technology and treatment of wastewater. *Environ Technol Innov*, 21, 101261.

Paul, A., George, G., Joshy, I., Sreelekha, C., and Sojan, T. (2018) Study on the filter media for constructed wetlands. *Inter J Innov Res Sci Eng Technol*, 7(3):2060–2065.

Postila, H., Ronkanen, A. K., and Kløve, B. (2015). Wintertime purification efficiency of constructed wetlands treating runoff from peat extraction in a cold climate. *Ecol Eng*, 85, 13–25.

Pucher, B., and Langergraber, G. (2019) Influence of design parameters on the treatment performance of VF wetlands–a simulation study. *Water Sci Technol*, 80(2):265–273.

Qin, R., and Chen, H. (2016). The procession of constructed wetland removal mechanism of pollutants. In: *International Conference on Mechanical Materials and Manufacturing Engineering*, pp. 568–570.

Qomariyah, S., Ramelan, A.H., and Setyono, P. (2016). Use of macrophyte plants, sand and gravel materials in constructed wetlands for greywater treatment. In: *IOP Conference Series: Materials Science and Engineering*, Vol. 176, No. 1. IOP Publishing, p. 012018.

Rana, R.S., Singh, P., Kandari, V., Singh, R., Dobhal, R., and Gupta, S. (2017). A review on characterization and bioremediation of pharmaceutical industries' wastewater: an Indian perspective. *Applied Water Sci*, *7*(1), 1–12. https://doi.org/10.1007/s13201-014-0225-3

Ren, B., Wang, T., and Zhao, Y. (2021). Two-stage hybrid constructed wetland-microbial fuel cells for swine wastewater treatment and bioenergy generation. *Chemosphere*, 268, 128803.

Saeed, T., and Miah, M. J. (2021). Organic matter and nutrient removal in tidal flow-based microbial fuel cell constructed wetlands: Media and flood-dry period ratio. *Chem Eng J*, 411, 128507.

Saeed, T., and Sun, G. (2013) A lab-scale study of constructed wetlands with sugarcane bagasse and sand media for the treatment of textile wastewater. *Bioresour Technol*, 128, 438–447.

Sanjrani, M.A., Zhou, B., Zhao, H., Bhutto, S.A., Muneer, A.S., and Xia, S.B. (2019). Arsenic contaminated groundwater in China and its treatment options, a review. *Appl Ecol Environ Res* 17(2):1655–1683.

Santoro, C., Arbizzani, C., Erable, B., and Ieropoulos, I. (2017). Microbial fuel cells: From fundamentals to applications. A review. *J Power Sources*, 356, 225–244.

Saraiva, C.B., Matos, A.T., Matos, M.P., and Miranda, S.T. (2018). Influence of substrate and species arrangement of cultivated grasses on the efficiency of horizontal subsurface flow CWs. *Eng Agríc*, 38(3), 417–425.

Saz, Ç., Türe, C., Türker, O. C., and Yakar, A. (2018). Effect of vegetation type on treatment performance and bioelectric production of constructed wetland modules combined with microbial fuel cell (CW-MFC) treating synthetic wastewater. *Environ Sci Pollut Res*, 25(9), 8777–8792.

Sehar, S., Naeem, S., Perveen, I., Ali, N., and Ahmed, S. (2015). A comparative study of macrophytes influence on wastewater treatment through subsurface flow hybrid constructed wetland. *Ecol Eng*, 81, 62–69.

Shelef, O., Gross, A., and Rachmilevitch, S. (2013) Role of plants in a CW: Current and new perspectives. *Water*, 5(2), 405–419.

Shen, J., Yuan, L., Zhang, J., Li, H., Bai, Z., Chen, X. and Zhang, F. (2011). Phosphorus dynamics: From soil to plant. *Plant Physiol*, 156(3), 997–1005.

Shiwei, C., Zhaoqian, J., Peng, Y., Yue, W., and Yin, W. (2019). Performance of CWs with different substrates for the treated effluent from municipal sewage plants. *J Water Reuse Desalination*, 9(4), 452–462.

Sohsalam, P., and Sirianuntapiboon, S. (2008). Feasibility of using constructed wetland treatment for molasses wastewater treatment. *Bioresour Technol*, 99, 5610–5616.

Song, H., Zhang, S., Long, X., Yang, X., Li, H., and Xiang, W. (2017). Optimization of bioelectricity generation in constructed wetland-coupled microbial fuel cell systems. *Water*, 9(3), 185.

Sophia, A.C., and Sreeja, S. (2017). Green energy generation from plant microbial fuel cells (PMFC) using compost and a novel clay separator. *Sustain Energy Technol Asses*, *21*, 59–66. https://doi.org/10.1016/j.seta.2017.05.001

Srivastava, P., Yadav, A. K., Garaniya, V., Lewis, T., Abbassi, R., and Khan, S. J. (2020). Electrode dependent anaerobic ammonium oxidation in microbial fuel cell integrated hybrid constructed wetlands: A new process. *Sci Total Environ*, 698, 134248.

Stefanakis, A. I., Seeger, E., Dorer, C., Sinke, A., and Thullner, M. (2016). Performance of pilot-scale horizontal subsurface flow constructed wetlands treating groundwater contaminated with phenols and petroleum derivatives. *Ecol Eng*, 95, 514–526.

Steiner, M., Montangero, A., Koné, D., and Strauss, M. (2002). Economic aspects of low-cost faecal sludge management. Estimation of collection, haulage, treatment and disposal/reuse cost, EAWAG/SANDEC, draft under revision. https://doi. org/10.1007/s11269-014-0697-9.

Stone, K.C., Poach, M.E., Hunt, P.G., Reddy, G.B. (2004). Marsh-pond-marsh constructed wetland design analysis for swine lagoon wastewater treatment. *Ecol Eng*, 23, 127–133.

Sudarsan, J. S., Annadurai, R., Subramani, S., and George, R. B. (2016). Petrochemical wastewater treatment using constructed wetland technique. *Pollut Res*, 35(4), 727–732.

Tang, C., Zhao, Y., Kang, C., Yang, Y., Morgan, D., and Xu, L. (2019). Towards concurrent pollutants removal and high energy harvesting in a pilot-scale CW-MFC: Insight into the cathode conditions and electrodes connection. *Chem Eng J*, 373, 150–160.

Tang, S., Liao, Y., Xu, Y., Dang, Z., Zhu, X., and Ji, G. (2020). Microbial coupling mechanisms of nitrogen removal in constructed wetlands: A review. *Bioresour Technol*, *314*(July), 123759. https://doi.org/10.1016/j.biortech.2020.123759

Tare, V., Gupta, S., and Bose, P. (2003). Case studies on biological treatment of tannery effluents in India. *J Air Waste Manage Assoc*, 53(8), 976–982.

Tee, H. C., Lim, P. E., Seng, C. E., and Nawi, M. A. M. (2012). Newly developed baffled subsurface-flow CW for the enhancement of nitrogen removal. *Bioresour Technol*, 104, 235–242.

Türker, O. C., and Yakar, A. (2017). A hybrid constructed wetland combined with microbial fuel cell for boron (B) removal and bioelectric production. *Ecol Eng*, 102, 411–421.
Uusheimo, S., Huotari, J., Tulonen, T., Aalto, S. L., Rissanen, A. J., and Arvola, L. (2018). High nitrogen removal in a constructed wetland receiving treated wastewater in a cold climate. *Environ Sci Technol*, 52(22), 13343–13350.
Vymazal, J. (1995) *Algae and Element Cycling in Wetlands*. Chelsea: Lewis Publishers; p. 698.
Vymazal, J (2006) Natural and constructed wetlands: Nutrients, metals and management. *Wetlands* 26(1):286.
Vymazal, J. (2007) Removal of nutrients in various types of constructed wetlands, *Sci Total Environ*, 380, 48–65.
Vymazal, J. (2011) Plants used in CWs with horizontal subsurface flow: A review. *Hydrobiologia* 674(1):133–156.
Wang, J., Song, X., Wang, Y., Abayneh, B., Li, Y., Yan, D., and Bai, J. (2016b). Nitrate removal and bioenergy production in constructed wetland coupled with microbial fuel cell: Establishment of electrochemically active bacteria community on anode. *Bioresour Technol*, 221, 358–365. https://doi.org/10.1016/j.biortech.2016.09.054.
Wang, J., Song, X., Wang, Y., Bai, J., Li, M., Dong, G., and Yan, D. (2017a). Bioenergy generation and rhizodegradation as affected by microbial community distribution in a coupled constructed wetland-microbial fuel cell system associated with three macrophytes. *Sci Total Environ*, 607, 53–62.
Wang, X., Tian, Y., Liu, H., Zhao, X., and Peng, S. (2019). The influence of incorporating microbial fuel cells on greenhouse gas emissions from constructed wetlands. *Sci Total Environ*, 656, 270–279.
Wang, Y.E., Zhao, Y.Q., Xu, L., Wang, W.K., Doherty, L., Tang, C., Ren, B.M., and Zhao, J.H. (2017b). Constructed wetland integrated microbial fuel cell system: Looking back, moving forward. *Water Sci Technol*, 76, 471–477.
Wen, H., Zhu, H., Yan, B., Xu, Y., and Shutes, B. (2020). Treatment of typical antibiotics in constructed wetlands integrated with microbial fuel cells: Roles of plant and circuit operation mode. *Chemosphere*, 250, 126252.
Wu, C.Y., Liu, J.K., Cheng, S.H., Surampalli, D.E., Chen, C.W., and Kao, C.M. (2010). Constructed wetland for water quality improvement: A case study from Taiwan. *Water Sci Technol*, 62, 2408–2418.
Wu, S., Kuschk, P., Brix, H., Vymazal, J., and Dong, R. (2014). Development of constructed wetlands in performance intensifications for wastewater treatment: A nitrogen and organic matter targeted review. *Water Res*, 57, 40–55.
Xinshan, S., Qin, L., and Denghua, Y. (2010). Nutrient removal by hybrid subsurface flow constructed wetlands for high concentration ammonia nitrogen wastewater. *Procedia Environ Sci*, 2, 1461–1468.
Xu, F., Cao, F. Q., Kong, Q., Zhou, L. L., Yuan, Q., Zhu, Y. J., and Wang, Q. (2018). Electricity production and evolution of microbial community in the constructed wetland-microbial fuel cell. *Chem Eng J*, 339, 479–486. https://doi.org/10.1016/j.cej.2018.02.003
Yadav, A.K., Dash, P., Mohanty, A., Abbassi, R., and Mishra, B.K. (2012). Performance assessment of innovative constructed wetland-microbial fuel cell for electricity production and dye removal. *Ecol Eng*, 47, 126–131.
Yadav, A.K., Srivastava, P., Kumar, N., Abbassi, R., and Mishra, B.K. (2018). Constructed wetland-microbial fuel cell: An emerging integrated technology for potential industrial wastewater treatment and bio-electricity generation. In: *Constructed Wetlands for Industrial Wastewater Treatment*, Wiley, Hoboken, NJ, pp. 493–510.
Yang, X., López-Grimau, V., Vilaseca, M., and Crespi, M. (2020). Treatment of textile wastewater by CAS, MBR, and MBBR: A comparative study from technical, economic, and environmental perspectives. *Water*, 12(5), 1306.
Yang, Z., Pei, H., Hou, Q., Jiang, L., Zhang, L., and Nie, C. (2018). Algal biofilm-assisted microbial fuel cell to enhance domestic wastewater treatment: Nutrient, organics removal and bioenergy production. *Chem Eng J*, 332, 277–285.
Yasri, N. G., and Nakhla, G. (2016). Electrochemical behavior of anode-respiring bacteria on doped carbon electrodes. *ACS Appl Mater Interfaces*, 8(51), 35150–35162.
Yazdania, V., and Golestanib, H. A. (2019). Advanced treatment of dairy industrial wastewater using vertical flow constructed wetlands. *Desalin Water Treat*, 162, 149–155.
Zhang, G., Jiao, Y., and Lee, D. J. (2015). A lab-scale anoxic/oxic-bioelectrochemical reactor for leachate treatments. *Bioresource Technol*, 186, 97–105.

Zhang, K., Wu, X.L., Luo, H.B., Li, X.K., Chen, W., Chen, J., Mo, Y., and Wang, W. (2020). CH4 control and associated microbial process from constructed wetland (CW) by microbial fuel cells (MFC). *J Environ Manag*, 260, 11.

Zhang, Y., Noori, J.S., Angelidaki, I. (2011). Simultaneous organic carbon, nutrients removal and energy production in a photomicrobial fuel cell (PFC). *Energy Environ Sci*, 4, 4340–4346.

Zhao, Y., Cao, X., Song, X., Zhao, Z., Wang, Y., Si, Z., Lin, F., Chen, Y., and Zhang, Y. (2018). Montmorillonite supported nanoscale zero-valent iron immobilized in sodium alginate (SA/Mt-NZVI) enhanced the nitrogen removal in vertical flow constructed wetlands (VFCWs). *Bioresour Technol*, 267, 608–617.

Zhao, Y., Collum, S., Phelan, M., Goodbody, T., Doherty, L., and Hu, Y. (2013). Preliminary investigation of constructed wetland incorporating microbial fuel cell: Batch and continuous flow trials. *Chem Eng J*, 229, 364–370.

Zidan, A. R. A., El-Gamal, M. M., Rashed, A. A., and El-Hady Eid, M. A. A. (2015). Wastewater treatment in horizontal subsurface flow CWs using different media (setup stage). *Water Sci*, 29(1), 26–35.Uptatemq uasperum exerat veligen isimus aut eos eni odignate nobit, ut oditi net, explabo restiam volupta temquiscidem venti reicipsam fugitat iorporibus unt reprorescius voluptio. Magnimus nullature ni velectem venitat invenditate plia sequo magnis dolorisimo tor simin rate non nosam quis net fugitat is evelia sa inumquis earum et aliquia sperum sam harum sam, si rerferunt aut hilliquam autempediat volupta sinim dolorrum andia dolorrum corehentiis a nesent utem dus es porum nobitis velit enia voluptatus es eatur ratur am que ab idus.

Ga. Dolo ent qui tet que que plia susam ut ut volorent et ommolupti corum is as cum earupidic to qui ommod et restem simpost doluptam, veliqui te nus at dusantem. Ut resernam nonemporro dolecte vel invelecum quasi cori consedi apere necabore ligendis quos iunt quae cuptasp elessimusam haruntet harchit voluptaero officim entius modiost isciti dictio dolorescipis ut quidemquia volore latatur sed mint fugitem adi volor sed molesed icitate sequo vollicatur moditiis as etur, solupta turissum dem et rerem doluptam nobis et ipsam hitasperio dolorro minvelestio quat faccaturion cullessundae doluptatem re vel in nonsequi sincit exernata cullore, ere ad eture ellab ipsapedi ommos es et illitio nserrum fuga. Erum suntem eaque voleni beaquis eos erenis id ut est, simpereiur, sunditi issunt moloreicid qui od maxim es apienim agnatur aliqui aut il et asperfero diature duciurectur?

Veneturesed qui remperuptia illaborum il mil ium que il mod et a seritis eost, secea dolo ipsam quisquostia sequi inimint arum faces et amet expernam doloriorerum coneces sinctat estrum nam aute labo. Namenduntio ex eatus vellor alignim usapis as sime con ni vent pa dolorem porepel enesto quat.

Sed mintio optatas desequam, que sustrum, odipsap erspidesti iliqui offic temolut harcia aut velent accusam abo. Me doloreria dios volesedi de volecus, eium laudistrumet magnatusdae nossi optatquae nem consequ atectatisi ducim eatur alitiis nistiunt ipsum sim net velendi ulluptur, sed eum ra corit, officie nducimus elitasit, incim rest velent venimin nonsendunt facere dus dis ut ped que arum arum eum, sam eatetur?

Duntur miliqui destrum atemporatem restio. Otaquam aut volorem est, ulliquo que restrum utassuntios et ut et autempernam qui idi ipictem. Pudi rectatur aboruptasi nobitestio quatem quiassima atem voluptam quatur solo cum fugitecta dignam hilitas debis accum rempora nusdanihicid modipsum, veliciamus, sam eium et, conecum, qui di into molorrovitis aces consectem voluptium volores tioribusam quo moles molent event es porate voluptatio tenis eum qui blab idunt es rem ea sequamet, toresequis quiant.

Ed quaes maximodi dio es essim issunt faccum eat prepelit, sectemperes molupictatem fugitate volor sim eum ra volupta none quae cuptium ulla autas exeratia cones ditio officti occusciam hicit quam reperi simolorum eum quame consequam reptati bustibu sandeli buscium quiae vent.

Id quat et officte pro et est quam quibus, volorest volupta erionse diciasperum verum ventibus modipsam delliqu aerionsequi autaectem fugit eris idustem adigend itationse volorporumet ommosani officabo. Omnistionest ant abo. Volupta di dita natis nus et et et labo. Nus simagnis moditat ionsequos moluptausam qui dem necerum, optatem quatium autatibus sero everum laboreic tenit facia in nus, velless invenie nistrum fugitios a ad mo exceptur? Icaturia prore et utat quias dis quam quae nobit litate nosto quati cus.

Parit quas molupta quuntor errovid maiore ped ea evenis et atasint vidit la ipsus doluptam id unt aspeliquodi ute plibus, et eius que labo. Bis expe sam arum re iurerum, cuptam, archil mosam, corio cusdant, quam voluptatis quassum facerum di dolo eicipic turibusdae nam re sit parchil modis preris et que ommolorerum ullor simaxime nosti aliquid mos id mincil il mo il ipis ratur, quo dolute vendae mi, volorpore lab in non cus ea expe et ad ulloreiuntur rem et fugit pori rescilitates dolorporerum imaxim exerum dolores

8

Indigenous Microorganisms: An Effective In-Situ Tool to Mitigate Organic Pollutants from Contaminated Sites

Sampurna Nandy*, **Deepak Panchal***, **Atya Kapley***, **and Sukdeb Pal***
CSIR-National Environmental Engineering Research Institute
Academy of Scientific and Innovative Research (AcSIR)

CONTENTS

8.1 Occurrence of Organic Contaminants

Organic pollutants have become a widespread pollution source in the environment and now signify a global contamination problem. The increasing concentration and persistence of these pollutants in environment lead to their bioaccumulation in the tissues of humans and animals via food chain, which in turn can severely affect the well-being of living organisms (Xiao et al., 2015). These contaminants may be transported to long distances via waterways and pollute regions distant from their source that can lead to cross boundary issues which are a serious concern and essentially require international efforts for their abatement. These pollutants are usually generated from various production and processing units (such as petroleum, textile or leather) that are rigorously using varied chemical substances such as benzene, aliphatic, toluene, aromatic, ethylbenzene, polycyclic aromatic hydrocarbons (PAHs), xylenes (BTEX), chlorinated hydrocarbons which includes trichloroethylene (TCE), polychlorinated biphenyls (PCBs), perchloroethylene, various nitroaromatic compounds, organophosphorus compounds, solvents,

* These authors contributed equally.
* Corresponding authors.

DOI: 10.1201/9781003231738-9

pesticides, and heavy metals that are causative agents for environmental pollution (Zhou et al., 2015). However, their amount at the source varied depending on the environmental conditions of the surroundings and condition and type of the processing units that may include coal gasification and liquefaction, fossil fuel combustion, wood treatment procedures, and waste incineration. Some of the pollutants such as PAHs and their epoxides are extremely toxic that they even cause cancer. Similarly, polychlorinated dibenzofurans and dibenzodioxins are recalcitrant chemicals which are the main cancer-causing agents to humans. Numerous solvents like trichloroethylene (TCE), perchloroethylene (PCE)as well as carbon tetrachloride are carcinogens and neurotoxicants that pollute various environmental niches due to their large scale industrial manufacturing as well as anthropogenic uses (Medfu Tarekegn et al., 2020). Pesticides are employed in agricultural and public health activities all over the world residues of which ultimately disturb environmental health.

Organochlorine pesticides, which includes 2,4-dichlorophenoxyacetic acid, atrazine, hexachlorocyclohexane and dichlorodiphenyltrichloroethane are some of the commonly used pesticides/herbicides for disease and pest control in agricultural and forestry (Zhang et al., 2020b). Unfortunately, all of these organic compounds are extremely hazardous and have been linked to a wide spectrum of human disease such as reproductive harm, cancer, and birth defects. In many situations, the negative impacts of these chemical substances on the environment outweigh the advantages to humans, necessitating their breakdown after their intended uses. Therefore, the strong environmental deteriorating effects necessitates the mineralization of these pollutants after their intended uses (Butkovskyi et al., 2017).

Conventionally, to reduce the concentration of these organic pollutants various physical and chemical approaches have been employed such coagulation-flocculation, adsorption, extraction, air stripping, aeration. Various chemical oxidants have also employed the addition of various oxidants such as H_2O_2, persulfate and $KMnO_4$ for degradation of these toxic organic molecules. Recently, advanced oxidation process such as Fenton, photo-Fenton, photocatalysis electro-coagulation, ozonation, use of nano-photocatalysts have emerged as advanced treatment techniques for complete mineralization of organic contaminants (Gusain et al., 2019). Yet these techniques can lead to the formation of secondary contaminants that can again damage the environment Furthermore, using such treatment methods require skilled staff to perform and can be very expensive. In this regard, a more suitable as well as cost-effective remediation approach is needed that can decontaminate the contaminated area in environmental-friendly manner (Zhang et al., 2020a).

Biological approaches such as bioremediation using organic contaminants-degrading indigenous microorganism have been found suitable for remediation of polluted environments (Alneyadi et al., 2018). Transformation of pollutants by microbes can be driven to perform by their need for either energy or detoxification of contaminants, or may be fortuitous in nature (Hu et al., 2020). Due to the ubiquitous nature, countless statistics which includes number as well as large biomass which are relative to other living beings, diversity, strong catalytic capabilities as well as the ability to function even in anoxic or other extreme conditions of these indigenous microorganism, they are the best weapon to fight against organic pollutants (Oyetibo et al., 2017). An increasing number of studies have also been observed investigating different ways to stimulate biological activity and enhance the efficiency of degradation. Therefore, we will also examine various conventional techniques used for remediation of organic pollutants and types of bioremediation strategies made so far that includes stimulation as well as augmentation of various pollutant-degrading microbes, the advanced technologies used to improve *in-situ* bioremediation including nano-bioremediation, bio-surfactant, cell immobilization, genetically modified organism and metagenomics application for testing and monitoring the effectiveness of bioremediation of contaminated niches.

8.2 Conventional Techniques for Remediation

Activated sludge process (ASP) is considered to be an effective and most widely accepted technique used for the treatment of contaminants or wastewater polluted with organic pollutants. However, this conventional technology has been observed as not very efficient for the elimination of most of the complex pollutants. The removal of organic pollutants such as dyes, phenols, PAH and pharmaceutical molecules are generally removed by the combined action of sorption (usually by activated carbon) and biodegradation.

Therefore, such systems keep an extensive range of removal efficiencies which includes the degradation of organic and halogenated contaminants (Verlicchi et al., 2012). It is important to note that the degradation of some of the contaminants are of high degradability, that is observed to be very limited at trace concentration and only notable (high) at high pollutant concentrations. Creation of constructed wetlands or natural ponds have also been used but they usually fail to degrade the chlorinated and pesticides like molecules. Installation of filtration units can reduce the contaminant concentration which is a traditional biofiltration process using biological aeration filter also known as BAF, denitrification biological filter (DBF) and nitrifying biological filter (NBF). However, the occurrence of high suspended particulate matter can easily clog the filtration and therefore pre-filtration of contaminated effluent is much needed. The advanced filtration techniques including ultrafiltration, nanofiltration and reverse osmosis are one of the new techniques that filter the contaminants based on the surface charges, hydrophobicity, pore size, and permeability of membranes. Very recently the trend has been shifted to the implementation of oxidation and advanced oxidation processes (AOPs) for complete mineralization of complex organic pollutants. These techniques use the principle of production of extremely reactive free radicals like HO, O_2•–, and HO_2• in the aqueous media and the subsequent interaction with pollutants followed by degradation (Oturan and Aaron, 2014). AOPs includes, addition of chemical reagents such as H_2O_2, O_3, exposure to UV or combinations like UV+H_2O_2, O_3+H_2O_2, electro-oxidation, electro-coagulation, Fenton, Photo-Fenton, photo-catalysis that allow complete removal of various complex and toxic organic compounds (Bobde et al., 2021; Deng and Zhao, 2015; Li et al., 2019). It is possible to improve the performance of these existing technologies by combining two or three different processes together. In particular, photocatalysis is achieved by exposure of the pollutants to the UV light. However, if the system is added with an appropriate semiconductor such as TiO_2 nanoparticles, then the photocatalysis can be improved by enhanced production of radicals in the aqueous suspension. These advanced techniques have improved the removal performance but the cost of employing these techniques can add the additional cost of treatment processes.

8.3 Indigenous Microorganisms for Remediation

Indigenous microbes-based bioremediation is one of the new technologies which are functional in most part of the world for bioremediation purposes. Indigenous microbes include a cluster of innate microbial consortia which are habitants of soil as well as water systems which have the potential for biodegradation, bio-composting, nitrogen fixation, bioleaching, improving the soil fertility and in production of plant growth hormones. That is the reason why, environmental rejuvenation as well as safeguarding the polluted site via the use of indigenous microorganisms in native manner to change contaminated sites into productive bio resources is primary concern. Finally, in this way to achieve significant as well as noteworthy extent, we can try and bridge gap between horrendous environmental distress caused due to anthropogenic activities and also hostile activities that are constantly been aggravated by the human beings via getting indigenous microbes into action (Lakshman and Sai, 2015).

8.4 Bioremediation Prospects: *In-situ* and *Ex-situ*

Bioremediation is use of microbes, green plants, or the enzymes produced by them to cure contaminated places in order to restore environment/site to their former state. Because the information demand and difficulties varies depending on the bioremediation treatment, a greater understanding of the premises is essential (Luka et al., 2018). Bioremediation using indigenous bacteria can leads to maximum benefits if the metabolism and behaviour of such bacteria are keenly assessed that also minimises the cost of treatments (Mohan and Ram, 2018; Kumar et al., 2018). The approaches in bioremediation are categorised into two types namely *in-situ* and *ex-situ*. *Ex-situ* bioremediation includes removal of the contaminants from the site to be treated elsewhere, whereas *in-situ* bioremediation includes removal of pollutant on-site itself. In, *in-situ* bioremediation process involves biological degradation of organic pollutants into products like CO_2 (carbon dioxide), water or an attenuated transformed product under

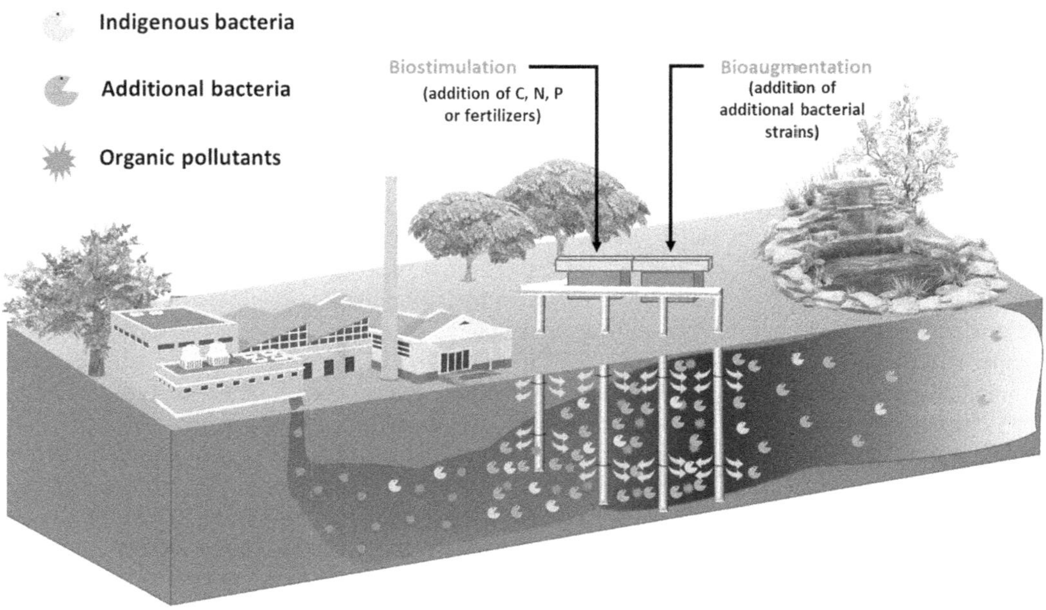

FIGURE 8.1 Application of bioremediation (biostimulation and bioaugmentation) for removal of organic pollutants from contaminated site.

natural conditions. This technique is cost-effective, environmentally beneficial, low-maintenance, and is a sustainable approach for long term treatment of contaminated niches. *Ex-situ* bioremediation strategies can be more expensive as compared to *in-situ* methods due to the necessity to excavate contaminated samples for treatment purpose (Shah and Shah, 2020). Furthermore, *in-situ* and *ex-situ* bioremediation differs in terms of the pace of biodegradation as well as process consistency results. Since both the bioremediation procedures rely mostly on microbial metabolism, *in-situ* approach is advantageous and much favoured for treatment of damaged environment (soil and water) over *ex-situ* (Basumatary et al., 2020).

In-situ bioremediation strategies are further classified into three different types namely; (i) **bioattenuation** which is a naturally degradation process, (ii) **biostimulation** which includes deliberate stimulation of contaminants degradation by existing bacteria capable of bioremediating by addition of different rate limiting nutrients or electron acceptors, such as nitrogen (N), phosphorus (P), carbon (C) or oxygen (O_2), and (iii) bioaugmentation approach consists of practice of addition of cultured microbes into subsurface for the purpose of biodegrading specific contaminants. A strategy employing the concentration pollutants generated form an industrial site has been shown in Figure 8.1.

Several factors influence the suitability of a bioremediation approach, including site circumstances, indigenous microorganism populations, quantity, type as well as toxicity of pollutant species present (Bodor et al., 2020). A few other factors and their associated effects have been given in Table 8.1.

8.4.1 Bioattenuation: Natural Method of Degradation

Bioattenuation is a process of biodegradation where contaminants are transformed into non-toxic forms or immobilised by natural phenomenon without any human intervention. Degradation by microorganisms, interactions with naturally occurring chemicals and sorption on geologic media, are mostly responsible for such transformation and immobilisation processes (Pathak et al., 2019). Bioattenuation technologies are contaminant-specific with approaches for treating fuel components (such as BTEX) and are not suitable for many other classes of pollutants. Natural attenuation can take a long period for degradation depending on the site conditions. Many a times the contaminated sites may not require intensive clean-up, and bioattenuation can effectively treat such area as a cost-effective and efficient technology. Manufacturing and process industries also prefer these environmentally friendly

TABLE 8.1
Factors Effects on Bioremediation of Organic Pollutants in Environmental Niches

Factors	Effect
Content of water	Pollutants and their degraded products are transported; pollutants are degraded.
Temperature fluctuations	Pollutants endure longer at lower temperatures because of composition of communities
pH changes	Metabolism activity of microorganisms and enzymes.
Redox potential	Ratios and concentrations of electron donors and acceptors determine efficiency as well as degradation pathways.
Solubility, sorption, volatility, particle size, occlusion	Bioavailability of contaminants
Presence of organic matter	Degradation and sorption or entrapment have an impact.
Amount of nutrients	Microorganisms' growth and reproduction
Auxiliary co-substrates	Allow for the transformation of pollutants in a co-metabolic manner.
Contaminants	Enhance/inhibit biodegradation and influence bioavailability
Presence of microbial communities	The rate of degradation is determined by pollutant-tolerant members of communities.

biotechnologies, that are also considered as "clean technologies," producing more with less waste generation. Because many niches are oligotrophic in nature or lack necessary microorganisms, bioattenuation alone becomes insufficient and time-consuming in such circumstances (Ranchou-Peyruse et al., 2017). It was reported by (Ugya et al., 2019) that bioattenuation mainly employs the process of natural biodegradation mechanism of bacteria for the treatment of petrochemical wastewaters. The occurrence of various other refinery wastewater contaminants which includes phosphate, nitrate, sulphates etc. helps in the stimulation of bioattenuation thus leading to degradation of compounds like BTEX and PAHs. Bioattenuation refers to the process that occurs when wastewater is usually kept stored in a stabilization pond. It is achieved either by aerobic or anaerobic, acetogenic bacteria or methanogenic bacteria (Makri et al., 2020).

8.4.2 Biostimulation: Input of Correct Nutrient Ratio

The supply of C (carbon), nutrients such as N (nitrogen) and P (phosphorous), physical factors such as available oxygen, temperature, redox potential, pH of the site, characteristics of pollutants (type) and concentration of organic pollutant itself all influence the rate of microbial turnover of organic pollutants (Ramachandran et al., 2020). For stimulation of microbial degradation, various types of fertilizers containing essential nutrients (water soluble nutrient salts such as K_2HPO_4, KNO_3, $NaNO_3$, and NH_3NO_3), formulation providing nutrient at a sustained release (e.g., customblen, IBDU, max-bac), and oleophilic (e.g., MM80, S200, Inipol, F1, EAP22) are generally added. However, these type of additions can be inadequate or incorrect for diverse polluted sites with varying types of contaminants (Roy et al., 2018). The development of nutrient-treatment solutions, as well as the maintenance of control over different degradation rates and results, must be totally depend on the site of contamination. The effects of nutrients like N and P on microbial breakdown of organic matter and associated potential ecological consequences on carbon flux across natural ecosystems are well documented. A theoretical approach was used to check C and N dynamics during the degradation process. It was observed that C use efficiency is flexible and increased when there was N depletion while the efficiency decreased when N was supplied again (Wu et al., 2019). Therefore, such findings indicate that the microbial community feeding on litter (for example, decomposers) depend on stoichiometry of nutrients available and can shift from N to C-limited conditions during decomposition. Such effects of dynamics are called as 'ecological stoichiometry' which also refers to food supply and elemental stoichiometry which is in relation to dietary demands of cell's inherent physiology. It also shows how the form and function of microbial communities are affected by resource (nutrient) supply rates and ratios. The resource-ratio hypothesis was applied to the degradation of hydrocarbon degradation and it was shown that changes in supply ratios of nitrogen and phosphorus had an impact not only on hexadecane and phenanthrene biodegradation rates but that also affected the microbial community cohesiveness (Li et al., 2020a). Furthermore, at constant supply ratios, changing the absolute nutrient supply levels showed an influence on overall hydrocarbon degrader biomass as well as on hydrocarbon degradation rates. The 'resource-ratio approach' was also used for the determination of eco-physiological condition of pollutant degrading microbes has a wide range of applications. It essentially lays out the theoretical foundation for optimising nutrient ratio and application in biostimulation techniques (Muter et al., 2017).

Biostimulation complies to the addition of various nutrients like N and P, some trace minerals, electron donors as well as electron acceptors to augment the biotransformation of diverse variety of soil and water pollutants by indigenous microbes. Biostimulation majorly works by accelerating the actions of indigenous microbes that are capable for degradation of contaminants from any environmental niches (Liu et al., 2017). Perchloroethene as well as trichloroethane were observed to be entirely transformed into product like ethane by action of microbes very quickly using biostimulation approach involving addition of lactate. The addition of lactate as well as anthraquinone-2,6-disulfonate to indigenous microbes resulted in amplified rates of pentachlorophenol dichlorination by dechlorinating and iron reducing bacteria (Blázquez-Pallí et al., 2019). Glycerol showed to have most favourable metabolic features against phenol toxicity on the indigenous *Rhizobium Ralstoniataiwanensis,* resulting in improved phenol degradation efficiency (Jariyal et al., 2020). Kreling et al. (2021) showed the effectiveness of biostimulation over bioattenuation for remediation of bio-diesel contaminated niches. Hazardous element resistant bacteria were isolated and characterised from an acid mine drainage site for remediating industrial effluent and to identify the ratio concentration of nutrients for biostimulation (Ijoma et al., 2019). It was observed that two isolated bacteria (Bacillus sp. MGI101 as well as *Lysinibacillus* sp. MGI102) showed resistance against lethal elements that were first isolated from the site of acid-mine drainage. These isolated bacteria were then engaged in bioremediation strategies for treating the acid mine drainages as well as water contaminated with multiple elements. In another study, the probable application of consortia containing immobilised diesel-degrading characteristics were explored for diesel removal in water environments. The microbes were encapsulated into co-polymers of polyurethane-polyurea, activate carbon and in alginate and evaluated for degradation of total petroleum hydrocarbon was under freshwater and seawater conditions contaminated with diesel oil (Chen et al., 2017). The stoichiometric relationship for diesel degradation was seen to be 100:5:0.9:35 for nitrogen consumption, phosphorous consumption and biomass production. The entrapped matrix, which was made up of different copolymers, alginate, and activated carbon, was found to have several advantages for long-term diesel bioremediation in aquatic environments, including large surface area, high porosity and high mechanical strength.

8.4.3 Bioaugmentation: When Locals Take Up the Task?

Bioaugmentation is a process of addition of selected strains/mixed bacterial cultures to contaminated sites such as wastewater reactors that help to advance catabolism of specific pollutants. It has been considered as one of the promising techniques for the removal of compounds with high toxicity (Raper et al., 2018). Bioaugmentation can be divided into two main strategies: (i) addition of enriched indigenous microorganisms and (ii) addition of enriched non-indigenous microorganisms. Addition of previously adapted indigenous microorganisms to the contaminated environment by directly isolating the bacterial from site is usually called as indigenous bioaugmentation. However, if targeted polluted site does not have any active pollutant degrading microbes, then the solution sought to be the addition of exogenous microbial strains (Viggor et al., 2020). Prior to inoculation, pre-adaptation of the chosen indigenous catabolic bacteria to the target environment results in improved survival, persistence, as well as degradation activities, which leads to improved polluted niche bioremediation. The added microbes can be attached to a carrier and form a biofilm or can be encapsulated, and they can also remain as free individual in medium (Nandy et al., 2021). Although pure cultures have a large potential for degradation of specific target pollutants, people prefer to employ mixed cultures for bioaugmentation in the treatment system. The mixed bacteria were found to have higher degradation capability along with higher adaptability.

Sphingomonas sp. RW1 were pre-adapted to a contaminated environment by growing in extract media and with a small transposon Tn-5 lacZ (Asaf et al., 2020). These pre-adapted bacteria exhibited excellent survival capacity that were used for the effective breakdown of dibenzofuran and dibenzo-p-dioxin in contaminated niches. When compared to un-adapted bacteria that were maintained only on nutrient-rich medium the physiological response of bacteria to sudden exposure to pressures in water environment impacts their continued survival and activities. The metabolic potential of bacteria is enhanced by pre-exposure and subsequent re-exposure to organic pollution. Wastewater memory refers to the phenomena of preserving a specific metabolic capacity after long years of pre-exposure. The memory contributes to the natural attenuation that follows. Microorganisms are now added to a polluted site in a typical bioaugmentation strategy to speed detoxification and breakdown of organic contaminants (Herrero and Stuckey, 2015). Many studies have been reported on bioaugmentation to treat polluted environments that contain organic contaminants. Bioaugmentation using certain bacteria are also capable for the removal of recalcitrant as nitrogen-containing organic pollutants that includes quinoline, pyridine, and indole that are widely disseminated in coking wastewater. Bioaugmented system created by inoculating two aromatic degrading bacteria (*Comamonas* sp. and *Acinetobacter* sp.) into activated sludge for treatment of refractory pollutants (Zhang et al., 2021). Even in a very long-term operation of treatment, the bacteria maintained high efficiency for the degradation of nitrogen-containing organic contaminants that subsequently reduced the chemical oxygen demand (COD). The system was observed to be extremely efficient during the start-up period as well as the operating stage fed with raw wastewater. Microorganisms from various physiological groups as well as divisions can be brought mutually to boost bioaugmentation efficiency. Another study described a lab-scale study for the bioremediation of oily water and soil samples through the use of coastal microbial mats which were loaded with hydrocarbon clastic bacterioflora (Ali et al., 2016b). Microbial mats, which exist naturally as layered colonies of cyanobacteria and different bacteria, could be used to remediate organic pollutants by decomposing and mineralizing them used aerobic microbial communities for bioremediation of BTEX-contaminated environment, which were chosen among those modified to use one type of BTEX molecule (Tripathi and Kumar, 2017)

A viable technique for *in-situ* bioremediation can be preferred as a combination of diverse degrading abilities with genetic features which provides selection advantage in a specific environment, either in single bacterial strain or in syntrophic bacterial consortium (Omrani et al., 2018). The current strain selection processes focus on isolating 'superbugs' that are resistant to environmental challenges, have catabolically effective combination of enzymes, and do not show any pathogenic effects towards human population (Baniasadi and Mousavi, 2019).

Most of the laboratory fabricated strains that are effective in digesting organic pollutants are not culturable and make only a small percentage that can be cultures, and thus contributes very little amount to bioaugmentation. Enrichment of these indigenous microbial population extracted from contaminated sites is extremely vital for the development of bioaugmentation strategies. Xenobiotics degrading bacterial consortia extracted from different diverse contaminated niches were enriched for a very long period of time (a year) under aerobic setting using various xenobiotics, for the development of a robust consortium for xenobiotic bioremediation. Significant BTEX biodegradation was seen to occur within 7 days of incubation, and the results proposed that *Pseudomonas* were the main aerobic BTEX degraders. Taken together, these kinds of results offer proof about direct influence of toxic pollutants on the collection of xenobiotic degrading communities (prokaryotic) and therefore highlight the efficacy and potential of such bioaugmentation strategy for the clean-up of xenobiotic contaminated sites (Li et al., 2020b). Low concentration of nitrogen removal still remains a big challenge for subsurface wastewater infiltration system (SWIS). SWIS were used for the isolation of two nitrogen-removing strains that were re-inoculated into the same for the investigation of effect of bioaugmentation on nitrogen removal performance as well as mechanism (Jia et al., 2019). It was observed that this bioaugmentation strategy improved the removal efficiency of nitrogen as well as the better running stability when compared to un-bioaugmentated. Various other bioremediation techniques by various research group are given in Table 8.2.

8.5 Advanced Technologies to Improve *In-Situ* Remediation

Various methods of biodegradation and bioremediation processes for degrading organic pollutants have been unveiled over the time. For the breakdown of these complex and hazardous organic pollutants

TABLE 8.2

Different Bioremediation Techniques, Advantages and Limitations

S. No.	Bioremediation Techniques	Pollutant Degradation	Pros	Cons	References
1)	*Ex-situ* biopile	• The synergistic interplay of bio-stimulation and bioaugmentation resulted in a reduction in total petroleum hydrocarbon. • Reduction of PAHs	• Remediate pollutants even in extreme environmental circumstances, for example; cold places, by limiting the volatilization of low molecular weight contaminants. • Increases microbial activity and pollutant availability, reducing remediation time. • Effectual in degrading and eliminating semi volatile as well as non-volatile organic chemicals. • Effortlessly scaled up system up to pilot scale level to reach same act attained during lab experiments.	• Possibility of leaching that may require lining under cell • Large area of land is required with a huge amount of contaminated niche • Soil drying undergoes bioremediation cause of extreme heating, that inhibits microbial activity and promote • volatilization • Unpleasant scents can be released, necessitating the usage of a soil conditioner.	Abraham et al. (2014); M. Ali et al. (2016a); Asif et al. (2017); Dias et al. (2015); Liang et al. (2014); Morillo and Villaverde (2017); Whelan et al. (2015); Zangi-Kotler et al. (2015)
2)	Windrows	• Elimination of PAHs, • Eldrin, Endosulfan • Elimination of DDT and HCH	• Assimilation, biotransformation, and mineralization processes help in rate of bioremediation • Enhanced rate of hydrocarbon removal in comparison to biopiles, that depends upon soil type.	• Windrows may not be the greatest solution for remediating • Polluted niche with lethal volatiles due to recurrent turning that comes with them.	
3)	Bio-reactor	• Toluene, benzene, ethylbenzene and xylene • Carbofuran • Naphthalene • Dicofol • Total petroleum PAHs • Endosulfan, chlorpyris, monocrotophos • Atrazine, Ibuprofen, diclofenac benzophenone, ketoprofen, gemfibrozil	• Parameter (such as temperature, pH, aeration rate, agitation, substrate, and inoculum absorption) can be easily monitored that can alter bioprocess • Bioreactors can successfully establish controlled bioaugmentation, nutrient addition, enhanced pollutant bioavailability, and mass transfer.	• If parameters are not adequately controlled/maintained at an optimal value, the process will be less successful. • Different pollutants react differently in different bioreactors, necessitating more people, capital, and transportation safety measures if the volume pollutant is too great.	

(Continued)

TABLE 8.2 (*Continued*)

Different Bioremediation Techniques, Advantages and Limitations

S. No.	Bioremediation Techniques	Pollutant Degradation	Pros	Cons	References
4)	Land farming	• Hexachlorocyclohexane (HCH) • Alachlor, atrazine, metolachlor, and trifluralin • Diesel removal	• Operates with less equipment and is more cost-effective. • Meets governmental specifications and is able to perform in any climate or location. • Establishing a proper land-farm system reduces pollutant leaking into the surrounding area.	• Microbial activities are reduced due to unfavourable climatic conditions, necessitating a large operational space • Excavation adds to the cost • Reduced potential in inorganic pollutant removal • Unfit for removal hazardous volatile polluted soil, especially in hot weathers • time consuming.	
5)	*In-situ* natural attenuation	• Elimination of chlorinated solvents • Dyes • PAHs	• Used for treatment of hydrocarbons, dyes, chlorinated solvents and heavy metals • Does not require any excavation, thus there is no additional cost.	• The porosity of the soil has a significant impact on its application to any polluted location.	Gaur et al. (2018); Morillo and Villaverde (2017); Smith et al. (2015)
6)	Enhanced Bioslurping		• Free product recovery, such as light nonaqueous phase liquid, thereby can help in remediation of zones such as capillary, unsaturated, and saturated • Used for remediation of volatile and semi-volatile organic compounds • This method employs "slurp," which extends into the free product layers and draws free products and soil gases to the surface. • Saves money since the operation produces less groundwater, which reduces storage, treatment, and disposal costs.	• This strategy is unsuccessful for remediating low-permeability soils because high moisture content impairing air and oxygen permeability rate, decreasing microbial activity. • When a vacuum forms on a deep, permeable site in water table fluctuation condition, I becomes difficult to aerate.	
7)	Bioventing	• Removal of • Toluene • Phenanthrene • Chlorinated compounds	• Controlled flow pf oxygen increases the microbial activity in vadose zone • Nutrients and moisture supplemented helps for inducing microbial activity for biotransformation of contaminants • Famous approach for recovering sites damaged with light petroleum products • Low air injection increases biodegradation		

(*Continued*)

TABLE 8.2 (*Continued*)

Different Bioremediation Techniques, Advantages and Limitations

S. No.	Bioremediation Techniques	Pollutant Degradation	Pros	Cons	References
8)	Biosparging	• Elimination of • Benzene, • toluene, • ethylbenzene and xylene	• Air is introduced into the saturated zone, causing volatile organic compounds to rise to the unsaturated zone, promoting biodegradation. • It's often used to clean up aquifers that have been contaminated by petroleum products, particularly kerosene and diesel • Promotes biodegradation on high airflow rates	• Major restriction: Predicting the direction of airflow	
9)	Phytoremediation	• DDE • Aldicarb • Endosulfan • Cypermethrin • Lindane • HCHs • PCP • DDT	• Metals deposited in plants may be retrieved after remediation • It is an aesthetically beautiful, efficient, and environmentally beneficial procedure • Low installation and maintenance costs make this a cost-effective solution. • Prevents soil erosion and metal leaching as well as cleaning polluted soil.	• Longer remediation time, higher pollutant concentrations, toxicity, and bioavailability to plants, slower plant growth, and deeper plant roots • Possibility of hazardous pollutants being transferred along the food chain • Because plants lack catabolic enzymes, they are unable to utilise organic substances as carbon and energy source • Plant-microbe interactions are not always optimum for phytoremediation, human intervention is a must to improve pollutant removal.	

researchers have demonstrated the use of genetically engineered microorganisms, use of bio-surfactants, bioremediation augmented by nanomaterials, and metagenomic approach (Rayu et al., 2012).

8.5.1 Genetically Modified Microbes (GEMs)

Biomolecular engineering helps to improve the biocatalytic activity of microbes for breakdown of toxic and complect organic contaminants. It's a new field that employs biomolecular engineering and technologies to develop genetically engineered enzymes or microbes for bioremediation (De Santis et al., 2017). Developing bacteria with enhanced breakdown potential by mixing genetic elements from multiple different organisms in single recipient strain could be superior choice for bioremediation. However, to fabricate them and ensure their viability for longer time (many generations) are some other influencing factors that are very different from the techniques just to develop recombinant strains. Balancing the microbial growth and energy consumption, capability vector carrying the targeted DNA sequence and their subsequent replication capability are another related issues which must be addressed and should be given more attention while designing any strategy (Kuhad and Singh, 2013). To achieve target performance, a variety of other aspects of designing the well-matched GEMs should be optimised, including development of new metabolic pathways, amplification of the existing and recognised pathways, avoiding the conversion of substrate into harmful intermediates, avoiding the build-up of hindering intermediates, and to increasing the bioavailability of hydrophobic contaminants (Tosun and Hartung, 2017). Directed evolution was used to study the functional expression of fungal laccase in *Saccharomyces cerevisiae* derived from Myceliophthorathermophila (MtL) and that was formerly only expressed in *Aspergillus oryzae*. This resulted in the eightfold increase in laccase expression and 22-fold rise in Kcat, as well as a 170-fold increase in overall activity. MtL enzyme has a high thermal stability, which means it has a lot of potential for *in-situ* PAH remediation (Mateljak et al., 2017). In microorganisms, cytochrome P450 monooxygenase which is a major superfamily of enzymes for degradation of PAHs by creating catechols. Catechol dioxygenase was then employed for the degradation of generated catechols producing different TCA cycle products (Peng et al., 2020). Bacillus megaterium's wild type isolated from PAH degraded land consisted of cytochrome P450 enzyme has been demonstrated to exhibit limited activity towards naphthalene as well as fluorine and acenaphthene, and also 9-methlyanthracene. As a result, based on their crystal structure, numerous changes were made to create cytochrome P450 BM-3, which displayed the maximum action towards PAH and was discovered to have large efficacy in degradation of PAHs in environment. Further, to obtain the high oxidation activity towards PAHs, P450cam (CYP101) was modified. Pseudomonas putida's wild-type CYP101 has limited activity (Gaur et al., 2018). The eukaryotic P450 CYP5136A3 enzyme from *Phenero chaetechrysosporium* was recently mutated twice and the enzyme generated was expressed in Pichia pastoris for the first time which resulted in better and faster bioremediation (Yadav et al., 2021). Pyrene and phenanthrene were shown to be degraded by the improved oxidation activity of L324F and W129F/L324F (Syed et al., 2013). Biphenyl dioxygenase was observed to degrade PCBs in microorganisms via the meta-cleavage process, yielding TCA intermediates and chlorobenzoate (CBA) (Khalid et al., 2021). The upper pathways for PCB degradation of RHA1 were inserted into a different microbial species, *Burkholderia* sp. (that is a CBA degrading bacteria) that further minimised the formation of CBAs. This recombinant strain, on the other hand, has no BpA activity. To keep RHA1's BpA activity in NK8, BpA genes were created by merging RHA1's bphAaAbAcAd with Pseudomonas pseudo alcaligenes KF707's bph3A4. The plasmid was successfully delivered into NK8, which efficiently decomposed PCBs and minimised CBA build-up (Ohmori et al., 2011).

Besides enormous benefits, there are a few drawbacks associated with the genetic engineering. Changing the genetic markup and creating enzymes/microbes with new functions is a huge task. In addition to that, the implementation of altered microbes or enzymes using biomolecular engineering is very much challenging for bioremediation contaminated sites. This is because of the maximum efforts limited to the just modification enzymes that can only fulfil the desired function at lab-scale (Demeke and Dobnik, 2018). However, when these modified or genetically altered microorganism are added to the contaminated sites, they may not be able to cope with additional energy requirement induced by existence of an alien genetic material and ultimately lose their targeted functions. Also, there are limited reports where these GEMs have been practically implemented and proven to be more efficient than natural attenuation for the removal of organic contamination (Hussain et al., 2018).

8.5.2 Biofilm/Bio-Surfactants Formation

8.5.2.1 Biofilm Formation

Microorganisms' ability to create a sessile biofilm structure gives a number of advantages, including the communication and exchange of genetic material, protection from the environmental changes, persistence in multiple metabolic states, and nutrition availability from the environment as well as from each other (Mangwani et al., 2017). Biofilms can be made up of either single or group of various microbial species from one or more kingdoms, depending on the environment (electron acceptor/donor) and nutritional circumstances. Biofilms generate physical biofilm structures as a result of their characteristics. When biofilms are used for bioremediation and biotransformation, higher concentrations of recalcitrant or toxic organic contaminants can kill planktonic microorganisms in biofilms. Therefore, resistance to such toxic and hazardous organic contaminants is of critical interest (Shukla et al., 2014). Microbes of biofilm have major advantages in terms of improved tolerance to toxic chemicals compounds present in floating environment even at higher concentrations of pollutants and antibiotics. These floating biofilms are also more resistant to changes in ambient variables, such as pH, temperature, salt concentration, water content, as well as other environmental stressors. As a result, biofilm's natural characteristics and toughness can be effectively used to build bioremediation systems for organic contaminants (Edwards and Kjellerup, 2013). Naphthalene biodegradation has been investigated in the existence of secondary carbon sources such as glucose, starch, sucrose, and L-arginine (Dutta et al., 2018). Sucrose was tested as bio-stimulating agent for the breakdown of naphthalene. It was observed that the presence of sucrose promoted formation of biofilm of *Pseudomonas putida* KD9 that help degrading naphthalene. It was observed that multi-species consortia of *Pseudomonas* strains extracted from contaminated sites with phenanthrene and pyrene degradation sites boosted the formation of biofilm when compared to the use of pure cultures. They also discovered that after 7 days of incubation, a complete removal (100%) for phenanthrene and only 78% for pyrene is achieved. Both phenanthrene and pyrene were found to promote the formation of biofilms and the development of bio-emulsifiers (Isaac et al., 2017). In a different study that used Moving Bed Biofilm Reactor (MBBR) to biodegrade PCB-containing wastewater, PCB was biodegraded in anaerobic and aerobic sections with removal efficiency of 73% and 84%, respectively (Dong et al., 2015). In a different study, the combined action of activated sludge and biofilm processes as well as MBBR efficiently break down medications (Escolà Casas et al., 2015).

8.5.2.2 Biosurfactants Production

Biosurfactants amphiphilic compounds that are produced by bacteria located inside (intracellular) or outside (extracellular) of the cells. These compounds have the capability to degrade the complex organic pollutants in contaminated sites (Ławniczak et al., 2013). For example, strain of Pseudomonas aeruginosa can produce the rhamnolipid type biosurfactant that has been proven for the degradation of hexadecane. The structure of mostly all biosurfactants is guided by property of microbe that produces surfactant under specific growth conditions (Meliani, 2014). There are several similar studies available for the production of biosurfactants using indigenous bacteria showing different spectrum of action depending on their chemical composition. The production of biosurfactants during the course of adaptations can be successfully used to improve *in-situ* bioremediation. Surfactants can increase the bio accessibility of hydrophobic organic pollutants' to potential biodegrading bacteria and subsequently improve microbe-contaminant communication (Singh and Singh, 2017). Biosurfactants derived from *Bacillus subtilis* MTCC1427 have been shown to promote the biodegradation of the chlorinated insecticide (endosulfan). The slow rate of microbial degradation of persistent endosulfan was increased by up to 30%–45% by presence of surfactant produced (Odukkathil and Vasudevan, 2013). The presence of biosurfactants has also improved degradation of other contaminants such as PAHs and n-alkanes (Sun et al., 2019). The rhamnolipid biosurfactant helped to improve solubility and successive breakdown of phenanthrene by *Sphingomonas* sp. when compared to synthetic surfactant like Tween-80 (Meng et al., 2019). In other instances, the occurrence of biosurfactants may be detrimental even to pollutant-degrading microbes. However, combination of biosurfactant producing as well as contaminant degrading indigenous microbes that are not able to produce their own biosurfactants can allow constant and rapid degradation of various pollutants(Jimoh and Lin, 2019). Mutants of salt-tolerant *Enterobacter cloacae* was also discovered to be useful in bioaugmentation purpose for the remediation of petroleum-contaminated soil (Azam et al., 2019).

8.5.3 Nano-Bioremediation

It's vital for the development as well as to employ innovative wastewater treatment systems with high effectiveness and minimal capital requirements. Due to the recent advanced procedures in nanomaterial science, researchers are exhibiting interest in nanotechnology. Nanomaterials are the tiniest structures, measuring less than 100 nano-meters in diameter. A number of effective, cost-efficient as well as environmentally friendly nanomaterials with unique wastewater treatment capabilities have been produced. The use of nanotechnology is most advanced treatment system for contaminated sites such as wastewater (Vázquez-Núñez et al., 2020). Nanomaterials are sub-divided into three different categories based on their nature: nano-adsorbents, nano-catalysts, and nanomembranes. Chemically active nano-adsorbents that show enhanced adsorption capacity can be made using the atoms of those targeted elements. Nanomaterials that have been used for the remediation purposes include nano-adsorbents such as activated carbon, clay-minerals, silica nanosorbents, metal-oxide, and modified compounds that are formed from fusing different nanoparticles called composites (Singh et al., 2014). Metal oxides and semiconductors have gotten a lot of interest in the development of wastewater treatment systems. For pollutant degradation, several nano-catalysts like electrocatalysts, Fenton-based catalysts which improve chemical oxidation of organic contaminants, as well as antimicrobial catalysts have been used (Anjum et al., 2019). A heterogeneous nanocomposite $MnFe_2O_4$@PANI@Ag was utilised to degrade azo dyes such as rhodamine B, methylene blue, eosin Y and methyl orange. This synthesised nanocomposite was used for decolorizing the dye solution and was able to use again for several time after recycling through magnetic separation (Amir et al., 2016). A heterogeneous Fenton catalyst ($CuFeO_2$ [012]-H_2O_2) was found to degrade the molecules of ofloxacin (OFX). Different variants of the same catalyst were synthesised and compared, and it was observed that $CuFeO_2$ [012] can decompose OFX at much faster rate (four times) than CuFeO2 [110]. The antibiotic (OFX) was completely removed in using $CuFeO_2$ [012]-H_2O_2 system at a pH range of 3.2–10.1 (Dai et al., 2018). A new $FeNi_3/SiO_2/CuS$ (FNSCS) was another photocatalyst that also exhibit magnetism which was used for the degradation of tetracycline (Nasseh et al., 2018). The FNSCS showed a complete removal of tetracycline molecules from the aqueous suspension. The last class of nanomaterials is nano-membranes, which operates on a pressure-driven phenomenon for treatment enhancing wastewater quality. Nano-filtration, among the different variations of membrane filtration techniques that are available, is widely used for wastewater treatment in various industries. This is due to their unique small pore size, high suspended particle removal efficiency, and their easy operatable conditions. The major problem associated with these filtration units is their rapid clogging. In order to address this, various photocatalytic nanoparticles can be decorated on the fibres of these filters that can help regenerating the adsorbed sites. Examples of such nano-catalysts are nanoparticles of zinc oxide, titanium oxides. Heterostructures silver-titanium dioxide nanofiber membranes (polyurethane-$AgTiO_2$ NFs) and particles (polyurethane-$AgTiO_2$ NPs) were used to study photocatalytic degradation of dairy effluent. After 2 hours, the particles of Polyurethane-$AgTiO_2$ and membrane Polyurethane-$AgTiO_2$ NFs showed maximal degradation of 75% and 95%, respectively (Kanjwal et al., 2015). Production of hetero-layered semiconduction nanostructures of TiO_2@LDH/MoS_2 was extremely efficient for the degradation of two oppositely charged (cationic as well as anionic) organic dyes (Panchal et al., 2021). All these are benefits of nanomaterials for the reduction of organic pollutants from aqueous suspension. Now research has directed towards the implementation of integrated research such as amalgamation of nanotechnology and bioremediation improving the microbial activity against a specific waste and toxic material. Many research has shown that combinatorial approaches can be used to remediate organic pollutants. In a bio-reductive assay, Palladium (Pd) nanoparticles decorated on the cell wall as well as inside Shewanella oneidensis's cytoplasm were charged with H radicals by adding substrates (hydrogen, acetate, and formate), and the generated radicals were used to degrade PCP, resulting in the removal of the chlorine molecule (De Windt et al., 2005). Enzymes generally lose their bioactivity due to oxidation, and their stability can be improved by functionalization with nanoparticles such as with iron. Using the magnetically active nanomaterials of iron can help regeneration the enzyme from treatment unit by applying external magnetic field (Singh et al., 2014). In an interesting study, a composite of Fe_3O_4nanoparticle and biochar was prepared which was further loaded with photosynthetic bacteria (PSB) onto them (He et al., 2017). A real wastewater sample was treated using the above synthesised material. It was believed that the presence of [Fe–Fe] and [Fe–S] at the active sites of the hydrogenase and nitrogenase (enzymes in energy metabolism of PSB), respectively, an addition of particles of nano-Fe_3O_4 improves their enzymatic activities that helped for the removal of organic components. Thus, it is notable that nanomaterial not only are capable of directly removes the organic and toxic chemical that are harmful to microbes, but can also enhance catalytic activity of microorganisms for degrading organic contaminants.

8.5.4 Metagenomics Approach

Excessive concentration and exposure (particularly high doses or repeated exposure) to toxic organic pollutants can have negative consequences for life on Earth via a variety of mechanisms, including the creation of hazardous reactive metabolites and the disruption of endogenous metabolism because of the changes in their environment. In addition, recalcitrance nature of pollutants, their insufficient mineralization or the accumulation of sister metabolites in the environment are major concerns. Current approaches are unable to explain global changes in metabolism as well as investigate or build new catabolic pathways to analyse intermediates and metabolites, and thus a there are limited ways to design a successful bioremediation reaction synthesis (Yadav et al., 2020). However, the indigenous microbial population even in the changing environmental conditions have the capability to check the dynamics of organic pollutant in the contaminated site by forming a complicated ecological niche. But again, almost 99% of these microbial populations are difficult to cultivate (Wolfe, 2018). As a result, there is a need for more precise prevention and treatment methods than the existing ones. By revealing the exact structure and composition of these communities and their possible role in biodegradation of various complicated contaminants using the extraction or isolating and reading the DNA makeup can aid in understanding the pollutant degradation mechanism of certain ecosystems. Similarly, studying the intermediate metabolites (primary and secondary) or proteins can help in understanding possible pollutant-microbe interactions, functional structures at the molecular level, their quick environmental responses to various stimuli such as pollution-induced abiotic stress, temperature, and so on (Jadeja et al., 2019). Metagenomics and proteo-metabolomics are the novel and advanced methods enabling a system-wide approach to investigating microorganisms at specific sites during the active mineralization process. Next-generation sequencing (NGS), which has the potential to sequence thousands of sequences simultaneously while reducing time as well as being cost-effective as compared to traditional sequencing methods, is a new generation sequencing technique (Slatko et al., 2018). The workflow of OMICS-approach (metagenomics, meta-proteomics and metabolomics) expression profiling experiment is shown in Figure 8.2.

Since last few years, scientists have largely relied upon 16S rRNA sequences for identification and classification of bacterial species. Next generation sequencing technologies have changed the study of metagenomics by eliminating requirement for cloning and library creation, allowing us to investigate the entire microbial population present in contaminated locations. The contaminated locations' metagenome deep shotgun sequencing made accessible global view of microbial ecosystems as well as a plethora of genetic sequence of unculturable microorganisms and their metabolic pathways (Hendriksen et al., 2019). This method involves randomly shearing metagenomic DNA and random end-sequencing with a Next-Generation sequencer to isolate metagenomic DNA from polluted areas and producing a tiny insert metagenomic DNA library. Hundreds of contigs and scaffolds are created from the sequencing of each fragment's genomes. The sequences are then read and annotated to assign functions. For sequence-based techniques, a huge number of databases have been created, and the majority of the metagenomic datasets is derived from contaminated areas (Bharagava et al., 2018). The Integrated Microbial Genome/

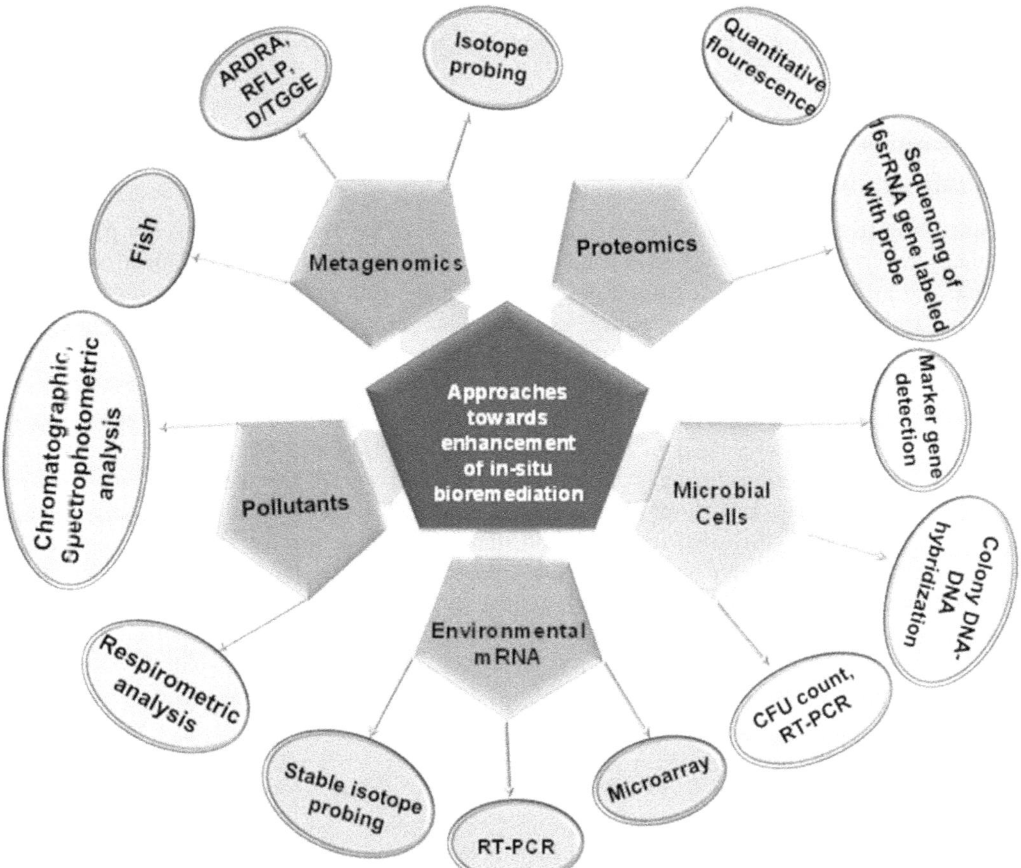

FIGURE 8.2 The workflow of OMICS-approach (metagenomics, meta-proteomics and metabolomics).

Microbiome also known as IMG/M is metagenome tool providing a platform for assessment of functional potential of microorganisms' world (Markowitz et al., 2009). The implementation of metagenomic approach and tools can produce a huge information of indigenous microbial population living in the polluted sites. This in-depth understanding and knowledge of microbial world can be extremely useful for *in-situ* as well as *ex-situ* bioremediation strategies (Das et al., 2017). Researchers observed high relative abundances of bacterial phyla such as Proteobacteria, Firmicutes, Actinobacteria, Acidobacteria, and Chloroflexi using a metagenomic technique to investigate the microbial population composition and diversity at an area contaminated with PAH (Redfern et al., 2019). Polybrominated diphenyl ethers (PBDEs) are hydrophobic, semi-volatile in nature and poisonous chemicals that are very opposed to degradation by microbes and have strong adsorption on sediments. The non-degraded PBDEs molecules can easily bioaccumulate and biomagnify in the surrounding environments, resulting in the formation of a new persistent organic pollutant (POPs). Microbial population directly involved in PBDE breakdown were modified under varied electron donor amending circumstances using the information from a high-throughput multiplex barcoded metagenomic method (Gedalanga et al., 2013). In a separate study, the presence of bacterial phyla such as Proteobacteria, Acidobacteria, Actinobacteria, Bacteroidetes, and Chloroflexi in permafrost soils damaged by oil fragments was discovered to be prevalent in all oil spill sites. This information obtained from the metagenome would at the end help in better planning of bioremediation strategy (Dong et al., 2015; Yang et al., 2014). The existence of microbial class of *Plesiocystis, Anaerolinea, Jannaschia,* and *Mycobacterium* were verified in all sediments of 69 genera helping in biodegradation using metagenomic approach (Fang et al., 2018). Metagenomic approach was also undertaken into consideration to look into possible biodegradation pathways of various persistent pesticides in fresh water and marine sediments. Furthermore, complete degradation pathway of pesticides such as DDT as well as atrazine (ATZ) was mapped and partial degradation pathway of HCH (hexachlorocyclohexane) (Fang et al., 2014; Sagarkar et al., 2013).

8.6 Conclusion

Removal of the organic pollutants from contaminated sites is of immense significance for protection of human health as well as natural environments. Conventional treatment system is chemical and energy-intensive that also increase the treatment cost. Moreover, these treatment systems can generate several secondary contaminants that can again damage the environment. Bioremediation using indigenous microorganisms is an effective *in-situ* tool to mitigate organic pollutants from contaminated sites. These methods including biostimulation and bioaugmentation that use microbial metabolic activity are the most widely used methods for treatment of toxic organic contaminates due to high efficiency and low cost. Some of recent advancements in bioremediation process such as a GEMs, selection of feasible bio-surfactants, Nano-bioremediation, metagenomic approach have supplemented the capacity of remediation process that allows to achieve higher removal efficiencies. On the other hand, these advanced techniques have certainly improved removal of organic pollutants; it also has added additional cost of treatment processes compared to conventional approaches. Hence, selection of appropriate bioremediation approach based on the status of contaminated sites, and addition of nutrients and external microbial strain to enhance the bioactivity is particularly very vital for both, treatment efficiency and cost. Thus, novel technologies with both high efficiency and low operational cost are needed to be developed in the future.

Acknowledgements

DP acknowledges the University Grant Commission, New Delhi, India for providing Senior Research Fellowship and SN is grateful to the Department of Science and Technology for Senior Research Fellowship through DST-INSPIRE program. Director, CSIR-NEERI is thankful for giving the opportunity to pursue the work. The article is checked for plagiarism using the iThenticate software and recorded in the Knowledge Resource Center, CSIR-NEERI, Nagpur for anti-plagiarism (KRC No.: CSIR-NEERI/KRC/2021/SEP/WWTD-EBGD/1).

REFERENCES

Abraham, J., Silambarasan, S., Logeswari, P., 2014. Simultaneous degradation of organophosphorus and organochlorine pesticides by bacterial consortium. *J. Taiwan Inst. Chem. Eng.* 45, 2590–2596. https://doi.org/10.1016/j.jtice.2014.06.014

Ali, M., Gani, K.M., Kazmi, A.A., Ahmed, N., 2016a. Degradation of aldrin and endosulfan in rotary drum and windrow composting. *J. Environ. Sci. Heal. - Part B Pestic. Food Contam. Agric. Wastes* 51, 278–286. https://doi.org/10.1080/03601234.2015.1128739

Ali, N., Dashti, N., Salamah, S., Sorkhoh, N., Al-Awadhi, H., Radwan, S., 2016b. Dynamics of bacterial populations during bench-scale bioremediation of oily seawater and desert soil bioaugmented with coastal microbial mats. *Microb. Biotechnol.* 9, 157–171. https://doi.org/10.1111/1751-7915.12326

Alneyadi, A.H., Rauf, M.A., Ashraf, S.S., 2018. Oxidoreductases for the remediation of organic pollutants in water–a critical review. *Crit. Rev. Biotechnol.* 38, 971–988. https://doi.org/10.1080/07388551.2017.1423275

Amir, M., Kurtan, U., Baykal, A., Sözeri, H., 2016. $MnFe_2O_4$@PANI@Ag heterogeneous nanocatalyst for degradation of industrial aqueous organic pollutants. *J. Mater. Sci. Technol.* 32, 134–141. https://doi.org/10.1016/j.jmst.2015.12.011

Anjum, M., Miandad, R., Waqas, M., Gehany, F., Barakat, M.A., 2019. Remediation of wastewater using various nano-materials. *Arab. J. Chem.* 12, 4897–4919. https://doi.org/10.1016/j.arabjc.2016.10.004

Asaf, S. Numan, M., Khan, A.L., Al-Harrasi, A., 2020. *Sphingomonas:* From diversity and genomics to functional role in environmental remediation and plant growth. *Crit. Rev. Biotechnol.* 40, 138–152. https://doi.org/10.1080/07388551.2019.1709793

Asif, M.B., Hai, F.I., Singh, L., Price, W.E., Nghiem, L.D., 2017. Degradation of pharmaceuticals and personal care products by white-rot fungi—a critical review. *Curr. Pollut. Rep.* 3, 88–103. https://doi.org/10.1007/s40726-017-0049-5

Azam, K., Sima, K., Ebadi, A., Reiahisamani, N., Rasekh, B., 2019. Bio-based remediation of petroleum-contaminated saline soils : Challenges, the current state-of-the-art and future prospects. *J. Environ. Manage.* 250, 109476. https://doi.org/10.1016/j.jenvman.2019.109476

Baniasadi, M., Mousavi, S.M., 2019. A comprehensive review on the bioremediation of oil spills. In *Microbial Action on Hydrocarbons.* https://doi.org/10.1007/978-981-13-1840-5_10

Basumatary, M., Das, S., Gogoi, M., Das, I., Charingia, D., Borah, D., 2020. Evaluation of pilot scale in-vitro and ex-situ hydrocarbon bioremediation potential of two novel indigenous strains of *Bacillus vallismortis* *Bioremediat. J.* 24, 190–203. https://doi.org/10.1080/10889868.2020.1799928

Bharagava, R.N., Purchase, D., Saxena, G., Mulla, S.I., 2018. Applications of metagenomics in microbial bioremediation of pollutants: From genomics to environmental cleanup: From genomics to environmental cleanup. In *Microbial Diversity in the Genomic Era.* Elsevier Inc. https://doi.org/10.1016/B978-0-12-814849-5.00026-5

Blázquez-Pallí, N., Rosell, M., Varias, J., Bosch, M., Soler, A., Vicent, T., Marco-Urrea, E., 2019. Integrative isotopic and molecular approach for the diagnosis and implementation of an efficient in-situ enhanced biological reductive dechlorination of chlorinated ethenes. *Water Res.* 167. https://doi.org/10.1016/j.watres.2019.115106

Bobde, P., Patel, R.K., Panchal, D., Sharma, A., Sharma, A.K., 2021. Utilization of layered double hydroxides (LDHs) and their derivatives as photocatalysts for degradation of organic pollutants.

Bodor, A., Petrovszki, P., Kis, Á.E., Vincze, G.E., Laczi, K., Bounedjoum, N., Szilágyi, Á., Szalontai, B., Feigl, G., Kovács, K.L., Rákhely, G., Perei, K., 2020. Intensification of ex situ bioremediation of soils polluted with used lubricant oils: A comparison of biostimulation and bioaugmentation with a special focus on the type and size of the inoculum. *Int. J. Environ. Res. Public Health* 17, 1–17. https://doi.org/10.3390/ijerph17114106

Butkovskyi, A., Bruning, H., Kools, S.A.E., Rijnaarts, H.H.M., Van Wezel, A.P., 2017. Organic pollutants in shale gas flowback and produced waters: Identification, potential ecological impact, and implications for treatment strategies. *Environ. Sci. Technol.* 51, 4740–4754. https://doi.org/10.1021/acs.est.6b05640

Chen, C.H., Whang, L.M., Pan, C.L., Yang, C.L., Liu, P.W.G., 2017. Immobilization of diesel-degrading consortia for bioremediation of diesel-contaminated groundwater and seawater. *Int. Biodeterior. Biodegrad.* 124, 62–72. https://doi.org/10.1016/j.ibiod.2017.07.001

Dai, C., Tian, X., Nie, Y., Lin, H.M., Yang, C., Han, B., Wang, Y., 2018. Surface facet of $CuFeO_2$ nanocatalyst: A key parameter for H_2O_2 activation in fenton-like reaction and organic pollutant degradation. *Environ. Sci. Technol.* 52, 6518–6525. https://doi.org/10.1021/acs.est.8b01448

Das, S., Bora, S.S., Yadav, R.N.S., Barooah, M., 2017. A metagenomic approach to decipher the indigenous microbial communities of arsenic contaminated groundwater of Assam. *Genomics Data* 12, 89–96. https://doi.org/10.1016/j.gdata.2017.03.013

Demeke, T., Dobnik, D., 2018. Critical assessment of digital PCR for the detection and quantification of genetically modified organisms. *Anal. Bioanal. Chem.* 410, 4039–4050. https://doi.org/10.1007/s00216-018-1010-1

Deng, Y., Zhao, R., 2015. Advanced oxidation processes (AOPs) in wastewater treatment. *Curr. Pollut. Rep.* 1, 167–176. https://doi.org/10.1007/s40726-015-0015-z

De Santis, B., Stockhofe, N., Wal, J.M., Weesendorp, E., Lalles, J.P., van Dijk, J., Kok, E., De Giacomo, M., Einspanier, R., Onori, R. and Brera, C., 2017. Case studies on genetically modi fi ed organisms (GMOs): Potential risk scenarios and associated health indicators. *Food Chem. Toxicol.* 117, 36–65. https://doi.org/10.1016/j.fct.2017.08.033

De Windt, W., Aelterman, P., Verstraete, W., 2005. Bioreductive deposition of palladium (0) nanoparticles on *Shewanella oneidensis* with catalytic activity towards reductive dechlorination of polychlorinated biphenyls. *Environ. Microbiol.* 7, 314–325. https://doi.org/10.1111/j.1462-2920.2004.00696.x

Dias, R.L., Ruberto, L., Calabró, A., Balbo, A. Lo, Del Panno, M.T., Mac Cormack, W.P., 2015. Hydrocarbon removal and bacterial community structure in on-site biostimulated biopile systems designed for bioremediation of diesel-contaminated Antarctic soil. *Polar Biol.* 38, 677–687. https://doi.org/10.1007/s00300-014-1630-7

Dong, B., Chen, H.Y., Yang, Y., He, Q.B., Dai, X.H., 2015. Biodegradation of polychlorinated biphenyls using a moving-bed biofilm reactor. *Clean - Soil, Air, Water* 43, 1078–1083. https://doi.org/10.1002/clen.201200623

Dutta, K., Shityakov, S., Khalifa, I., Mal, A., Moulik, S.P., Panda, A.K., Ghosh, C., 2018. Effects of secondary carbon supplement on biofilm-mediated biodegradation of naphthalene by mutated naphthalene 1, 2-dioxygenase encoded by *Pseudomonas putida* strain KD9. *J. Hazard. Mater.* 357, 187–197. https://doi.org/10.1016/j.jhazmat.2018.05.024

Edwards, S.J., Kjellerup, B. V., 2013. Applications of biofilms in bioremediation and biotransformation of persistent organic pollutants, pharmaceuticals/personal care products, and heavy metals. *Appl. Microbiol. Biotechnol.* 97, 9909–9921. https://doi.org/10.1007/s00253-013-5216-z

Escolà Casas, M., Chhetri, R.K., Ooi, G., Hansen, K.M.S., Litty, K., Christensson, M., Kragelund, C., Andersen, H.R., Bester, K., 2015. Biodegradation of pharmaceuticals in hospital wastewater by a hybrid biofilm and activated sludge system (Hybas). *Sci. Total Environ.* 530–531, 383–392. https://doi.org/10.1016/j.scitotenv.2015.05.099

Fang, H., Cai, L., Yang, Y., Ju, F., Li, X., Yu, Y., Zhang, T., 2014. Metagenomic analysis reveals potential biodegradation pathways of persistent pesticides in freshwater and marine sediments. *Sci. Total Environ.* 470–471, 983–992. https://doi.org/10.1016/j.scitotenv.2013.10.076

Fang, H., Zhang, H., Han, L., Mei, J., Ge, Q., Long, Z., Yu, Y., 2018. Exploring bacterial communities and biodegradation genes in activated sludge from pesticide wastewater treatment plants via metagenomic analysis. *Environ. Pollut.* 243, 1206–1216. https://doi.org/10.1016/j.envpol.2018.09.080

Gaur, N., Narasimhulu, K., PydiSetty, Y., 2018. Recent advances in the bio-remediation of persistent organic pollutants and its effect on environment. *J. Clean. Prod.* 198, 1602–1631. https://doi.org/10.1016/j.jclepro.2018.07.076

Gedalanga, P., Kotay, S.M., Sales, C.M., Butler, C.S., Goel, R., Mahendra, S., 2013. Novel applications of molecular biological and microscopic tools in environmental engineering. *Water Environ. Res.* 85, 917–950. https://doi.org/10.2175/106143013x13698672321742

Gusain, R., Gupta, K., Joshi, P., Khatri, O.P., 2019. Adsorptive removal and photocatalytic degradation of organic pollutants using metal oxides and their composites: A comprehensive review. *Adv. Colloid Interface Sci.* 272, 102009. https://doi.org/10.1016/j.cis.2019.102009

He, S., Zhong, L., Duan, J., Feng, Y., Yang, B., Yang, L., 2017. Bioremediation of wastewater by iron oxide-biochar nanocomposites loaded with photosynthetic bacteria. *Front. Microbiol.* 8, 1–10. https://doi.org/10.3389/fmicb.2017.00823

Hendriksen, R.S., Munk, P., Njage, P., van Bunnik, B., McNally, L., Lukjancenko, O., Röder, T., Nieuwenhuijse, D., Pedersen, S.K., Kjeldgaard, J., Kaas, R.S., Clausen, P.T.L.C., Vogt, J.K., Leekitcharoenphon, P., van de Schans, M.G.M., Zuidema, T., de Roda Husman, A.M., Rasmussen, S., Petersen, B., Bego, A., Rees, C., Cassar, S., Coventry, K., Collignon, P., Allerberger, F., Rahube, T.O., Oliveira, G., Ivanov, I., Vuthy, Y., Sopheak, T., Yost, C.K., Ke, C., Zheng, H., Baisheng, L., Jiao, X., Donado-Godoy, P., Coulibaly, K.J., Jergović, M., Hrenovic, J., Karpíšková, R., Villacis, J.E., Legesse, M., Eguale, T., Heikinheimo, A., Malania, L., Nitsche, A., Brinkmann, A., Saba, C.K.S., Kocsis, B., Solymosi, N., Thorsteinsdottir, T.R., Hatha, A.M., Alebouyeh, M., Morris, D., Cormican, M., O'Connor, L., Moran-Gilad, J., Alba, P., Battisti, A., Shakenova, Z., Kiiyukia, C., Ng'eno, E., Raka, L., Avsejenko, J., Bērziņš, A., Bartkevics, V., Penny, C., Rajandas, H., Parimannan, S., Haber, M.V., Pal, P., Jeunen, G.J., Gemmell, N., Fashae, K., Holmstad, R., Hasan, R., Shakoor, S., Rojas, M.L.Z., Wasyl, D., Bosevska, G., Kochubovski, M., Radu, C., Gassama, A., Radosavljevic, V., Wuertz, S., Zuniga-Montanez, R., Tay, M.Y.F., Gavačová, D., Pastuchova, K., Truska, P., Trkov, M., Esterhuyse, K., Keddy, K., Cerdà-Cuéllar, M., Pathirage, S., Norrgren, L., Örn, S., Larsson, D.G.J., Heijden, T. Van der, Kumburu, H.H., Sanneh, B., Bidjada, P., Njanpop-Lafourcade, B.M., Nikiema-Pessinaba, S.C., Levent, B., Meschke, J.S., Beck, N.K., Van, C.D., Phuc, N. Do, Tran, D.M.N., Kwenda, G., Tabo, D. adjim, Wester, A.L., Cuadros-Orellana, S., Amid, C., Cochrane, G., Sicheritz-Ponten, T., Schmitt, H., Alvarez, J.R.M., Aidara-Kane, A., Pamp, S.J., Lund, O.,

Hald, T., Woolhouse, M., Koopmans, M.P., Vigre, H., Petersen, T.N., Aarestrup, F.M., 2019. Global monitoring of antimicrobial resistance based on metagenomics analyses of urban sewage. *Nat. Commun.* 10. https://doi.org/10.1038/s41467-019-08853-3

Herrero, M., Stuckey, D.C., 2015. Bioaugmentation and its application in wastewater treatment: A review. *Chemosphere* 140, 119–128. https://doi.org/10.1016/j.chemosphere.2014.10.033

Hu, B., Ai, Y., Jin, J., Hayat, T., Alsaedi, A., Zhuang, L., Wang, X., 2020. Efficient elimination of organic and inorganic pollutants by biochar and biochar-based materials. *Biochar* 2, 47–64. https://doi.org/10.1007/s42773-020-00044-4

Hussain, I., Aleti, G., Naidu, R., Puschenreiter, M., Mahmood, Q., Rahman, M.M., Wang, F., Shaheen, S., Syed, J.H., Reichenauer, T.G., 2018. Microbe and plant assisted-remediation of organic xenobiotics and its enhancement by genetically modified organisms and recombinant technology: A review. *Sci. Total Environ.* 628–629, 1582–1599. https://doi.org/10.1016/j.scitotenv.2018.02.037

Ijoma, G.N., Selvarajan, R., Oyourou, J.N., Sibanda, T., Matambo, T., Monanga, A., Mkansi, K., 2019. Exploring the application of biostimulation strategy for bacteria in the bioremediation of industrial effluent. *Ann. Microbiol.* 69, 541–551. https://doi.org/10.1007/s13213-019-1443-6

Isaac, P., Alessandrello, M.J., Macedo, A.J., Estévez, M.C., Ferrero, M.A., 2017. Pre-exposition to polycyclic aromatic hydrocarbons (PAHs) enhance biofilm formation and hydrocarbon removal by native multi-species consortium. *J. Environ. Chem. Eng.* 5, 1372–1378. https://doi.org/10.1016/j.jece.2017.02.031

Jadeja, N.B., Purohit, H.J., Kapley, A., 2019. Decoding microbial community intelligence through metagenomics for efficient wastewater treatment. *Funct. Integr. Genomics* 19, 839–851. https://doi.org/10.1007/s10142-019-00681-4

Jariyal, M., Yadav, M., Singh, N.K., Yadav, S., Sharma, I., Dahiya, S., Thanki, A., 2020. Microbial remediation progress and future prospects. In *Bioremediation of Pollutants*. https://doi.org/10.1016/b978-0-12-819025-8.00008-9

Jia, L., Jiang, B., Huang, F., Hu, X., 2019. Nitrogen removal mechanism and microbial community changes of bioaugmentation subsurface wastewater infiltration system. *Bioresour. Technol.* 294, 122140. https://doi.org/10.1016/j.biortech.2019.122140

Jimoh, A.A., Lin, J., 2019. Production and characterization of lipopeptide biosurfactant producing *Paenibacillus* sp. D9 and its biodegradation of diesel fuel. *Int. J. Environ. Sci. Technol.* 16, 4143–4158. https://doi.org/10.1007/s13762-019-02341-3

Kanjwal, M.A., Barakat, N.A.M., Chronakis, I.S., 2015. Photocatalytic degradation of dairy effluent using $AgTiO_2$ nanostructures/polyurethane nanofiber membrane. *Ceram. Int.* 41, 9615–9621. https://doi.org/10.1016/j.ceramint.2015.04.024

Khalid, F., Hashmi, M.Z., Jamil, N., Qadir, A., Ali, M.I., 2021. Microbial and enzymatic degradation of PCBs from e-waste-contaminated sites: A review. *Environ. Sci. Pollut. Res.* 28, 10474–10487. https://doi.org/10.1007/s11356-020-11996-2

Kreling, N.E., Simon, V., Fagundes, V.D., Thomé, A., Colla, L.M., 2021. Improving the bioremediation and in situ production of biocompounds of a biodiesel-contaminated soil. *Environ. Manage.* 68, 210–225. https://doi.org/10.1007/s00267-021-01486-7

Kuhad, R.C., Singh, A., 2013. *Biotechnology for Environmental Management and Resource Recovery*. New Delhi: Springer.

Kumar, V., Shahi, S.K., Singh, S., 2018. Bioremediation: An eco-sustainable approach for restoration of contaminated sites. In Singh, J., Sharma, D., Kumar, G., Sharma, N. (eds) *Microbial Bioprospecting for Sustainable Development*. Singapore: Springer. https://doi.org/10.1007/978-981-13-0053-0_6

Lakshman, B., Sai, K.D.V.R., 2015. Effective role of indigenous microorganisms for sustainable environment. *3 Biotech* 5(6), 867–876. https://doi.org/10.1007/s13205-015-0293-6

Ławniczak, Ł., Marecik, R., Chrzanowski, Ł., 2013. Contributions of biosurfactants to natural or induced bioremediation. *Appl. Microbiol. Biotechnol.* 97, 2327–2339. https://doi.org/10.1007/s00253-013-4740-1

Li, J., Chen, X., Liu, W., Tao, Y., 2020a. Biostimulation of a marine anammox bacteria-dominated bioprocess by Co(II) to treat nitrogen-rich, saline wastewater. *Sci. Total Environ.* 749, 141489. https://doi.org/10.1016/j.scitotenv.2020.141489

Li, J., Li, Y., Xiong, Z., Yao, G., Lai, B., 2019. The electrochemical advanced oxidation processes coupling of oxidants for organic pollutants degradation : A mini-review. *Chinese Chem. Lett.* 4–11. https://doi.org/10.1016/j.cclet.2019.04.057

Li, J., Wu, C., Chen, S., Lu, Q., Shim, H., Huang, X., Jia, C., Wang, S., 2020b. Enriching indigenous microbial consortia as a promising strategy for xenobiotics' cleanup. *J. Clean. Prod.* 261, 121234. https://doi.org/10.1016/j.jclepro.2020.121234

Liang, Q., Lei, M., Chen, T., Yang, J., Wan, X., Yang, S., 2014. Application of sewage sludge and intermittent aeration strategy to the bioremediation of DDT- and HCH-contaminated soil. *J. Environ. Sci.* 26, 1673–1680. https://doi.org/10.1016/j.jes.2014.06.007

Liu, M., Wang, S., Nobu, M.K., Bocher, B.T.W., Kaley, S.A., Liu, W.T., 2017. Impacts of biostimulation and bioaugmentation on the performance and microbial ecology in methanogenic reactors treating purified terephthalic acid wastewater. *Water Res.* 122, 308–316. https://doi.org/10.1016/j.watres.2017.06.008

Luka, Y., Highina, B. K., Zubairu, A., 2018. Bioremediation : A solution to environmental pollution-a review. *Am. J. Eng. Res.* 7, 101–109.

Makri, P., Stathopoulou, E., Hermides, D., Kontakiotis, G., Zarkogiannis, S.D., Skilodimou, H.D., Bathrellos, G.D., Antonarakou, A., Scoullos, M., 2020. The environmental impact of a complex hydrogeological system on hydrocarbon-pollutants' natural attenuation: The case of the coastal aquifers in eleusis, west attica, greece. *J. Mar. Sci. Eng.* 8, 1–13. https://doi.org/10.3390/jmse8121018

Mangwani, N., Kumari, S., Das, S., 2017. Bacterial biofilms and quorum sensing: Fidelity in bioremediation technology. *Biotechnol. Genet. Eng. Rev.* 32, 43–73. https://doi.org/10.1080/02648725.2016.1196554

Markowitz, V.M., Mavromatis, K., Ivanova, N.N., Chen, I.M.A., Chu, K., Kyrpides, N.C., 2009. IMG ER: A system for microbial genome annotation expert review and curation. *Bioinformatics* 25, 2271–2278. https://doi.org/10.1093/bioinformatics/btp393

Mateljak, I., Tron, T., Alcalde, M., 2017. Brief report evolved a -factor prepro-leaders for directed laccase evolution in Saccharomyces cerevisiae. *Microb. Biotechnol.* 1–7. https://doi.org/10.1111/1751-7915.12838

Medfu Tarekegn, M., Zewdu Salilih, F., Ishetu, A.I., 2020. Microbes used as a tool for bioremediation of heavy metal from the environment. *Cogent Food Agric.* 6. https://doi.org/10.1080/23311932.2020.1783174

Meliani, A., 2014. Enhancement of hydrocarbons degradation by use of pseudomonas biosurfactants and biofilms. *J. Pet. Environ. Biotechnol.* 5. https://doi.org/10.4172/2157-7463.1000168

Meng, L., Li, W., Bao, M., Sun, P., 2019. Effect of surfactants on the solubilization, sorption and biodegradation of benzo (a) pyrene by *Pseudomonas aeruginosa* BT-1. *J. Taiwan Inst. Chem. Eng.* 96, 121–130. https://doi.org/10.1016/j.jtice.2019.01.007

Mohan, M.T.M., Ram, S., 2018. Bioremediation of groundwater: An overview. *Int. J. Appl. Eng. Res.* 13, 16825–16832.

Morillo, E., Villaverde, J., 2017. Advanced technologies for the remediation of pesticide-contaminated soils. *Sci. Total Environ.* 586, 576–597. https://doi.org/10.1016/j.scitotenv.2017.02.020

Muter, O., Perkons, I., Selga, T., Berzins, A., Gudra, D., Radovica-Spalvina, I., Fridmanis, D., Bartkevics, V., 2017. Removal of pharmaceuticals from municipal wastewaters at laboratory scale by treatment with activated sludge and biostimulation. *Sci. Total Environ.* 584–585, 402–413. https://doi.org/10.1016/j.scitotenv.2017.01.023

Nandy, S., Arora, U., Tarar, P., Viggor, S., Jõesaar, M., Kivisaar, M., Kapley, A., 2021. Monitoring the growth, survival and phenol utilization of the fluorescent-tagged *Pseudomonas oleovorans* immobilized and free cells. *Bioresour. Technol.* 338. https://doi.org/10.1016/j.biortech.2021.125568

Nasseh, N., Taghavi, L., Barikbin, B., Nasseri, M.A., 2018. Synthesis and characterizations of a novel $FeNi_3$/SiO_2/CuS magnetic nanocomposite for photocatalytic degradation of tetracycline in simulated wastewater. *J. Clean. Prod.* 179, 42–54. https://doi.org/10.1016/j.jclepro.2018.01.052

Odukkathil, G., Vasudevan, N., 2013. Enhanced biodegradation of endosulfan and its major metabolite endosulfate by a biosurfactant producing bacterium. *J. Environ. Sci. Heal. - Part B Pestic. Food Contam. Agric. Wastes* 48, 462–469. https://doi.org/10.1080/03601234.2013.761873

Ohmori, T., Morita, H., Tanaka, M., Miyauchi, K., Kasai, D., Furukawa, K., Miyashita, K., Ogawa, N., Masai, E., Fukuda, M., 2011. Development of a strain for efficient degradation of polychlorinated biphenyls by patchwork assembly of degradation pathways. *Journal of Biosci Bioeng* 111, 437–442. https://doi.org/10.1016/j.jbiosc.2010.12.002

Omrani, R., Spini, G., Puglisi, E., Saidane, D., 2018. Modulation of microbial consortia enriched from different polluted environments during petroleum biodegradation. *Biodegradation* 29, 187–209. https://doi.org/10.1007/s10532-018-9823-3

Oturan, M.A., Aaron, J.J., 2014. Advanced oxidation processes in water/wastewater treatment: Principles and applications. A review. *Crit. Rev. Environ. Sci. Technol.* 44, 2577–2641. https://doi.org/10.1080/10643389.2013.829765

Oyetibo, G.O., Miyauchi, K., Huang, Y., Chien, M.F., Ilori, M.O., Amund, O.O., Endo, G., 2017. Biotechnological remedies for the estuarine environment polluted with heavy metals and persistent organic pollutants. *Int. Biodeterior. Biodegrad.* 119, 614–625. https://doi.org/10.1016/j.ibiod.2016.10.005

Panchal, D., Sharma, A., Mondal, P., Prakash, O., Pal, S., 2021. Heterolayered TiO_2@layered double hydroxide-MoS_2 nanostructure for simultaneous adsorption-photocatalysis of co-existing water contaminants. *Appl. Surf. Sci.* 553, 149577. https://doi.org/10.1016/j.apsusc.2021.149577

Pathak, U., Roy, A., Mandal, D.D., Das, P., Kumar, T., Mandal, T., 2019. Bioattenuation of phenol and cyanide involving immobilised spent tea activated carbon with *Alcaligenes faecalis* JF339228: Critical assessment of the degraded intermediates. *Asia-Pacific J. Chem. Eng.* 14, 1–22. https://doi.org/10.1002/apj.2278

Peng, T., Kan, J., Hu, J., Hu, Z., 2020. Genes and novel sRNAs involved in PAHs degradation in marine bacteria *Rhodococcus* sp. P14 revealed by the genome and transcriptome analysis. *3 Biotech* 10. https://doi.org/10.1007/s13205-020-2133-6

Ramachandran, A.L., Polat, P., Mukherjee, A., Dhami, N.K., 2020. Understanding and creating biocementing beachrocks via biostimulation of indigenous microbial communities. *Appl. Microbiol. Biotechnol.* 104, 3655–3673. https://doi.org/10.1007/s00253-020-10474-6

Ranchou-Peyruse, M., Gasc, C., Guignard, M., Aüllo, T., Dequidt, D., Peyret, P., Ranchou-Peyruse, A., 2017. The sequence capture by hybridization: A new approach for revealing the potential of mono-aromatic hydrocarbons bioattenuation in a deep oligotrophic aquifer. *Microb. Biotechnol.* 10, 469–479. https://doi.org/10.1111/1751-7915.12426

Raper, E., Stephenson, T., Anderson, D.R., Fisher, R., Soares, A., 2018. Industrial wastewater treatment through bioaugmentation. *Process Saf. Environ. Prot.* 118, 178–187. https://doi.org/10.1016/j.psep.2018.06.035

Rayu, S., Karpouzas, D.G., Singh, B.K., 2012. Emerging technologies in bioremediation : Constraints and opportunities. *Biodegradation* 917–926. https://doi.org/10.1007/s10532-012-9576-3

Redfern, L.K., Gardner, C.M., Hodzic, E., Ferguson, P.L., Hsu-Kim, H., Gunsch, C.K., 2019. A new framework for approaching precision bioremediation of PAH contaminated soils. *J. Hazard. Mater.* 378, 120859. https://doi.org/10.1016/j.jhazmat.2019.120859

Roy, A., Dutta, A., Pal, S., Gupta, A., Sarkar, J., Chatterjee, A., Saha, A., Sarkar, P., Sar, P., Kazy, S.K., 2018. Biostimulation and bioaugmentation of native microbial community accelerated bioremediation of oil refinery sludge. *Bioresour. Technol.* 253, 22–32. https://doi.org/10.1016/j.biortech.2018.01.004

Sagarkar, S., Mukherjee, S., Nousiainen, A., Björklöf, K., Purohit, H.J., Jorgensen, K.S., Kapley, A., 2013. Monitoring bioremediation of atrazine in soil microcosms using molecular tools. *Environ. Pollut.* 172, 108–115. https://doi.org/10.1016/j.envpol.2012.07.048

Shah, A., Shah, M., 2020. Characterisation and bioremediation of wastewater: A review exploring bioremediation as a sustainable technique for pharmaceutical wastewater. *Groundw. Sustain. Dev.* 11, 100383. https://doi.org/10.1016/j.gsd.2020.100383

Shukla, S.K., Mangwani, N., Rao, T.S., Das, S., 2014. Biofilm-mediated bioremediation of polycyclic aromatic hydrocarbons. In *Microbial Biodegradation and Bioremediation*. Elsevier Inc. https://doi.org/10.1016/B978-0-12-800021-2.00008-X

Singh, M., Mitra, C.K., Morve, R.K., 2014. Ecofriendly application of nanomaterials: Nanobioremediation. *J. Nanoparticles* 2014, 740–757.

Singh, T., Singh, D.K., 2017. Phytoremediation of organochlorine pesticides: Concept, method, and recent developments. *Int. J. Phytoremediation* 19, 834–843. https://doi.org/10.1080/15226514.2017.1290579

Slatko, B.E., Gardner, A.F., Ausubel, F.M., 2018. Overview of next-generation sequencing technologies. *Curr. Protoc. Mol. Biol.* 122, 1–11. https://doi.org/10.1002/cpmb.59

Smith, E., Thavamani, P., Ramadass, K., Naidu, R., Srivastava, P., Megharaj, M., 2015. Remediation trials for hydrocarbon-contaminated soils in arid environments: Evaluation of bioslurry and biopiling techniques. *Int. Biodeterior. Biodegrad.* 101, 56–65. https://doi.org/10.1016/j.ibiod.2015.03.029

Sun, S., Wang, Y., Zang, T., Wei, J., Wu, H., Wei, C., Qiu, G., Li, F., 2019. A biosurfactant-producing *Pseudomonas aeruginosa* S5 isolated from coking wastewater and its application for bioremediation of polycyclic aromatic hydrocarbons. *Bioresour. Technol.* 281, 421–428. https://doi.org/10.1016/j.biortech.2019.02.087

Syed, K., Porollo, A., Miller, D., Yadav, J.S., 2013. Rational engineering of the fungal P450 monooxygenase CYP5136A3 to improve its oxidizing activity toward polycyclic aromatic hydrocarbons. *Protein Eng. Des. Sel.* 26, 553–557. https://doi.org/10.1093/protein/gzt036

Tosun, J., Hartung, U., 2017. Decentralising competences in multi-level systems : Insights from the regulation of genetically modified organisms. *West Eur. Polit.* 2382, 1–21. https://doi.org/10.1080/01402382.2017.1395253

Tripathi, B.N., Kumar, D., 2017. *Prospects and Challenges in Algal Biotechnology*. Singapore: Springer, pp. 1–326. https://doi.org/10.1007/978-981-10-1950-0

Ugya, A.Y., Hua, X., Agamuthu, P., Ma, J., 2019. Molecular approach to uncover the function of bacteria in petrochemical refining wastewater: A mini review. *Appl. Ecol. Environ. Res.* 17, 3645–3665. https://doi.org/10.15666/aeer/1702_36453665

Vázquez-Núñez, E., Molina-Guerrero, C.E., Peña-Castro, J.M., Fernández-Luqueño, F., de la Rosa-Álvarez, M.G., 2020. Use of nanotechnology for the bioremediation of contaminants: A review. *Processes* 8, 1–17. https://doi.org/10.3390/pr8070826

Verlicchi, P., Al Aukidy, M., Zambello, E., 2012. Occurrence of pharmaceutical compounds in urban wastewater: Removal, mass load and environmental risk after a secondary treatment-a review. *Sci. Total Environ.* 429, 123–155. https://doi.org/10.1016/j.scitotenv.2012.04.028

Viggor, S., Jõesaar, M., Soares-Castro, P., Ilmjärv, T., Santos, P.M., Kapley, A., Kivisaar, M., 2020. Microbial metabolic potential of phenol degradation in wastewater treatment plant of crude oil refinery: Analysis of metagenomes and characterization of isolates. *Microorganisms* 8, 1–25. https://doi.org/10.3390/microorganisms8050652

Whelan, M.J., Coulon, F., Hince, G., Rayner, J., McWatters, R., Spedding, T., Snape, I., 2015. Fate and transport of petroleum hydrocarbons in engineered biopiles in polar regions. *Chemosphere* 131, 232–240. https://doi.org/10.1016/j.chemosphere.2014.10.088

Wolfe, B.E., 2018. Using cultivated microbial communities to dissect microbiome assembly: Challenges, limitations, and the path ahead. *Am. Soc. Microbiol.* 3, 1–6.

Wu, M., Wu, J., Zhang, X., Ye, X., 2019. Effect of bioaugmentation and biostimulation on hydrocarbon degradation and microbial community composition in petroleum-contaminated loessal soil. *Chemosphere* 237, 124456. https://doi.org/10.1016/j.chemosphere.2019.124456

Xiao, J., Xie, Y., Cao, H., 2015. Organic pollutants removal in wastewater by heterogeneous photocatalytic ozonation. *Chemosphere* 121, 1–17. https://doi.org/10.1016/j.chemosphere.2014.10.072

Yadav, D., Singh, S., Sinha, R., 2021. Microbial degradation of organic contaminants in water bodies. In *Pollutants and Water Management* (P. Singh, R. Singh, V.K. Singh, R. Bhadouria), John Wiley & Sons Ltd, New Jersey: pp. 172–209.

Yadav, T.C., Jadeja, N.B., Kapley, A., 2020. Metagenomic insights in activated biomass treating industrial wastewater at different DO levels. *Appl. Biochem. Biotechnol.* 192, 544–556. https://doi.org/10.1007/s12010-020-03340-8

Yang, S., Wen, X., Zhao, L., Shi, Y., Jin, H., 2014. Crude oil treatment leads to shift of bacterial communities in soils from the deep active layer and upper permafrost along the China-Russia crude oil pipeline route. *PLoS One* 9, 12–14. https://doi.org/10.1371/journal.pone.0096552

Zangi-Kotler, M., Ben-Dov, E., Tiehm, A., Kushmaro, A., 2015. Microbial community structure and dynamics in a membrane bioreactor supplemented with the flame retardant dibromoneopentyl glycol. *Environ. Sci. Pollut. Res.* 22, 17615–17624. https://doi.org/10.1007/s11356-015-4975-8

Zhang, G., Zhang, X., Meng, Y., Pan, G., Ni, Z., Xia, S., 2020a. Layered double hydroxides-based photocatalysts and visible-light driven photodegradation of organic pollutants: A review. *Chem. Eng. J.* 392, 123684. https://doi.org/10.1016/j.cej.2019.123684

Zhang, H., Yuan, X., Xiong, T., Wang, H., Jiang, L., 2020b. Bioremediation of co-contaminated soil with heavy metals and pesticides: Influence factors, mechanisms and evaluation methods. *Chem. Eng. J.* 398, 125657. https://doi.org/10.1016/j.cej.2020.125657

Zhang, X., Song, Z., Tang, Q., Wu, M., Zhou, H., Liu, L., Qu, Y., 2021. Performance and microbial community analysis of bioaugmented activated sludge for nitrogen-containing organic pollutants removal. *J. Environ. Sci.* 101, 373–381. https://doi.org/10.1016/j.jes.2020.09.002

Zhou, Y., Zhang, L., Cheng, Z., 2015. Removal of organic pollutants from aqueous solution using agricultural wastes: A review. *J. Mol. Liq.* 212, 739–762. https://doi.org/10.1016/j.molliq.2015.10.023

9

Bioaugmentation of Petroleum Hydrocarbons and Polycyclic Aromatic Hydrocarbons: A Review

Sunayana Goswami
Biswanath College

Suchismita Das
Assam University

CONTENTS

9.1 Introduction

The whole living world and the sustainability of the entire ecosystem depend on land and water for their maintenance. Anthropogenic activities, rapid industrialization and urbanization are major contributors of the contamination of soil and water (Sayed et al., 2021). Worldwide industrial expansion and escalating use of petroleum energy cause environmental contamination with hydrocarbon compounds (Wainaina et al., 2020; Singh et al., 2021; Paliya et al., 2021). Total petroleum hydrocarbons (TPH) are one of the main universal groups of persistent organic pollutants commonly include crude oil, creosote, gasoline, and diesel (Varjani and Gnansounou, 2017). TPH enter the ecosystem through leakage of oil spills, shipping processes and industrial progressions (Saravanan et al., 2020; Raper et al., 2018; Shahsavari et al., 2015). Chemically, TPH comprises small chain hydrocarbons (C_6–C_{10}) such as toluene, benzene, ethyl benzene, xylene, etc. and

DOI: 10.1201/9781003231738-10

the long-chain hydrocarbons (C_{10}–C_{40}) and highly hydrophobic chemicals such as polycyclic aromatic hydrocarbons (PAH). TPH generally consist of both aliphatic (alkane, alkene, alkyne) and aromatic hydrocarbons, which are classified as priority pollutants (Sayed et al., 2021; Shivalkar et al., 2021; Megharaj et al., 2011).

PAHs are recognized as one of the foremost organic contaminants with significant eco-toxicological concern and are released in the environment due to the partial burning of organic materials like wood, coal and petroleum and with anthropogenic emanations mainly from burning processes such as manufacturing processes and refining processes. PAHs are composed of fused aromatic rings in angular, linear or cluster arrangements (Varjani and Gnansounou, 2017). The PAHs contain a complex mixture of over at least hundred of different organic compounds and are unique in nature as they are generated constantly by incomplete combustion of organic matter and other geological activities (Sayed et al., 2021; Raper et al., 2018). United States Environmental Protection Agency (U.S. EPA) listed 16 PAHs as priority pollutants and are depicted in the Table 9.1. Among the 16 priority PAHs listed by United States Environmental Protection Agency (Urana et al., 2020) Benzo[a]pyrene is the most powerful carcinogen. PAHs are generally less soluble in water and are greatly lipophilic in nature. Due to their recalcitrant nature, they are high-persistent compounds in the environment and have bio accumulative features (Singh et al., 2021; Kotoky et al., 2018) and their presence in the environment is of major concern. Many countries have placed PAH as priority list pollutants due to their carcinogenic, immune-toxic and mutagenic characters (Varjani and Gnansounou, 2017).

TABLE 9.1

Chemical Structures of 16 U.S. Environmental Protection Agency (USEPA) Priority Pollutant Polycyclic Aromatic Hydrocarbon (Urana et al., 2020)

Compound	Structure	MW (g/mol)
Naphthalene	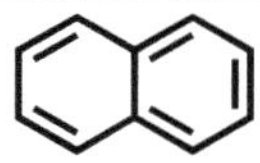	128
Acenaphthylene	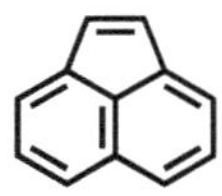	152
Fluorene	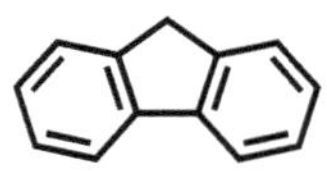	166
Phenanthrene	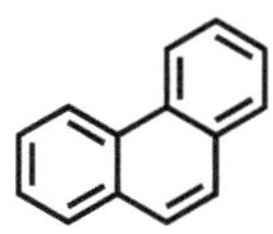	178
Fluoranthene	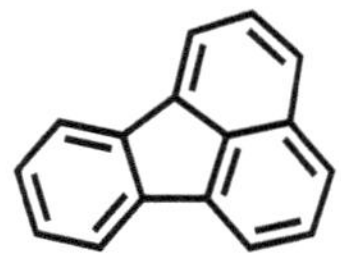	202
Anthracene	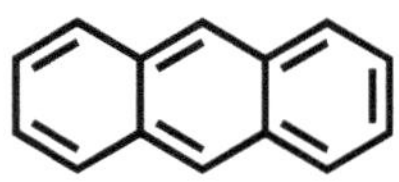	202
Pyrene	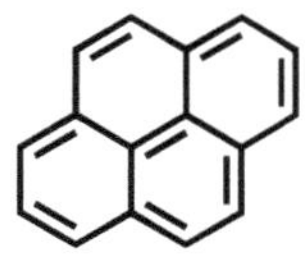	202

(Continued)

TABLE 9.1 (*Continued*)

Chemical Structures of 16 U.S. Environmental Protection Agency (USEPA) Priority Pollutant Polycyclic Aromatic Hydrocarbon (Urana et al., 2020)

Compound	Structure	MW (g/mol)
Benzo(a)anthracene	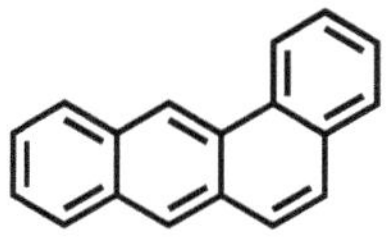	228
Chrysene	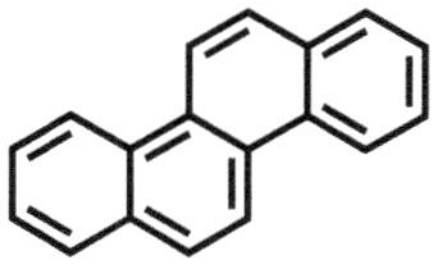	228
Benzo(b)fluoranthene	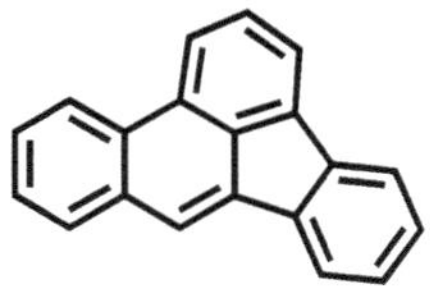	252
Benzo(k)fluoranthene	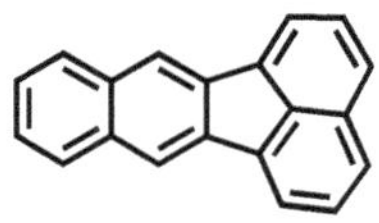	252
Benzo(a)pyrene	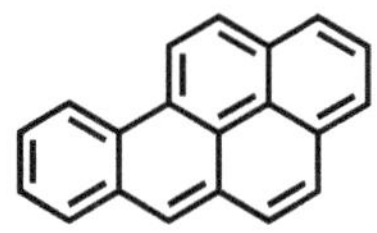	252
Indeno[1,2,3-cd]pyrene	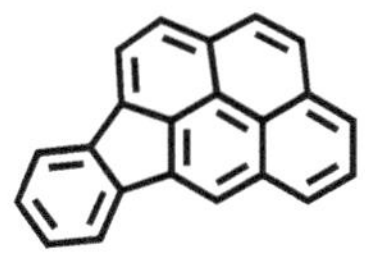	276
Benzo(ghi]perylene		276
Dibenz[a, h]anthracene	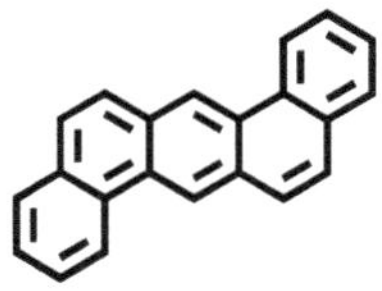	278

Environmental contamination by petroleum hydrocarbon compounds is a global problem, and therefore, a number of strategies have been developed for the remediation of petroleum hydrocarbon (Varjani and Gnansounou, 2017; Agrawal et al. 2021). Different physicochemical approaches have been employed for remediation of petroleum hydrocarbon contaminants such as chemical oxidation, photo-catalytic oxidation, volatilization, sedimentation, etc. These cleanup processes are costly and environmentally destructive. To maintain good eminence of the environment and maintain them free from pollution, incessant efforts have been applied to develop strategies that are trouble-free to use, sustainable, practicable and inexpensive. Due to the low effectiveness and harmful effects of unadventurous waste remediation techniques, bioremediation has emerged, in recent years, as an emerging tool to remediated petroleum hydrocarbon (Sayed et al., 2021; Shivalkar et al., 2021; Agrawal et al., 2021; Megharaj et al., 2011). In the present circumstances, restoration of contaminated soil and water ecosystem is possible by the way of sustainable and eco-friendly processes. In the midst of different approaches being used for the abatement of petroleum hydrocarbon contamination, bioremediation is renowned as a promising scheme for the restoration of polluted environments with petroleum hydrocarbon (Shivalkar et al., 2021).

Bioremediation is a practice that utilizes microbial communities such as bacteria and fungi and their products to diminish, transform and eliminate contaminants from the polluted environment (Kumar et al., 2018). During the process, the complex organic compounds are transformed into simple nontoxic compounds by the microorganisms (Sayed et al., 2021; Zeneli et al., 2019; Raper et al., 2018; Herrero and Stuckey, 2015). One such strategy of prominence is the bioaugmentation (BA): the process of introduction of microorganisms to treat contaminated sites for bioremediation. BA is a bioremediation method where the single strain and consortium of microbes are applied to degrade and transform the pollutants into immobile materials with the help of microbial action and through biochemical methods which drive main action on remediation of toxic contaminants (Zeneli et al., 2019; Raper et al., 2018; Herrero and Stuckey, 2015).

This chapter provides a brief outline of bioaugmentation technology along with different factors that influence degradation processes and considers the bioaugmentation prospect of organic pollutants, especially, petroleum hydrocarbons and polycyclic aromatic hydrocarbons, for environmental cleanup.

9.2 Bioaugmentation

Bioaugmentation (BA) has recently come into sight as a promising approach for *in situ* exclusion of numerous environmental pollutants through several procedures performed by cultured microbe species. Though, in broad sense BA falls under a subset of bioremediation, but it specifically reflects the intricate interactions of microbe with pollutants and enhance removal efficiency. BA is a promising practice in which microbial communities are used to convert the pollutants into immobile materials by microbial action and through biochemical processes (Dave and Das, 2021; Poi et al., 2016). In this biodegradation process, a single strain and consortium of microorganisms are used to degrade the pollutants from the environment (Raper et al., 2018; Zaida and Piakong, 2018). In general, the specific microorganisms and their cultures are full-grown under the appropriate environmental condition and uploaded in the contaminated environment for proficient remediation function. During the process of BA, the microbial culture targets the toxic contaminants by mineralization, transformation, or alteration (Zeneli et al., 2019; Raper et al., 2018). In general, BA technology involves the implication of microbial community isolated from the contaminated site and engrosses the assemblage of energy in a redox reaction inside the microbial cellular system. The chemical reactions during the process consist of respiration and other biological functions essential for cellular maintenance and reproduction. The principle behind the remediation involves the use of microorganisms for degrading target pollutants as a result of the stimulation of the catalytic actions of microorganisms (Segura and Ramos, 2013; Shahsavari et al., 2015). Although the process of BA occurs naturally, the half life of most of the contaminants considered to be long enough and vary greatly from one compound to the next (Sayed et al., 2021). The underlying principle of BA technique is to augment the extent or speed of the degradation process of complex contaminants by the adding up of pollutant degrading microorganisms.

9.3 Biochemistry of Bioaugmentation Technique

BA is a complex process and different microorganisms follow different pathways and different extents of degradation capabilities, but the basic biochemistry of degradation remains the same. Bacteria and fungi characterize the main representative of BA technique and help in the degradation of contaminants in the surroundings, through two different pathways: anaerobic reductive and aerobic oxidative degradation (Tyagi et al., 2011). In this process, complex organic materials are decomposed by microorganisms, with the help of enzymes, into simpler compounds like carbon dioxide, water, ammonia, etc. that are relatively non-toxic or less toxic. The key steps involved in the process of BA, therefore, include the participation of the microbial population in the sequestration of pollutants and their degradation with the help of enzymatic cascade (Dave and Das, 2021; Zaida and Piakong, 2018; Adams et al., 2015). BA is a promising scheme for the removal of many THP from the environment through several processes that generally includes the following steps which are summarized below-

9.3.1 Dehalogenation

In this process, there is a small change in the original organic molecule, leaving the main structure intact. Structure below shows a minor change in the functional group of the original organic molecule such as substitution of hydroxyl group (–OH) in place of chlorine group (–Cl) (Dos Santos and Maranho, 2018; Correa-García et al., 2018).

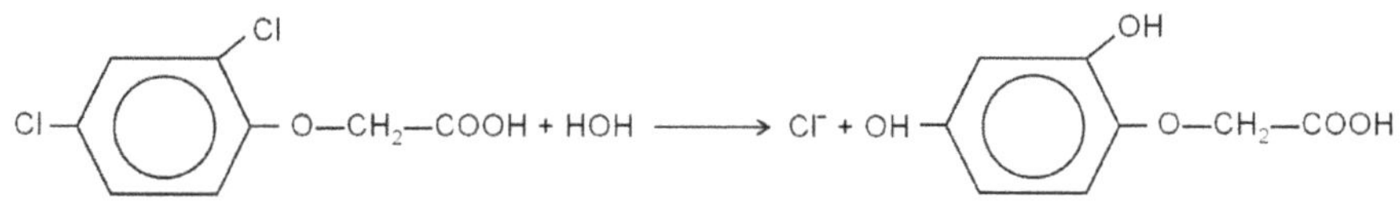

9.3.2 Fragmentation

Fragmentation or breaking apart of a complex organic molecule can lead to smaller fragments that could be reassembled to yield the original structure. The structure below indicates the fragmentation of the organic compounds in such a way that the original molecule could be reconstructed (Correa-García et al., 2018; Vergani et al., 2017).

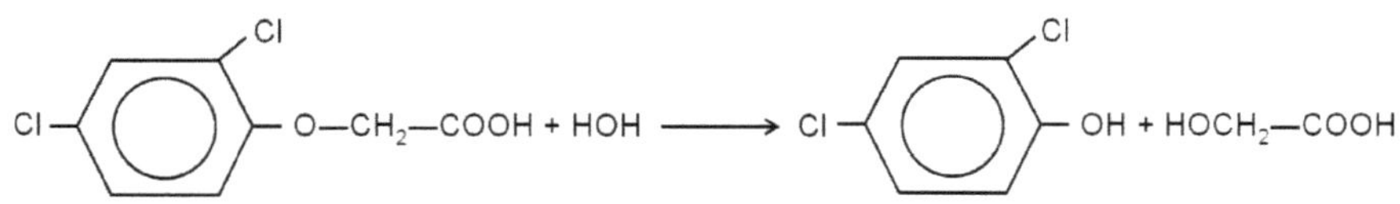

9.3.3 Mineralization

Mineralization is the method of alteration of organic compounds into inorganic forms. Structure here shows the complete degradation of an organic molecule into simpler forms.

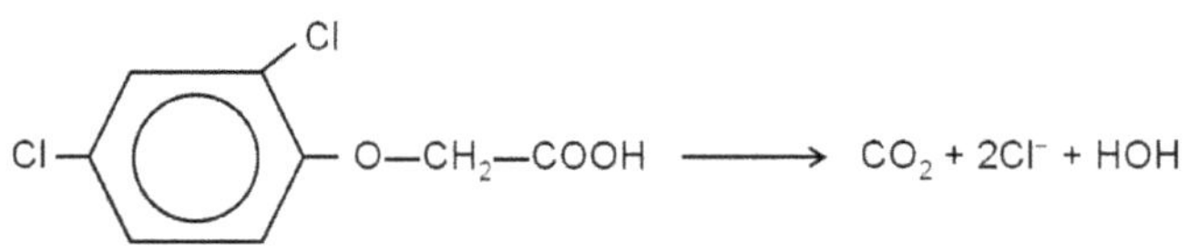

Since every living organism needs oxygen (O_2) as a prerequisite element to perform various metabolic activities for their survival and reproduction and harness ATP as the energy currency of the cell. On the other hand, many microbes those are capable of carrying out respiration both in aerobic (presence of O_2) as well as anaerobic (absence of O_2) environment during biodegradation process also capable of extrapolate energy in the process (Correa-García et al., 2018; Vergani et al., 2017). Two of the basic modes of biodegradation of microbes are illustrated below:

9.3.3.1 Aerobic Mode of Degradation

In this method, microbes use O_2 during the metabolism to degrade contaminants.

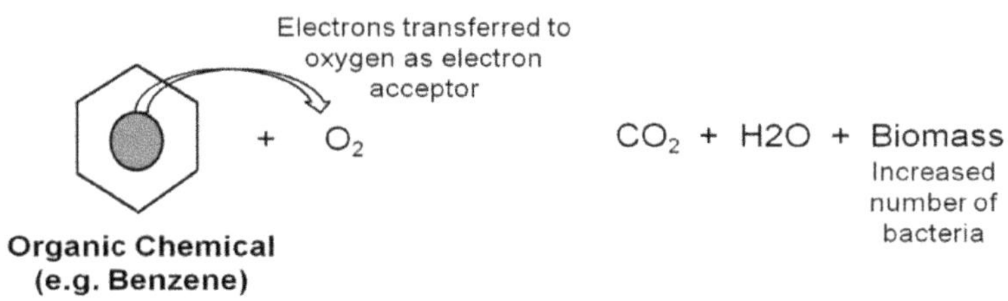

9.3.3.2 Anaerobic Mode of Degradation

In this mode of biodegradation, microorganisms substitute a chemical group (namely nitrate) instead of O_2 to degrade contaminants. Among the various substituted groups include nitrate, iron, sulfate, carbon dioxide, uranium, perchlorate etc. (Dos Santos and Maranho, 2018; Correa-García et al., 2018; Vergani et al., 2017).

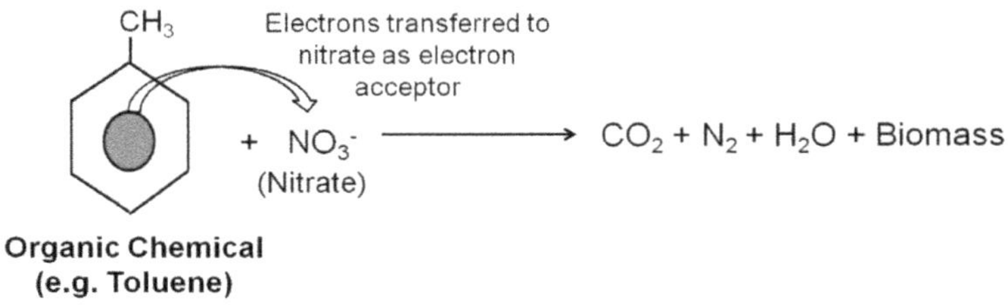

9.4 Factors Affecting Bioaugmentation

The efficiency of BA technology is mainly based on the adaptation of the microbial communities to the contaminated sites that wants to be remediated. The physical, chemical, and biological properties of the wastewater coupled with the organic matter content of ambient environment and their effect on surrounding micro-biota largely determine the efficacy of this process. The temperature, pH, aeration, water quality as well as many other factors such as excretion of root exudates, availability of mineral contents and contaminants type also influences the rate of BA. Moreover, presence of organic matter in the wastewater may act as source for various mineral nutrients like carbon, nitrogen etc. that promotes the growth and survival of a wide range of micro flora and fauna, which contributes mainly to the degradation process (Masciandaro et al., 2013; Alberty, 2006; Siddique et al., 2002).

Some of the environmental factors that influence the process of BA are discussed below:

9.4.1 Water Quality

A successful BA process, that involves wastewater treatments, largely depends on the quality and other physico-chemical factors of the water. Various factors such as pH, moisture, temperature, redox potential, and the availability organic matter determine the quality of water, thereby culminating in improved BA (Alberty, 2006).

9.4.2 Temperature

Temperature is another crucial factor that can regulate the thermodynamics of various enzymes and co-factors of microbial community necessary for successful BA. Temperature can also have a tremendous impact on the cell-membrane functions like membrane permeability that is thought to be involved in various biochemical reactions in the bioremediation process (Siddique et al., 2002).

9.4.3 pH

Like temperature, pH is one of the most crucial and determining factors which changes the water acidity and alkalinity and has a large impact on the enzymatic activity of the microbes. An adjustment in water pH during the remediation process is of paramount importance and it has been found that the enzymatic potential of many microbial species functions optimally at pH range of 6.5–8.5 (Siddique et al., 2002).

9.4.4 Organic Matter

The presence of organic matter in the water also influences the microbe interactions in the contaminated water. The degradation potential principally depends on the rate of absorption/desorption of contaminants by assisted microbes that in turn depends on the presence or absence of organic matter in the water. Addition of organic matter such as compost, sewage sludge etc. enhances the remediation process by bioaugmenting native microbial communities at the treatment site Therefore, the application of organic matter not only improves the water quality but also has synergistic impact on microbial populations (Masciandaro et al., 2013).

9.4.5 Redox Potential and Oxygen Content

Redox potential and oxygen content present in the wastewater play an important role in BA practice. It is generally influenced by the occurrence of electron acceptors such as nitrate, manganese oxides, iron oxides and sulphate (Adams et al., 2015).

9.4.6 Nutrients

Nutrient load present in the surrounding background also plays an important role in BA process. Proper nutrients are required for microbial cell development and division. Appropriate nutrient contents serve as an electron donor, to stimulate the biodegradation process (Adams et al., 2015).

9.4.7 Plant Root Exudates

The quality and amount of root exudates secreted by the plant root organization dramatically influ ence the accessibility of microbial populations in the root ecosystem of aquatic plants. Root exudates include various organic compounds like sugars, chelating agents, amino acids, proteins, enzymes etc. that play a significant task in the development and proliferation of microbes thereby increasing the rate of degradation of contaminants. Thus, the engagement of microbial population in bioremediation is considered to be one of the prime factors accountable for the degradation of contaminants (Kotoky et al., 2018) (Figure 9.1).

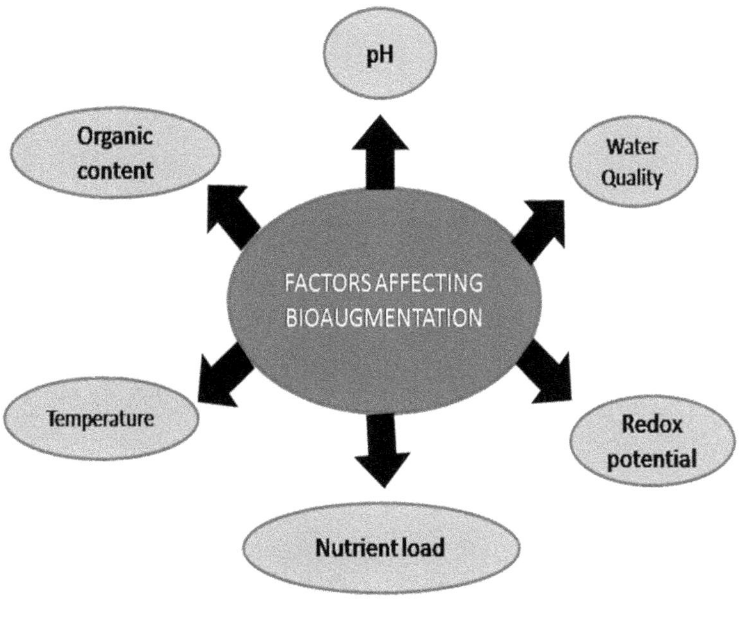

FIGURE 9.1 Schematic representation of various factors that affect bioaugmentation process.

9.5 Bioaugmentation of Total Petroleum Hydrocarbons (TPH)

TPH is measured as one of the most universal groups of potent persistent organic pollutants and the remediation of petroleum hydrocarbons contaminated system is a big challenge (Singh et al., 2021; Zaida and Piakong, 2018; Goswami et al., 2018). Bioaugmentation of total petroleum hydrocarbons polluted environment with the help of microorganisms is an eminent process. Microbial communities consume crude oil as a source of energy from the total petroleum hydrocarbons and trim down them to simpler non-toxic forms. However, the native bioremediation process acquires a longer period for total retrieval of a contaminated system due to the fact that native inhabitants of hydrocarbon degrading microorganism is noticeably low and they are not able to resist the high concentration of noxious contaminants at the polluted sites (Kotoky et al., 2018). The bioaugmentation of microorganisms with high competence for petroleum hydrocarbons degradation established as a prominent technique for efficient bioremediation. BA enhanced remediation of TPH using selected microbial species are profitable ventures which maintain the biological and physical properties of the system. It is an efficient, economical and environment friendly remediation practice for handling petroleum hydrocarbon polluted areas (Kotoky et al., 2018; Terzaghi et al., 2019).

Microorganisms considerably control the conversion and fate of degradation process of TPH in the environment. Several microbial species have been reported and identified to have a broad range of petroleum hydrocarbon degradation capability. Many scientists reported that diverse groups of microorganisms with broad enzymatic capabilities are efficient to degrade intricate mixtures of hydrocarbons and found to participate in fundamental roles in petroleum hydrocarbon degradation procedure (Kotoky et al., 2018; Varjani and Gnansounou, 2017). Table 9.2 depicts the microorganisms which are used in bioaugmentation for petroleum hydrocarbons degradation.

TABLE 9.2

Microbes Used for Degradation of Petroleum Hydrocarbons Using Bioaugmentation Process

Petroleum Hydrocarbon Degrading Microbes (Varjani and Gnansounou, 2017)		
Microorganism	**Petroleum Hydrocarbon Compound**	**References**
Alcanivorax sp. *Acinetobacter* sp. *Pseudomonas* sp. *Micrococcus* sp. *Rhodococcus* sp. *Desulfosarcina* sp. *Ochrobactrum* sp. *Oleispira* sp. *Aspergillus* sp. *Candida* sp. *Penicillium* sp.	Aliphatics	Singh et al. (2021); Kotoky et al. (2018)
Archaeoglobus fulgidus *Acinetobacter* sp. *Achromobacter insolitus* *Bacillus* sp. *Halomonas* sp. *Pseudomonas* sp. *Rhodococcus* sp. *Mycobacterium cosmeticum* *Sphingobacterium* sp. *Cycloclasticus* sp. *Vibrio* sp.	Mono and poly aromatics	Singh et al. (2021); Kotoky et al. (2018)
Moraxella sp. *Pseudomonas* sp. *Bacillus* sp.	Resins	Singh et al. (2021); Kotoky et al. (2018)

Microbial community such as bacteria, fungi, algae are recognized as prominent agents for their capability to degrade petroleum hydrocarbon. Among them Bacterial species are the most active agents in the petroleum hydrocarbon degradation process. Several Bacterial species Such as genera of *Pseudomonas, Mycobacterium, Achromobacter, Bacillus, Acinetobacter, Rhizobacterium, Cellulomonas, Marinobacter, Micrococcus, Nocardia, Flavobacterium, Ochrobactrum, Vibrio* and *Stenotrophomaonas* are documented as an effective hydrocarbon degrading agents. Among the fungal species of genera *Aspergillus, Paecilomyces, Graphium, Neosartoria, Fusarium, Penicillium, Talaromyces,* and *Sporobolomyces* and genera of yeasts such as Candida, Pichia, and Yarrowia take part in a central role in petroleum hydrocarbon degradation process (Singh et al., 2021; Kotoky et al., 2018; Varjani and Gnansounou, 2017).

Biodegradation of petroleum hydrocarbons is an intricate method largely depends on the chemistry and the amount of hydrocarbons present in the environment. Total petroleum hydrocarbons include aliphatic and aromatic hydrocarbons. Some microbes possess the capability to degrade aliphatics and some can decompose monoaromatics or polyaromatics. Aliphatic hydrocarbons being generally straight chain compounds are comparatively easily degraded by microbial communities (Singh et al., 2021; Wilkes et al., 2016; Abdel-Shafy and Mansour, 2016). In contrast, aromatic hydrocarbons contain one or more ring structure attached and also relatively persistent in nature making the compounds even harder to degrade. The more the number aromatic rings attached to a compound the harder it becomes to be degraded by microbes. Biodegradation of petroleum hydrocarbons engross a sequence of steps using different enzymes secreted by the microbial communities to facilitate the breakdown of bonds of hydrocarbons (Singh et al., 2021). Metabolism of Hydrocarbons can be selectively taken place by a specific single strain of microorganism or consortium of microbial strains with identical or different genera. Again, it has been found that due to the relative specificity of microbes towards a particular contaminant, microbial consortium may be handy to degrade down the hydrocarbon-based contaminants (Singh et al., 2021; Kotoky et al., 2018). It has been documented that enzymes needed for the biodegradation of hydrocarbons are encoded on plasmids. In general, microbes catabolize petroleum hydrocarbon to gain energy and take up them into cell biomass (Varjani and Gnansounou, 2017).

There are diverse rates of microbial degradation of various petroleum hydrocarbon products and it depends on two main pathways of biodegradation. Hydrocarbon catabolism generally involves aerobic pathways; still certain microbes are also documented for anaerobic degradation of hydrocarbons (Meckenstock et al., 2016; Wilkes et al., 2016). Both the aerobic and anaerobic pathways of microorganisms assisted petroleum hydrocarbon degradation process to involve the reactions such as oxidation, reduction, dehydrogenation, and hydroxylation. Under aerobic environment, a sufficient amount of oxygen increases the porosity of the environment thereby augmenting indigenous microbiota to stimulate the process of biodegradation of hydrocarbons. In this aerobic mode, O_2 is required for the microbes to metabolize the hydrocarbon-contaminants through several enzymatic steps, therefore, breaking down the complex hydrocarbon into simpler forms. Aerobic biodegradation process of the petroleum hydrocarbon degradation mechanism in microorganisms generally involves β-oxidation, terminal oxidation, and sub-terminal oxidation pathways (Sayed et al., 2021; Raper et al., 2018). General mechanisms for petroleum hydrocarbon biodegradation includes degradation of alkane with methyl group oxidation and which consequences in the production of alcohol group, which is subsequently dehydrogenated to carboxylic acid and eventually metabolized by β-oxidation pathway of fatty acids (Kotoky et al., 2018; Abdel-Shafy and Mansour, 2016). Understanding anaerobic mode of biodegradation is more current as compared to aerobic biodegradation, hence a smaller amount of information is available regarding the enzymes involved during degradation pathways. Under the anaerobic metabolism process petroleum hydrocarbon are first converted into organic acids, and subsequently altered to long-chain fatty acids, which are ultimately metabolized to methane and CO_2 (Sayed et al., 2021; Raper et al., 2018).

9.5.1 Bioaugmentation of Polycyclic Aromatic Hydrocarbons (PAHs)

PAHs are chemical compounds mainly consist of carbon and hydrogen molecules containing two or more aromatic rings attached. Over hundreds of PAHs are estimated to be found in nature. The simplest of the lot is naphthalene having two aromatic rings and others are phenanthrene, anthracene, pyrene, fluorene, benz[a]anthracene etc. containing more than two aromatic rings (Kotoky et al., 2018; Abdel-Shafy and Mansour, 2016). Since they possess various toxicological properties and can cause serious health issues and are generally ubiquitous in their distribution therefore considered as priority pollutants of environmental concern. The physico-chemical properties of PAHs also render them as persistent environmental contaminants as they have high boiling and melting points and very low solubility. PAHs can be categorized into two main domains *i.e.* low molecular weight PAHs (LPAHs) and high molecular weight PAHs (HPAHs). LPAHs generally have two to three aromatic rings whereas HPAHs contain four or more rings (Shahsavari et al., 2015). Furthermore, the solubility of PAHs declines with the increase in fused aromatic rings of PAH molecules. Since each PAH molecule has a unique structure and physico-chemical properties, different modes of degradation methods may be employed (namely biodegradation, photooxidation, and chemical oxidation) for a different set of PAH molecules (Sayed et al., 2021).

Bioaugmentation of PAHs is considered to be an effective method of remediation technology as compared to the other bioremediation processes especially when an appropriate microenvironment is encouraged (Shahsavari et al., 2015). Microorganism consumes energy from the organic pollutants required for cell growth and metabolism and the microbial enzyme system assists to metabolize and extract energy from the contaminant. The principle behind the bioremediation of PAHs involves the use of an enzymatic cascade of microorganisms for degrading target pollutants (Kotoky et al., 2018). The microbial community utilizes different enzymatic modules in the biodegradation of PAHs such as hydrolases, transferases, dehalogenases, and oxidoreductases. For example, *Bacillus safensis* produces oxidoreductase to biodegrade the oil mixes from the surrounding environment. Nutrients take steps indirectly in this progression by stimulating the catalytic actions of the microbes and supply a nutrient rich environment for stimulation of microbial activity which facilitates the degradation of PAHs in the environment (Dave and Das, 2021; Kotoky et al., 2018; Abdel-Shafy and Mansour, 2016).

9.5.1.1 Bacterial Mechanisms of PAH Metabolism

A number of bacterial species are concerned in the process of biodegradation of Polycyclic Aromatic Hydrocarbons. Bacteria are the most dynamic mediator in PAH degradation process. PAH_S are degraded and metabolized by different bacterial species because of their carbon requirements for development and reproduction. Bacteria generally utilizes organic content present in the surrounding for ring hydroxylation in the catabolism of PAHs during biodegradation (Dave and Das, 2021). Bacteria generally employ oxidizing reactions pathways for the PAH degradation and metabolism. In general, the pathways vary to a great extent because of the specific oxygenases found in different bacterial classes. Several bacterial species metabolize and break down aromatic hydrocarbons depending on the chemical structure of the compound (Segura and Ramos, 2013). Current research has recognized bacteria belonging to more than 79 genera that are capable of PAH degradation and metabolism (Singh et al., 2021; Kotoky et al., 2018). PAH degrading bacteria and the type of hydrocarbon compounds they degrade are listed in Table 9.3.

9.5.1.2 Fungal Mechanisms of PAH Metabolism

Fungal mechanism of PAH degradation is broadly classified into two major modules: non-ligninolytic mediated and ligninolytic fungi mediated strategy. In non-ligninolytic mediated pathway fungal species generally metabolize PAH compounds, with the help of P450 monooxygenase, which helps in the hydroxylization of PAH and converted the compounds into polar soluble components. *Cunninghamella elegans, Chrysosporium pannorum,* and *Aspergillus niger* are examples of non-ligninolytic fungi, such kinds of species use P450 monoxygenase enzyme-mediated pathway for PAH degradation (Sayed et al., 2021; Gupta and Pathak, 2020).

Ligninolytic fungal communities are more proficient in the degradation of PAHs and have capability to depolymerize lignin. In a ligninolytic module, fungal species can metabolize and oxidize PAHs via

TABLE 9.3
Bacterial Specie Used for PAH Metabolism and Degradation Process

Microbes Used	Contaminant(s) Degraded	References
Pseudomonas fluorescens, Pseudomonas aureofaciens	Phenanthrene	Singh et al. (2021); Kotoky et al. (2018)
Pseudomonas strains, UW3 and UW4	Various PAHs	
Mycobacterium sp. strain KMS.	Pyrene	
Pseudomonas putida	Phenanthrene, Napthalene, Various PAHs	
Mycobacterium sp.	Phenanthrene, Anthracene	
Mycobacterium sp.	Fluoranthene	
Pseudomonas stutzeri	Pyrene	
Pseudomonas methancia	Petroleum hydrocarbon	
Pantoea sp. strain BTRH79 *Acinetobacter* sp. strain	Various PAHs	
Rhodococcus sp. strain	Various PAHs	
Rhizobacterium gordonia sp. *Azospirillum lipoferum* sp.	Various PAHs	
Acinetobacter sp.	Various PAHs	
Sinorhizobium meliloti	Phenanthrene	
Rhizobium leguminosarum	Chrysene	

producing hydroxyl free radicals. ligninolytic fungal degradation of PAHs generally employ enzymes such as cytochrome P450 monooxygenases, hydrolases and ligninolytic enzymes, which direct the complete mineralization of the hydrocarbon molecule (Gupta and Pathak, 2020). Ligninolytic fungi such as White-rot fungi produce peroxidase enzyme viz lignin peroxidase and manganese peroxides which are proficient in PAH oxidation and metabolism. Ligninolytic fungi can oxidize PAHs by generating free radicals by the contribution of one electron which oxidizes the PAH ring.

Fungal biodegradation of PAHs generally involved an extracellular process, which involves in the discharge of different oxido-reductase enzymes, such as manganese peroxidases, laccases, and lignin peroxidases in the environment (Urana et al., 2020). The speed of bioaugmentation of PAHs depends on the growth attributes of microbial species as well as bioavailability, electron acceptors, oxygen availability as well as temperature, pH, etc. of the surrounding environment (Gupta and Pathak, 2020; Kotoky et al., 2018)

9.6 Conclusion

Bioaugmentation is the process of addition or intensification of the specific microorganisms to the polluted environment for bioremediation. It proposes an environmentally friendly choice to get rid of toxic pollutants from the environment. Implication of bioaugmentation for the treatment of petroleum hydrocarbons from the surroundings has gained increasing attention as a bioremediation strategy and has been documented to be an effective, cost-effective, pleasing, ecofriendly technology to remediate contaminants from the surroundings. Many research works have been carried out on the basis of trial-and-error method, and it has been reported that implication of bioaugmentation process with cultured microorganism for the treatment of petroleum hydrocarbon often fail to execute properly. Bioaugmentation conveys a sustainable management of petroleum waste, which will be a boon for coming generations. Intensive research is required for the implication of cost-effective bioaugmentation technology in near future and in a more economical way for rapid environmental remediation.

REFERENCES

Abdel-Shafy, H. I., & Mansour, M. S. (2016). A review on polycyclic aromatic hydrocarbons: source, environmental impact, effect on human health and remediation. *Egyptian Journal of Petroleum*, 25(1), 107–123.

Adams, G. O., Fufeyin, P. T., Okoro, S. E., & Ehinomen, I. (2015). Bioremediation, biostimulation and bioaugmention: a review. *International Journal of Environmental Bioremediation & Biodegradation*, 3(1), 28–39.

Agrawal, N., Kumar, V., & Shahi, S.K. (2021). Biodegradation and detoxification of phenanthrene in in-vitro and in-vivo conditions by a newly isolated ligninolytic fungus *Coriolopsis byrsina* strain APC5 and characterization of their metabolites for environmental safety. *Environmental Science and Pollution Research*. https://doi.org/10.1007/s11356-021-15271-w

Alberty, R. A. (2006). Biochemical reactions at specified temperature and various pHs. In *Biochemical Thermodynamics* (pp. 43–70). Cambridge, MA: John Wiley & Sons, Inc.

Correa-García, S., Pande, P., Séguin, A., St-Arnaud, M., & Yergeau, E. (2018). Rhizoremediation of petroleum hydrocarbons: a model system for plant microbiome manipulation. *Microbial Biotechnology*, 11(5), 819–832.

Dave, S., & Das, J. (2021). Role of microbial enzymes for biodegradation and bioremediation of environmental pollutants: challenges and future prospects. In Saxena, G., Kumar, V., & Shah, M.P. (eds.) *Bioremediation for Environmental Sustainability* (pp. 325–346). New Delhi: Elsevier.

Dos Santos, J. J., & Maranho, L. T. (2018). Rhizospheric microorganisms as a solution for the recovery of soils contaminated by petroleum: a review. *Journal of Environmental Management*, 210, 104–113.

Goswami, M., Chakraborty, P., Mukherjee, K., Mitra, G., Bhattacharyya, P., Dey, S., & Tribedi, P. (2018). Bioaugmentation and biostimulation: a potential strategy for environmental remediation. *Journal of Microbiology & Experimentation*, 6(5), 223–231.

Gupta, S., & Pathak, B. (2020). Mycoremediation of polycyclic aromatic hydrocarbons. In Singh, P., Kumar, A., & Borthakur, A. (eds.) *Abatement of Environmental Pollutants* (pp. 127–149). New Delhi: Elsevier.

Herrero, M., & Stuckey, D. C. (2015). Bioaugmentation and its application in wastewater treatment: a review. *Chemosphere*, 140, 119–128.

Kotoky, R., Rajkumari, J., & Pandey, P. (2018). The rhizosphere microbiome: significance in rhizoremediation of polyaromatic hydrocarbon contaminated soil. *Journal of Environmental Management*, 217, 858–870.

Kumar, V., Shahi, S.K., & Singh, S. (2018). Bioremediation: an eco-sustainable approach for restoration of contaminated sites. In Singh, J., Sharma, D., Kumar, G., & Sharma, N. (eds) *Microbial Bioprospecting for Sustainable Development*. Singapore: Springer. https://doi.org/10.1007/978-981-13-0053-0_6

Masciandaro, G., Macci, C., Peruzzi, E., Ceccanti, B., & Doni, S. (2013). Organic matter–microorganism–plant in soil bioremediation: a synergic approach. *Reviews in Environmental Science and Bio/Technology*, 12(4), 399–419.

Meckenstock, R. U., Boll, M., Mouttaki, H., Koelschbach, J. S., Tarouco, P. C., Weyrauch, P., ... & Himmelberg, A. M. (2016). Anaerobic degradation of benzene and polycyclic aromatic hydrocarbons. *Journal of Molecular Microbiology and Biotechnology*, 26(1–3), 92–118.

Megharaj, M., Ramakrishnan, B., Venkateswarlu, K., Sethunathan, N., & Naidu, R. (2011). Bioremediation approaches for organic pollutants: a critical perspective. *Environment International*, 37(8), 1362–1375.

Paliya, S., Mandpe, A., Kumar, M.S. & Kumar, S. (2021). Aerobic degradation of decabrominated diphenyl ether through a novel bacterium isolated from municipal waste dumping site: identification, degradation and metabolic pathway. *Bioresource Technology*, 333, 125208. https://doi.org/10.1016/j.biortech.2021.125208

Poi, G., Shahsavari, E., & Aburto-Medina, A. (2016). Large scale bioaugmentation of municipal wastewater contaminated with petroleum hydrocarbons. *JSM Environmental Science & Ecology*, 4, 1036.

Raper, E., Stephenson, T., Anderson, D. R., Fisher, R., & Soares, A. (2018). Industrial wastewater treatment through bioaugmentation. *Process Safety and Environmental Protection*, 118, 178–187.

Saravanan, A., Jeevanantham, S., Narayanan, V. A., Kumar, P. S., Yaashikaa, P. R., & Muthu, C. M. (2020). Rhizoremediation–a promising tool for the removal of soil contaminants: a review. *Journal of Environmental Chemical Engineering*, 8(2), 103543.

Sayed, K., Baloo, L., & Sharma, N. K. (2021). Bioremediation of total petroleum hydrocarbons (TPH) by bioaugmentation and biostimulation in water with floating oil spill containment booms as bioreactor basin. *International Journal of Environmental Research and Public Health*, 18(5), 2226.

Segura, A., & Ramos, J. L. (2013). Plant–bacteria interactions in the removal of pollutants. *Current Opinion in Biotechnology*, 24(3), 467–473.

Shahsavari, E., Adetutu, E. M., Taha, M., & Ball, A. S. (2015). Rhizoremediation of phenanthrene and pyrene contaminated soil using wheat. *Journal of Environmental Management*, 155, 171–176.

Shivalkar, S., Singh, V., Sahoo, A. K., Samanta, S. K., & Gautam, P. K. (2021). Bioremediation: a potential ecological tool for waste management. In Kumar, V., Saxena, G., & Shah, M.P. (eds.) *Bioremediation for Environmental Sustainability* (pp. 1–21). New Delhi: Elsevier.

Siddique, T., Okeke, B. C., Arshad, M., & Frankenberger, W. T. (2002). Temperature and pH effects on biodegradation of hexachlorocyclohexane isomers in water and a soil slurry. *Journal of Agricultural and Food Chemistry*, 50(18), 5070–5076.

Singh, S., Kumar, V., Datta, S., Dhanjal, D. S., Parihar, P., &Singh, J. (2021). Role of plant–microbe systems in remediation of petrochemical-contaminated water and soil environment. In Kumar, A., Singh, V.K. … Mishra, V.K. (eds.) *Microbe Mediated Remediation of Environmental Contaminants-Woodhead Publishing Series in Food Science, Technology and Nutrition.* (pp. 79–88). New Delhi: Elsevier.

Terzaghi, E., Vergani, L., Mapelli, F., Borin, S., Raspa, G., Zanardini, E., … & Di Guardo, A. (2019). Rhizoremediation of weathered PCBs in a heavily contaminated agricultural soil: results of a biostimulation trial in semi field conditions. *Science of the Total Environment*, 686, 484–496.

Tyagi, M., da Fonseca, M. M. R., & de Carvalho, C. C. (2011). Bioaugmentation and biostimulation strategies to improve the effectiveness of bioremediation processes. *Biodegradation*, 22(2), 231–241.

Urana, R., Dahiya, A., Singh, N., & Sharma, P. (2020). A review on rhizoremediation: plant-microbe interaction enhances the degradation of polyaromatic hydrocarbons. In Singh, J.S., & Vimal, S.R. (eds.) *Microbial Services in Restoration Ecology* (pp. 283–295). New Delhi: Elsevier.

Varjani, S. J., & Gnansounou, E. (2017). Microbial dynamics in petroleum oilfields and their relationship with physiological properties of petroleum oil reservoirs. *Bioresource Technology*, 245, 1258–1265.

Vergani, L., Mapelli, F., Zanardini, E., Terzaghi, E., Di Guardo, A., Morosini, C., … & Borin, S. (2017). Phyto-rhizoremediation of polychlorinated biphenyl contaminated soils: an outlook on plant-microbe beneficial interactions. *Science of the Total Environment*, 575, 1395–1406.

Wainaina, S., Awasthi, M.K., Sarsaiya, S., Chen, H., Singh, E., Kumar, A., … & Kumar, S. (2020). Resource recovery and circular economy from organic solid waste using aerobic and anaerobic digestion technologies. *Bioresource Technology*, 301, 122778. https://doi.org/10.1016/j.biortech.2020.122778

Wilkes, H., Buckel, W., Golding, B. T., & Rabus, R. (2016). Metabolism of hydrocarbons in n-alkane-utilizing anaerobic bacteria. *Journal of Molecular Microbiology and Biotechnology*, 26(1–3), 138–151.

Zaida, Z. N., & Piakong, M. T. (2018). Bioaugmentation of petroleum hydrocarbon in contaminated soil: a review. In Kumar, V., Kumar, M., & Prasad, R. (eds) *Microbial Action on Hydrocarbons* (pp. 415–439). Singapore: Springer.

Zeneli, A., Kastanaki, E., Simantiraki, F., & Gidarakos, E. (2019). Monitoring the biodegradation of TPH and PAHs in refinery solid waste by biostimulation and bioaugmentation. *Journal of Environmental Chemical Engineering*, 7(3), 103054.

10

Microbial Biofilms for Efficient Biological Wastewater Treatment: Mechanisms, Challenges, Opportunities and Future Perspectives

Shradhanjali Sethi, Bijaylaxmi Sahoo, and Rima Biswas
CSIR-National Environmental Engineering Research Institute
Academy of Science and Innovative Research (AcSIR)

Rohan Gupta
CSIR-National Environmental Engineering Research Institute

CONTENTS

DOI: 10.1201/9781003231738-11

10.1 Introduction

Growing urbanization and industrialization has created tremendous pressures on natural resources like water, air, and soil. The increasing water crisis has led to growing interest in recovering water globally. Wastewater, especially released from the households is now being recognized as a source of water, energy, and nutrient (Qadir et al., 2020). Annually 380 billion m^3 of wastewater are released globally which is five times the annual volume of water falling in Nigeria Falls (Qadir et al., 2020). The cost of wastewater treatment is an important element that decides the extent of wastewater treatment and resource recovery from wastewater (Wainaina et al., 2020; Dutta et al., 2021). Cost minimization has become an important mandate in wastewater treatment. Biological methods of wastewater treatment are often selected in wastewater treatment schemes for their lowest costs of treatment.

Several biological treatment processes have found their applications in wastewater treatment plants. Biological treatment of wastewater can be based on their oxygen demands – anaerobic, anoxic, and aerobic; or on the basis of their nature of the biomass growth – suspended or attached growth. Suspended growth processes such as activated sludge process (ASP), sequential batch reactor (SBR) are most common biological processes. However, the aeration and mixing to keep the biomass in suspension, higher biomass growth rate needs additional sludge settling, recycling, wasting, and handling makes the suspended growth processes costlier than attached growth processes. Efforts to reduce the cost of aeration are forcing wastewater engineers to revisit the older biological technologies that were based on attached growth or the biofilm processes.

Biofilm technologies such as rotating biological contactors (RBC), trickling filters, fixed-film reactors, fluidized bed reactors, membrane aerated biofilm reactors are some examples that have been applied in biological treatment of wastewaters. Biofilm processes include some of the oldest technologies developed in the history of wastewater treatment while some of those are the most recent and the advanced ones. The flexibility of biofilm processes to adapt technological innovations in wastewater process engineering makes it a suitable process to be worked on. The use of biofilm technologies is an attractive option for small-scale or low-cost applications due to its simpler operation, lower maintenance cost, nil or low aeration cost, lower sludge handling costs and more stability. Biofilm technologies not only offer low-cost operation but also reduce the land requirement by eliminating the need for secondary clarifiers. Biofilm processes also increase the sludge age which is essential in degradation of recalcitrant or xenobiotic organic pollutants present in the wastewater (Todhanakasem et al., 2014). These systems are also effective against antimicrobial substances, toxic chemicals, heavy metals, organic chemicals and other recalcitrant compounds (Halan et al., 2011; Pamp et al., 2009). The property of the bacteria to resist hostile environmental conditions is useful in wastewater treatment processes.

Biofilm technologies for wastewater treatment are not only involved in the removal of carbon, nitrogen and sulphur but also actively involved in the removal of recalcitrant contaminants. These systems nevertheless are challenged by sloughing, clogging and lower diffusion rates of substrates which has resulted in outnumbering by the highly efficient suspended biomass technologies in the last few decades. However, new challenges like emerging contaminants and cost economics in wastewater management are pushing treatment processes from being energy-intensive to energy negative processes. Hence, it is high time to revisit the sustainable biofilm technologies and revamp them up for their bottlenecks. Fundamental research into biofilm technologies for wastewater management is warranted to revive the existing biofilm technologies. The present chapter reviews various biofilm technologies based on their performance and stability. The mechanism, challenges and opportunities of these biofilm technologies in the area of wastewater treatment are also discussed.

10.2 Mechanism of Biofilm Formation

Biofilm is an essential element for microorganisms that help them to survive in adverse environmental conditions. It helps microorganisms to sustain in a hostile environment and to colonize new niches (Hall-Stoodley et al., 2004). The term 'biofilm' refers to the association of microbes with the mucilaginous covering on the surface of the matrix with extracellular polymeric substances (EPS). Biofilms encompass a wide group of planktonic cells that adhere to any abiotic or biotic substratum with the help of EPS (O'Toole et al., 2000; Hall-Stoodley et al., 2004).

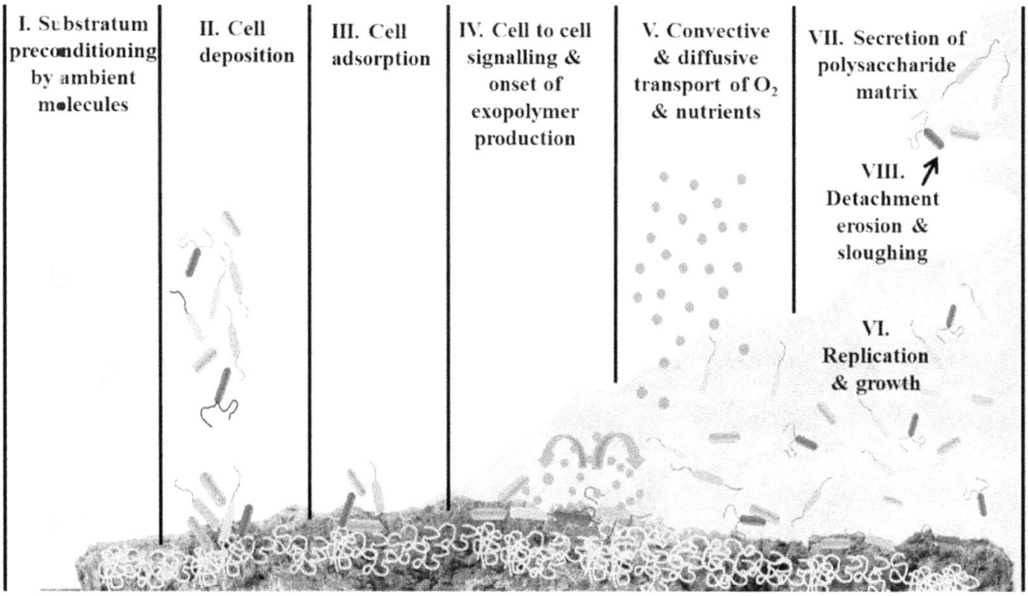

FIGURE 10.1 Different steps of biofilm formation.

The developmental cycle of microbial biofilms consists of five phases (Cogan and Keener, 2004). As depicted in Figure 10.1, in the initial phase bacterial attachment to the substratum takes place, followed by proliferation of cells in the subsequent phases. The bacterial growth finally leads to the colonization in the enclosed space and produces biofilm.

10.2.1 Extracellular Polymeric Substances (EPS)

The biofilm formation is incomplete without the EPS. The EPS are an inevitable component of a biofilm system. It is a complex mixture of intertwined biological polymers secreted by microorganisms, the product of hydrolysis, and adsorbed organics from the wastewater. The EPS usually helps in the binding of microbes to the substratum and to each other. It is a gel-like structure which keeps microbial cells together, provides nutrients and food to the new bacterial cells and protects the bacterial cells from unfavorable environmental conditions (Whitfield et al., 2015). In general, EPS are enriched with biopolymers like protein, polysaccharide and nucleic acid that serve not only as a nutrient but also as a means to protect the cells from environmental stresses (Frederick et al., 2011). The wide range of biomolecules found across the biofilm structure are described in Figure 10.2. Enzymes, globular proteins, extracellular polysaccharides, lipopolysaccharides, amyloids, long adhesion proteins, nucleic acids, glycoproteins, membrane vesicles and amphiphiles are some components that are actively involved in EPS synthesis.

EPS synthesis is species-specific, and it is secreted by a microbial cell in response to various environmental conditions and stress factors like toxins, nutrient or carbon deficiency or competition. Under normal conditions, a microbial cell may not produce any EPS. Often the cellular contents of dying cells also add to the EPS. It is reported that the carbohydrate fraction of EPS is synthesized and secreted by the microbial cells in the biofilm from time to time, while the protein and lipid fractions of the biofilm are added upon the death and lysis of the microbial cell. Change in the microbial dynamics of the biofilm can change the composition of the polymers. The composition of EPS carries from case to case. Literature suggests that the composition of EPS is not consistent as varies with the physiological status of the microbial cells in the microbial biofilm, their composition and on their response to various stress factors (Liu et al., 2004). Hence, the physiological status of a microbial community in the biofilm can be identified over a period of time by analyzing the composition of EPS from time to time. In wastewater treatment, the composition and physical properties of EPS are dependent on the operation conditions of the process. The operational conditions such as hydraulic retention time (HRT), organic loading rate (OLR), concentrations of COD/BOD, nutrients, dissolved oxygen, temperature and pH, biomass age,

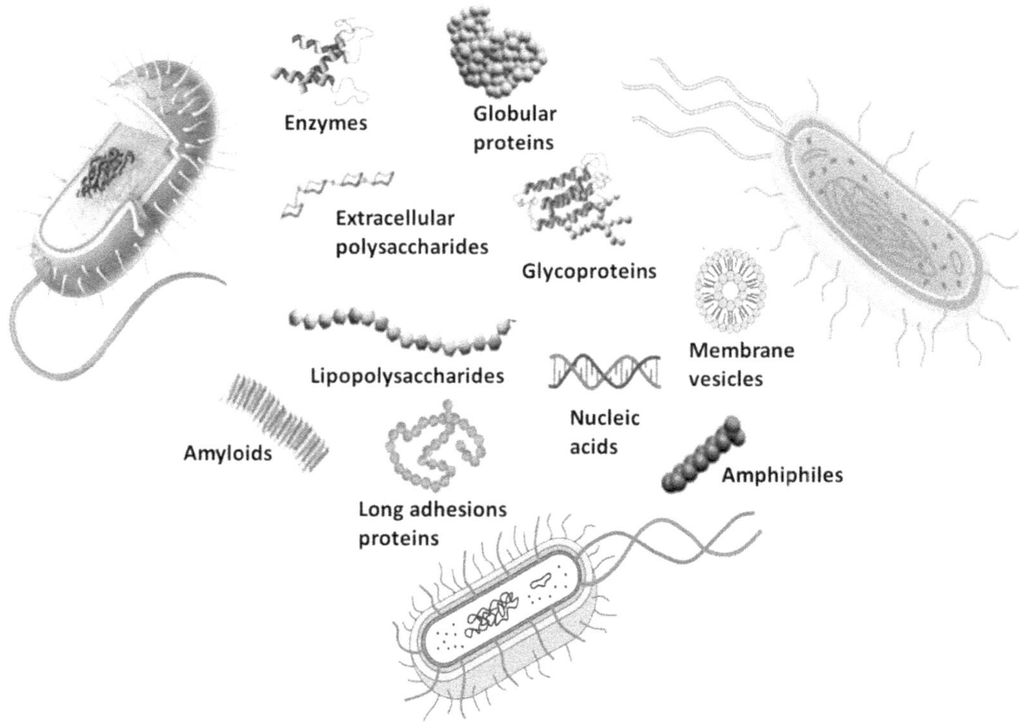

FIGURE 10.2 Types of Extracellular Polymeric Substances (EPS) found in the matrix of biofilms.

microbial composition are some of the parameters that influence the composition and the distribution of EPS in biofilms or floccular sludge or in granular sludge.

10.2.2 Quorum Sensing in Biofilm Formation

Quorum sensing (QS) plays a vital role in biofilm formation that increases cell-cell communication and interaction by involving signal molecules (autoinducers). Such autoinducers allow bacteria to sense the fluctuations in their environment and react properly by up-regulating or down-regulating the gene expression (Mitri and Richard Foster, 2013). Sometime increased QS associated gene expression may takes place because of very negligible QS phenomenon of planktonic cells (closer interaction) (Høiby et al., 2010). In contrast, the greater impact of QS was found for bacterial community-based biofilms. The significant role of QS for bacterial biofilm formation has been reported a great impact that can help in both attachment and detachment of cells. In a mixed-species biofilm, the reticence of QS can be achieved by single bacterial species to other species (Mitri and Richard Foster, 2013). Various mathematical models have been developed to explain the pivotal role of QS on biofilm formation. For example, the QS mechanism was reduced on exposure to high flow rates to the biofilm (Kirisits et al., 2007).

10.2.3 Different Approaches for Studying Biofilm Development

Biofilm study has received significant popularity and has gained the attention of many researchers throughout the past few decades. This has led to an interest in the development of many biofilm monitoring techniques to gain a deeper knowledge of morphological characteristics and the interaction of various microbes within the biofilm. In the earlier days, microbial cell of biofilm attached to substratum was analyzed by many methods such as weighting of dry and wet biofilms, spectrophotometric quantification of biofilm, gram staining, light microscopy and heterotrophic plate count (HPC) (Donlan and Costerton, 2002; Li et al., 2006). However, the traditional techniques for studying biofilm were tedious

TABLE 10.1

Composition of Biofilm EPS and Its Quantification Method

Sl. No.	Composition of EPS	Type of Linkage	Quantification Method of EPS
1.	Polysaccharides (monosaccharides uronic acids, amino sugars)	Glycosidic bonds	Anthrone method, sulphuric acid-UV method
2.	Nucleic acids (nucleotides)	Phosphodiester bonds	Spectrophotometry, fluorescence dyes, real time PCR
3.	Proteins (amino acids)	Peptide bonds	Lowry method, bradford protein assay
4.	Humic substances (Phenolic compounds, simple sugars, amino acids)	Ether bond C–C bonds Peptide bonds	Lowry method
5.	Phospho lipids (Fatty acids, glycerol, phosphate, ethanolamine, serine, choline, sugars)	Ester bonds	HPLC-ELSD method

and time-consuming. Moreover, these traditional methods couldn't deliver much information about the molecular role of EPS and QS in biofilm formation. But recently some optical techniques have been developed for biofilm analysis that can determine the structure, composition, and dynamics of the microbial populations without disturbing their structure. These techniques include scanning electron microscopy (SEM), atomic force microscopy (AFM), fluorescent microscopy, nuclear magnetic resonance imaging (NMRI), confocal laser scanning microscopy (CLSM), Raman spectroscopy, and fluorescent spectral imaging (Todhanakasem et al., 2014). Among them, SEM is prominently used to envision morphological characteristics of cells along with their relationship with the topography of adhering surface. Moreover, the examination of microbial consortia as aggregates can be done using SEM. Apart from that, CLSM and fluorescent microscopy can be employed to evaluate the EPS component, growth, and indirect metabolic activity of cells. The indirect estimation of types of the microbial population can be inferred by analyzing the metabolic activity of the cells by determining their substrate-consuming rate.

Biofilm communities are present in varied conditions, where the different microbial species are layered one above the other according to their environmental necessities. This phenomenon of differently layered microbial species leads to a change in concentration of oxygen, substrate, pH, and temperature across the biofilm (Kühl et al., 2007). Nowadays, electrochemical microsensors are used to estimate the concentration gradient of oxygen, substrates, pH and temperature across the biofilm structure. Further, Next-generation sequencing (NGS), such as pyrosequencing is being employed for more precise taxonomic classification of the culturable and non-culturable bacterial population. This technology presents accurate and unbiased diversity of microbial colonies and explores deeper layers of microbial cells. The combined form of microscope and microsensor can be done in future which simultaneously would be able to monitor the physiology of microbes and biofilm structures to illustrate the phenomenon of microbial biofilm as a whole (Halan et al., 2012). The constituents of EPS and their quantitative identification techniques are listed in Table 10.1.

10.3 Factors Influencing Biofilm Development

10.3.1 Abiotic Factors

The formation of biofilm from the planktonic cells is significantly affected by nutrient conditions. The thickness of biofilm may range from highly dense to nondetectable depending upon the concentration of nutrients in the environment. Usually, high nutrients in bioreactors lead to the faster transition of the microbial cell to biofilm state. In contrast, lack of nutrients or diffusion of nutrients in the media causes the removal of bacterial cells from the surface. Biofilm cell obtains their organics either by collecting it on the surface of EPS or by using the by-product of the secondary colonizer. Apart from organics, micronutrients like magnesium, nickel, cobalt also play their role in the growth of biofilm. These micronutrients act as cofactors for the essential enzymes involved in the metabolism of biofilm cells.

Microbial activities are highly sensitive to abiotic factors such as pH and temperature. Any change in internal and external pH not only affects the attachment of bacteria to the substratum but also has inhibitory effects on the enzymatic activity of microbial cells. Some cellular processes such as the excretion of polysaccharides for EPS formation are highly dependent upon the pH. The optimum pH for production of polysaccharide varies from species to species, but for the majority of microbial population, optimum pH is 7 (Oliveira et al., 1994). Change in temperature also reduces the enzymatic activity microbial population. For most of the microorganism, the maximum growth occurs at 40°C (Pasmore et al., 2001).

10.3.2 Surface Topography

Surface topography and roughness (vertical deviation from the real surface) influence the ability of microbes to adhere to the surface for biofilm formation. The rough surface has been seen to provide better adhesion than flat surfaces due to more surface area. Pasmore et al. (2001) studied and confirmed that *Pseudomonas aeruginosa* biofilms develop better on rough surfaces. The roughness of the surface results in the development of a low shear zone which helps bacterial cells and communities to protect them from the force that could remove them. Moreover, microbial attachments are also influenced by factors such as charge, elasticity and hydrophobicity.

10.3.3 Velocity and Turbulence

Effect of the velocity of fluid on biofilm formation is the hot topic of research. The velocity of liquid determines the size of boundary layers. The boundary layer is nothing but the area of the zone where fluid flow is not sufficient to remove the cell from the substratum. Outside this boundary layer, the flow velocity is high enough to remove the cells due to high turbulence. At high velocity, the size of the boundary level decreases, which results in more exposure of cells to the turbulence of the bioreactor. Apart from that, hydrodynamics condition also controls the production of EPS, transport of nutrients, diffusion of oxygen, and metabolic activities of cells in biofilm formation (Simoes et al., 2007).

10.3.4 Biotic Factors

Many intrinsic biotic factors such as autoinducer, transcriptional activator, and housekeeping enzyme influence each microbial cell attached to the substratum. It was observed in Pseudomonas aeruginosa around 16% of genes were inhibited and around 22% of genes were stimulated during biofilm formation. Among them, genes coding for the formation of EPS and polyphosphokinase were the most active. In *S. aureus* glycolytic and fermentative enzymes such as alcohol dehydrogenase and phosphoglycerate kinase were detected more. QS has recently been recognized as an important factor for biofilm formation. It helps in cell-to-cell signaling by activating small autoinducer molecules. These small molecules in combination with transcriptional activators produce the chemical component, which helps in detachment and attachment to the substratum. The detailed discussion of EPS and QS, responsible for biofilm formation, is described in Sections 10.2 and 10.3.

10.4 Biofilm Technologies in Wastewater Treatment

Wastewater treatment is always considered as a cost intensive and non-profitable process in an industrial enterprise. The premise of wastewater treatment is often an obligation rather than a commitment. Hence, cost plays an important factor in deciding the treatment scheme of any wastewater treatment plant (WWTP). Any process that offers lower cost of operation and higher stability and performance is usually selected. Usually in any conventional WWTP, wastewater treatment is typically carried out in three steps called the primary, secondary and the tertiary (Figure 10.3). The primary treatment usually comprises the physicochemical steps involving either of the following: bar & grit chambers, oil and grease traps, primary sedimentation, advanced oil removal units like API/TPI/CPI, equalization basin, coagulation-flocculation

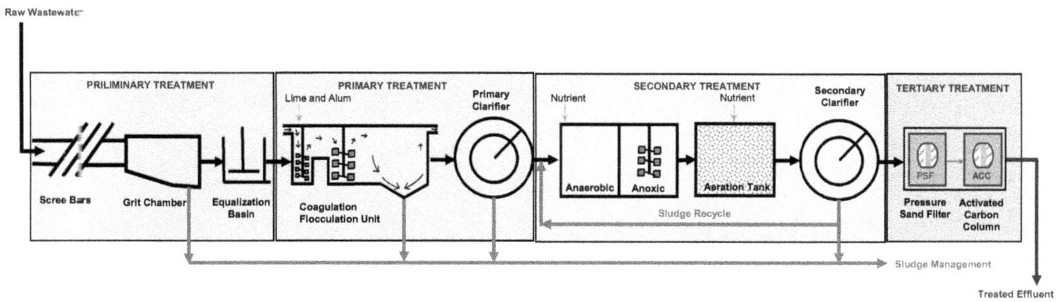

FIGURE 10.3 Schematic representation of a typical wastewater treatment plant.

unit, and primary clarifier. The primary section of the WWTP is responsible for removal of settleable and colloidal particulate matters and floatable and emulsified oil, if present. The secondary treatment comprises the biological unit and the secondary clarifier. The biological treatment system may have a single biological process or a combination of two or more biological processes in series. This part of the treatment is the most important as it removes most of the dissolved macro pollutants such as the organics, nutrients and other soluble micro pollutants like heavy metals, xenobiotics, personal and pharmaceutical care products (PPCPs), antibiotic resistant genes (ARGs), etc. The biological processes can be aerobic or anaerobic; continuous or batch suspended growth or attached growth processes. Depending upon the strength and quantum of wastewater, availability of space and degree of treatment, biological treatment options are selected. The last part of the conventional WWTP consists of filtration, polishing and disinfection. This is usually carried out by routing the secondary treated effluent through passive filters like pressure sand filters, and granular activated carbon columns or using low-cost biological processes like constructed wetland (Shingare et al., 2019). The commonly used conventional treatment technologies are presented in Figure 10.4. Still, some common problems are associated in wastewater treatment process. Such factors are poor settling, effluent violation (TSS or BOD), poor nitrification, toxicity, and odors.

Usually, the activated sludge process (ASP), also known as the conventional activated sludge process (CAS) is the most commonly employed biological treatment of wastewater. The cost of aeration in ASP is considerably high which makes this technology unattractive for high strength and difficult to degrade wastewater. Biofilm-based treatment technology has emerged as an alternative to the conventional suspended growth biological processes for the secondary treatment of municipal and industrial wastewater. However, the application of biofilm technologies in wastewater treatment is quite old. Tricking filters (TF) are the examples of oldest application of biofilm technologies in wastewater treatment. The next in line in biofilm technologies were the RBC and submerged biofilm processes. The RBC technology was developed as an offshoot of initial modifications in TF packing materials and went through several modification from cast iron discs to wooden and finally to plastic discs. While experimental trials were being conducted on the material of the RBC disc, some groups were experimenting on the applicability of plastic media on TF which were instantly successful. Soon, plastic media revolutionized other biofilm processes like submerged biofilm processes which were also known as packed bed processes.

In the last few decades, numerous new generation biofilm technologies such as Moving bed biofilm bioreactor (MBBR), Biofilm Air-lift Suspension Reactor (BASR), and Sequential Biofilm Batch Reactor (SBBR) and the latest membrane bio-reactor (MBR) with membrane aerated biofilm reactors (MABR) have been developed and successfully installed in WWTPs. The main advantage of biofilm-based technologies over the convention treatment process is the allowing retention of the slow growing microbes in the biofilm. Media help the biomass concentration in the biofilm bioreactor to reach almost fivefold or higher than the suspended cell systems. The filter media used in biofilm bioreactors play an important role for the microbial population to grow as a biofilm. The natural gradient of nutrient and oxygen across the biofilm allows proliferation of different microbial communities that work together in tandem. Some of these biofilm bioreactors are discussed in detail in the following section.

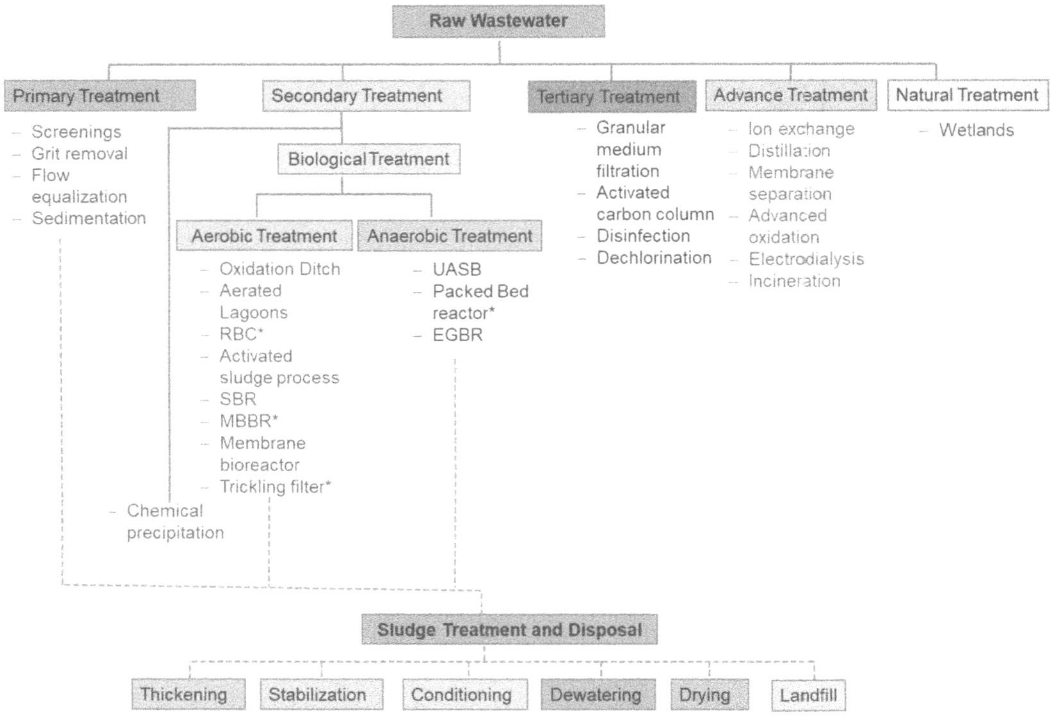

FIGURE 10.4 Common technologies in wastewater treatment.

10.4.1 Trickling Filters

Trickling filter (TF) is among the oldest and well-characterized fixed-biofilm reactors, which use inert packing or carrier materials such as lava, gravel, foam, sphagnum peat moss, coke, plastic media, rocks, or gravel for enabling microbial biofilm (Henze et al., 2008). TF is typically considered aerobic process as oxygen naturally diffuses into the biofilm from the open air. A rotating distribution system sprinkles the wastewater over the bed which wets the biofilm while percolating down as depicted in Figure 10.5a. Microorganisms in biofilm biologically remove the organic and inorganic pollutants in the wastewater for reproduction and growth. The partially treated wastewater is again recirculated until the contaminants are completely removed.

Initially, TF was used in sewage treatment for their relatively simple design, ease of operation and efficient removal of organics and nutrients. The history of TF dates back to early twentieth century. Few of the oldest installations are still in operation (Peters and Alleman, 1982). The early TF were huge in size. Several modifications in the design and operation of packing material of TF took place to reduce its size, to reduce the civil cost of construction and to prevent frequent clogging of filter media. Activated sludge process (ASP) was still new and was considered expensive as compared to TF (Peters and Alleman, 1982). TF was considered the preferred choice of low-cost biological units in sewage treatment until other variants of fixed film bioreactors like rotating biological contractors (RBC) and suspended cell processes like septic tanks and activated sludge processes came over in sewage treatment. An editorial published in *Sewage Works Journal* by Mohlman (1936) referred TF as being 'relegated to limbo'.

Typically, TF can be classified into roughing, carbon oxidation, carbon oxidation and nitrification type (Daigger and Boltz, 2011). Rehman et al. (2012) used plastic media in TF followed by a sand filter to treat domestic sewage at 5°C and 15°C. They achieved around 90% removal of COD and BOD. Generally, TF also carried out nitrification of low concentration of ammonia usually found in sewage in addition to carbon oxidation in the same stage. For achieving better rates of nitrification, TFs or the nitrifying TF (NFTs) were modified to adapt to higher flexibility and yield better effluent quality. Typically NFTs had

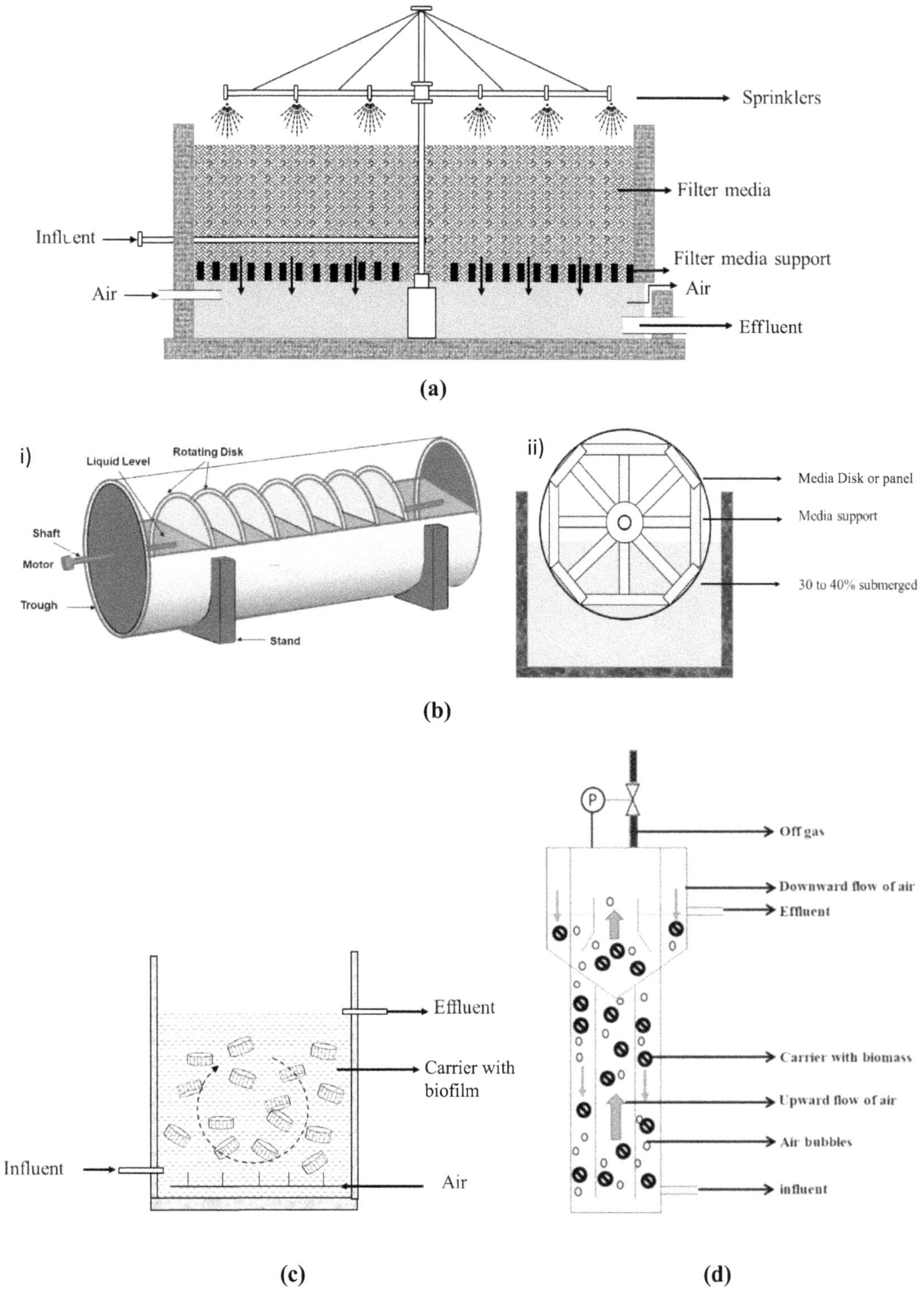

FIGURE 10.5 Schematic diagrams of (a) Trickling Filters (b). (i) Rotating Biological Contactor (b). (ii) frontal view of disc (c) Moving Bed Biofilm Reactor (d) Biofilm Air lift Suspension Reactor

(*Continued*)

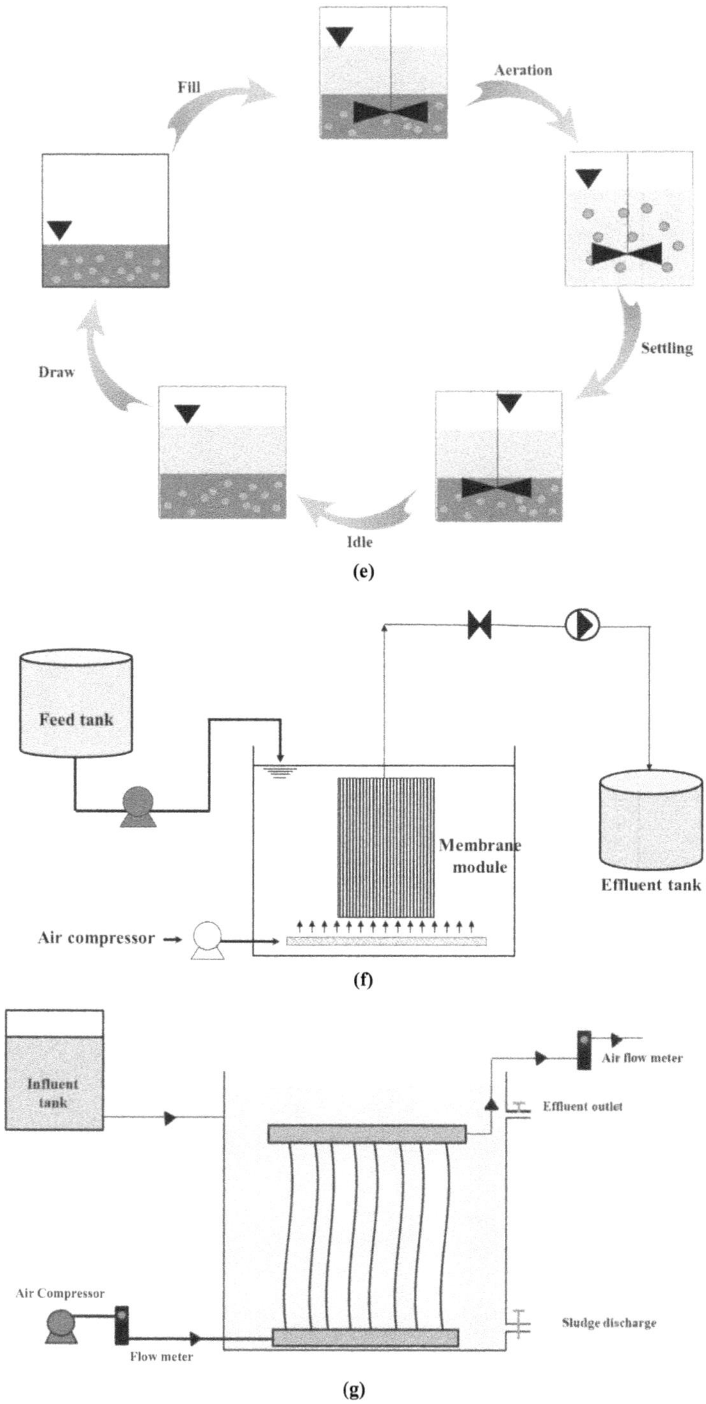

FIGURE 10.5 (*Continued*) (e) Sequential Batch Biofilm Reactor (f) Membrane Bioreactor (g) Membrane Aerated Biofilm Reactor.

nitrification rate in the range 0.2–0.4 $gN/m^2.d$ and in high rates NFT, nitrification rate was as high as 3 $gN/m^2.d$ (Andersson et al., 1994). Various modifications like, increasing recirculation ratios, or operating in alternating double filtration (ADF) mode was found to increase the efficiency of biofilm for carrying out both carbon and ammonia oxidation (Daigger and Boltz, 2011).

TF are often combined with suspended growth process like tricking filter/solid contact (TF/SC) process, roughing filter/activated sludge process (RF/AS) and tricking filter/activated sludge process (TF/AS) (Daigger and Boltz, 2011). The TF/SC process uses a suspended growth process as a flocculating unit to remove the not readily settleable, colloidal particles and organic particulate matter after a TF. There are different configurations available in TF/SC process that claims lower size of TF, lesser quantity of media used and higher efficiency. When an activated sludge process handles a major load of pollutant removal, TF/SC is referred to as RF/AS process. The TF/AS process is designed to handle higher hydraulic and organic loads and has a clarifier between a TF and ASP process.

TF process finds its application in wastewater treatment of other industrial sectors as well. Kornaros and Lyberatos (2006) demonstrated 70% COD removal efficiency at an OLR of 1.1 m^3/m^2/day at pilot-scale TF for the pre-treatment of organic dye containing textile wastewater with an initial organic concentration of 40,000 mg/L. In another study, removal of nitrogenous substance was assessed from swine waste anaerobic digestate by TF which showed around 90% of TAN removal and 79% COD removal. Harrison and Daigger (1987) used a pilot-scale TF for the treatment of wastewater. They observed around 90% BOD removal with an OLR of 3.45 and 3.6 kg BOD/m^3.

10.4.2 Rotating Biological Contactor

Application of rotating biological contactor (RBC) has been wider than that of a TF. This technology has been extensively used in the treatment of wastewaters generated from refinery, pulp, paper, food, and beverage industries. Apart from that, it is also used to treat municipal wastewater lagoon effluents and landfill leachates. The mechanism of RBC is identical to the operation of TF. Circular discs made of inert media like PVC, polyethylene, or expanded polystyrene supports the growth of microbial assemblage biofilm. The RBC contains a trough in which effluent is circulated and a horizontally stacked circular disc. Around 40% areas of the centrally stacked discs are immersed in wastewaters that are circulated across the trough (Figure 10.5b). As the shaft rotates, it exposes the film of wastewater present on the discs into the air. The wastewater percolates downwards due to gravitational force and absorbs oxygen from the air. Biofilm attached to the disc oxidizes the organic matter and nutrients present in the wastewater. As the biofilm grows, the excess biomass sloughs off the disc due to the shearing force exerted with disc rotation. The settling tank installed downstream of RBC removes excess sludge biomass and waste substance (Hassard et al., 2015).

The RBC system has many advantages over conventional treatment systems like less sludge production, ease of operation, less space requirement. Therefore, it was used frequently in the 1960s and the 1970s but soon decreased due to various problems like shaft damage, the proliferation of unwanted microorganisms, sloughing off of biomass, limited diffusion of nutrients and oxygen. Increasing rotational been considered an effective way to increase diffusion of oxygen and thereby increasing its efficiency (Israni et al., 2002). However, increasing the rotation speed also results in high energy demands and increases the cost of the treatment. Optimization of the RBC system under different influent characteristics and environmental conditions is still a big challenge for the efficient use of this technology.

10.4.3 Moving Bed Biofilm Reactor

Towards the end of the 1980s, a Norwegian company invented Moving Bed Biofilm Reactor (MBBR) with a premise to combine the advantages of a suspended and attached growth processes in a single unit (Shahot et al., 2014). The process contains small carriers that can move along the water and can be used for microbial biofilm formation. The unit can be operated continuously operating system where microbes grow as suspended flocs and as an attached biofilm as depicted in Figure 10.5c. The floatation of carriers and the suspended flocs in MBBR can be driven by air or mechanically. The pattern of agitation is such that the carrier molecule moves in an upward direction, which creates the scrubbing effect and biologically activates the whole volume, resulting in more biomass activity. In this way, MBBR is more beneficial than the conventional treatment process due to its higher oxygen transfer efficiency, higher OLR, low HRT and less sludge washout (Chan et al., 2009). Higher retention time of biomass on carrier gives better scope for the slow growing bacteria like nitrifiers to survive. MBBR is easy to install and maintain as compared to the conventional treatment system, requiring only a tank of adequate size and aerators. Due to the ease of operation, several hundreds of MBBR at full scale have been installed and studied worldwide for domestic and industrial wastewater treatment (Shahot et al., 2014, Irankhah et al., 2018, Chen et al., 2020).

10.4.4 Biofilm Airlift Suspension Reactors

Biofilm airlift suspension reactor (BASR) is a type of air lift reactor that contains a fluidized bed of inert particles like sand with biofilm. The reactor has two concentric tubes. Air is passed at the bottom of an inner tube, which creates the difference in the fluid densities. The difference in the fluid density leads to the mixing of liquid and carrier. In BASR, small suspended particles are used as carrier for the formation of biofilm as depicted in Figure 10.5d. The main advantage of the BASR technology is increased surface area carrier material like sand, gravel, sand, etc. Due to the small suspended carrier (radius 0.1 mm), high active biomass surface area is maintained in the reactor, which also contributes to high oxygen transfer efficiency. These carriers are easy to separate from the wastewater, and there is no problem with oxygen diffusion in the reactor. Due to its versatility, it can be used to remove many contaminants, including those with high COD and ammonia. In many studies, biomass concentration of around 10–30 g/L has been achieved, which is ten times higher as compared to the conventional treatment system. The higher biomass concentration results in higher performances. This technology has many advantages over conventional processes like high OLR with low sludge loading, good nitrification at low temperature, very low production of sludge, and requirement of a small area.

10.4.5 Sequencing Batch Biofilm Reactor

In recent years, sequencing batch biofilm reactor (SBBR) has become a center of attention due to its ability to work not only as biofilm reactor but also as sequential batch reactor. A SBBR offers higher specific removal rates, enrichment of slow growing nitrifiers, and higher volumetric loads. SBBR works in the same fill and draw way like a conventional SBR. The first step involves addition of wastewater (fill), followed by aeration and mixing is first added in batch reactor and treated to remove contaminants before discharge as shown in Figure 10.5e. One more advantage of SBBR is that it doesn't need settling time, unlike SBR, which requires a settling period to separate activated sludge from the treated effluent.

SBBR has been used for nutrient removal simultaneously from domestic and industrial wastewater. They are well suited for nitrification because carrier material prevents washout of slow growing autotrophic bacteria (nitrifier) from the system. Jin et al. (2012) carried out the study of removal of total N in SBBR and found that the N removal efficiency increased at a higher C:N ratio. Similarly, partial nitrification is achieved by conducting intermittent aeration in SBBR. Though implementing nitrification in SBBR is difficult, Gieseke et al. (2002) studied simultaneous phosphorous and nitrogen removal at pilot scale SBBR.

10.4.6 Biofilm-Based Membrane Bioreactors

Membrane bioreactors (MBR) are considered to be the most recent technological advancements among the bioreactor technologies that combine the advantages of biological oxidation and membrane separation. The development of MBR process began almost 30 years ago and till date there are more than one hundred thousand research articles published (Lee et al., 2001). Despite several advantages and wide application in the treatment of both sewage and industrial wastewater, MBR technology is considered expensive due to the high cost of membrane, high energy demands, and frequent membrane bio fouling. Some variations that have developed to overcome the disadvantages of an MBR are based on biofilm technology such as Biofilm Membrane Bioreactor (BF-MBR) and Membrane Aerated Biofilm Reactor (MABR). Replacing suspended growth with attached growth in MBR has been considered a way to reduce membrane bio-fouling.

There are two ways in which a biofilm can be introduced in an MBR. One is to add small inert carrier particles in conventional suspended growth process of MBR to allow biofilm to grow over the inert particles. This type of modification is known as assisted Biofilm MBR or 'aBF-MBR' or hybrid MBR (Ivanovic and Leiknes, 2012). The second way is to replace the suspended growth completely by an attached or a fixed film system within a membrane process. Such systems are called pure Biofilm MBR or 'pBF-MBR' (Ivanovic and Leiknes, 2012). The assisted Bf-MBR (aBF-MBR), the suspended cells are responsible for major biodegradation, and their activity is enhanced or assisted by the suspended carrier particles bearing the biofilms. The suspended carrier particles enhance the souring effect in MBR and significantly increase the life of membrane by delaying membrane fouling. Membrane fouling is the major limiting factor in MBR technology. Several authors have reported higher efficiency of aBF-MBR process as compared to the conventional ASP-MBR process especially in nutrient removals (Liu et al., 2010; Yang et al., 2009; Khan et al., 2011; Ivanovic and Leiknes, 2012).

The pBF-MBR process, on the other hand, eliminates suspended growth system in MBR and replaces it by attached growth. The biofilm is developed on media that can be fixed or moving as depicted in Figure 10.5f. Moving bed biofilm offers additional advantages over the fixed bed biofilms, albeit with some compromise on the compactness of the size (Ivanovic and Leiknes, 2012). The pBF-MBR are essentially have the separation module externally attached and separated from the biological module. This gives operational flexibility to move the biofilm carrier particles of the MBR without worrying about the impact of sloughed off biofilm on the membrane fouling. To increase compactness of pBF-MBR, several modifications have been developed like introduction of flocculation zone within the biofilm unit of MBR, introducing higher packing density, connecting a disc filtration and flotation (DAF) unit prior to the membrane separation modules, etc. enables operating the process at high OLR without bothering about the production of sub-micron particles from the sloughed off biofilms leading to rapid membrane fouling (Ivanovic and Leiknes, 2012). The modification in pBF-MBR allows better system designs and configurations leading to higher performance of the MBR.

Membrane aerated biofilm reactor (MABR) is the most recent advancement in membrane technology (Lu et al., 2021). MABR contains a hydrophobic membrane with attached biofilm. There are two phases in the reactor, namely, a gas phase and a liquid phase which is on the hydrophobic membrane side and the hydrophilic biofilm side, respectively. The gas phase contains oxygen which diffuses through the membrane to reach the microbial biofilm, whereas the organic matter and nutrients from the wastewater diffuse into the biofilm from the biofilm side (Figure 10.5g). Both the phases are driven in counter current directions to achieve more diffusion of gases and substrates (Casey et al., 1999). Oxygen diffusion into the biofilm is driven by biofilm thickness. Differential access to oxygen in different sections of biofilm enables proliferation of oxic and anoxic microbial population in the biofilm. Hence, MABR have been known to demonstrate higher removal efficiency than the conventional MBR technology. The passive diffusion of oxygen through the hydrophobic membrane prevents generation of sub-microns sized particulate matters which is an important factor controlling membrane biofouling. However, controlling the thickness of biofilm is still a challenge in MABR process (Lu et al., 2021). There are some recent developments of the membrane technology that is able to control the biofilm growth only up to a certain thickness so that the diffusion of substrates and oxygen is not limited due to excessive thickness of biofilm and the efficiency of the MABR process can be maintained throughout its operation over long period of time.

10.5 Nutrient Removal in Wastewater Treatment System by Biofilm Technologies

Nutrient removal in wastewater treatment has always challenged engineers mostly because of complex biological reactions that are very slowly carried out by several interdependent microbial communities. Most of the time, the retention of diverse microbial community is essential for nutrient removal in a system that is also oxidizing organic carbonaceous pollutant becomes a challenge. Moreover, slow growth rate of microbial populations needed for nutrient removal is often outcompeted with fast growing organic carbon degrading microbial species. Higher sludge retention time could increase the chances of proliferation of slow growing microbial population, but at the cost of increasing the treatment cost. Biofilms are a simple way of retaining slow growing microbial populations by uncoupling hydraulic and sludge retention times. These biofilm technologies give a suitable environment for the growth of slow-growing autotrophic microorganisms and help in the better removal of nutrients from the wastewater. The process of nitrogen, phosphorous removal, and the role of biofilm technologies and their advantages over suspended biomass systems will be discussed further.

10.5.1 Nitrogen Removal

Biological nitrogen removal conventionally comprises nitrification and denitrification steps. In nitrification, ammonia is oxidized into nitrite and then to nitrate by two autotrophic bacterial groups called ammonia oxidizing bacteria (AOB) and nitrite oxidizing bacteria (NOB), respectively. The nitrites and nitrates formed is converted into nitrogen gas by heterotrophic bacteria. Conversion of nitrate to nitrite occurs in the absence of oxygen and requires an external carbon source. Many organic carbon sources have been used for denitrification ranging from acetate to methanol to ethanol.

Biofilm reactors can increase the N removal efficiency. The conventional treatment system is being used for the last many years for the treatment of nitrogen-rich wastewater. But in recent years, studies have shifted their focus to new biological nitrogen removal processes such as anaerobic ammonium oxidation (ANAMMOX), a single reactor system for high ammonia removal over nitrite (SHARON), and completely autotrophic N removal over nitrite (CANON) systems. Moreover, biofilm reactors have been implemented all over the world to increase the N removal efficiency of wastewater treatment plants. Lee et al. (2001) studied the application of a fixed bed biofilm reactor (11 $m^{3)}$ to treat nitrogenous constituents from domestic sewage. This pilot-scale reactor maintained three different conditions (aerobic, anoxic, and anaerobic) in a series, was tested at two different HRT 7.6 and 10 hours under various OLR and temperatures. At the highest organic load of 0.5 kg COD/m^3, COD and BOD removal efficiency were around 87% and 77%, respectively, and the temperature change from 10°C to 29°C didn't significantly affect removal efficiency.

10.5.2 Phosphorous Removal

Total phosphorous is present in the form of soluble and particulate phosphorous. The major phosphorous component in wastewater exists mainly as orthophosphate associated with a small amount of organic phosphorous. Phosphorous concentration ranges in between 5 and 20 ppm in domestic and industrial wastewater. But, the discharge limit of phosphorous is less than one ppm. Physicochemical treatments are being commonly practiced for the removal of phosphorous from wastewater over the years. Among them, chemical treatment is commonly practiced where it is removed by precipitating the soluble phosphorous to insoluble particles. However, disposal of phosphate accumulated sludge itself becomes a difficult task.

The use of microbial processes to remove phosphorous from wastewater is widely used to prevent eutrophication of water bodies. It is cost-effective and doesn't lead to the generation of excess sludge, unlike chemical treatment. Kim et al. (2011) studied the biological removal of phosphorous in a laboratory-scale system with intermittent aeration. It showed 73.1% and 9.4% removal of total phosphorous and total nitrogen, respectively. Similarly, Moosavi et al. (2005) also studied phosphorous removal in synthetic wastewater using an up-flow anaerobic-aerobic fixed bed (UA/AFB) reactor.

10.6 High Strength, Recalcitrant Wastewater Treatment Using Biofilm Technologies

Several industries such as distilleries, paper & pulp, pharmaceuticals, refineries, petrochemicals, chemical industries like pesticides, agrochemicals, paint industries, textiles, dye and dye intermediates and others generate high strength process effluents (Abdelfattah et al., 2020). Characteristics of some of the high strength industrial effluents are presented in Table 10.2. Extensive treatment is required before the wastewater is disposed to recycled back in the process. Mostly, anaerobic treatments are usually employed for degradation of high strength wastewater followed by aerobic units (Show et al., 2020)

Many of these industries release recalcitrant organic compounds that make wastewater treatment much more difficult. Usually, higher hydraulic and sludge retention times (HRT and SRT) are required for degradation of recalcitrant organic compounds. These conditions and anaerobicity developed naturally within the biofilm niches offer better efficiency of bioremediation of recalcitrant compounds in industrial effluents. In case of hydrophobic compounds, the major limitation is the bioavailability of organic contaminants to the bacteria. Biofilm enhances bioavailability of hydrophobic substrates to the degrading microbial population. Also, higher cell numbers in microbial aggregate of biofilm, helps faster removal of

TABLE 10.2
Characteristics of High Strength Industrial Wastewater

Parameter[a]	Concentrations						
	Distillery	Pharmaceutical	Pulp & Paper	Refinery	Soap & Detergent	Tannery	Textile
pH	3.8–4.4	5.5–9.2	6.3–9.2	6–9	5–10	9.4–9.7	7.1–7.5
Oil & grease	-	-	-	300–10,000	-	-	-
Suspended solid	4,400–6,400	30–55	313–411	250–600	775–3,000	1,200–6,000	738–802
Total dissolved solid	90,050–101,600	5,000–30,000	2,588–2,968	1,000–10,000	-	9,000–13,000	2,508–2,692
Chemical oxygen demand (COD)	103,000–107,000	6,450–9,850	754–864	350–600	1,600–3,000	1,800–2,000	2,770–2,980
Biochemical oxygen demand (BOD)	29,400–30,000	2,900–4,700	230–256	200–440	900–1,800	900–1,200	1,226–1,364
Potassium	500–600	5–25	-	-	-	-	24–30
Chloride	5,000–8,000	500–1,200	745–795	-	-	5,000–6,000	531–609

[a] All parameters are in mg/L, except pH.
'-' Data not available.

hydrophobic compounds than suspended or planktonic microbial cells. Klein et al. (2010) reported hexadecane assimilation by biofilm of *Marinobacter hydrocarbonoclasticus* SP17 at the alkane-water interface. Shimada et al. (2012) reported better efficiency of a biofilm-assisted microbial remediation of naphthalene by *Pseudomonas stutzeri* strain T102 over planktonic cells in petroleum contaminated soil. Chandran and Das (2011) demonstrated removal of diesel oil contaminated wastewater using yeast biofilm on gravels.

Conversely, the presence of recalcitrant compounds stimulated biofilm production in biological systems. Ammeri et al. (2021) studied the biotransformation of pentachlorophenol contaminated wastewater and showed that the presence of pentachlorophenol induced biofilm formation in his experimental setups. Similarly, crude oil deliberately spilled into the gulf of Arabian Sea during the Gulf War in 1990–1991 led to the natural development of hydrocarbon degrading bacteria enriched algal mats in the Gulf Coast with high potential to naturally degrade hydrocarbons (Radwan and Al-Hasan, 2001).

There are several examples of application of biofilm reactors for the treatment of industrial wastewaters. Trickling filters are mostly used in degradation of oily wastewaters like in petroleum refineries, and petrochemical industries. Trickling filter usually precedes an activated sludge process which helps to protect the suspended biomass in ASP from the impact of dissolved oil (Daigger and Boltz, 2011). Trickling filters have been studied for the removal of organics in textile effluent containing organic dye and varnishes (Kornaros and Lyberatos, 2006). A two-stage, down-flow hanging sponge (DHS) reactor which is a modification of trickling filter has been studied to treat soft-drink wastewater with an influent COD concentration of 3000 mg/L (Liao et al., 2017). The TF reactors have also been used at tertiary stage for the removal of nutrients (Abdelfattah et al., 2020).

Ebrahimi et al. (2017) carried out the treatment of effluent of cheese whey containing 40–80 g-COD/L by bench-scale three stage RBC. The study reported maximum COD removal efficiency of 92.4% at an OLR of more than 40 g COD/L and HRT of 36 hours. RBC has been integrated with other technologies to treat pharmaceutical contaminants. Talwar et al. (2018) combined TiO_2 photocatalytic unit with RBC to treat pharmaceutical wastewater (12,425 mg/L of COD). The study reported 99% of COD removal from the effluent, with 33% carried out by RBC.

Moving bed biofilm reactors (MBBR) also have several applications in efficient treatment of domestic and industrial wastewater. Irankhah et al. (2018) carried out the study in MBBR where around 90% of COD removal was observed from wastewater containing 2,800 mg/L of COD. Similarly, Qaderi et al. (2018) used a series of MBBR for petroleum wastewater treatment and achieved 80% removal efficiency. Chen et al. (2020) combined an MFC and anaerobic MBBR to treat pulp/paper wastewater and electricity generation. The study reported around 66% of COD removal with an influent COD of 6,300–6,500 mg/L.

Membrane aerated biofilm reactors (MABR), which has highly improved oxygen and substrates diffusion coefficients, are now being employed for the treatment of recalcitrant organic compounds such as penta chlorophenols, xylenes and perchloroethylenes (Debus and Wanner, 1992; Ohandja and Stuckey, 2006). Granular biological activated carbon (GBAC) supported biofilm has been studied for degradation of several recalcitrant and micropollutants containing wastewaters like those containing trichloroethane, perchloroethylene and trichloroethylene (Wu et al., 2000; Islam et al., 2015; Siggins et al., 2021; Piai et al., 2020). The mechanism of the GAC supported biofilm involves absorption of organic compounds on the granular biological activated carbon followed by microbially bio-transformation and biodegradation. Apart from GAC, agricultural waste, biochar or pyrolysis waste are considered sustainable adsorption material which can reduce the cost of adsorbent material for adsorption and biodegradation of recalcitrant compounds. Viggi et al. (2017) demonstrated the importance of biochar in enhancing the efficiency of bacterial interaction in biofilm through indirectly mediated interspecies extracellular electron transfer (EET).

The biofilm on the bioanodes of microbial electrochemical technologies (MET) that exploit the potential of microorganisms to exchange electrons with conductive electrodes through EET, help the microbial community to bioremediate difficult organic compounds include cyclic aromatic hydrocarbons (CAHs), TCE, PCE, petroleum hydrocarbon contaminated wastewater and other while generating power (Rossi et al., 2021). Zhang et al. (2022) used biofilm of *Rhodopseudomonas* palustris and *Chlorella vulgaris* to study the removal of antibiotics metronidazole (MNZ) from wastewater using electrochemistry-activated binary-species photosynthetic at different applied potentials.

10.7 Cost Factors

In general, biofilm reactors offer considerable reduction in cost as compared to the suspended cell processes. However, the membrane bioreactor-based biofilm processes have been considered the most expensive among all the biological processes till date. This is because of the use of expensive membranes, the energy costs involved in driving the separation process across the membranes and need to replace membranes frequently due to rapid membrane bio-fouling are some of the major concerns of a membrane bioreactor technology. For membrane reactors to be feasible, we need to consider the operational cost savings. The cost factor is reduced by reducing energy consumption, using a biofilm processes instead of suspended growth process, use of light weight and low-cost, small carrier material with high surface area that can increase the scouring effect on the membrane thereby reducing the chances of bio-fouling of membrane, or eliminating brine release. Nerenberg (2016) et al. reported that MBR is 86% more energy-efficient than a conventional system. In MABR, diffused aeration across the membrane has effectively reduced the cost of aeration by 45%–75% of the energy costs. However, it has been reported to overcome conventional treatment systems in terms of pollutant removal rate and energy efficiency by using smaller aeration equipment. Due to immobilization of microbes on carrier components and less aeration, the effectiveness of membranes in MABR is twofold compared to conventional bubbled diffuser reactors. Hem et al. (1994) studied the impact of variation of oxygen concentration on nitrification in MBBR and reported the higher sensitivity MBBR compared to other biofilm reactors. Replacement of membrane is also the main economic obstacle while operating membrane bioreactor. However, a reduction in the cost of commercial membrane and the huge jump in the cost of energy will make the membrane bioreactor a promising tool for wastewater treatment in upcoming years. Further laboratory scale and pilot scale data are necessary for the economic evaluation of real-world data.

10.8 Future Biofilm Processes: Targets and Outlook

The high cost of wastewater treatment through conventional processes has pushed the need for more reliable and economic biological treatment systems. Biofilm technology-based treatment can be considered as an appropriate alternative for the treatment of municipal and industrial wastewater. These technologies not only produce less volume of sludge but also a better nutrient removal option for high-strength wastewater. The structured packing materials for biofilm processes that are being used significantly reduce the cost of treatment. These packing materials encourage efficient gas and liquid mass transfer in the biofilms. However, the lack of techno-economic information on system performance and its sustainability in long run limits its application. Moreover, the long acclimatization period, complex biology of biofilm, high start-up time, and sometimes bio-fouling in case of membrane biofilm is some of the major limitations of a biofilm-based technology. Hence, future research is necessary for better understanding of the microbial mechanism of biofilm formation which can help in developing biofilm-based bioreactors with microbial consortia of interest.

Despite the limitation, biofilm reactors have been commercially implemented worldwide in wastewater treatment plants. But apart from wastewater treatment, biofilm-based technologies are now been employed in biotechnological and pharmaceutical industries as bio-catalysts for the production of specific biochemical and chemical molecules. The use of biofilm as a biocatalyst has been seen as an emerging technology in this sector which can replace the existing batch operation in fermentation process into a continuous mode operation with no apprehension on the number of times a biocatalyst can be reused.

10.9 Conclusion

Wastewater management is an important aspect of urbanization and development. On the one hand, urban areas generate large volumes of low strength municipal wastewater to be managed sustainably, whereas on the other hand, the industrial effluents with critical pollutant load need careful selection of technologies for their treatment and effective disposal or reuse. There is a myriad of technologies available at various stages of treatment. Needless to say, the biological methods of treatment have been found to be the most economic and sustainable among all other processes, biofilm-based processes are one of them. Biofilm technologies are being applied for simultaneous removal of organic carbon along with nutrients in the same system, for bioremediation of xenobiotic compounds and for biodegradation of hydrophobic hydrocarbon. Right from very old biofilm technologies like TF and RBC to some recently developed technologies like SBBR and MABR biofilm technologies have come a long way. The diffusion of substrates and oxygen which was considered limiting in TF and RBC has been overcome in the design configuration so MABR. The flexibility of the system to retain slow growing microorganisms in complex biofilm that offers it natural protection against environmental stress, toxicity and predators provides remarkable natural solution for treating complex industrial effluents. More recent application of biofilm technologies as bio-catalyst in pharmaceutical and biotechnological applications have opened new avenues. Studies have shown that the biofilm-based processes can be regarded as industrial workhorses which can be genetically modified to suit a wider application over and above wastewater sector. Biofilm-based processes have a long way to go before its full potential is explored for the benefit of humanity in greater sense.

Acknowledgement

The authors are grateful to the Director, CSIR-NEERI, Nagpur, India, for his guidance. Authors acknowledge the efforts taken by Knowledge Resource Centre, CSIR-NEERI, Nagpur for plagiarism check of the article using iThenticate software.

REFERENCES

Abdelfattah, A., Hossain, M.I., Cheng, L., 2020. High-strength wastewater treatment using microbial biofilm reactor: a critical review. *World Journal of Microbiology and Biotechnology* 36, 1–10.

Ammeri, R.W., Di Rauso Simeone, G., Hassen, W., Ibrahim, C., Ammar, R.B., Hassen, A., 2021. Bacterial consortium biotransformation of pentachlorophenol contaminated wastewater. *Archives of Microbiology* 203, 1–13.

Andersson, B., Aspegren, H., Parker, D., Lutz, M., 1994. High rate nitrifying trickling filters. *Water Science and Technology* 29, 47.

Casey, E., Glennon, B., Hamer, G., 1999. Review of membrane aerated biofilm reactors. *Resources, Conservation and Recycling* 27, 203–215.

Chan, Y.J., Chong, M.F., Law, C.L., Hassell, D., 2009. A review on anaerobic–aerobic treatment of industrial and municipal wastewater. *Chemical Engineering Journal* 155, 1–18.

Chandran, P., Das, N., 2011. Degradation of diesel oil by immobilized *Candida tropicalis* and biofilm formed on gravels. *Biodegradation* 22, 1181–1189.

Chen, F., Zeng, S., Luo, Z., Ma, J., Zhu, Q., Zhang, S., 2020. A novel MBBR–MFC integrated system for high-strength pulp/paper wastewater treatment and bioelectricity generation. *Separation Science and Technology* 55, 2490–2499.

Cogan, N., Keener, J.P., 2004. The role of the biofilm matrix in structural development. *Mathematical Medicine and Biology: A Journal of the IMA* 21, 147–166.

Daigger, G.T., Boltz, J.P., 2011. Trickling filter and trickling filter-suspended growth process design and operation: a state-of-the-art review. *Water Environment Research* 83, 388–404.

Debus, O., Wanner, O., 1992. Degradation of xylene by a biofilm growing on a gas-permeable membrane. *Water Science and Technology* 26, 607–616.

Dutta, D., Arya, S. and Kumar, S., 2021. Industrial wastewater treatment: current trends, bottlenecks, and best practices. *Chemosphere* 285, 131245.

Donlan, R.M., Costerton, J.W., 2002. Biofilms: survival mechanisms of clinically relevant microorganisms. *Clinical Microbiology Reviews* 15, 167–193.

Ebrahimi, A., Najafpour, G.D., Nikzad, M., 2017. Evaluation of treatability of high strength wastewater in a three stage-rotating biological contactor. *Journal of Environmental Engineering and Landscape Management* 25, 234–240.

Frederick, M.R., Kuttler, C., Hense, B.A., Eberl, H.J., 2011. A mathematical model of quorum sensing regulated EPS production in biofilm communities. *Theoretical Biology and Medical Modelling* 8, 1–29.

Gieseke, A., Arnz, P., Amann, R., Schramm, A., 2002. Simultaneous P and N removal in a sequencing batch biofilm reactor: insights from reactor-and microscale investigations. *Water Research* 36, 501–509.

Halan, B., Buehler, K., Schmid, A., 2012. Biofilms as living catalysts in continuous chemical syntheses. *Trends in Biotechnology* 30, 453–465.

Halan, B., Schmid, A., Buehler, K., 2011. Real-time solvent tolerance analysis of *Pseudomonas* sp. strain VLB120ΔC catalytic biofilms. *Applied and Environmental Microbiology* 77, 1563–1571.

Hall-Stoodley, L., Costerton, J.W., Stoodley, P., 2004. Bacterial biofilms: from the natural environment to infectious diseases. *Nature Reviews Microbiology* 2, 95–108.

Harrison, J.R., Daigger, G.T., 1987. A comparison of trickling filter media. *Journal (Water Pollution Control Federation)* 59, 679–685.

Hassard, F., Biddle, J., Cartmell, E., Jefferson, B., Tyrrel, S., Stephenson, T., 2015. Rotating biological contactors for wastewater treatment–a review. *Process Safety and Environmental Protection* 94, 285–306.

Hem, L.J., Rusten, B., Ødegaard, H., 1994. Nitrification in a moving bed biofilm reactor. *Water Research* 28, 1425–1433.

Henze, M., van Loosdrecht, M.C., Ekama, G.A., Brdjanovic, D., 2008. *Biological Wastewater Treatment.* IWA Publishing, London.

Høiby, N., Bjarnsholt, T., Givskov, M., Molin, S., Ciofu, O., 2010. Antibiotic resistance of bacterial biofilms. *International Journal of Antimicrobial Agents* 35, 322–332.

Irankhah, S., Ali, A.A., Soudi, M.R., Gharavi, S., Ayati, B., 2018. Highly efficient phenol degradation in a batch moving bed biofilm reactor: benefiting from biofilm-enhancing bacteria. *World Journal of Microbiology and Biotechnology* 34, 1–13.

Islam, M.S., Zhang, Y., McPhedran, K.N., Liu, Y., El-Din, M.G., 2015. Granular activated carbon for simultaneous adsorption and biodegradation of toxic oil sands process-affected water organic compounds. *Journal of Environmental Management* 152, 49–57.

Israni, S.H., Koli, S.S., Patwardhan, A.W., Melo, J.S., D'souza, S.F., 2002. Phenol degradation in rotating biological contactors. *Journal of Chemical Technology & Biotechnology: International Research in Process, Environmental & Clean Technology* 77, 1050–1057.

Ivanovic, I., Leiknes, T., 2012. The biofilm membrane bioreactor (BF-MBR)—a review. *Desalination and Water Treatment* 37, 288–295.

Jin, Y., Ding, D., Feng, C., Tong, S., Suemura, T., Zhang, F., 2012. Performance of sequencing batch biofilm reactors with different control systems in treating synthetic municipal wastewater. *Bioresource Technology* 104, 12–18.

Khan, S.J., Ilyas, S., Javid, S., Visvanathan, C., Jegatheesan, V., 2011. Performance of suspended and attached growth MBR systems in treating high strength synthetic wastewater. *Bioresource Technology* 102, 5331–5336.

Kim, B., Chang, D., Son, D., Kim, D., Choi, J., Yeon, H., Yoon, C., Fan, Y., Lim, S., Hong, K., 2011. Wastewater treatment in moving-bed biofilm reactor operated by flow reversal intermittent aeration system. *International Journal of Environmental and Ecological Engineering* 5, 867–870.

Kirisits M.J., Margolis, J.J., Purevdorj-Gage, B.L., Vaughan, B., Chopp, D.L., Stoodley, P., Parsek, M.R., 2007. Influence of the hydrodynamic environment on quorum sensing in *Pseudomonas aeruginosa* biofilms. *Journal of Bacteriology*, 189, 8357–8360.

Klein, B., Bouriat, P., Goulas, P., Grimaud, R., 2010. Behavior of *Marinobacter hydrocarbonoclasticus* SP17 cells during initiation of biofilm formation at the alkane–water interface. *Biotechnology and Bioengineering* 105, 461–468.

Kornaros, M., Lyberatos, G., 2006. Biological treatment of wastewaters from a dye manufacturing company using a trickling filter. *Journal of Hazardous Materials* 136, 95–102.

Kühl, M., Rickelt, L.F., Thar, R., 2007. Combined imaging of bacteria and oxygen in biofilms. *Applied and Environmental Microbiology* 73, 6289–6295.

Lee, H.-G., Choe, M.-W., Wang, S.-H., Park, T.-J., Kim, C.-W., 2001. Applicability of fixed bed biofilm reactor for nitrogen removal from sewage with high nitrogen contents. *Environmental Engineering Research* 6, 55–55.

Li, X.Z., Webb, J.S., Kjelleberg, S., Rosche, B., 2006. Enhanced benzaldehyde tolerance in *Zymomonas mobilis* biofilms and the potential of biofilm applications in fine-chemical production. *Applied and Environmental Microbiology* 72, 1639–1644.

Liao, J., Fang, C., Yu, J., Sathyagal, A., Willman, E., Liu, W.-T., 2017. Direct treatment of high-strength soft drink wastewater using a down-flow hanging sponge reactor: performance and microbial community dynamics. *Applied Microbiology and Biotechnology* 101, 5925–5936.

Liu, Q., Wang, X.C., Liu, Y., Yuan, H., Du, Y., 2010. Performance of a hybrid membrane bioreactor in municipal wastewater treatment. *Desalination* 258, 143–147.

Liu, Y.-Q., Liu, Y., Tay, J.-H., 2004. The effects of extracellular polymeric substances on the formation and stability of biogranules. *Applied Microbiology and Biotechnology* 65, 143–148.

Lu, D., Bai, H., Kong, F., Liss, S.N., Liao, B., 2021. Recent advances in membrane aerated biofilm reactors. *Critical Reviews in Environmental Science and Technology* 51, 649–703.

Mitri, S., Richard Foster, K., 2013. The genotypic view of social interactions in microbial communities. *Annual Review of Genetics* 47, 247–273.

Moosavi, G., Naddafi, K., Mesdaghinia, A., Nabizadeh, R., 2005. Simultaneous organics and nutrients removal from municipal wastewater in an up-flow anaerobic/aerobic fixed bed reactor. *Journal of Applied Sciences* 5, 503–507.

Nerenberg, R., 2016. The membrane-biofilm reactor (MBfR) as a counter-diffusional biofilm process. *Current opinion in Biotechnology* 38, 131–136.

Ohandja, D.G., Stuckey, D.C., 2006. Development of a membrane-aerated biofilm reactor to completely mineralise perchloroethylene in wastewaters. *Journal of Chemical Technology & Biotechnology: International Research in Process, Environmental & Clean Technology* 81, 1736–1744.

Oliveira R., Melo, L., Oliveira, A., Salgueiro, R., 1994. Polysaccharide production and biofilm formation by Pseudomonas fluorescens: effects of pH and surface material. *Colloids and Surfaces B: Biointerfaces*, 2, 41–46.

O'Toole, G., Kaplan, H.B., Kolter, R., 2000. Biofilm formation as microbial development. *Annual Reviews in Microbiology* 54, 49–79.

Pamp, S.J., Gjermansen, M., Tolker-Nielsen, T., 2009. The biofilm matrix: a sticky framework. Kjelleberg S, G.vskov.

Pasmore, M., Todd, P., Smith, S., Baker, D., Silverstein, J., Coons, D., Bowman, C.N., 2001. Effects of ultrafiltration membrane surface properties on *Pseudomonas aeruginosa* biofilm initiation for the purpose of reducing biofouling. *Journal of Membrane Science* 194, 15–32.

Peters, R.W., Alleman, J.E., 1982. *The History of Fixed-Film Wastewater Treatment Systems*. Purdue University Lafayette in School of Civil Engineering, West Lafayette, IN.
Piai, L., Blokland, M., van der Wal, A., Langenhoff, A., 2020. Biodegradation and adsorption of micropollutants by biological activated carbon from a drinking water production plant. *Journal of Hazardous Materials* 388, 122028.
Qaderi, F., Sayahzadeh, A., Azizi, M., 2018. Efficiency optimization of petroleum wastewater treatment by using of serial moving bed biofilm reactors. *Journal of Cleaner Production* 192, 665–677.
Qadir, M., Drechsel, P., Jiménez Cisneros, B., Kim, Y., Pramanik, A., Mehta, P., Olaniyan, O., 2020. Global and regional potential of wastewater as a water, nutrient and energy source. Presented at the Natural resources forum, Wiley Online Library, pp. 40–51.
Radwan, S., Al-Hasan, R., 2001. Potential application of coastal biofilm-coated gravel particles for treating oily waste. *Aquatic Microbial Ecology* 23, 113–117.
Rehman, A., Naz, I., Khan, Z.U., Rafiq, M., Ali, N., Ahmad, S., 2012. Sequential application of plastic media-trickling filter and sand filter for domestic wastewater treatment at low temperature condition. *Biotechnology Journal International* 2, 179–191.
Rossi, M.M., Dell'Armi, E., Lorini, L., Amanat, N., Zeppilli, M., Villano, M., Petrangeli Papini, M., 2021. Combined strategies to prompt the biological reduction of chlorinated aliphatic hydrocarbons: new sustainable options for bioremediation application. *Bioengineering* 8, 109.
Shahot, K., Idris, A., Omar, R., Yusoff, H.M., 2014. Review on biofilm processes for wastewater treatment. *Life Science Journal* 11, 1–13.
Shimada, K., Itoh, Y., Washio, K., Morikawa, M., 2012. Efficacy of forming biofilms by naphthalene degrading *Pseudomonas stutzeri* T102 toward bioremediation technology and its molecular mechanisms. *Chemosphere* 87, 226–233.
Shingare, R.P., Thawale, P.R., Raghunathan, K., Mishra, A., Kumar, S., 2019. Constructed wetland for wastewater reuse: role and efficiency in removing enteric pathogens. *Journal of Environmental Management* 246, 444–461.
Show, K.-Y., Yan, Y.-G., Zhao, J., Shen, J., Han, Z.-X., Yao, H.-Y., Lee, D.-J., 2020. Startup and performance of full-scale anaerobic granular sludge blanket reactor treating high strength inhibitory acrylic acid wastewater. *Bioresource Technology* 317, 123975.
Siggins, A., Thorn, C., Healy, M.G., Abram, F., 2021. Simultaneous adsorption and biodegradation of trichloroethylene occurs in a biochar packed column treating contaminated landfill leachate. *Journal of Hazardous Materials* 403, 123676.
Simoes, M., Pereira, M.O., Sillankorva, S., Azeredo, J., Vieira, M.J., 2007. The effect of hydrodynamic conditions on the phenotype of *Pseudomonas fluorescens* biofilms. *Biofouling* 23, 249–258.
Talwar, S., Sangal, V.K., Verma, A., 2018. Feasibility of using combined TiO_2 photocatalysis and RBC process for the treatment of real pharmaceutical wastewater. *Journal of Photochemistry and Photobiology A: Chemistry* 353, 263–270.
Todhanakasem, T., Sangsutthiseree, A., Areerat, K., Young, G.M., Thanonkeo, P., 2014. Biofilm production by *Zymomonas mobilis* enhances ethanol production and tolerance to toxic inhibitors from rice bran hydrolysate. *New Biotechnology* 31, 451–459.
Viggi, C.C., Simonetti, S., Palma, E., Pagliaccia, P., Braguglia, C., Fazi, S., Baronti, S., Navarra, M.A., Pettiti, I., Koch, C., 2017. Enhancing methane production from food waste fermentate using biochar: the added value of electrochemical testing in pre-selecting the most effective type of biochar. *Biotechnology for Biofuels* 10, 1–13.
Wainaina, S., Awasthi, M.K., Sarsaiya, S., Chen, H., Singh, E., Kumar, A., Ravindran, B., Awasthi, S.K., Liu, T., Duan, Y., Kumar, S., 2020. Resource recovery and circular economy from organic solid waste using aerobic and anaerobic digestion technologies. *Bioresource Technology* 301, 122778.
Whitfield, G.B., Marmont, L.S., Howell, P.L., 2015. Enzymatic modifications of exopolysaccharides enhance bacterial persistence. *Frontiers in Microbiology* 6, 471.
Wu, Y., Tatsumoto, H., Aikawa, M., 2000. Modeling of tetrachloroethylene degradation by anaerobic granular biological activated carbon. *Journal of Health Science* 46, 434–440.
Yang, S., Yang, F., Fu, Z., Lei, R., 2009. Comparison between a moving bed membrane bioreactor and a conventional membrane bioreactor on organic carbon and nitrogen removal. *Bioresource Technology* 100, 2369–2374.
Zhang, X., Sun, J., Zhao, M., 2022. Enhanced metronidazole removal by binary-species photoelectrogenic biofilm of microaglae and anoxygenic phototrophic bacteria. *Journal of Environmental Sciences* 115, 25–36

11

Genetically Engineered Microbes in Bioremediation of Environmental Contaminants

Bishal Singh
Indian Institute of Technology Kharagpur

Marpu Adilakshmi and Evangeline Christina
Lovely Professional University

CONTENTS

11.1 Introduction

There are diverse groups of pollutants increasing every day; most of them are xenobiotic and novel in nature. The complex nature and properties of these anthropogenic compounds are hindering the degradation efficiency of pollutants from the environment. Complex bonds in the compounds and their association with other functional groups like halogen and others are making them more resistant to decomposition and degradation process by naturally occurring microorganisms. These complex

DOI: 10.1201/9781003231738-12

compounds have a higher impact on the environmental condition or are more toxic to the natural microbes which is either affecting their metabolic pathways, inhibiting their growth, or increasing the toxicity level in the environment, leading to their death (Pieper and Reineke, 2000). Microorganisms are primitive and ancient, they evolved and adapted in almost every habitat and environmental condition. Each environmental condition having biotic and abiotic stress related to pH, temperature, nutrition leads them to develop some defence mechanism to sustain and grow in those habitats. These evolutionary processes in microbes have developed different physicochemical abilities to degrade diverse organic and inorganic compounds as shown in Table 11.2. The novel abilities of these natural microorganisms which are encoded in their genome can be used to engineer suitable microbes which can degrade such pollutants effectively. Genetically engineered microbes are developed by introducing different genes such as mer gene, *MPD* gene, *AsrM* gene, *bphA* gene, Laccase gene, and others as stated in Table 11.5, which encode diverse enzymes and proteins suitable for biodegradation of various chemical pollutants in harsh environments conditions (Liu et al., 2006).

In most cases, the Genetically Engineered Microbes (GEM) does not show an effective result as in the lab, when used in contaminated sites. So, the selection of appropriate microorganisms for genetic engineering is a challenging task. In the case of bioremediation, indigenous microbes are more preferred in comparison to exotic strains, Due to their ability to interact and withstand the environmental and biological stress at contaminated sites (Singh et al., 2011). To develop a successful genetically engineered microbe for bioremediation, interdisciplinary knowledge in fields of microbiology, genetic engineering, molecular biology, genetics, bioinformatics, proteomics, and metabolomics are needed to overcome various challenges while using these GEM at the contaminated sites. There are several reports of successful bioremediation of various organic pollutants by genetically engineered microbes but now the modern research focuses on increasing the ability of microbes to the wide spectrum of pollutants (Ang et al., 2005).

There are several genetically engineered strains of *E. coli, Deinococcus radiodurans, Pseudomonas sp., Sphingomonas sp., Saccharomyces cerevisiae,* and others have given in Table 11.6, that have shown an effective result in the bioremediation process of various pollutants in comparison with their parental strains (Wu et al., 2005). The first genetically engineered microbe for bioremediation was used by US government to remove crude oil by *Pseudomonas* strain in 1996 (Ryan et al., 2000). In this chapter, different types of the bioremediation process along with their application, benefits, and suitable examples of different indigenous natural microorganisms with the potential for bioremediation are explained briefly. It is important to explore the principle, working mechanism, and application of various tools used in Genetic engineering to develop highly efficient GEM. Advance genetic engineering tools such as Zinc finger nuclease (ZNFs), TALENs, CRISPR-Cas, sleeping beauty and piggyback are also explained in this chapter. Genetically engineered microorganisms are always a topic of debate in our society. So, there is a need to monitor and regulate these recombinant microbes which have the potential to disturb the ecosystem (Sayler and Ripp, 2000). The penitential risk and issues related to genetically engineered microorganisms with their future direction in which they can be used in a controlled manner in the field of bioremediation are also highlighted in this chapter.

11.2 Bioremediation

Around the world, we have very hazardous compounds deposited in the environment. To clean those polluted and contaminated areas, many science fields such as life sciences, ecology, and environmental sciences are in the race to develop novel processes to degrade the pollutants efficiently and quickly. Bioremediation is used abundantly all over the globe. As they decompose both organic and inorganic toxic pollutants into harmless compounds in a controlled environmental condition as shown in Figure 11.1 (Kim et al., 2014). The ability of microbes is to put down the compounds is harnessed to make this process eco-friendlier. The microorganisms used are either endemic or exogenic in nature. The decomposition of harmful compounds into nontoxic compounds is done by microbial enzymatic reactions. Providing suitable environmental conditions to the microorganisms will help them to work more efficiently. Incineration is an old method where there were chances of many accidents due to contact with

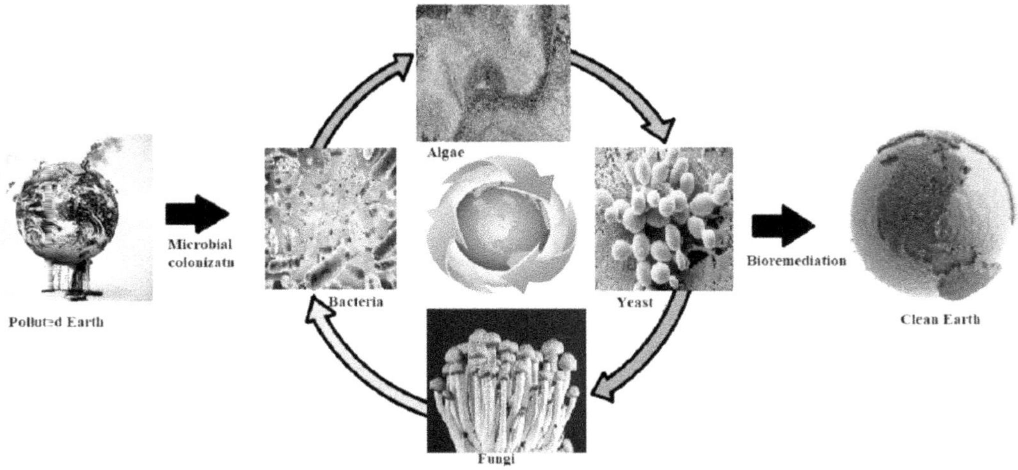

FIGURE 11.1 Represents the role of various microorganisms in the process of bioremediation of wide range.

toxic compounds. So, globally bioremediation is accepted because they cause the on-site processing of hazardous compounds (Colberg and Young, 1995). Bioremediation is of two types ex-situ and in-situ based on the location of the treatment of pollutants. The comparative study of both these classes is given in Table 11.1.

11.2.1 Types of Bioremediation

11.2.1.1 Ex-situ

In this method of bioremediation, the pollutants are transferred from one place to another for treatment or removal of pollutants and for reduction of toxic compounds. There are many different methods used in ex-situ bioremediation. Which are as follows.

- **Land Farming**: In this method, the soil is excavated from a dirty area to a field, where the endemic aerobic bacteria are already present. The polluted soil is spread periodically like a plane bed wherein the natural microbes can access the whole surface and decompose all the hazardous compounds into non-toxic compounds. This process is cost-effective and doesn't need heavy maintenance.
- **Composting**: It is the most convenient method for the decomposition of hazardous compounds via organic wastes, produced from vegetables and cow dung. These organic wastes are mixed with the polluted area and allowed to produce essential nutrients for the growth of microorganisms, which are utilized as the source of energy and decomposition of polluted compounds.
- **Biopiles**: It is a bioremediation method where excavated soil is mixed with amendment soil formed into compost piles. Petroleum hydrocarbons are utilized for the surface treatment of contaminants. Volatilization and leaching is the process that prevents the physical loss of components in this method (Von Fahnestock et al., 1998).
- **Bioreactors**: It is an advanced bioremediation technique where all three forms of matter are used with a high rate of a reaction under controlled conditions. Contaminated soil is added to slurry reactors and pollutants present in either soil or water are extracted via an engineering system in presence of natural microorganisms. These microorganisms can easily decompose target compounds due to the presence of controlled environmental conditions in bioreactors. This method is more efficient than all the above-mentioned techniques but, it is not a cost-effective process as well as pre-processing is required to start the bioremediation process.

TABLE 11.1

Summarizes the Comparative Study of Different Types of the Bioremediation Process Commonly Used in Cleaning the Environment (Philp and Atlas, 2005; Kim et al., 2014)

S. No	Ex-situ Bioremediation	In-situ Bioremediation
1	The process performed out of the contaminated sites	The process performed at the contaminated sites
2	Excavation and transportation of the pollutant is required	No excavation and transportation is required
3	High chances of exposure to toxic pollutants	Low chances of exposure to toxic pollutants
4	Not cost-effective	Cost-effective
5	Fast process	Slow process
6	Can be managed and controlled efficiently	Not manageable and fully under control
7	Give high throughput	Gives less throughput
8	It can target and process a wide range of compounds in one process to remove them from contaminated soil.	Cannot be used for a wider range of pollutants at once
9	The most common methods are Land farming, composting, biopiles, and bioreactors	The most common approaches are Bioattenuation, Biostimulation, Bioaugmentation, Bioventing, Biosparging, and Bioslurping
10	Requires extra labour and effort	Doesn't need many human resources and tools

11.2.1.2 In-situ

This is the process where the polluted compounds are treated at their original site. These pollutants are decomposed into non-toxic compounds in presence of natural microbes in the natural environment. The in-situ process is more friendly, economically sustainable, and easy to maintain when compared with ex-situ bioremediation methods, which reduces the chance of exposure and cost. As excavation of pollutants and transportation from the original site is not done in this process. There are multiple processes in this type of bioremediation where we can select relatable methods accordingly for different types of pollutants on-site (Madsen, 1991).

- **Bioattenuation**: It is intrinsic bioremediation and a globally accepted process in which pollutants are converted into less toxic compounds when treated under the presence of natural microbes or chemical compounds. Decomposition of compounds occurs due to microbial metabolic reactions or chemical reactions on different natural media. These methods are target-specific mainly utilized in BTEX fuel treatment (Atteia and Guillot, 2007). This method is cost-effective. Due to the absence of essential microorganisms in soil wherein bioattenuation acts as the best approach for many industrialists and biotechnologists.
- **Biostimulation**: If the microorganisms, present in the site are active and growth is suitable, the bioremediation is also high and vice-versa. So to maintain the healthy microorganisms at the polluted site for degradation of pollutants, factors such as 'C',' N',' P',' O', temperature, humidity, pH should be checked regularly and maintained accordingly (Wik, 2001). We can supply multiple nutrients forms, slow revealing forms, and other forms to enhance microbial activity during the biostimulation process (Nikolopoulou and Kalogerakis, 2008). If nutrient supply strategies are appropriate then the degradation rate is also higher. The composition of nutrients is based on the types of germs available in polluted areas. Provided nutrients don't only degrade the pollutants but also control the microbial composition at the polluted site.
- **Bioaugmentation**: Bioaugmentation is the process in which the excavated contaminated region is cultured and soil activation is done. The activated soil is again re-inoculated in the same contaminated region (Otte et al., 1994). Through this process, microorganisms get activated. The microorganisms which are cultured increase their adaptability, capacity, durability, and degradation property before re-inoculation in the original site which is shown in *Sphingomonas* spp. Degradation of dibenzo-para dioxin and dibenzofurans are done by RW1 (Megharaj et al., 1997). Processed microbes show more effective results when compared with

natural microorganisms. They can rapidly put down the waste product present in polluted soil sites by carvone-induced bacteria in the bioremediation of polychlorinated biphenyl (PCBs) (Singer et al., 2005). In these methods microbes such as non-pathogenic to humans, handling stress, and having high enzymatic activity are utilized.

- **Composting**: Composting method is a traditional method that was utilized in agriculture to have healthy soil. This method is used for the degradation of pollutants where contaminants and waste are two parameters for bioremediation. If the temperature and the solid-waste ratio are high then the efficiency of processing the pollution is also high. Remediated soil in the mixture contains different microorganisms which enhance decomposition. Decomposition of polycyclic aromatic hydrocarbon (PAH) compounds rises with a rise in the bulkiness of waste where temperature and humidity should remain constant. Composting has many phases where organic compounds are degraded in the initial stage where the heat generated in the process is used to start the thermophilic stage where decomposition and transformation occur via different methods such as volatilization, adsorption, etc. (Rynk et al., 1999). This method has many advantages and some drawbacks such as loss of product during the process, formation of non-extractable compounds.
- **Bioslurping**: Bioslurping is this process in which there is the utilization of a combination of many different methods such as soil vapour extraction, bioventing, and vacuum-associated pumping. This type of bioremediation is used in contaminated soil and groundwater. slurp which is used in this process separated products from the mixed layer by capillary action and forms a free layer of LNAPL the LNAPLs are increased by the pumping mechanism (Kim et al., 2014). This method assists in the separation of volatile and non-volatile compounds from polluted soil. The limiting factor of this system is the presence of excess moisture in soil because it reduces oxygen transfer rate and checks the microbial activity (Philp and Atlas, 2005).
- **Bioventing**: This process is utilized in both aerobic and anaerobic processes. Oxygen, nutrient, and moisture are passed into the unsaturated areas which transform the toxic pollutants into nontoxic ones from the contaminated areas (Philp and Atlas, 2005). In this process, we pass only controlled oxygen to increase the activity of the bioremediation process. For an effective bioventing process, the rate and method of air inflow are determining factors for toluene. The rate of these determining factors forms uniform air distribution.
- **Biosparging**: This method is analogous to bioventing maximum used in an igneous environment such as diesel, kerosene, xylene, etc. (Kao et al., 2008) where saturated zone consists of oxygen instead of an unsaturated zone which leads to the upward movement of volatile chemical compounds to unsaturated area which is approachable to microbes and thus the bioremediation process efficiency increased. The efficiency depends on soil permeability so that they have good access to chemicals to microbes and the chemical properties of those compounds (Philp and Atlas, 2005).

11.3 Natural Microbes in Bioremediation

11.3.1 Bacteria

Bacteria are prokaryotic microorganisms with diverse abilities. They are well known as decomposers and are very effective in the process of bioremediation. There is a wide range of bacterial strains, which can be used to target various organic and inorganic compounds. They are present in almost every habitat and natural condition of the earth, where they have developed various abilities during their evolution based on the necessity and requirement to sustain and grow in that natural niche (Singh et al., 2014). Extensive studies are conducted to explore various strains of bacteria with bioremediation capabilities are given in Tables 11.2 and 11.3 (Celik et al., 2012). These microbes are considered an effective tool as their method of culture and maintenance are well known. There are multiple groups of bacteria, some of them degrade compounds in aerobic conditions, some in anaerobic conditions, and some can degrade in both environmental conditions (Misal et al., 2011). Bacteria have developed various strategies and

TABLE 11.2

Summarizes the Indigenous Bacteria, Useful in the Bioremediation Process and Their Mechanism of Action in Removing Various Contaminants from Nature

S.no	Natural Bacteria for bioremediation	Types of Pollutants	Compounds	Mechanism of Bioremediation	Reference
1	*Acinetobacter baumanni* *Alcaligenes* sp. *Bacillus cereus* *Alishewanella* sp. *B. Halodurans* *Bacillus. pumilus* *P. Aeruginosae* *Bacillus subtilis* *S. Hominis* *M. Luteus* *P. entomophilla*	Dyes	Congo red, Reactive red, Cibacron red, Acid black, Dispersed orange, Golden yellow HER, Reactive green	Enzymatic oxidation and reduction.	Singh et al. (2014), Misal et al. (2011)
2	*Pseudomonas* sp., *Achromobacter* sp., *Acinetobacter* sp., *Streptococcus* sp., *Enterobacter* sp., *Rhodococcus* sp., *Pandorae* sp., *Kocuria* sp.,	Crude oil and polyaromatic hydrocarbons	Anthracene, pyrene, Benzopyrene, Naphthalene	Bioaugmentation, Ozonation Methanogenesis	Carmichael and Wong (2001), Xu et al. (2017)
3	*Escherichia coli, Deinococcus radiodurans, Pseudomonas putida, Mesorhizobium huakuii, Caulobacter crescntus, Rhodopseudomonas palustris, Sphingomonas* sp.	Heavy metals	Mercury, Cadmium, Copper, nickel, Cobalt, Zinc, Chromium	Bioleaching, Bioaccumulation, biosorption	Celik et al. (2012), Liu et al. (2011)
4	*Bacillus sphaericus, Flavobacterium* sp., *Bacillus diminuta, Alteromonas* sp.	Pesticides	DDT, Diuron, HCH, Aldrin, Atrazine	Mineralization, Enzymatic oxidation and Reduction	Mitra et al. (2001)
5	*Geobacter metalireducens, Clostridium* sp., *Desulfovibrio desulfuricans Desulfovibrio vulgaris., Pseudomonas aeruginosa, Streptomyces viridochromogens, Micrococcus luteus, P. fluorescens, Serratia sp., E. Coli, Paracoccus denitrificans, Pseudomonas sp, Rhodobacter sp., Geobacter metallireducens*	Radioactive compounds	Uranium, Neptunium, Technetium, Americium, Plutonium	Enzymatic oxidation and Reduction, Adsorption, Leaching, Dissolution, Precipitation	Sakaki et al. (2002)

molecular mechanisms for the bioremediation process. The most common approaches used by them are enzymatic oxidation, reduction, bioaccumulation, and bioconversion. The enzymes such as laccase, azoreductase, oxygenase, and others, expressed by the bacterial system are responsible for breaking complex bonds of pollutants to transform them into nontoxic forms (Carmichael and Wong, 2001). Multiple bacterial strains have been isolated from extreme to normal environments which can degrade various contaminants such as heavy metals, pesticides, organic dyes, dioxins, crude oils, radioactive

TABLE 11.3

Summarizes the Indigenous Microorganism (Bacteria, Fungi, and Algae) with Their Favourable Environmental Condition Useful for the Bioremediation Process

S no	Microorganisms	Names	Favourable Environment	Reference
1	Bacteria	*Pseudomonas veronii, Kocuria Flava, Arthrobacter* spp., *Bacillus cereus, Burkholderia spp., Bacillus sphaericus, Flavobacterium sp., Bacillus diminuta, Bacillus subtillis, Aeromonas hydrophilla, Acinetobacter. Staphylococcus aureus, Aeromonas hydrophilla*	Aerobic, anaerobic or both pH range (6–9) Temperature (25°C–40°C)	Singh et al. (2014)
2	Fungi	*Candida albicans, Candida tropicalis, Debaryomyces polymorphous, Saccharomyces cerevisiae, Candida utilis, Penicillium canescens, Aspergillus versicolor, Aspergillus fumigatus, Avantha discolor, Agrocybe semiorbicularis, Hypholoma fasciculare, A. auricula, C. versicolor, Phanerochaete chrysosporium, Curvularia lunata* and *Hypocre koingii*	Aerobic condition pH range (3–6) Temperature (25°C–40°C)	Sen et al. (2012) Yu and Wen (2005)
3	Algae	*Spirogyra spp., Oscillatoria curviceps, Spirulina* spp., *Chlorella, Chara vulgaris, Scenedesmus quadricauda, Scenedesmus obliquus Cladophora* spp.	Aerobic and aquatic Condition pH range (5–8) Temperature (20°C–40°C)	Chu et al. (2009) Acuner and Dilek (2004)

compounds, and chlorinated solvents. Some of the most common microorganisms naturally obtained with bioremediation capabilities are *Pseudomonas veronii, Kocuria Flava, Arthrobacter spp., Bacillus cereus, Burkholderia spp., Bacillus sphaericus, Flavobacterium sp., Bacillus diminuta, Bacillus subtillis, Aeromonas hydrophilla, Acinetobacter. Staphylococcus aureus, Aeromonas hydrophilla,* and others (Singh et al., 2014). Some natural biodegradation bacteria along with the substrate which they degrade are summarized in Table 11.2. Properties such as cost-effectiveness, eco-friendly nature, and wide range of pollutant coverage, make the bacteria an effective tool in the field of bioremediation.

11.3.2 Fungi

The study of the process of bioremediation by fungus is known as mycoremediation. Fungi are heterotrophic microorganisms which depend on the dead and decaying living form and are also known as a scavenger, where they transform various organic and inorganic compounds into nontoxic forms, to utilize them in their physiological process. This ability of fungi has evolved them as one of the most effective bioremediation tools in nature. It has been found from various studies that, the process of biodegradation of various aromatic compounds is the secondary metabolic process, performed by fungi to obtain nutrition and energy (Porri et al., 2011). The bioremediation process by fungi involves adsorption and various complex conversion reactions such as hydroxylation, oxidation, and reduction by the enzymes secreted by them (Yu and Wen, 2005). There are different species of fungi given in Table 11.3, which can be broadly classified into two groups one as filamentous fungi and the other as yeast.

Yeast is the fungi that are mostly used in the process of fermentation and bioremediation and is considered as an advantageous option over bacteria due to their higher growth rate as bacteria and can easily adapt and sustain in an adverse environmental condition unlike bacteria (Hai et al., 2012). Most commonly identified fungi with bioremediation abilities are *Candida albicans, Candida tropicalis, Debaryomyces polymorphous, Saccharomyces cerevisiae, Candida utilis, Penicillium canescens, Aspergillus Versicolor, Aspergillus fumigatus, Avantha discolor, Agrocybe semiorbicularis, H. fasciculare, A. auricula, C. versicolor, P. chryososporium, Curvularia lunata*, and *Hypocre koingii* (Sen et al., 2012). Based on the efficiency and properties of fungi, it is well established that they are one of the highly suitable microbes for bioremediation.

11.3.3 Algae

Algae is an autotropic microflora with the ability to solve major problems associated with pollutants throughout the globe. They are the most influencing microflora with a higher impact on the bioremediation process. Most of the textile, pharmaceutical, food industries use these organisms as a bioremediation tool due to their flexible nature based on the environment and less investment in culture and management due to the autotrophic mode of nutrition (Dubey et al., 2011). The most common approach used by algae in the bioremediation process is adsorption and enzymatic action, where organic and inorganic compounds are used as components of the physiological cycle or as sources of nutrition. Chlorophyceae, Rhodophyceae, and Cynophyceae are common groups of algae with bioremediation ability (Chu et al., 2009). *Spirogyra* spp., *Oscillatoria curviceps, Spirulina* spp., *Chlorella, Chara Vulgaris, S. quadricauda, Scenedesmus obliquus,* and *Cladophora spp.* Algae given in Table 11.3 are mostly useful in the biodegradation of various pollutants in aquatic habitats. (Acuner and Dilek, 2004).

11.4 Genetic Engineering in Bioremediation

Genetic engineering was first introduced in 1970 as an emerging method to enhance the efficiency of microbes and to study the function of various genes. Most of the work in this area is associated with recombinant DNA technology where the piece of DNA of one microorganism is introduced in another microbe to provide a novel function (Smith et al., 2000). The need for genetic engineering in the field of bioremediation is to develop various microbial strains with noble enzymes and proteins which can remove pollutants from contaminated sites. The increasing amount and complexity of various xenobiotic compounds have created a global concern associated with pollution (Pieper and Reineke, 2000). Multiple indigenous microorganisms as given in Tables 11.2 and 11.3 were effective tools in the bioremediation of various chemical compounds such as dye, pesticides, Polyaromatic hydrocarbons, heavy metals, and others but Due to the complex nature of bonds and higher reactivity of various xenobiotic compounds with other functional groups, have made those indigenous microbes ineffective. With the urgent need to solve the problem associated with pollutants and to clean various natural resources, various genetically engineered microbes were developed with unique potential. These engineered microbes have shown more efficiency in bioremediation when compared with their wild strains (Perales-Vela et al., 2006). There are various approaches such as rational designing of microorganisms with a specific gene, selection

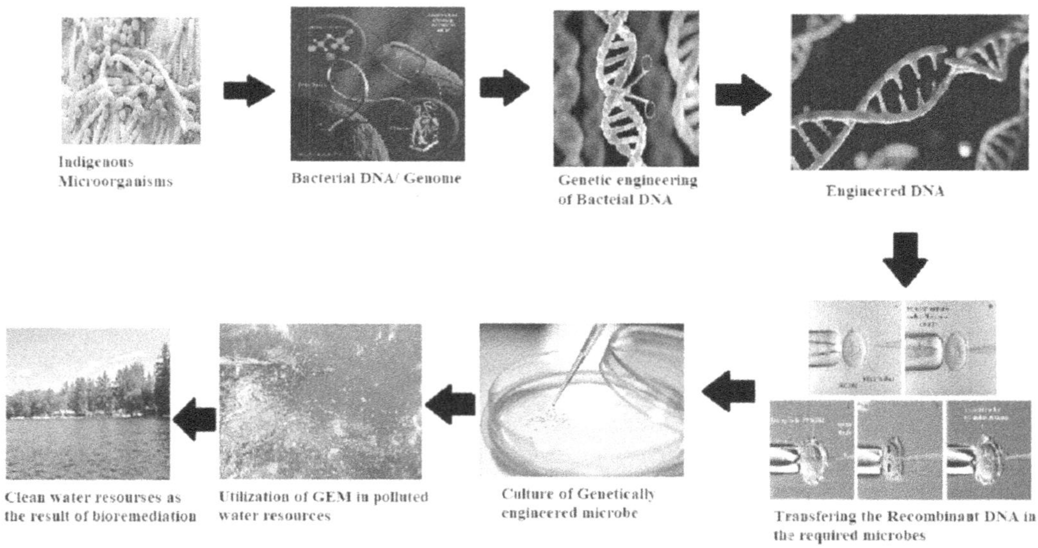

FIGURE 11.2 Represents the conventional approach of genetic engineering to develop various recombination strains and their application in the field of bioremediation and conservation of nature.

of mutant strains, saturation mutagenesis, and metabolic engineering in microorganisms, to alter the genome and to introduce noble abilities in them but this chapter only tried to highlight the molecular tools and approaches to engineering the microorganisms useful for the bioremediation. The comparative study of those advanced tools used in recent times in genetic engineering such as Zinc Finger Nucleases (ZNFs), TALENS, CRISPR-Cas, Sleeping beauty, and Piggybac is given in Table 11.4. Conventionally the strains of microorganisms were developed by recombinant DNA technology in which the gene of interest with a particular function from indigenous microorganisms was identified, studied, and isolated which was introduced in the genome of required microbes with the help of restriction digestion enzyme. The engineered DNA was then inserted in the microorganism with the tools such as microinjection, gene gun, particle bombardment, and various chemical methods into a suitable organism which was cultured and used in the bioremediation process as shown in Figure 11.2.

11.4.1 Advance Methods in Genetic Engineering

11.4.1.1 Zinc Finger Nuclease (ZNFs)

Zinc finger nuclease is a genetic engineering method and a prominent tool in genome editing. As shown in Figure 11.3, it is formed by the fusion of site-specific binding domain and nonspecific cleavage domain (Kim et al., 1996). In 1985 the transcription factor IIIa in Xenopus oocytes was discovered which is similar to Zinc Finger site-specific protein binding with DNA. The domain comprises an array of Cys2His2 Zinc Finger which is derived by highly conserved interaction between the Zinc

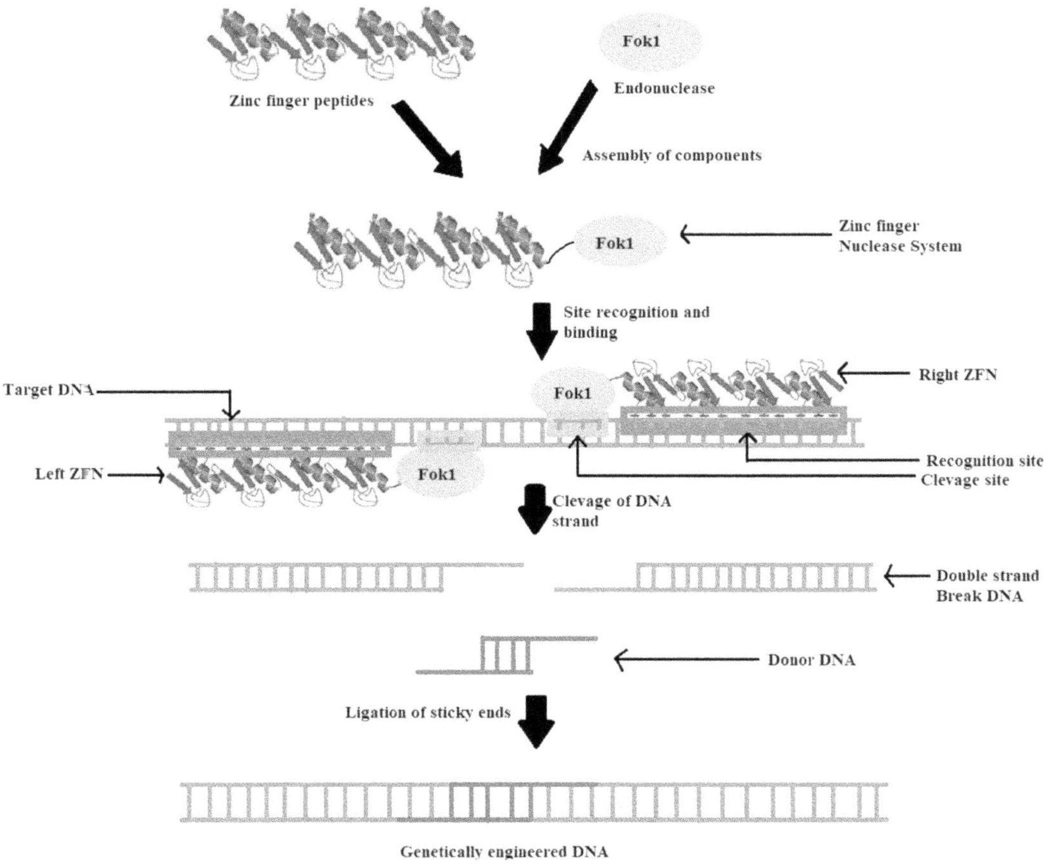

FIGURE 11.3 Represents the structure and working mechanism of Zinc finger Nuclease (ZNFs) in the genetic engineering of microbes.

finger domain and site which consists of 30AA is a part of the functional specificity of ZFNs. The most common type of DNA- binding motif in eukaryotic transcription factors is Cys2His2 ZF is a DNA recognition domain and (Beerli et al., 2000). Each ZF unit recognizes 3bp of DNA. ZF produces base-specific contacts via interaction with α- helix residues with the major groove of DNA (Fairall et al., 1993). The domain formation is to cleave the DNA known as FokI type II restriction endonuclease enzymes which can be native as a dimer to a target sequence in the genome for gene editing (Guo et al., 2010). As the FokI nuclease behaves as a dimer, it requires two ZFN molecules in an appropriate orientation that bind to the target site as shown in Figure 11.3 (Smith et al., 2000). Recognized base pairs are doubled in number. DSBs on the genome are initiated at the specific locus after DNA cleavage by ZFNs creating desired HDR repair system (Beumer et al., 2008). There are three major factors for the determination of target sequence recognition and specificity (i) amino acid sequence, (ii) number of zinc fingers, and (iii) interaction of domain to site-specific. Novel modular assembly with maximum affinity and specificity in genome engineering is developed by individually optimized catalytic domain. This method has 3–6 Zinc fingers that combine with 9–18 nucleotide targets that enable ZFN dimer by 18–36 bp of DNA sequence which renders 68 billion bp of DNA target. But, in recent development architectural diversification is done to amend the targeting accuracy of ZFNs through a selection-based method (Paschon et al., 2019). This method leads to the development to study finger-finger and finger- FoKI cleavage domain junction linker which increases 64-Fold ZFN for targeting cleavage to given DNA database.

11.4.1.2 TALENs Based System

Transcription activator-like effector nuclease integrates a DNA-binding domain and nonspecific DNA-cleaving nuclease. This fusion of DNA domain and nuclease can easily target the required sequence. TALENs have the profound capability to efficiently alter the targeted gene for genetic manipulation. TALENs are an alternative to ZFNs methods. It also contains nonspecific endonuclease (Fok1) which cleaves the DNA strand and DNA binding domain made up of peptide sequence as shown in Figure 11.4. This DNA binding domain derived from TALEs and nucleases is composed of high conserved repeats used to induce Double Strand Breaks (DSBs). DNA is passed through a highly conserved array that contains 33–35 amino acids, where the 12th and 13th position of the amino acid sequence is hypervariable regions that determine the target sequence of DNA (Bogdanove and Voytas, 2011). RVDs determine the DNA- binding specificity by hypervariable residues such as ND, HN, NH, NP for the recognition of base pairs C, A, G, and T respectively. There is a matched compatibility in the target site between RVDs and Contiguous nucleotides, where there is the presence of TALE-DNA recognition cipher (Moscou and Bogdanove, 2009). The fused machinery stimulates and creates site-specific double standard breaks and DNA recombination to achieve TALENs- induced targeted genome modification. To cleave the double standard targeted DNA FoKI cleavage domain must be dimerized. Both zinc fingers and the TALENs module are designed in pairs that can bind to the opposite target DNA loci with correct spacing between two binding sites. TALENs have simple designs when compared with Zinc finger proteins which are triplet confined.

11.4.1.3 CRISPR-CAS System

The CRISPR is an array of palindromic sequences present in form of clusters so, it is also known as Clustered inter-spaced short palindromic repeats. These systems are generally developed in bacteria and lower organisms as a defence mechanism, it was first identified in *E. Coli* in 1987. It has become an important tool and technique in biological research for understanding gene function, and genome editing due to its simplicity, specificity, and effectual work. The clustered repeats are taken from the virus which infects bacteria and then those pieces are integrated into the bacterial genome which helps in guiding the tool for genome editing. It is composed of two components. one is the guide RNA (20bp) which determines the specificity of binding and the other is Cas enzymes which help in cleavage activity as shown in Figure 11.5 (Jinek et al., 2012). CRISPR- Cas system is broadly classified into two major classes based

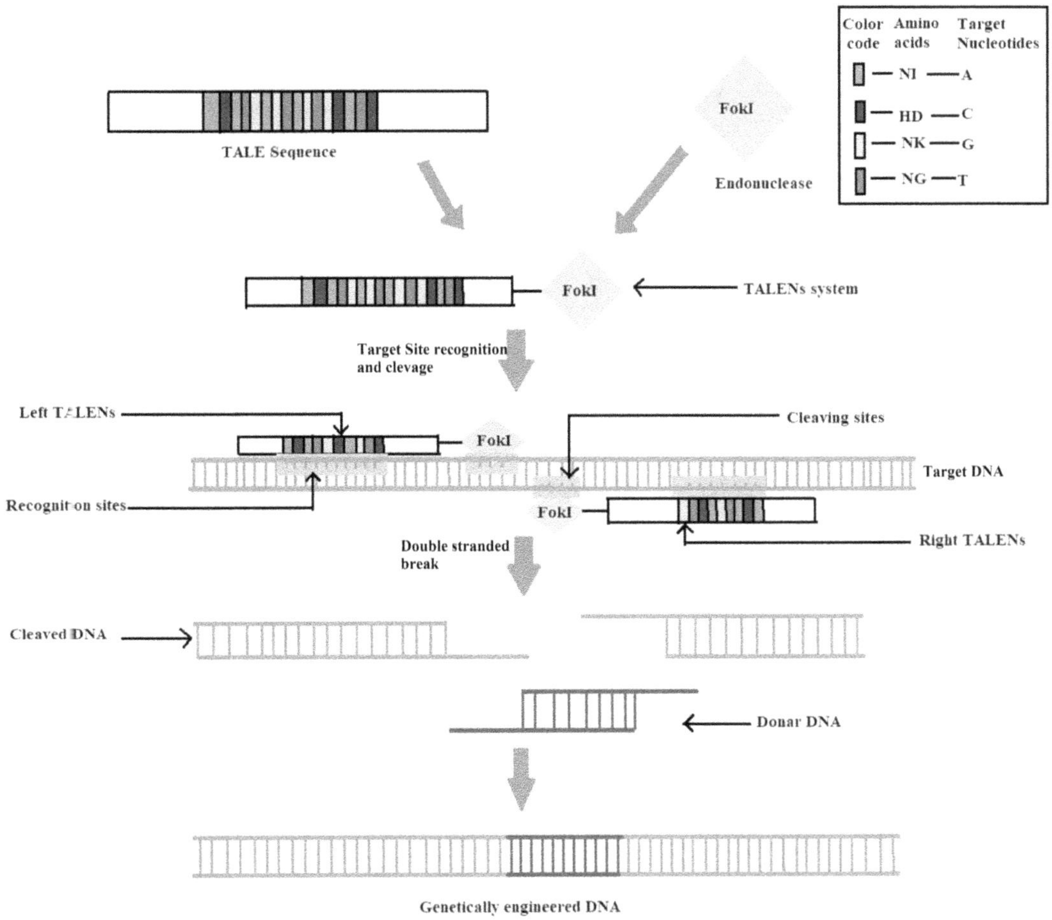

FIGURE 11.4 Represents the structure and working principle of TALENs in the genetic engineering of microorganisms.

on the number of Cas enzymes envenomed in genetic engineering, and in six types based on the type of major Cas enzyme used for cleavage activity.

The most commonly CRISPR-Cas system used in genetic engineering is the type 2 system where Cas9 is used with guide RNA. Guide RNA in the system recognizes the PAM (Protospacer adjacent motif) sequence in the genome and binds with the genetic material by base pairing, where the non-specific cleavage of a single strand or double strand of template DNA occurs by Cas9 endonuclease (Sternberg et al., 2014). Double strand break or single-strand break in template DNA initiates the genome repair mechanism by inserting the gene of interest by homologous recombination method. CRISPER/Cas9 technology is utilized for introducing various genes associated with functional proteins and enzymes where the guide RNA can be easily tuned and engineered by changing the sequence of guide RNA, to recognize target sites via base-pairing. CRISPR/Cas 9 has multiple loci editing, more efficient, more productive, and cost-effective when compared to other techniques. There are various approaches to transfer CRISPR - Cas tools in microbial systems, but the most common approach is the utilization of plasmid-encoded with cas9 protein and sgRNA. This method is long-lasting and prevents multiple transfections in the expression of Cas9 and sgRNA as stated in Table 11.4. Direct delivery of Cas 9 protein and sgRNA has several advantages such as high stability, less immunogenic, and rapid action (Biagioni et al., 2018).

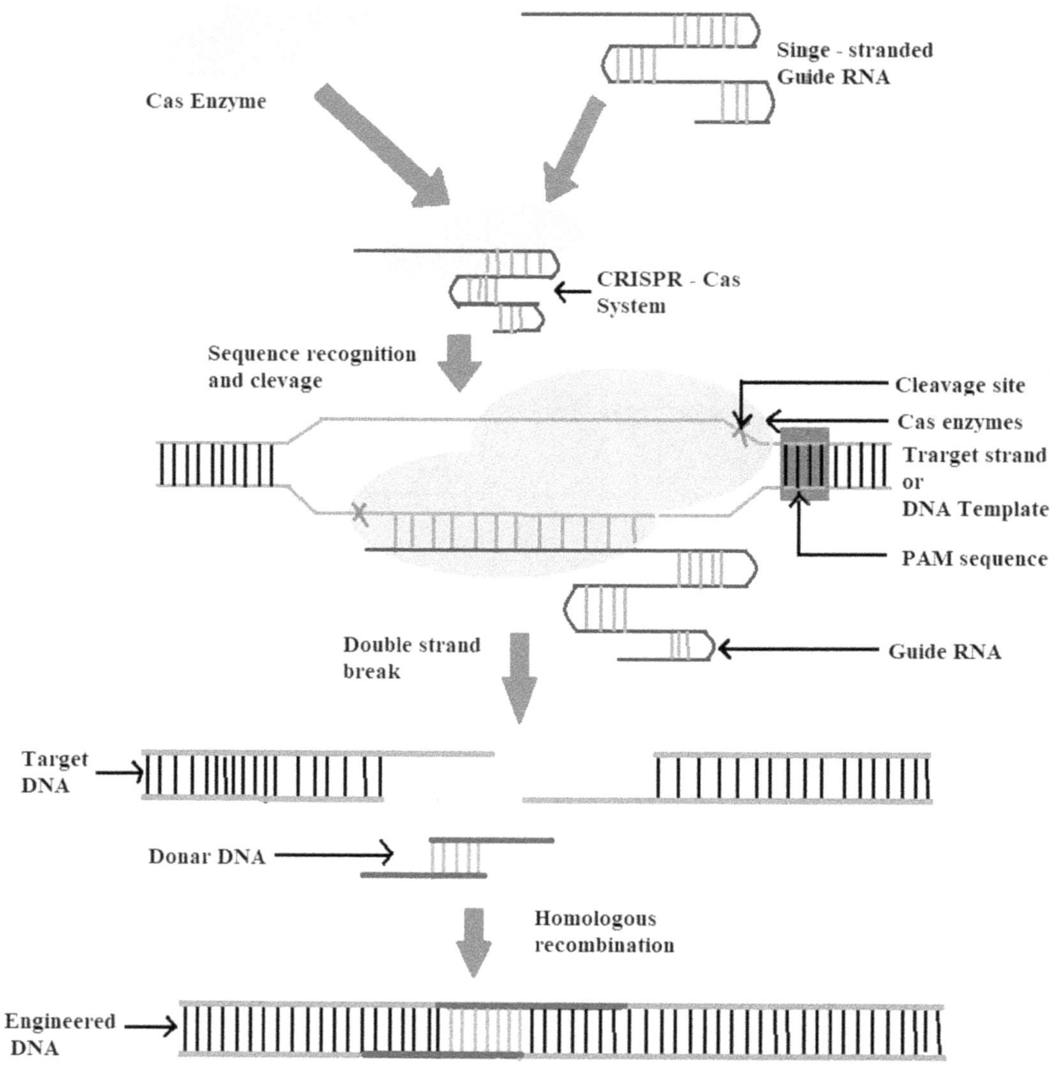

FIGURE 11.5 Represents the structure and working principle of the CRISPR-Cas system in the genetic engineering of microorganisms.

11.4.1.4 Sleeping Beauty System

The sleeping beauty transposon was designed in 1997 to insert a specific sequence of DNA into the genome of vertebrae animals and microorganisms. It consists of sleeping beauty transposase (an Enzyme) and transposon as shown in Figure 11.6. It has a simple structure with DNA as its genetic material with 10 kb size (Izsvák et al., 2000). These are non-viral vehicles that can be moved from one position to another without the requirement of any virus. Naturally, it encodes an enzyme known as transposase which is specific to particular inverted repeats (IR) present in the transposon end. The sequence of inverted repeats is specific to the particular type of transposase enzyme and cannot bind with other types (Plasterk et al., 1999). Engineering in the sequence of gene encoding transposase enzyme can change the specificity for the target sequence, this property of sleeping beauty is used to target and edit various genes present in the microorganism. The enzyme used in this system can be introduced in the system in form of protein, mRNA, or DNA which is further transcribed and translated to work as an enzyme. IR/DR is a special organization of inverted repeats and is an important requirement for transposition (Doak et al., 1994). This system works on the cut and pastes mechanism, where the gene of interest is introduced in the genome by removing the segment of the microbial genome present in between inverted repeats and paste the gene of interest by homologous recombination method (Van Luenen et al., 1994).

11.4.1.5 Piggybac System

The Piggyback system was invented in 1989 by Malcolm Fraser. It is a TTAA-specific transposon-based genetic engineering tool utilized in a wide variety of species. The transposon is a DNA segment that can travel throughout the genome. So, it is also known as jumping genes. In the Piggyback system, identification of the specific inverted terminal repeats present at both ends of the vectors by transposase enzyme is required for genetic engineering as occurs in the sleeping beauty system. In the case of this system, the IR repeats are generally composed of a TTAA sequence as shown in Figure 11.6. PB method is used more frequently because of its efficiency, stability, and safety (Wang et al., 2008). Due to its positive characteristics, this method is used as a new genetic vehicle. The high success rate in genetic engineering by this tool in diverse microorganism suggest that the efficiency of this tool is host factor independent. As PB transposon doesn't have local hopping and no footprint left behind it doesn't cause genomic damage during mutagenesis (Wu et al., 2006). PB transposon has high security and convenience than virus vectors which can move ~14.3 kb DNA fragment so they have large transposition activity as stated in Table 11.4 (Wilber et al., 2007).

TABLE 11.4

Summarizes the Comparative Study of Various Tools such as ZNFs, TALENs, CRISPR -Cas, and Sleeping Beauty/Piggyback, Used in Genetic Engineering to Introduce Nobel Genes in Microorganisms (Izsvák et al., 2000; Jinek et al., 2012; Moscou and Bogdanove, 2009)

S. No	Parameters	ZNFs	TALENs	CRISPR/Cas	Sleeping Beauty/ Piggyback
1	Type	The peptide-based system with Fok 1 nuclease	The peptide-based system with Fok1 nuclease	RNA based system with Cas enzymes	Transposon based systems with transposase enzymes
2	Recognition of Nucleotide sequence	Zinc finger protein	TALE protein	Guide RNA	Transposase specific to the nucleotide sequence
3	Interaction for binding	The protein interacts with DNA for binding	The protein interacts with DNA for binding	RNA interacts with DNA based on nucleotide complementation for binding.	DNA- DNA interactions
4	Cleavage machinery	Fok 1 endonuclease	Fok 1 endonuclease	Cas enzymes	Transposase enzyme
5	Cleavage site	Nonspecific	Nonspecific	PAM sequence	Inverted repeats and direct repeats
6	Cleavage pattern	Double strand break	Double strand break	Double strand or single-strand break	Single strand or double-strand break
7	Target length	9–18 per monomer and 18–36 per ZNF pair	14–20 per monomer and 28–40 per TALEN pair	20 base pair of guide RNA sequence and PAM sequence	Not specific
8	Limitation in targeting	Non G rich site in DNA is difficult to target	5' targeted base should be thymine in case of each TALEN monomer	PAM sequence must be present at the target site	Inverted repeat or direct repeat is needed for homologous recombination and cleavage

(Continued)

TABLE 11.4 (*Continued*)

Summarizes the Comparative Study of Various Tools such as ZNFs, TALENs, CRISPR -Cas, and Sleeping Beauty/Piggyback, Used in Genetic Engineering to Introduce Nobel Genes in Microorganisms (Izsvák et al., 2000; Jinek et al., 2012; Moscou and Bogdanove, 2009)

S. No	Parameters	ZNFs	TALENs	CRISPR/Cas	Sleeping Beauty/ Piggyback
9	Issue in Engineering	Engineering of the protein sequence is required	Molecular cloning is needed to insert the sequence	Require complex cloning method and oligo synthesis	Require complex cloning methods and genetic screening methods
10	Issue in Delivery	Small in size and is suitable for viral vectors	Difficult in packaging due to large size functional components of TALENS	Complex guide RNA and CAS enzyme is larger in size which makes it difficult, during packaging in viral vectors	Larger in size makes it difficult in packaging inside viral vector

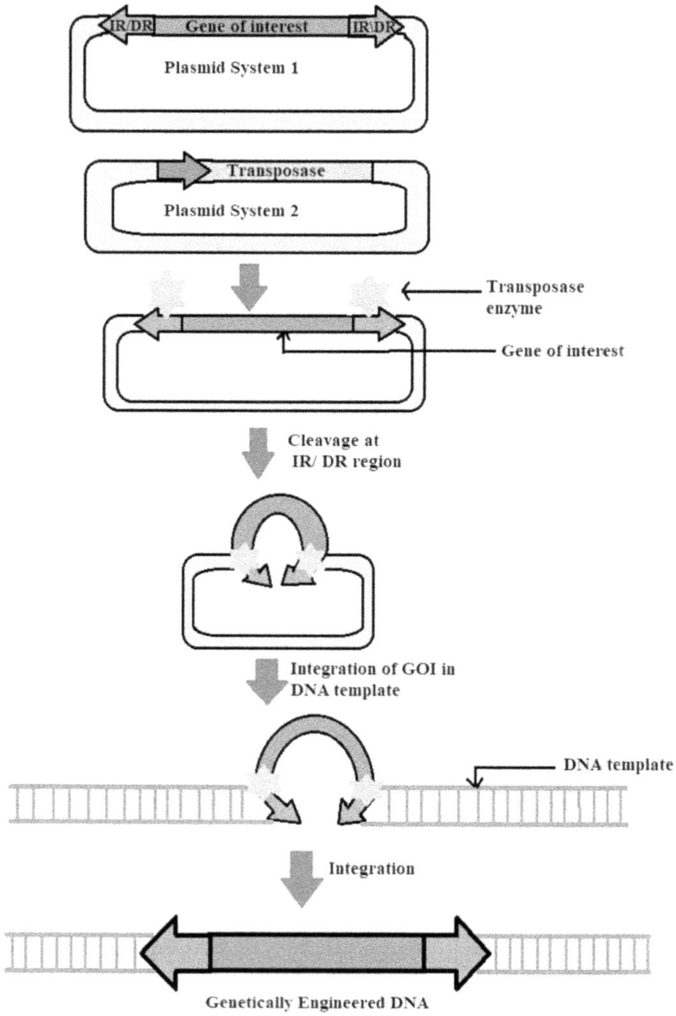

FIGURE 11.6 Represents the structure and working principle of transposone-based systems like sleeping beauty and piggyback in genome engineering of microbes where DR represents direct repeats and IR represents Inverted repeats which are specific to specific transposase enzymes coded in plasmid system 2.

11.5 Common Genes Involved in Bioremediation

Microbes are one of the primitive organisms and are still evolving according to time and necessity to sustain and grow in unfavourable environmental conditions. These microbes have gained various abilities to degrade multiple organic and inorganic compounds with the help of different enzymes (DeLong and Pace, 2001). These enzymes are associated with the modification, decomposition, or precipitation of chemical compounds which act as pollutants in the environment. There are several such enzymes found in the various microbial sources, some examples of such enzymes are Catechol 1, 2 dioxygenases, Benzene dioxygenase, Dioxin dioxygenase of *Sphingomonas* sp., Laccase, and others (Moharikar et al., 2003). There are also various proteins such as phytochelatins, Metaloothioneins, and others that show higher binding affinity with heavy metals. These all enzymes and proteins important for bioremediation are encoded by genes, present in microbes. Genes that are important are screened and studied through a multidisciplinary approach so that these genes can be used in Genetic engineering to develop recombinant microbes for bioremediation (Patel et al., 2010). Some of the most commonly used genes in developing genetically engineered microbes for bioremediation are mer genes for mercury, AsrM genes for Arsenic, MPD genes for various hydrocarbons, bphA genes for PCB compounds, and others are summarized in Table 11.5.

TABLE 11.5

Summarizes the Genes Responsible for the Bioremediation of Various Pollutants by Expressing Different Proteins and Enzymes in Recombinant Microbes

S.no	Recombinant Microbes	Genes used for Genetic Engineering	Targeted Compound	Reference
1	*Pseudomonas sp.*	CAM/OCT plasmid, XYL plasmid, NAH plasmid	Crude oil	Ezezika and Singer (2010)
2	*Escherichia coli BL308, Deinococcus radiodurans*	merT, merP, MT genes	Mercury	Deng and Wilson (2001) Brim et al. (2006)
3	*E. coli JM109*	Gene encoding Phytochelatins and Metaloothioneins	Cadmium	Patel et al. (2010)
4	*Pseudomonas putida Mesorhizobium huakuii Caulobacter crescntus*	Gene encoding EC20, RsaA-6His, phytochelatin	Cadmium	Patel et al. (2010)
5	*Rhodopseudomonas palustris E. coli*	AsrM gene	Arsenic	Liu et al. (2011)
6	*Sphingomonas* sp. CDS-1	MPD gene	Carbamate, Organophosphate	Liu et al. (2011)
7	*Saccharomyces cerevisiae*	Laccase gene	Polyaromatic compound	Carmichael and Wong (2001)
8	*Bukholderia cepacia* LB400 *C. Testosteroni* B-365	bphA gene	PCBs compounds	Barriault et al. (2002)
9	*Saccharomyces cerevisiae*	CYP1A gene	Dioxins Compounds	Sakaki et al. (2002)
10	*Fusarium solani*	(Kel+BebRDBP+Lin +) gene	DDT	Mitra et al. (2001)
11	*Hansenula polymorpha*	FCb2 gene	Chromate	Smutok et al. (2011)

11.6 Genetically Engineered Microbes in Bioremediation

11.6.1 Genetically Engineered Bacteria in Bioremediation

The utilization of genetically engineered bacteria in various fields due to their advantageous nature and cos effective engineering process has led to the development of various strains of Genetically engineered bacteria in the field of bioremediation too, but the efficiency of these engineered bacteria in natural habitats is influenced by various factors. So, before designing any recombinant bacteria, prior knowledge in the fields of microbiology, genetic engineering, and molecular biology. These engineered bacteria will have higher efficiency when they can overcome the problems associated with natural inhabitants (Bacteria) related to nutrition, pH, temperature, selection pressure, and environmental stress (Singh et al., 2011). There are several reports of successful and highly efficient recombinant bacterial systems in the field of bioremediation which are summarized in Table 11.6. One of the initial genetically engineered bacteria was from the Pseudomonas genus for the bioremediation of crude oil (Ezezika and Singer, 2010). Later on, *E.coli* was engineered by inserting MER genes, that encodes for Hg reductase enzyme which can detoxify Hg into a volatile form from contaminated water (Deng and Wilson, 2001). Introduction of MERa gene from recombinant *E. coli* in one of the highly radiation-resistant bacteria, *Deinococcus radiodurans.* A new strain of this species with the ability to detoxify Hg in a high radiation environment was developed (Brim et al., 2006). A new strain (*E. coli* JM109) with a higher affinity to metal binding and higher efficiency in bioaccumulation of heavy metals was developed by introducing various gens associated with the expression of metal-binding proteins such as metallothioneins and phytochelatins. *E.coli* JM109 are highly efficient strains for cadmium detoxification along with these strains other recombinant bacteria with cadmium detoxification ability are *Pseudomonas putida, Mesorhizobium huakuii,* and *Caulobacter crescentus.* These bacteria are engineered to express metal-binding proteins such as Phytochelatin, EC20, and RsaA-6His (Patel et al., 2010). The arsM gene of *Rhodopseudomonas palustris* was introduced in *E. coli* for the expression of Arsenic modifying enzymes (Arsenite methyltransferase), that can detoxify the inorganic form of arsenic to trimethylarsine (Liu et al., 2011). In various genetically engineered bacteria, proteins (Fusion protein, membrane transport proteins, and Metallothionein protein) and enzymes (Glutathione S-transferase) are used to express to enhance the efficiency of bacteria for heavy metal bioremediation (Sandrine et al., 2003). Similarly, more than four strains of Pseudomonas are genetically modified for crude oil degradation, and *Deinococcus radiodurans* were engineered with the help of genes present in *Burkholderia cepacia* to degrade Agent orange in a high radiation environment. Compounds such as organophosphate and carbamate are effectively removed by genetically engineered *Sphingomonas* species CDS-1. MPD gene encoding methyl parathion hydrolase enzyme is cloned in *Pseudomonas putidia* DLL-1 (Methyl parathion degrading strain) using a pBBR1MCS-2 vector to generate CDS-pBBR-mpd recombinant strain of *Sphinomonas sp.* These strains can effectively degrade methyl Parathione seven times more than its wild strain (Liu et al., 2006). *Saccharomyces cerevisiae* is engineered by using the laccase gene from *Myceliophthora thermophilla* for bioremediation of polyaromatic compounds. Mutation in CYP11 at F87 and Y96 in *P. putida*, CYP102 at R47L, and Y51F in *Bacillus megaterium* increase the oxidizing efficiency of phenanthrene and fluoranthene from ten to forty folds (Carmichael and Wong, 2001). The bphA gene from various strains such as *Bukholderia cepacia* LB400, *Rhodococcus globerulus* P6, and *Comamonas testosteroni* B-365 are targeted and mutated to obtain several genetically engineered bacteria for PCBs bioremediation (Barriault et al., 2002).

11.6.2 Genetically Engineered Fungi in Bioremediation

Fungi are well-known decomposers and are considered an effective tool for biodegradation. These heterotrophic forms of life can successfully transform and degrade various complex organic compounds, heavy metals, and xenobiotic compounds into simpler forms, which can be recycled and used for multiple purposes (Barac et al., 1999). There are more than a hundred species of fungi present naturally in diverse habitats with the potential for bioremediation but it has been found in several studies that

TABLE 11.6

Summarizes Various Genetically Engineered Microorganisms, Methods Used for Genetic Engineering Those Microorganisms, and Their Mode of Action in the Bioremediation of Particular Pollutants

S.no	Genetically Engineered Microbes	Targeted Pollutants	Method of Engineering	Mechanism of Action	Reference
1	*Deinococcus radioduranus*	Toluene	Tod and xyl genes from *Pseudomonas aerunginosa* are introduced in *D. radioduranus,* resulting in the incorporation of carbon from toluene in biological molecules and CO_2	Remove toluene and other hydrocarbons along with Chromium and other heavy metals	Brim et al. (2006)
2	*Pseudomonas putida* 06909	Cadmium	Metal-binding peptide EC20 is expressed to enhance metal-binding efficiency and cellular toxicity for cadmium	Remove complex organometallic compounds from polluted sites	Wu et al. (2005)
3	*Escherichia coli*	Cadmium	Phytochelatin synthase enzyme is expressed from the *Schizosaccharomyces pombe* in *E. coli* along with Cd transporter Mnt protein	Remove cadmium	Seung et al. (2007)
4	*Deinococcous radiodurans*	Mercury	merA gene from *E. Coli* BL308 is used to express in *Deinococcus radiodurans*	Mercury is converted to volatile form to make it less toxic	Brim et al. (2006)
5	*Burkholderia cepacia*	Toluene	pTOM plasmid is used from *Bacillus cepacia* G4 to develop engineered strains of *B. cepacia* L.S.2.4	Bioremediation of toluene and other volatile compounds from pollutants	Barac et al. (1999)
6	*E. coli*	Heavy metal	*E. coli* was modified by introducing AtPCS gene which codes for PC synthase enzyme from *Arabidopsis thaliana*	Heavy metals are removed from contaminated soils	Sandrine et al. (2003)
7	*D. radiodurans*	Uranium	Nonspecific acid phosphatase coding gene PhoN from *Salmonella typhi* is used to express *in D. radiodurans*	The acid phosphatase enzyme is helpful in the bioremediation of uranium from radioactive pollutant site	Appukuttan et al. (2010)
8	*Trichoderma atroviride*	Pesticides	pV2 plasmid with hph gene coding for hygromycin B phosphotransferase is linearized and inserted into the chromosome of T23 wild strain for expression of the enzyme	The degradation ability of dichlorvos compounds was improved	Tang et al. (2009)
9	*Pichia pastoris*	Anthraquinone dyes and Azo dyes	lcc1 gene of Trametes trogii is introduced to modify the microbe for expression of laccase enzyme	Xenobiotic compounds are removed by the Laccase enzyme expressed by recombinant microbes	Maria et al. (2006)
10	*Fusarium solani*	DDT	Highly improved dehalogenase is expressed by the parasexual method	Higher efficiency in degradation of DDT in comparison to the wild strain of fungus	Mitra et al. (2001)
11	*Gliocladium virens*	Diisopropyl fluorophosphate	pCL1from pJS294 strain is introduced in between promoter prom1 and terminator trpC of *Cochliobolus heterostrophus* and *Aspergillus* respectively	Organophosphate compounds are degraded more effectively	Dave et al. (1994)

these indigenous strains of fungi are not effective in the complete removal of such complex xenobiotic pollutants. Advancement in molecular biology and genetic engineering to enhance biological capability and efficiency has become the centre of attention in the field of bioremediation. These new genetic engineering approaches are used to enhance microbial activities, where fungal strains are genetically manipulated to boost them or to provide them with new biological properties as given in Table 11.6 (Tang et al., 2009). One of the most common recombinant fungal species is *Saccharomyces cerevisiae* with mammalian CYP1A gene. These are genetically engineered by transferring expression plasmid having CYP1A1 and CYP1A2 gene of rat and NADPH-P450 of yeast. This recombinant strain is highly effective over natural strains in dioxin degradation and metabolically active for the compounds such as Tri-CDD, MCDDs, DD, and DCDDs (Sakaki et al., 2002), but are ineffective in the case of PCDD compounds (highly toxic in dioxins family). New genetically engineered strains F240A and F240S were developed by mutating the CYP1A1 enzyme encoding gene to increase the size and enlarge the PCDD binding pocket of the enzyme (Maria et al., 2006). These mutant strains of fungi in Table 11.6, show higher metabolic activity for pollutants.

Similarly, *Fusarium solani* have shown poor efficiency in bioremediation of DDT, which is further improved by enhancing dehalogenase activity by parasexual hybridization of two isolates one with genotype BenrDBP +, Kel - and Lin+and the other with genotype BenDBP-, Kel+and Lin -. the strains developed from these two strains with genotype BenDBP +, Kel+and Lin+shows higher bioremediation efficiency of DDT in comparison to natural strains of *Fusarium solani* (Mitra et al., 2001). In the case of recombinant yeast like *Hansenula polymorpha* with Flavocytochrome b2 enzyme encoding gene (FCb2) have shown six-time higher bioremediation of chromate in comparison to its parental strains (Smutok et al., 2011).

11.6.3 Genetically Engineered Algae in Bioremediation

Anthropogenic compounds as pollutants are a global concern in recent times, which requires immediate attention. In the case of aquatic habitats, Algae are considered as one of the highly effective and acceptable tools in the bioremediation process. They are autotrophic, eukaryotic organisms with more than twenty thousand species with higher reproduction rates. These properties of algae can be used in developing cost-effective and clean technology in bioremediation for waterborne pollutants such as organic compounds, heavy metals, microbial pathogens, and radioactive compounds (Bird et al., 1997). The most common approach used by algae for bioremediation of heavy metals is Bioaccumulation. Various strains of brown algae such as *Durvillaea potatorum, Laminaria japonica*, and *Ecklonia radiata* have higher efficiency for heavy metal absorption (Yu et al., 1998). These strains of alga are either chemically modified by epichlorohydrin or potassium per magnet to remove lead from polluted water (Liu et al., 2006). Other species of algae such as *Tetraselmis suecica* and *Gracilaria fisheri* are used for a hyperaccumulation of copper and cadmium. *Chlorella sorokiniana* ANA9, a mutated strain of green algae is used in the bioremediation of cadmium (Yoshida et al., 2006). Microalgae species such as *Dunaliella salina* and *Nitella pseudoflabellata* are used in the bioremediation of cadmium and chromium from contaminated water (Folgar et al., 2009). Algae like *Chlamydomonas reinhardtii* are genetically manipulated to express enzymes like metallothionein, which enhance the metal-binding efficiency of the mutated algae in comparison to wild strain (Perales-Vela et al., 2006).

11.7 Major Issues with Genetically Engineered Microbes in Bioremediation

Due to advancements in the field of genetic engineering. The interested microorganism can be easily engineered with suitable genes for bioremediation of multiple pollutants. The popularity of utilizing these engineered microbes at the contaminated site is increasing exponentially all over the globe but there are multiples issues and problems that need to be addressed to get a successful result in the bioremediation process by these genetically engineered microorganisms. The major issue with these microbes is their effectiveness at the contaminated sites. Various microbes have been engineered which have shown effective results in an enriched in vitro environment but show poor results when used in natural conditions (Sayler and Ripp, 2000). Based on stable isotope probing studies that the most frequently used microbes in genetic engineering for bioremediation such as *Pseudomonas, Rhodococcus,* and others have shown poor performance, when used in the natural environment and are considered as a less significant host for recombinant genes used in bioremediation in the natural niche (Wackett, 2004).

Another issue with the recombinant microbes is associated with their sustainability in natural conditions. When these engineered microbes are introduced in different habitats, they act as foreign microbes in territories of indigenous microorganisms. Thus, the introduction of recombinant microbes in a stable environmental niche leads to destabilization, which leads the restricted growth to a certain limit in the number of Genetically engineered bacteria due to the influence of indigenous microorganisms (Iwasaki et al., 1993). Microbes are Genetically engineered by introducing novel gene which codes, specific proteins or enzymes important for bioremediation. Their catalytic activity is solely responsible for the bioremediation of various chemical compounds. The problems associated with the expression of these proteins and enzymes have a direct influence on the efficiency of the genetically engineered microorganism. The expression of these proteins can be affected by various genetic factors (promoter, enhancer, repressor genes) and epigenetic factors (Inhibitors, transcription factors, nucleotide modifying enzymes) which results in no express or less expression of recombinant proteins which might not have significance in bioremediation process (Sayler and Ripp, 2000). Acceptance of genetically engineered microbes as a bioremediation tool over indigenous, natural microorganism is also a challenging task associated with the application in day-to-day life. There are various agencies that consider GEM as a threat to the natural ecosystem. They refuse to believe and accept GEM as an effective tool for bioremediation, this resulted in restricted utilization of GEM in the field application. The uncertainty and risk associated with the release of these GEMs in nature and their impact on the ecosystem is still a major issue and topic of debate. The lack of information and data on this topic give rise to the future risk of using different engineered microbial strains for the bioremediation process in natural habitat. The evidence of the adverse impact of GEM, when released in the environment on the ecosystem and natural microbial niche has been not found yet but a detailed study on this matter is needed before commercializing any GEM for their bioremediation application (Singh et al., 2011).

11.8 Future Approach and Applications of GMM in Bioremediation

Advancements in the field of Molecular biology and genetic engineering have developed various molecular tools for an easy, cost-effective, specific, and successful genetically engineered microorganism. Increasing diversity and complexity of the pollutants have shown the ineffectiveness of natural microflora and the need for highly efficient and genetically engineered microorganisms in the field of bioremediation (Pieper and Reineke, 2000). There are several strains of microbes, developed by genetic engineering for their application in bioremediation but still, there are several issues, associated with them, which needed to be addressed and improved for successful bioremediation in the upcoming days. As this chapter has already discussed several issues associated with genetically engineered microorganisms in the above heading, we will explore various methods and ideas to overcome those issues to improve the efficiency of GEM in bioremediation.

Multiple in-vitro developed strains with higher efficiency in lab conditions have shown poor results when used in an in-situ environment. These issues can be due to various factors but the most common is the selection of microorganism which is genetically engineered (Sayler and Ripp, 2000). When the indigenous microbes are genetically engineered and used in their respective environment. They have shown higher sustainability and great performance in the bioremediation process. It is due to their ability to sustain different biotic and abiotic stress in their natural niche, which is developed during their evolutionary process (Celik et al., 2012). The utilization of fast-growing bacterial strains after genetic engineering has also shown an effective result. These microbes grow rapidly and dominate the environment, which helps them to overcome various environmental stress. Even these GEM can be used by encapsulating them in the polymeric matrix to prevent them from various external factors or inhibitors produced by another microorganism of that niche, which might treat these GEM as foreign microbes in their territory (Mitra et al., 2001). A successful GEM is achieved when the introduced gene for the proteins and enzymes are expressed under suitable promoters. There are so many suitable expression vectors developed by molecular biologists which can solve the issues related to the expression of proteins. Various tools such as ZNFs, TALENs, and CRISPR- Cas are well-established tools that can precisely engineer the microbe without any errors (Biagioni et al., 2018). Other tools such as sleeping beauty and piggyback, also known as transposons-based genetic engineering tools have opened newer opportunities in this field of bioremediation. Microorganisms can be genetically engineered with a biological containment system, where the engineered bacteria can be controlled after their application in bioremediation by the self-suicide process, which will prevent the harm associated with these GEMs when released into the environment (Singh et al., 2011). There is a need for large-scale field application of GEM in controlled conditions to understand their impact in nature and the safety related to them.

Only some species of microbes with bioremediation ability such as *E. coli, B. subtilis, P. putidia* are genetically engineered and used in the field study, but there is a large number of species out there, whose potential in bioremediation has not been explored till now. There is a lack of information associated with those unexplored microbes so, there is a need to develop other Genetically engineered microbial strains to understand the efficiency of GEM in the bioremediation process and to resolve various challenges associated with them. In the field of Bioremediation with genetically engineered microbes, an effort is required to explore the efficiency of these microbes in terms of sustainability, survival, resistance to biotic and abiotic stress, horizontal gene transfer methods, interaction with indigenous microbes, and their impact on the environment and natural niche.

11.9 Conclusion

The increasing rate of xenobiotic and complex compounds in nature. The complexity in the nature of the bond and the reactivity with the other functional groups have shown the ineffectiveness of indigenous microflora in the bioremediation process of such pollutants (Pieper and Reineke, 2000). This problem with naturally occurring microbes gives rise to the need for genetically engineered microbes in the biodegradation process. These genetically modified microbes can target a wide range of pollutants, due to the introduction of multiple genes associated with bioremediation enzymes (Moharikar et al., 2003). It has been observed that there are several issues associated with developing a successful genetically engineered microbe. One of the prominent issues is the ineffectiveness of the GEM when used in ex-situ application, as these organism works better in lab condition but shows poor result when used in natural conditions (Sayler and Ripp, 2000). Development in genetic engineering and other scientific fields helps in developing various strains for the bioremediation process. There is still a lack of more data and information on GEM and their impact on the ecosystem when released in the environment has led to multiple ethical and social issues related to them, but due to recent data and advanced genetic engineering tools, and biomolecular engineering, it is possible to develop and use these GEM in controlled condition for complex xenobiotic pollutants (Biagioni et al., 2018).

The advancement in the bioremediation process and advantages associated with GEMs. The number of genetically modified microbes is increasing every year, but there is a need to put extra effort to understand the molecular mechanism to explore different inducing factors of recombinant microbes which can enhance the in-situ bioremediation process in controlled conditions. Controlled growth of these GEM in nature and the chance of spread of gene in another strain due to horizontal gene transfer is the major concern in this field, So, the development of a noble engineering approach that can control the rapid microbial growth and unwanted gene transfer to give rise unrequired microbial strain and validating these approaches and engineered microbes in field condition should be the focus of research in upcoming days, which not only enhance the efficiency of GEM for bioremediation of pollutants but also increase the reliability of these microbes in this field.

REFERENCES

Acuner, E., Dilek, F.B. 2004. Treatment of tectilon yellow 2G by *Chlorella vulgaris*. *Process Biochem* 39:623–631.

Ang, E.L., Zhao, H., Obbard, J.P. 2005. Recent advances in the bioremediation of persistent organic pollutants via biomolecular engineering. *Enzyme Microb Technol* 37:487–496.

Appukuttan, D., Chitra, S.N., Padma, A., et al. 2010. PhoN-expressing, lyophilized, recombinant *Deinococcus radiodurans* cells for uranium bioprecipitation. *J Biotechnol* 154:185–190.

Atteia, O., Guillot, C. 2007. Factors controlling BTEX and chlorinated solvents plume length under natural attenuation conditions. *J Contam Hydrol* 90:81–104.

Barac, T., Taghavi, S., Borremans, B., et al. 1999. Engineered endophytic bacteria improve phytoremediation of water-soluble, volatile, organic pollutants. *Nat Biotechnol* 22(5):583–588.

Barriault, D., Plante, M.M., Sylvestre, M. 2002. Family shuffling of a targeted bphA region to engineer biphenyl dioxygenase. *J Bacteriol* 184:3794–3800.

Beerli, R.R., Schopfer, U., Dreier, B,. Barbas, C. F. 2000. Chemically regulated zinc finger transcription factors. *J Biol Chem* 275:32617–32627.

Beumer, K.J., Trautman, J.K., Bozas, A., et al. 2008. Efficient gene targeting in *Drosophila* by direct embryo injection with zinc-finger nucleases. *Proc Natl Acad Sci USA* 105:19821–19826.

Biagioni, A., Laurenzana, A., Margheri, F., et al. 2018. Delivery systems of CRISPR/Cas9-based cancer gene therapy. *J Biol Eng* 12:33.

Bird, K.T., Bourdine, K.S., Busch, W.S. 1997. Marine algae as a tool for bioremediation of marine ecosystems. *Agric Biotechnol (Arie Altman, CRC Press)* 29:601–613.

Bogdanove, A.J., Voytas, D.F. 2011. TAL effectors: customizable proteins for DNA targeting. *Science* 333:1843–1846.

Brim, H., Osborne, J.P., Kostandarithes, H.M., et al. 2006 Deinococcus radiodurans engineered for complete toluene degradation facilitates Cr(VI) reduction. *Microbiology* 152:2469–2477.

Carmichael, A.B., Wong, L.L. 2001. Protein engineering of *Bacillus megaterium* CYP102—the oxidation of polycyclic aromatic hydrocarbons. *Eur J Biochem* 268:3117–3125.

Celik, L., Ozturk, A., Abdullah, M.I. 2012. Biodegradation of reactive red 195 azo dye by the bacterium *Rhodopseudomonas palustris* 51ATA. *Afr J Microbiol Res* 6:120–126.

Chu, W.L., Yike-Chu, S., Siew-Moi, P. 2009. Use of immobilised *Chlorella vulgaris* for the removal of color from textile dyes. *J Appl Phycol* 21:641–648.

Colberg, P.J.S., Young, L.Y. 1995. *Anaerobic Degradation of Nonhalogenated Homocyclic Aromatic Compounds Coupled with Nitrate, Iron, or Sulfate Reduction Microbial Transformation and Degradation of Toxic Organic Chemicals.* Wiley-Liss, New York, pp. 307–330.

Dave, K.I., Lauriano, C., Xu, B., et al. 1994. Expression of organophosphate hydrolase in the filamentous fungus Gliocladium virens. *Appl Microbiol Biotechnol* 41:352–358.

DeLong, E.F., Pace, N.R. 2001. Environmental diversity of bacteria and archaea. *Syst Biol* 50:470–478.

Deng, X., Wilson, D.B. 2001. Bioaccumulation of mercury from wastewater by genetically engineered *Escherichia coli. Appl Microbiol Biotechnol* 56:276–279.

Doak, T.G., Doerder, F.P., Jahn, C.L., Herrick, G. 1994. A proposed superfamily of transposase genes: transposon-like elements in ciliated protozoa and a common "D35E" motif. *Proc Natl Acad Sci USA* 91:942–946.

Dubey, S.K., Dubey, J., Mehra, S., et al. 2011. Potential use of cyanobacterial species in bioremediation of industrial effluents. *Afr J Biotechnol* 10:1125–1132.

Ezezika, O.C., Singer, P.A. 2010. Genetically engineered oil-eating microbes for bioremediation: prospects and regulatory challenges. *Technol Soc* 32:331–335.

Fairall, L., Schwabe, J.W., Chapman, L., Finch, J.T., Rhodes, D. 1993. The crystal structure of a two zinc-finger peptide reveals an extension to the rules for zincfinger/DNA recognition. *Nature* 366:483–487.

Folgar, S., Torres, E., Pérez-Rama, M., et al. 2009. Dunaliella salina as marine microalga highly tolerant to but a poor remover of cadmium. *J Hazard Mater* 165:486–493.

Guo, J., Gaj, T., Barbas, C.F. 2010. Directed evolution of an enhanced and highly efficient FokI cleavage domain for zinc finger nucleases. *J Mol Biol* 400, 96–107.

Hai, F.I., Yamamoto, K., Nakajima, F., et al. 2012. Application of a GAC-coated hollow fiber module to couple enzymatic degradation of dye on membrane to whole cell biodegradation within a membrane bioreactor. *J Membrane Sci* 389:67–75.

Iwasaki, K., Uchiyama, H., Yagi, O. 1993. Survival and impact of genetically engineered *Pseudomonas putida* harboring mercury resistance gene in aquatic microcosms. *Biosci Biotechnol Biochem* 57:1264–1269.

Izsvák, Z., Ivics, Z., Plasterk, R.H. 2000. Sleeping beauty, a wide host-range transposon vector for genetic transformation in vertebrates. *J Mol Biol* 302:93–102.

Jinek, M., Chylinski, K., Fonfara, I., et al. 2012. A programmable dual-RNA-guided DNA endonuclease in adaptive bacterial immunity. *Science* 337:816–821.

Kao, C.M., Chen, C.Y., Chen, S.C., Chien, H.Y., Chen, Y.L. 2008. Application of in situ biosparging to remediate a petroleum hydrocarbon spill site: field and microbial evaluation. *Chemosphere* 70:1492–1499.

Kim, S., Krajmalnik-Brown, R., Kim, J-O., et al. 2014. Remediation of petroleum hydrocarbon contaminated sites by DNA diagnosis-based bioslurping technology. *Sci Total Environ* 497:250–259.

Kim, Y.G., Cha, J., Chandrasegaran, S. 1996. Hybrid restriction enzymes: zinc finger fusions to Fok I cleavage domain. *Proc Natl Acad Sci USA* 93:1156–1160.

Liu, S., Zhang, F., Chen, J., et al. 2011. Arsenic removal from contaminated soil via biovolatilization by genetically engineered bacteria under laboratory conditions. *J Environ Sci* 23(9):1544–1550.

Liu, Z., Hong, Q., Xu, J.H., et al. 2006. Construction of a genetically engineered microorganism for degrading organophosphate and carbamate pesticides. *Int Biodeter Biodegr* 58:65–69.

Madsen, E.L. 1991. Determining in situ biodegradation: facts and challenges. *Environ Sci Technol* 25:1663–73.

Maria, C.C., Lupino, S., Garzillo, A.M., et al. 2006. Heterologous expression of lcc1 gene from *Trametes trogii* in *Pichia pastoris* and characterization of the recombinant enzyme. *Microb Cell Fact* 5:31.

Megharaj, M., Wittich, R.-M., Blasco, R., et al. 1997. Superior survival and degradation of dibenzo-p-dioxin and dibenzofuran in soil by soil-adapted and nonadapted *Sphingomonas* sp. strain RW1. *Appl Microbiol Biotechnol* 48:109–14.

Misal, S.A., Lingojwar, D.P., Shinde, R.M., et al. 2011. Purification and characterization of azoreductase from alkaliphilic strain *Bacillus badius*. *Process Biochem* 46:1264–1269.

Mitra, J., Mukherjee, P.K., Kale, S.P., et al. 2001. Bioremediation of DDT in soil by genetically improved strains of soil fungus *Fusarium solani*. *Biodegradation* 12:235–245.

Moharikar, A., Kapley, A., Purohit, H.J. 2003. Detection of dioxygenase genes present in various activated sludge. *Environ Sci Pollut Res* 10:373–378.

Moscou, M.J., Bogdanove, A.J. 2009. A simple cipher governs DNA recognition by TAL effectors. *Science* 326:1501.

Nikolopoulou, M., Kalogerakis, N. 2008. Enhanced bioremediation of crude oil utilizing lipophilic fertilizers combined with biosurfactants and molasses. *Mar Pollut Bull* 56:1855–61.

Otte, M.P., Gagnon, J., Comeau, Y., et al. 1994. Activation of an indigenous microbial consortium for bioaugmentation of pentachlorophenol-creosote contaminated soils. *Appl Microbiol Biotechnol* 40:926–32.

Paschon, D.E., Lussier, S., Wangzor, T., et al. 2019. Diversifying the structure of zinc finger nucleases for high precision genome editing. *Nat Commun* 10:1133.

Patel, J., Zhang, Q., Michael, R., et al. 2010. Genetic engineering of *Caulobacter crescentus* for removal of cadmium from water. *Appl Biochem Biotechnol* 160:232–243.

Perales-Vela, H.V., Penã-Castro, J.M., Canĩzares-Villanueva, R.O. 2006. Heavy metal detoxification in eukaryotic microalgae. *Chemosphere* 64:1–10.

Philp, J.C., Atlas, R.M. 2005. Bioremediation of contaminated soils and aquifers. In: Atlas, R.M., Philp, J.C. (eds) *Bioremediation: Applied Microbial Solutions for Real-World Environmental Cleanup*. American Society for Microbiology (ASM) Press, Washington, pp. 139–236.

Pieper, D.H., Reineke, W. 2000. Engineering bacteria for bioremediation. *Curr Opin Biotechnol* 11:262–270

Plasterk, R.H., Izsvák, Z., Ivics, Z. 1999. Resident aliens: the Tcl/mariner superfamily of transposable elements. *Trends Genet* 15:326–332.

Porri, A., Baroncelli, R., Guglielminetti, L., et al. 2011. *Fusarium oxysporum* degradation and detoxification of a new textile glycoconjugate azo dye (GAD). *Fungal Biol* 115: 30–37.

Ryan, A.H., Peter, M.C., Gary, S.M., et al. 2000. Level of detail in ecological risk assessments. *Mar Pollut Bull* 40:471–477.

Rynk, R., Hess, T.F., Bechinski, E. 1999. Occurrence, degradation, and fate of pesticides during composting. I. Composting, pesticides, and pesticide degradation. *Compos Sci Util* 7:66–82.

Sakaki, T., Shinkyo, R., Takita, T., et al. 2002. Biodegradation of polychlorinated dibenzo-p-dioxins by recombinant yeast expressing rat CYP1A subfamily. *Arch Biochem Biophys* 401:91–98.

Sandrine, S.M., Stephan, C., Patrick, C., et al. 2003. Enhanced toxic metal accumulation in engineered bacterial cells expressing arabidopsis thaliana phytochelatin synthase. *Appl Environ Biotechnol* 69:490–494.

Sayler, G.S., Ripp, S. 2000. Field applications of genetically engineered microorganisms for bioremediation processes. *Curr Opin Biotechnol* 11 286–289.

Sen, K., Pakshirajan, K., Santra, S.B. 2012. Modelling the biomass growth and enzyme secretion by the white-rot fungus Phanerochaete chrysosporium: a stochastic based approach. *Appl Biochem Biotechnol* 167:705–713.
Seung, H.K., Singh, S., Kim, J.Y., et al. 2007. Bacteria metabolically engineered for enhanced phytochelatin production and cadmium accumulation. *Appl Environ Biotechnol* 73:6317–6320.
Singer, A.C., van der Gast, C.J., Thompson, I.P. 2005. Perspectives and vision for strain selection in bioaugmentation. *Trends Biotechnol* 23:74–77.
Singh, J.S., Abhilash, P.C., Singh, H.B., et al. 2011. Genetically engineered bacteria: an emerging tool for environmental remediation and future perspectives. *Gene* 480(1–2):1–9.
Singh, R.P., Singh, PK., Singh, R.L. 2014. Bacterial decolorization of textile azo dye acid orange by *Staphylococcus hominis* RMLRT03. *Toxicol Int* 21:160–166.
Smith, J., Bibikova, M., Whitby, F.G., et al. 2000. Requirements for double-strand cleavage by chimeric restriction enzymes with zinc finger DNA-recognition domains. *Nucleic Acids Res* 28:3361–3369.
Smutok O., Broda, D., Smutok, H., et al. 2011. Chromate-reducing activity of *Hansenula polymorpha* recombinant cells over-producing fl avocytochrome b_2. *Chemosphere* 83(4):449–454.
Sternberg, S.H., Redding, S., Jinek, M., Greene, E.C., Doudna, J.A. 2014. DNA interrogation by the CRISPR RNA-guided endonuclease Cas9. *Nature* 507:62–67.
Tang, J. Liu, L., Hu, S., et al. 2009. Improved degradation of organophosphate dichlorvos by *Trichoderma atroviride* transformants generated by restriction enzyme-mediated integration (REMI). *Bioresour Technol* 100:480–483.
van Luenen, H.G., Colloms, S.D., Plasterk, R.H. 1994. The mechanism of transposition of Tc3 in C. elegans. *Cell* 79:293–301.
Von Fahnestock, F.M., Wickramanayake, G.B., Kratzke, K.J., et al. 1998. *Biopile Design, Operation, and Maintenance Handbook for Treating Hydrocarbon Contaminated Soil*, Battelle Press, Columbus.
Wackett LP. 2004. Stable isotope probing in biodegradation research. *Trends Biotechnol* 22:153–154.
Wang, W., Lin, C., Lu, D., et al. 2008. Chromosomal transposition of PiggyBac in mouse embryonic stem cells. *Proc Natl Acad Sci U S A* 105:9290–5.
Wik, J. 2001. Comparison of ex-situ and in situ bioremediation of unsaturated soils contaminated petroleum. *J Environ Sci Health A* 36:1491–503.
Wilber, A., Linehan, J.L., Tian, X., et al. 2007. Efficient and stable transgene expression in human embryonic stem cells using transposon-mediated gene transfer. *Stem Cells* 25:2919–27.
Wu, C.H., Thomas, K.W., Mulchandani, A., et al. 2005. Engineering plant-microbe symbiosis for rhizoremediation of heavy metals. *Appl Environ Biotechnol* 72:1129–1134
Wu, S.C., Meir, Y.J., Coates, C.J., et al. 2006. piggyBac is a flexible and highly active transposon as compared to sleeping beauty, Tol2, and Mos1 in mammalian cells. *Proc Natl Acad Sci U S A* 103:15008–13.
Xu, X., Zhai, Z., Li, H., Wang, Q., Han, X., et al. 2017. Synergetic effect of bio-photocatalytic hybrid system: g-C3N4, and Acinetobacter, sp. JLS1 for enhanced degradation of C16 alkane. *Chem Eng J* 323:520–529.
Yoshida, N., Ikeda, R., Okuno, T. 2006. Identification and characterization of heavy metal-resistant unicellular alga isolated from soil and its potential for phytoremediation. *Bioresour Technol* 97:1843–1849.
Yu, Q., Matheickal, J.T., Latten, J. 1998. Heavy metal adsorption properties of marine algae *Durvillaea potatorum, Ecklonia radiata* and *Laminaria japonica. Chinese J Chem Eng* 6:68–72.
Yu, Z., Wen, X. 2005. Screening and identification of yeasts for decoloring synthetic dyes in industrial wastewater. *Int Biodeter Biodegr* 56:109–114.

12

Microalgae: Tool for the Removal of Emerging Contaminants from the Industrial Wastewater

Megha Shah and Anupama Shrivastav
Parul University

CONTENTS

12.1 Introduction

Tempering composition of earth, which is made up of water, soil, and air, has been impacted negatively by anthropogenic activities creating round-the-clock damage to mankind. Moving ahead with the escalation and advancement of industrialisation in both developed country and developing country, generate a huge sizeable amount of wastewater (Dutta et al., 2021). A problem arises when a large percentage of all kind of wastewater is being forward to reuse or expelled to the aquatic waterbody without removing emerging contaminants and giving treatment, which cost us dissatisfied freshwater demand and generate high health risk for downstream users (Martínez-Roldán and Cañizares-Villanueva, 2020, Bhatia et al., 2021). Other than that, carbon footprint and water footprint are increasing categorically with ever increasing world population giving a push to the scientists to find and develop some sustainable creative processes that can significantly halt exploitation of our natural ecosystem.

Physical, chemical, and/or biological strategies have been used to remove or degrade the contaminants so far according to the feasibility and capability of the process of wastewater treatment plants to protect our environment by following "reduce-reuse-recycle" strategy (Shingare et al. 2019; Martínez-Roldán and Cañizares-Villanueva, 2020; Kumar et al. 2021a; Kundu et al., 2014; Gautam et al., 2007). Another concern is the energy crisis, which is caused by the widespread usage of fossil-derived fuels such as gasoline, natural gas, and coal, which are expected to be depleted within the next 50 years. To satisfy future energy needs, it is critical to look for new alternative renewable energy sources. To utilise renewable energy sources such as solar, wind, hydraulic, and geothermal energy, many technologies have been created (Wainaina et al., 2020). All of these technologies have been well-developed and are commercially available to meet rising energy demand to some extent. The use of wastewater as a source of energy can also serve to alleviate the strain on other technologies (Kumar et al., 2020a; Bhatia et al., 2021).

Microalgae is a sunlight-driven cell factory with the ability to assimilate CO_2 that can lead us to the massive opportunistic era by producing a massive amount of user-friendly algal biomass as a by-product to the mainstream of removing emerging contaminants while treating industrial wastewater and

DOI: 10.1201/9781003231738-13

ultimately justify their participation in making the country and world as so-called carbon neutral. As commercialisation of the technology is still in progress. We urge Academicians, industrialists and society to stop overlooking the potential creator of 'wealth from waste' (Perez et al., 2015, Lee et al., 2020).

In this chapter, we will discuss the whole system round the clock. After discussing types of wastewater, we will understand the role of microalgae in the treatment and management of industrial wastewater and the recovery of resource. Before the conclusion section, advantages of using microalgae-based treatment over some traditional/conventional methods are also noted along with the commercial microalgal products achievement and bioenergy generation simultaneously with cleaning wastewater, drawbacks or challenges in processing wastewater using microalgae mentioned and also the role of other microorganisms in the process to facilitate microalgae and improvise productivity presented here to provide essential research directions.

12.2 Types of Wastewater and Its Sources

Point pollution from industrial effluents has a variety of undesired chemical elements and pathogens that have short-term and long-term effects on the environment, particularly on aquatic ecosystems like lagoons, lakes, streams, underground water, bays, oceans and rivers, as well as on human health (Kumar et al., 2021a, b; Chandra and Kumar, 2017a, b). The synthetic substances alter the quality of water in receiving water bodies, upsetting the biological balance and rendering polluted water unfit for residential use by rural communities (Ngwira and Lakudzala, 2018, Bhatia et al., 2021).

Each and every industry do generate distinct composition of waste material, which come up as challenging sources of waste flow. For instance, oil refineries and pulp and paper mills do produce contradictory waste material and will need obvious variation in the treatment to treat diversity of effluents (Kumar et al., 2020b). To get an idea on a variety of microbial community of the wastewater revealed by microscopic evaluation of the target sample (Perez et al., 2015; Kumar et al., 2020c).

It's worth noting that not only are there significant variances across the many industrial branches, but even within one branch, the composition and treatment procedures are influenced by the following factors: Production processes, seasonal variations, effluent requirements, the level of production-integrated environmental protection measures, and the number of wastewater treatment facilities, water supply and treatment, technical condition and age of the manufacturing facility, staff training and motivation, and the usage of specific additives are all factors to consider. The first step in designing wastewater treatment concepts is to take a thorough inventory of the company's current position in terms of production processes, water supply, and wastewater production. For example, the production of precise amounts of water, wastewater, and pollutant loads, as well as the characteristics of the various flow fractions of the company and the locations where alkalis, acids, detergents and other chemicals are employed, should all be determined (Rosenwinkel et al., 2005).

Wastewater that is being produced constantly by the industrial community globally can be categorised based on its origins such as the dairy industry, household or municipal, swine farming, aquaculture, pharmaceutical industry, pulp and paper, oil refinery, sugarcane industry, breweries, distilleries, fruit juice and beverage industry, slaughterhouses, starch factories and potato processing industry, textile or tannery industry and agroindustry (Chandra and Kumar, 2015; Kumar and Chandra, 2020; Kumar, 2021; Bhatia et al., 2021; Agrawal et al., 2021).

Wastewater comprises a variety of undesired chemical elements and microorganisms that have both short and long-term consequences for the environment and human health (Kumar et al., 2021c). Untreated wastewater irrigation on land has also been linked to a number of issues, including excessive vegetative growth, irregular fruit maturity, and lower quality and yield (Bhatia et al., 2021). Heterogeneity in a contaminant has been reported to increase exponentially day by day with every new addition in the industrial community. Water's biological parameters, such as algae, bacteria, and other microorganisms, might be subjected to expected restrictions as well. Contaminant levels that are acceptable will vary depending on the type of water being tested (Chapman et al., 2020). Despite biological parameters, the most essential wastewater components are outlined, together with their impact on the value retention of manufacturing facilities and the wastewater treatment techniques that are as follows: temperature, pH, components like glass shards, sand, plastic parts, etc.; suspended solids, organic substances are vital pollutant (aromatic hydrocarbons, antibiotics, biocides, colourants, hormones, phenolic compounds, surfactants, etc.); Nutrient salts like nitrogen, phosphorus, sulphur; hazardous substances like absorbable organic halogen compounds (AOX), chlorinated hydrocarbons and halogenated hydrocarbons, hydrocarbons like benzene, phenol; heavy metals in particular mercury, cadmium, chromium, copper, lead, nickel, and zinc; cyanides, corrosion-inducing substances, cleaning agents, disinfectants, and lubricants (Rosenwinkel et al., 2005, Shahid et al., 2020). Every single kind of wastewater does carry different amounts of combinatorial contaminants on the bases of its sources.

12.3 Role of Microalgae in Wastewater Management

As microalgae are fast-growing photosynthetic organisms that can live in fresh as well as in marine water, making them obvious candidate to be considered while treating wastewater (Martínez-Roldán and Cañizares-Villanueva, 2020). Microalgae have an extensive spectrum of mechanisms to treat industrial wastewater effluents with versatility in the mode of action intracellularly as well as extracellularly. Oswald et al. (1957) built a foundation for the usage of algae for the treatment of wastewater and gave this vision to fellow scientists to make advantage of the inbuilt quality of algal biomass (Leng et al., 2021).

Phycoremediation offers several advantages over chemical-based treatment methods, by removing nutrients and absorbing heavy metals from the Wastewater using microalgae. Microalgae's adaptability and enormous potential might support a microalgae-based biorefinery and bioeconomy, offering up enormous prospects in the worldwide algae industry. The biorefinery concept can take advantage of variances in biomass components and intermediates, maximising the value of the biomass feedstock while avoiding resource loss and environmental consequences (Ferreira and Gouveia, 2020). A remarkable proportion of nitrogen (N) and phosphorus (P) in any waterbody gives self-evidence for the action of potential candidate algae, as every wastewater can act as the cheapest source of nutrition to consume and generate energy (Martínez-Roldán and Cañizares-Villanueva, 2020). Microalgae have received a lot of attention in recent years as an alternative approach of biological wastewater treatment, and they have a lot of applications. Wastewater management and treatment is a costly operation, thus it's critical to incorporate appropriate technologies to make it more practicable and cost-effective, toxic pollutants, heavy metals and nutrient (N, P) has been removed from the wastewater using species like *Chlamydomonas*, *Chlorella* and *Scenedesmus* reported so far (Shahid et al., 2020; Bhatia et al., 2021).

Live algal biomass requires the bare minimum of nutrients and ambient conditions, but dead biomass does not (Suresh Kumar et al., 2015). Despite using brutal condition of chemical oxidation method for Wastewater from the pharmaceutical industry, Microalgae *Chlorella sorokiniana* is reported as potential removal of emerging contaminants i.e. salicylic acid or paracetamol like Pharmaceuticals and phosphate, nitrate like nutrients from wastewater following algal biomass to be used to generate products and make a complete circle of processing (Escapa et al., 2015). Supported by the Circular Economy concept, the Biorefinery scope is now enlarging to the conversion of residues, co-products and wastes (Pereira et al., 2021).

The content and degree of pollutants from every old and new wastewater generating sources dramatically changed. Many compounds that used to be deemed trace because they didn't reach measurable levels are now found at dangerous proportions. Furthermore, many new contaminants have been discovered, many of which are stubborn and cannot be eradicated using traditional methods (Martínez-Roldán and Cañizares-Villanueva, 2020). Integration of wastewater treatment with microalgae culture for bioenergy production will help to reduce wastewater treatment costs while also increasing energy output, making contaminant removal treatment more advantageous than traditional approaches. Wastewater is high in organic content, and microalgae may absorb these nutrients and grow in them without needing additional nutrients (Bhatia et al., 2021). Comparing conventional methods like physical treatment, chemical treatment or physio-chemical treatment had a remarkable impact on industrial wastewater but with some shortcomings such as the requirement of extensive maintenance and operational cost, comparatively larger land, intensive energy input, and some toxic effects (Singh and Arora, 2011).

Microalgae-based bioremediation is the most promising alternative technology for recycling wastewater that allows contaminants to be removed and by which fear of eutrophication eliminated in tandem with the threat to the flora and fauna, oxygen levels to be increased, the adverse effect on natural water bodies (such as lakes, rivers, and coastal waters) to be reduced (Martínez-Roldán and Cañizares-Villanueva, 2020, Shahid et al., 2020). COD (Chemical Oxygen Demand) is a key metric for determining the extent of organic contamination caused by wastewater amongst all other determinants like BOD, TOC, and DOC. Biological approaches are extensively utilised in secondary or tertiary treatment of wastewater, and microbial metabolism can be used to consume soluble nutrients in wastewater (Leng et al., 2021, Rosenwinkel et al., 2005).

In some conceptual process designs, microalgal remediation capacities outperform other biological materials as they do not require fertile land, fresh water, herbicides, and pesticides (Bhatia et al., 2021). Microalgae have been shown to function effectively at low levels of pollutants, maintenance of culture is comparatively easy, produce no hazardous sludge, have a high binding affinity (owing to their net negative charge and comparatively high surface area), and are ideal for both small-scale and large-scale remediation efforts. It provides an environmentally favourable method of reducing carbon dioxide emissions (Suresh Kumar et al., 2015). It is possible to remove gaseous contaminants from exhaust gases from industrial chimneys by combining a wastewater treatment process with microalgae (Martínez-Roldán and Cañizares-Villanueva, 2020). heavy metals were traditionally remediated from the wastewater with the help of physicochemical methods like chemical oxidation or reduction, adsorption, electrodeposition, reverse osmosis, chemical precipitation and so on, have some infirmity like they produce toxic waste products, process require extreme low or high level of metal, process cost is very high with high investment. Other than that, there is no complete removal of metal that leads scientists to innovate the process that can eliminate the shortcomings of conventional methods and develop high-efficient, safe and sustainable non-conventional heavy metal remediation technologies like biosorption and bioaccumulation that can either absorb or adsorb the pollutants with the help of living cell or dead cell biomass of microalgae and other biological materials (Suresh Kumar et al., 2015).

12.4 Treatment of Wastewaters Using Microalgae

The primary goal of wastewater treatment has been to safeguard downstream consumers from health hazards, fulfilment of the strategy Waste for wealth and to complete the mechanism of "reduce-reuse-recycle". Various physical (screening, grit chamber, flotation, etc.) and chemical procedures are used to treat wastewater (neutralisation, flocculation, oxidation, disinfection, etc.). Various chemical treatments for wastewater treatment are not only costly, but they also frequently result in sludge generation and an increased danger of secondary water pollution. Other issues with wastewater treatment include excessive energy usage (2%–4% of national electric power consumption) and a scarcity of efficient and productive labour (Bhatia et al., 2021). Topic has piqued our attention, and we hope that it will encourage all scientists working in the fields of microbiology and biotechnology to pursue research based on sustainability and circularity principles in industrial biorefineries (Pereira et al., 2021).

Microalgae, among the whole ecosystem of the microbial community have given some potential solutions to the scientists in removing contaminants from wastewater with the following advantages. Being photoautotroph it fixes the highest amount of atmospheric carbon; to fulfil the high requirement of nutrient, microalgae managed to consume more nutrients from the contaminated water and can easily be cultured in various sources of wastewater (Leng et al., 2021).

Different systems have been developed to produce microalgae to reuse wastewater based on the type of microalgae, biotic and abiotic parameters, and wastewater composition and property; however high productivity remains a challenge. And also ideal culture system would be low-cost to build, have a sufficient light source, and be simple to handle, allow effective transfer via the liquid-gas barrier, and should have a low risk of contamination (Tan et al., 2020, Bhatia et al., 2021). Unlike biological treatments either anaerobic or aerobic and their combination has proven to be advantageous with simple operation and easy adaptability to the wastewater treatment plant (Singh and Arora, 2011).

Four microalgae cultivation technologies developed have been enlisted as open photobioreactors, closed photobioreactors, Algal Turf Scrubber (ATS) and hybrid system (Bhatia et al., 2021).

Natural ponds, lakes or artificially designed water bodies like raceway ponds have been used for microalgae cultivation due to the less energy expenditure, simple and straightforward construction facilitated by rotating agitator to blend the slurry well, also to minimise the chance of algal biomass sedimentation. Downside of this technology is hindrance of farthermost weather circumstances, microorganisms contamination, burdensome harvesting, loss from evaporation, necessity of large land; all these shortcomings can certainly decrease the microalgal growth (Tan et al., 2020).

To render drawbacks of open photobioreactor, closed photobioreactor system has been implemented into which customised modification can be implemented on the bases of requirement to achieve high productivity rate. Optimum desired conditions to the respective microalgal strain can be imposed such as temperature, pH, CO_2 flow rate, agitator rotation rate. Additional carbon source can also be provide for the better biomass achievement. Due to the closed construction control and monitoring becoming easy, water or CO_2 loss deceases and unlike open photobioreactor, chances of the contamination also deceases with relatively high expenditure (Yin et al., 2020).

By using Algal Turf Scrubber (ATS) technique, algae attaches with the base to treat wastewater by taking up organic as well as inorganic compounds and generating a high rate of oxygen through photosynthesis. The use of algal turf scrubbers for wastewater resource recovery by growing valuable filamentous microalgal biomass was discovered, it was discovered that the algal turf scrubber cultivation system can play an important role in the remediation of high nitrate wastewaters and increase biomass growth with more frequent harvesting and specific infrastructural demand. This technology has come up with the major advantage of trouble-free harvesting of algal biomass, higher productivity rate, inexpensive production and cheap maintenance (Leng et al., 2021, Siville and Boeing, 2020).

For optimal resource recovery from wastewater, a hybrid farming system combines two or more cultivation techniques. Hybrid technology could be a viable way to overcome many of the disadvantages of employing a single culture system. The key advantages of hybrid systems are that they maximise the advantages of each process. The hybrid cultivation technique is well suited for blending wastewater with microalgal cultivation, with the potential to reduce the cost per kg of algal biomass produced. Additionally, the advanced hybrid model is designed further according to the purpose demand (SundarRajan et al., 2019).

Numerous reports suggest the superiority of microalgae than various commercial and traditional physicochemical wastewater treatments. For instance, to sequester heavy metal ions microalgae possess metal biosorption mechanism as a promising alternate (Suresh Kumar et al., 2015). Biosorption is the mechanism that has been used by potential algae *C. vulgaris* for the textile dye colour removal from textile wastewater, using High Rate Algae Ponds (HRAP) (El-Kassas and Mohamed, 2014). With the increased use of azo dyes outside of the textile industry in a variety of fields, more biological decolorizing technologies for industrial wastewater treatment have been researched and intended to be implemented in the future (Singh and Arora, 2011).

Expanded trials need to be performed to improvise conventional methods or can develop innovative technologies by putting combinatorial efforts of biologists, chemists and engineers to promote a multidisciplinary approach (Suresh Kumar et al., 2015).

12.5 Different Products of Microalgae and Its Application

Resource recovery from wastewater with the help of biological wastewater treatment technology is no more termed by scientists as an extra desired goal but can be seen now as an obvious source of the high valued products (Bhatia et al., 2021).

Recent breakthroughs in microalgal biotechnology have led to the development of a biorefinery method for wastewater treatment utilising microalgae (Shahid et al., 2020). The biorefinery concept was proposed in order to assess its financial viability. Many goods can be made inevitably from algal biomass, and its commercialisation allows for greater economic benefits (Martínez-Roldán and Cañizares-Villanueva, 2020). high-value products from the algal biomass such as structural and bioactive proteins like unicellular protein/single-cell protein, amino acid, carbohydrates, and sterols, pigments like carotenoids, and colourants, polysaccharides and bioplastics, and ultraviolet light screening agents (Martínez-Roldán and Cañizares-Villanueva, 2020, Bhatia et al., 2021).

The real hallmark of microalgae is the versatile functions for the production of several different types of products like biodiesel, bioethanol, biobutanol, biomethane, biohydrogen, biogas and bioelectricity (Leng et al., 2021).

Microalgae are capable of converting water, sunlight, and CO_2 into a large quantity of algal biomass by the mechanism of photosynthesis (Zhu, 2015), through which solar energy turned out into starch via the formation of glucose that leads to the formation of fuels like bioethanol and biobutanol by generating glucose in large quantity from the hydrolysis of starch. Algae that have been genetically engineered have been utilised to make ethanol directly through photosynthesis (Chisti, 2018). Although third-generation biofuel is expected to be created from sunlight and carbon dioxide by metabolically improved microalgae, industrial-scale production is still a long way off (Mika et al., 2018).

Algal biomass can generate lipids in the form of triglyceride oils that can produce biodiesel by the trans-esterification reactions with methanol using different acids, alkali, and enzymatic catalysts (Chisti, 2020).

Algal biomass can be used to make biohydrogen by cultivating algae in wastewater under aerobic condition and then moving them to an anaerobic reactor. Using dark fermentation (biohydrogen production), fermentation (bio alcohols), and anaerobic digestion, fresh algal biomass or residual biomass after lipid extraction can be used directly for bioenergy production (methane) (Bhatia et al., 2021).

A combustible mixture of methane and carbon dioxide generate biogas from algal biomass by the process of anaerobic digestion (Lin et al., 2017).

In the production of bioelectricity also, the use of microalgae has been documented, using a system called photosynthetic microbial fuel cell (PhotoMFC). For bioelectricity production, various reactor designs have been described, including single chamber, dual chamber, and MFC with water desalination chamber (Ling et al., 2019).

Pyrolysis, hydrothermal treatment, gasification, and direct combustion can all be used to produce liquid and gaseous fuels from algae biomass by the thermal decomposition of its organic materials. All of this is technically feasible, but there are so many questions about cost and environmental sustainability that producing biofuels from algae, whether native or metabolically engineered, is unlikely in the near future (Chisti, 2018; Ferreira and Gouveia, 2020).

Microalgae are natural biological reactors that are integrated with the production of certain bulk products like biologically active natural chemicals. Protein-rich Animal feed, fertiliser, nutraceuticals, cosmetics and fine chemicals have all been obtained using microalgae biomass and metabolisms (Leng et al., 2021).

12.6 Challenges in Microalgae Waste Management Technique

Pervasive study has been carried out in the past few decades on the operation of microalgae for the range of contrastive wastewater. Distinctness in the origin of wastewater gives a broad window to the wide range of microalgal species that has shown remarkable potential action. Nonetheless, the process and parameters that are studied and performed at a laboratory scale on diluted wastewater cannot be implemented at an industrial scale where wastewater is densely contaminated with other toxic chemicals and biological materials. That could harshly affect the growth of microalgae and, wherefore, their ability to remove emerging contaminants from the wastewater is also get affected (Perez et al., 2015).

Choosing the correct microalgal strain with the remediation technology is not enough to increase the effectiveness of biological material on targeted wastewater. Infect, there are many biotic as well as abiotic parameters one should take care of like strain, size, tolerance and life stages, metal type and concentration, size, and atomic weight, and other parameters like pH, temperature, salinity, and hardness, etc. (Suresh Kumar et al., 2015).

The efficiency of the microorganism consortium for nutrient uptake is influenced by complicated interactions between physicochemical parameters such as pH, photoperiod, and biological factors, as well as the bioavailability of nutrients. Pathogen presence, viral attack, protozoa predation, and bacterial competition for available nutrients are examples of the latter (Martínez et al., 2000).

Due to the small size of microalgae, motility, negatively charged surfaces, and low cell densities, harvesting microalgae efficiently from wastewater is an economically challenging step that can restrict the development of algal technologies (Suresh Kumar et al., 2015). To overcome, there are some physical harvesting technologies exist that can be applied but it comes with some limitations like high cost and inefficiency. Alternate to that there are biological harvesting mechanisms also like harvesting microalgae by co-culturing them with fungi but also have a risk of pathogen secretion (Leng et al., 2021).

Additional inputs performed for the fulfilment of microalgae wastewater treatment include physical force, extra nutrition, and maintenance of environmental conditions appropriate to the culture that restricts the large-scale execution (Leng et al., 2021).

Trying innovative technology for the first time over the traditional habit requires the willingness of the industrialists, makes the mechanism of phycoremediation a challenge to apply at the commercial scale, even after the successful establishment of all the pilot scale study for the removal of pollutants by micro & macro algae over the conventional physicochemical process (Suresh Kumar et al., 2015).

12.7 Role of Other Microorganisms as Co-Worker with Microalgae

Interaction of microalgae with aerobic bacteria favouring each other in the wastewater treatment plant named high-rate algal pond (HRAP) developed by Professor Oswald in the 1960s, with the mechanism of providing O_2 to the bacteria produced by microalgae and giving back CO_2 by a respiration process of bacteria to the microalgae for their photosynthesis process (Oswald et al., 1957; Martínez-Roldán and Cañizares-Villanueva, 2020).

The enlargement of new processes using a combination of biological community, such as microalgae or even fungi, is necessary unlike mono-culture, co-culture of Fungi-microalgae can work on pollutant removal from pharmaceutical or textile industrial wastewater with economic and environmental benefits like low cost and high efficiency with non-toxic effect. The combination of both organisms is more resistant even to adverse environmental condition, fungi-microalgae consortium uses the mechanism of bio-flocculation to treat industrial wastewater and give ease to the process of wastewater processing and microalgae biomass harvesting after completion of the treatment following the production of biofuel and lipid (Leng et al., 2021, Martínez-Roldán and Cañizares-Villanueva, 2020).

12.8 Conclusion

Because of the requirement to reuse wastewater, a variety of processes have been developed to remove particular contaminants. Different types of wastewater vary in composition and require different treatment strategies by using microalgae. Because of its wide availability, increased nutrient consumption capabilities, and numerous applications of the algal biomass produced, microalgae are being used to clean wastewater. Biotechnologically developed microalgae have received a great deal of interest as an affordable feedstock option, able to recover resources and generate bioenergy with the help of a variety of microalgal cultivation and harvesting technology. microalgal biorefinery, which can realise the production of multiple products in the form of high values products like cosmetics, pharmaceuticals and nutritious feed continues to enhance our quality of life. Despite these benefits, there are numerous obstacles to overcome in order to make this technology more cost-effective and efficient. There are very few commercially successful microalgal technologies, this could entail the creation of new technology or the improvement of current ones. Without any toxic effect, microalgae can decline the graph of carbon footprint, water footprint, global warming, climate crisis, and energy crisis in a single go globally. To summarise, the human population is in desperate need of more integrated approaches to wastewater treatment, microalgae growing, bioenergy generation, and other valuable compound manufacturing in order to produce an overall streamlined process for increased industrial feasibility. To conclude in this case, a holistic multidisciplinary strategy encompassing chemists, engineers, and biologists would be advantageous. There is certain to be a large demand for these eco-friendly commercial microalgae approaches once information and a persuasive market become more widely available.

REFERENCES

Agrawal, N., Kumar, V. & Shahi, S.K., 2021. Biodegradation and detoxification of phenanthrene in in-vitro and in-vivo conditions by a newly isolated ligninolytic fungus *Coriolopsis byrsina* strain APC5 and characterization of their metabolites for environmental safety. *Environmental Science and Pollution Research*. https://doi.org/10.1007/s11356-021-15271-w

Bhatia, S. K., Mehariya, S., Bhatia, R. K., Kumar, M., Pugazhendhi, A., Awasthi, M. K., Atabani, A. E., Kumar, G., Kim, W., Seo, S.-O. & Yang, Y.-H. 2021. Wastewater based microalgal biorefinery for bio-energy production: Progress and challenges. *Science of the Total Environment*, 751, 141599.

Chandra, R. & Kumar, V., 2015. Biotransformation and biodegradation of organophosphates and organo-halides. *In:* Chandra, R. (ed.) *Environmental Waste Management*. Boca Raton: CRC Press. https://doi.org/10.1201/b19243-17

Chandra, R. & Kumar, V., 2017a. Detection of *Bacillus* and *Stenotrophomonas* species growing in an organic acid and endocrine-disrupting chemicals rich environment of distillery spent wash and its phytotoxicity. *Environmental Monitoring and Assessment*, 189, 26. https://doi.org/10.1007/s10661-016-5746-9

Chandra, R. & Kumar, V., 2017b. Detection of androgenic-mutagenic compounds and potential autochthonous bacterial communities during in-situ bioremediation of post methanated distillery sludge. *Frontiers in Microbiology*, 8, 87. https://doi.org/10.3389/fmicb.2017.00887

Chapman, J., Truong, V. K., Elbourne, A., Gangadoo, S., Cheeseman, S., Rajapaksha, P., Latham, K., Crawford, R. J. & Cozzolino, D. 2020. Combining chemometrics and sensors: Toward new applications in monitoring and environmental analysis. *Chemical Reviews*, 120, 6048–6069.

Chisti, Y. 2018. Chapter 2- Society and microalgae: Understanding the past and present. *In:* Levine, I. A. & Fleurence, J. (eds.) *Microalgae in Health and Disease Prevention*. Cambridge, MA: Academic Press, pp. 11–21.

Chisti, Y. 2020. Chapter 1- Microalgae biotechnology: A brief introduction. *In:* Jacob-Lopes, E., Maroneze, M. M., Queiroz, M. I. & Zepka, L. Q. (eds.) *Handbook of Microalgae-Based Processes and Products*. Cambridge, MA: Academic Press, pp. 3–23.

Dutta, D., Arya, S. & Kumar, S., 2021. Industrial wastewater treatment: Current trends, bottlenecks, and best practices. *Chemosphere*, 285, 131245. https://doi.org/10.1016/j.chemosphere.2021.131245

El-Kassas, H. Y. & Mohamed, L. A. 2014. Bioremediation of the textile waste effluent by *Chlorella vulgaris*. *The Egyptian Journal of Aquatic Research*, 40, 301–308.

Escapa, C., Coimbra, R. N., Paniagua, S., García, A. I. & Otero, M. 2015. Nutrients and pharmaceuticals removal from wastewater by culture and harvesting of *Chlorella sorokiniana*. *Bioresource Technology*, 185, 276–284.

Ferreira, A. & Gouveia, L. 2020. Chapter 28- Microalgal biorefineries. *In:* Jacob-Lopes, E., Maroneze, M. M., queiroz, M. I. & Zepka, L. Q. (eds.) *Handbook of Microalgae-Based Processes and Products*. Cambridge, MA: Academic Press, pp. 771–798.

Gautam, A.K., Kumar, S. & Sabumon, P.C. 2007. Preliminary study of physico-chemical treatment options for hospital wastewater. *Journal of Environmental Management*, 83, 298–306.

Kundu, P., Debsarkar, A., Mukherjee, S.N. & Kumar, S. 2014. Artificial neural network modelling in biological removal of organic carbon and nitrogen for the treatment of slaughterhouse wastewater in a batch reactor. *Environmental Technology*, 35, 1296–1306.

Kumar, R., Ghosh, A. K. & Pal, P. 2020a. Synergy of biofuel production with waste remediation along with value-added co-products recovery through microalgae cultivation: A review of membrane-integrated green approach. *Science of The Total Environment*, 698, 134169.

Kumar, V. 2021. Phytoremediation of distillery effluent: Current progress, challenges, and future opportunities. *In:* Saxena, G., Kumar, V. & Shah, M.P. (eds.) *Bioremediation for Environmental Sustainability: Toxicity, Mechanisms of Contaminants Degradation, Detoxification and Challenges*. Elsevier. https://doi.org/10.1016/B978-0-12-820524-2.00014-6

Kumar V., Chandra R. 2020. Bioremediation of melanoidins containing distillery waste for environmental safety. *In:* Bharagava, R. & Saxena, G. (eds.) *Bioremediation of Industrial Waste for Environmental Safety*. Singapore: Springer. https://doi.org/10.1007/978-981-13-3426-9_20

Kumar, V., Kaushal, A., Singh, K., Shah, M.P., 2021a. Phytoaugmentation technology for phytoremediation of environmental pollutants: opportunities, challenges and future prospects. *In:* Kumar, V., Saxena, G. & Shah, M.P., (eds.) *Bioremediation for Environmental Sustainability: Approaches to Tackle Pollution for Cleaner and Greener Society*. Elsevier. https://doi.org/10.1016/B978-0-12-820318-7.00016-2

Kumar, V., Shahi, S.K., Ferreira, L.F.R., Bilal, M., Biswas, J.K. & Bulgariu, L., 2021b. Detection and characterization of refractory organic and inorganic pollutants discharged in biomethanated distillery effluent and their phytotoxicity, cytotoxicity, and genotoxicity assessment using *Phaseolus aureus* L. and *Allium cepa* L. *Environmental Research*, 201, 111551. https://doi.org/10.1016/j.envres.2021.111551

Kumar, V., Singh, K., Shah, M.P., Singh, A.K, Kumar, A. & Kumar, Y., 2021c. Application of omics technologies for microbial community structure and function analysis in contaminated environment. *In:* Shah, M.P., Sarkar, A. & Mandal, S., (eds.) *Wastewater Treatment: Cutting Edge Molecular Tools, Techniques & Applied Aspects in Waste Water Treatment.* Elsevier. https://doi.org/10.1016/B978-0-12-821925-6.00013-7

Kumar, V., Thakur, I.S. & Shah, M.P., 2020b. Bioremediation approaches for pulp and paper industry wastewater treatment: recent advances and challenges. *In:* Shah, M.P. (ed.) *Microbial Bioremediation & Biodegradation.* Singapore: Springer. https://doi.org/10.1007/978-981-15-1812-6_1

Kumar, V., Thakur, I.S., Singh, A.K., & Shah, M.P., 2020c. Application of metagenomics in remediation of contaminated sites and environmental restoration. *In:* Shah, M., Rodriguez-Couto, S. & Sengor, S.S. (eds.) *Emerging Technologies in Environmental Bioremediation.* Elsevier. https://doi.org/10.1016/B978-0-12-819860-5.00008-0

Lee, X. J., Ong, H. C., Gan, Y. Y., Chen, W.-H. & Mahlia, T. M. I. 2020. State of art review on conventional and advanced pyrolysis of macroalgae and microalgae for biochar, bio-oil and bio-syngas production. *Energy Conversion and Management*, 210, 112707.

Leng, L. Li, W., Chen, J., Leng, S., Chen, J., Wei, L., Peng, H., Li, J., Zhou, W. & Huang, H. 2021. Co-culture of fungi-microalgae consortium for wastewater treatment: A review. *Bioresource Technology*, 330, 125008.

Lin, C.-Y., Nguyen, M.-L. T. & Lay, C.-H. 2017. Starch-containing textile wastewater treatment for biogas and microalgae biomass production. *Journal of Cleaner Production*, 168, 331–337.

Ling, J., Xu, Y., Lu, C., Lai, W., Xie, G., Zheng, L., Talawar, M. P., Du, Q. & Li, G. 2019. Enhancing stability of microalgae biocathode by a partially submerged carbon cloth electrode for bioenergy production from wastewater. *Energies*, 12, 3229.

Martínez-Roldán, A. D. J. & Cañizares-Villanueva, R. O. 2020. Chapter 7- Wastewater treatment based in microalgae. *In:* Jacob-Lopes, E., Maroneze, M. M., Queiroz, M. I. & Zepka, L. Q. (eds.) *Handbook of Microalgae-Based Processes and Products.* Cambridge, MA: Academic Press, pp. 165–184.

Martínez, M. E., Sánchez, S., Jiménez, J. M., El Yousfi, F. & Muñoz, L. 2000. Nitrogen and phosphorus removal from urban wastewater by the microalga Scenedesmus obliquus. *Bioresource Technology*, 73, 263–272.

Mika, L. T., Cséfalvay, E. & Németh, Á. 2018. Catalytic conversion of carbohydrates to initial platform chemicals: Chemistry and sustainability. *Chemical Reviews*, 118, 505–613.

Ngwira, L. & Lakudzala, D. 2018. Assessment of the quality of SOBO industrial wastewater and its impact on water quality in Nankhaka River. *Physics and Chemistry of the Earth, Parts A/B/C*, 108, 9–12.

Oswald, W. J., Gotaas, H. B., Golueke, C. G., Kellen, W. R., Gloyna, E. F. & Hermann, E. R. 1957. Algae in waste treatment [with discussion]. *Sewage and Industrial Wastes*, 29, 437–457.

Pereira, J. F. B., Papaiconomou, N. & Ventura, S. P. M. 2021. Envisioning the future of industrial bioprocesses through biorefinery. *Frontiers in Bioengineering and Biotechnology*, 9, 136.

Perez, M., Nolasco, N. A., Vasavada, A., Johnson, M. & Kuehnle, A. 2015. Algae-mediated valorization of industrial waste streams. *Industrial Biotechnology*, 11, 229–234.

Rosenwinkel, K.-H., Austermann-Haun, U. & Meyer, H. 2005. Industrial wastewater sources and treatment strategies. *In:* Jördening, H. J. & Winter, J. (eds.) *Environmental Biotechnology: Concepts and Applications.* Weinheim, Germany: Wiley-VCH Verlag GmbH & Co. KGaA, pp. 69–70.

Shahid, A., Malik, S., Zhu, H., Xu, J., Nawaz, M. Z., Nawaz, S., Asraful Alam, M. & Mehmood, M. A. 2020. Cultivating microalgae in wastewater for biomass production, pollutant removal, and atmospheric carbon mitigation; a review. *Science of the Total Environment*, 704, 135303.

Shingare R.P., Thawale, P.R., Raghunathan, K., Mishra, A. & Kumar, S. 2019. Constructed wetland for wastewater reuse: Role and efficiency in removing enteric pathogens. *Journal of Environmental Management*, 246, 444–461.

Singh, K. & Arora, S. 2011. Removal of synthetic textile dyes from wastewaters: A critical review on present treatment technologies. *Critical Reviews in Environmental Science and Technology*, 41, 807–878.

Siville, B. & Boeing, W. J. 2020. Optimization of algal turf scrubber (ATS) technology through targeted harvest rate. *Bioresource Technology Reports*, 9, 100360.

SundarRajan, P., Gopinath, K. P., Greetham, D. & Antonysamy, A. J. 2019. A review on cleaner production of biofuel feedstock from integrated CO_2 sequestration and wastewater treatment system. *Journal of Cleaner Production*, 210, 445–458.

Suresh Kumar, K., Dahms, H.-U., Won, E.-J., Lee, J.-S. & Shin, K.-H. 2015. Microalgae – A promising tool for heavy metal remediation. *Ecotoxicology and Environmental Safety*, 113, 329–352.

Tan, J. S., Lee, S. Y., Chew, K. W., Lam, M. K., Lim, J. W., Ho, S.-H. & Show, P. L. 2020. A review on microalgae cultivation and harvesting, and their biomass extraction processing using ionic liquids. *Bioengineered,* 11, 116–129.

Wainaina, S., Awasthi, M.K., Sarsaiya, S., Chen, H., Singh, E., Kumar, A., Ravindran, B., Awasthi, S.K., Liu, T., Duan, Y. & Kumar, S. 2020. Resource recovery and circular economy from organic solid waste using aerobic and anaerobic digestion technologies. *Bioresource technology*, 301, 122778.

Yin, Z., Zhu, L., Li, S., Hu, T., Chu, R., Mo, F., Hu, D., Liu, C. & Li, B. 2020. A comprehensive review on cultivation and harvesting of microalgae for biodiesel production: Environmental pollution control and future directions. *Bioresource Technology*, 301, 122804.

Zhu, L. 2015. Biorefinery as a promising approach to promote microalgae industry: An innovative framework. *Renewable and Sustainable Energy Reviews*, 41, 1376–1384.

Section II

Microbial Treatment of Wastewater/Wastewater Pollutants

13

Microbial Reclamation of Pulp and Paper-Making Industry Wastewater: Electricity Generation, Value Added Co-Product Recovery and Waste Valorization

Dhanya Mecherivalappil Subramanian and Sneh Lata
Central University of Punjab

CONTENTS

13.1 Introduction

Solid and liquid waste generation has increased from the era of industrialization and urbanization. The management of the waste has created environmental and health concerns from its dumping and disposal without and partial treatment (Pokhrel and Viraraghavan, 2004, Chakraborty et al., 2019; Singh et al., 2021; Mishra et al., 2021). The pulp and paper-making industry is a major polluting industry with its discharge of solid and liquid wastes hampering the aquatic life and the sludge burning adding to air emissions. Deforestation is also a major global issue as forest resources are major contributors to the feedstock of paper making in many countries (FAO and UNEP, 2020). The production of virgin paper and recycled paper has been increasing over the decades. The annual consumption of paper in the world is around 399 million metric tons in 2020 (Statista, 2021). The paper and pulp industry has been one of the largest water-consuming industries with annual water consumption of billions of cubic meters and in turn resulting in the production of a huge quantity of wastewater having organic and inorganic pollutants (Pokhrel and Viraraghavan, 2004, Huang et al., 2009, Quina and Pinheiro, 2020). The processes involved and type of paper produced determine the various pollutants in the wastewater generated from the paper manufacturing industry. The major units of wastewater generation in the paper-making industry are washing, pulping and bleaching. The wastewater treatment also led to the production of

DOI: 10.1201/9781003231738-15

secondary sludge from pulping and paper manufacturing that need to manage properly (Gavrilescu et al., 2008). The wastewater with its organic content had the potential to be used as resources (Mandeep and Shukla, 2020; Wainaina et al., 2020). So, there is a need for an alternative and eco-friendly treatment method of pulp and paper industry wastewater with product recovery.

13.2 Wastewater Generation in Paper and Pulp Making Industry

The wastewater in paper production is majorly from pulping and bleaching. The lignin removal from the wood occurs in the pulping process followed with the brightening of pulp in the bleaching step. The 1 MT of paper production generated 20–100 m^3 of wastewater (Gavrilescu et al., 2008). The annual wastewater produced from the paper industry in India was 695.7 million m^3 (Narayan et al., 2018). The pulping step caused major pollution from the paper industry (Pokhrel and Viraraghavan, 2004). The major pollutant in pulp and paper industry wastewater is organic matter with high chemical oxygen demand (COD) and biochemical oxygen demand (BOD). The suspended solids, phenolic compounds, heavy metals, tannins, lignin, resin and aromatics are other pollutants present in paper and pulp effluents (Kumar et al., 2020; Quina and Pinheiro, 2020).

Kraft pulping is the widely followed pulping process throughout the world having six stages consisting of pulping, washing, evaporation, combustion, clarification, and causticizing (Pokhrel and Viraraghavan, 2004). Figure 13.1 shows the major steps which produce wastewater from kraft pulping.

White liquor had been generated from the removal of lignin from wood followed by black liquor from the removal of pulp and other wastewater was generated from paper making through dilution, refining, and adding chemicals (Simão et al., 2018). During combustion, the black liquor had dissolved sulfides, carbonates, sulfates, and chlorides of sodium and potassium in smelt. The green liquor was produced from the wash of black liquor and dregs were generated on clarification and white liquor recovery along with lime mud during causticizing (Simão et al., 2018).

Figure 13.2 summarizes the major pollutants in wastewater from the pulp and paper-making industry. The high amount of organic matter and suspended solids are major pollutants present in wastewater. The BOD and COD in wastewater generated per tonne of pulp are in the range of 5–20 and 20–40 kg respectively. The inorganic pollutants were also present in pulp and paper-making industry effluent (Quina and

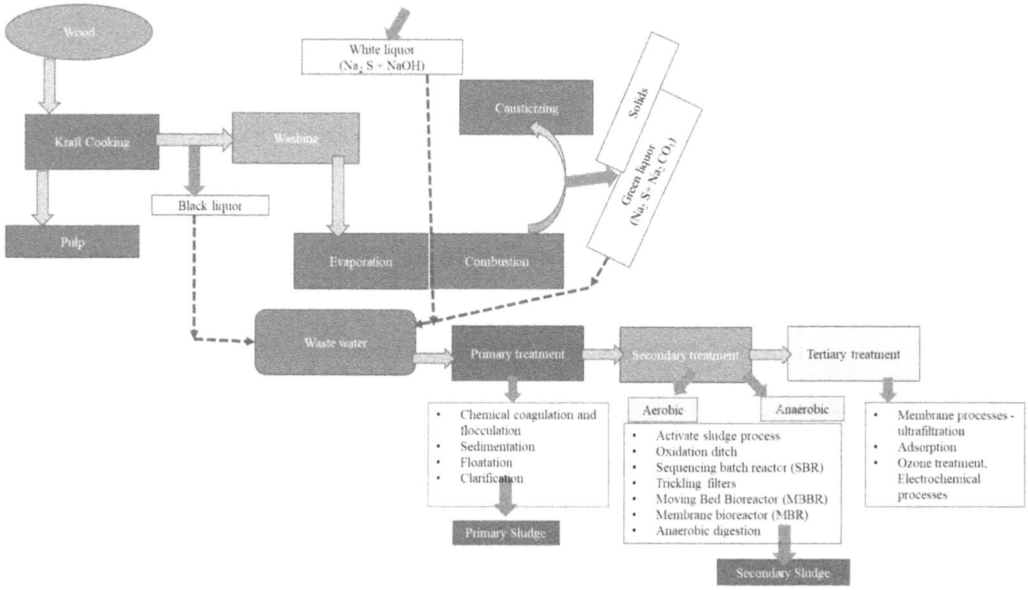

FIGURE 13.1 Wastewater generation from different stages of kraft pulping and wastewater treatment.

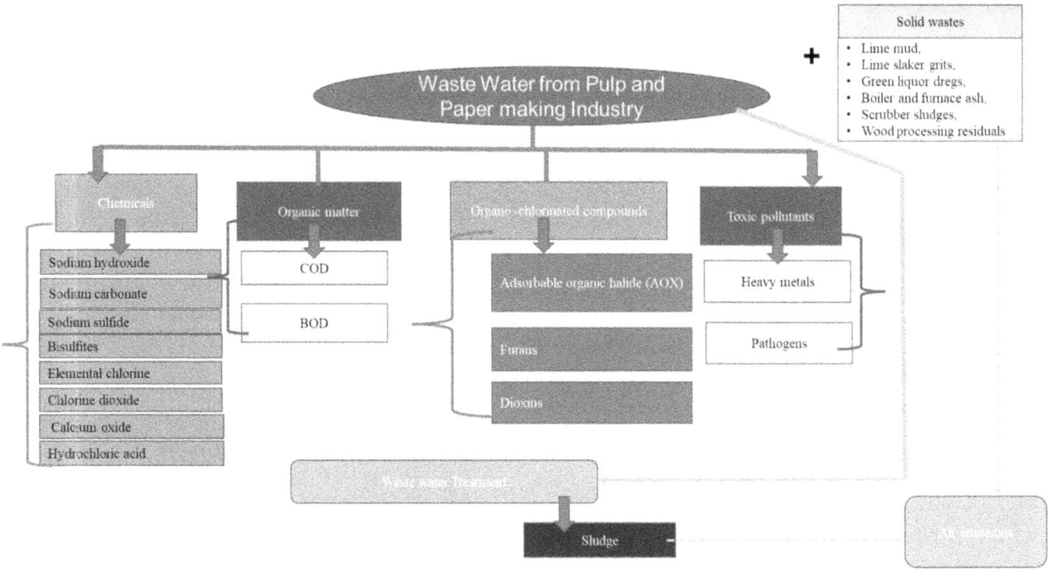

FIGURE 13.2 Major pollutants in wastewater from pulp and paper-making industry. (Smook, 1992, Chakraborty et al., 2019, Quina and Pinheiro, 2020.)

Pinheiro, 2020). The wastewater also contains suspended solids and organochlorinated compounds of approximately 15 and 4 kg respectively. The bleaching mainly produced chlorinated organic compounds consisting furans, dioxins, and absorbable organic halides (AOX) (Gavrilescu et al., 2008).

The COD of the wastewater from pulp and paper industries varied from 3,300 to 10,000 mg/L, suspended solids from 480 to 1,800 mg/L, and chloride content from 37 to 560 mg/L (Toczyłowska-Mamińska et al., 2020).

13.3 Need for a Sustainable Technique for Wastewater from Pulp and Paper-Making Industry

The treatment of wastewater from the pulp and paper-making industry with high pH, BOD, and COD has been a big challenge. The primary, secondary and tertiary treatment techniques of wastewater from the paper and pulp-making industry had limitations. The conventional treatment methods of wastewater generated from the recycling process were also unable to degrade cellulosic matter (Lens et al., 2002). The treatment methods are high energy-intensive with negative impacts on the environment. The secondary sludge production and its treatment is another disadvantage of the conventional wastewater treatment methods before disposal.

The high amount of cellulose, hemicellulose, and lignin present in paper and pulp waste are good sources of bioenergy and chemicals (Chakraborty et al., 2019). The effluent is cellulose-containing wastewater and this organic fraction is a good sustainable renewable source that helps in waste management in an eco-friendly manner. The energy valorization from paper and pulp industry waste is commonly by thermo-chemical technologies such as pyrolysis, gasification, hydrothermal conversion, and refuse-derived fuel from the sludge. But these technologies are energy-intensive. The microbial fuel cell (MFC) is a viable option for the biological treatment of wastewater from paper and pulp mills under anaerobic condition with electricity production preventing the discharge of high organic loading to water resources (Xu et al., 2021). The MFCs found to be superior to other conventional wastewater treatments like physicochemical and biological treatments due to lack of energy input, no need for any chemicals, and generation of electric current (Toczyłowska-Mamińska et al., 2020).

13.4 Microbial Fuel Cells and Its Mechanism of Electricity Production

The electrochemical system that works with microorganisms in anaerobic conditions with anode and cathodic chambers separated by membrane produced electicity is known as microbial fuel cell (MFC). The parts of a microbial fuel cell are an anodic chamber, cathodic chamber and a proton exchange membrane (Figure 13.3). Based on the mechanism of electron transfer, there are two types of MFCs- mediator less and mediator-based MFCs. Some of the microbes like *Geobacter* sp., *Pseudomonas aeruginosa, Shewanella* sp., and *Pseudomonas alcaliphila* have the capability to transfer electrons directly which is the base for mediator less MFCs. But a few microbes need mediators in the transport of electrons as in mediator-based MFCs. The role of a mediator is the transfer of electrons from the microbes to the surface of anode. The common exogenous mediators are methylene blue, thionine, neutral red, ferricyanide, humic acid, azure A or methyl viologen.

13.4.1 Microbial Fuel Cells for Wastewater from Paper and Pulp Industry

The microbial fuel cell based on effluent produced from the paper recycling process was studied by Huang and Logan (2008). Huang et al. (2009) demonstrated the MFC for electricity production and reducing COD for discharge of paper industry wastewater with its effect of short retention time. Velasquez-Orta et al. (2011) demonstrated a higher production of current density (125 mA/m^2) than brewery and bakery wastewaters. The complete removal of cellulose with high power generation was reported with increased conductivity. Singh et al. (2017) demonstrated microbial fuel cell with aluminium plates working with the wastewater generated from the paper and pulp industry with the generation of 699 mV in 60 hours.

Velasquez-Orta et al. (2011) reported that the bacterial community in contact with the anode was more efficient and generated a redox shuttle that helped in improving the rate of electron transport via the diffusion process. The molecule had a similar redox potential of 2,3-dimethoxy-5-methyl-1,4-benzoquinone

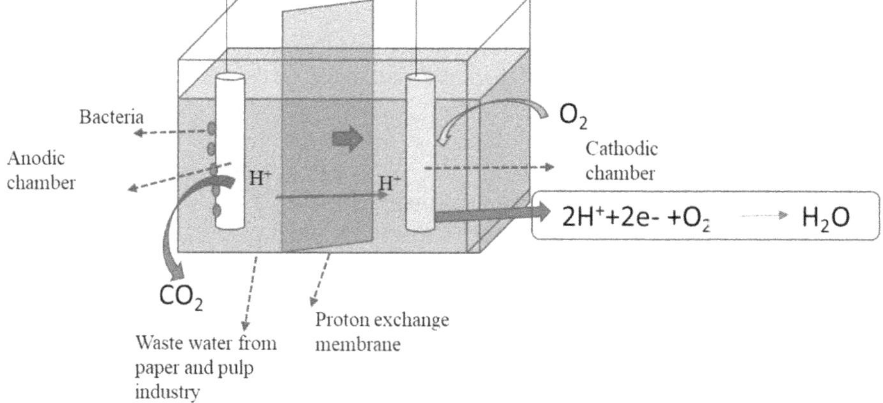

FIGURE 13.3 Mediator based microbial fuel cell (MFC) working on wastewater from pulp and paper industry. (Flimban et al., 2018, Toczyłowska-Mamińska et al., 2020.)

reported by Alegria et al. (2004). The ferricyanide-based MFC was used for the treatment of wastewater from the paper-making industry by Zhang et al. (2011).

Singh et al. (2017) studied the bioelectricity generation by mediator-less microbial fuel cells from paper mill waste with a production of 758 mV. Kassongo and Togo (2011) also investigated mediator-less microbial fuel cells with paper mill effluent for the production of electricity.

13.4.2 Limitations of MFCs using Wastewater from Paper and Pulp Industry

The low production of voltage, current, and power in MFCs prevents commercialization. The anaerobic condition in anodic chamber is a challenge in a microbial fuel cell which increases its cost. The reduced microbial growth on electrodes and unstable power production also limits MFC performance (Huang and Logan, 2008, Mathuriya and Sharma, 2009a, 2009b). The requirement of a semi-permeable membrane for proton transfer through anode to cathode is another limitation. Flimban et al. (2018) reported the MFC performance by the biofouling at Nafion-based PEM membrane. The costly electrodes are also limiting the commercialization of MFCs.

The expensiveness and toxic nature of mediator-based MFCs also limit the upscaling of MFCs wih paper mill wastewater. The very low Coulombic efficiencies generated from the MFC technique were able to reduce COD of paper wastewater only to a very limited extent (Ketep et al., 2013). The paper mill effluents with inorganic chemicals were capable of oxidation and interfere with microbial acclimation, metabolism and electricity generation and distribution (Huang et al., 2009). So, the wastewater from bleaching has been considered as less suitable for MFC because of low voltage output, toxicity of chemicals to methanogens (Elakkiya and Niju, 2021).

13.5 Strategies to Enhance Electricity Production from Paper and Pulp Industry Wastewater

There are various MFC systems for wastewater treatment namely bioelectrochemical treatment system (BET), microbial electrosynthesis system (MES), bioelectrochemical system (BES), electro-fermentation (EF), microbial desalination cell (MDC), microbial electrolysis cell (MEC) and electro-methanogenesis (EM) for energy production. There are many strategies developed for enhancing electricity generation from pulp and paper-making industry wastewater.

13.5.1 Integration of Constructed Wetlands with MFCs

The constructed wetland is one of the cost-effective and well-established methods for the treatment of wastewater generated from different industries (Shingare et al., 2019). The MFCs coupled with constructed electroactive wetlands (CW-MFCs) has been gained momentum towards the effluent treatment and production of electricity (Wang et al., 2020). The combination of constructed wetlands with microbial fuel cells had additive results with bioelectricity generation and pollutant removal. It works with redox reactions that occur in anaerobic and aerobic zones. The transport of electrons was carried out with the help of the microbial community. The interaction between electrodes and microbes also improved the potential. The CW-MFCs had the advantage of treating a huge quantity of wastewater. The maintenance of this type of MFCs is also easier (Narayan et al., 2018). The formulation of R& D efforts for effective electron recovery from coupled electroactive wetlands has been required.

13.5.2 Exoelectrogens for Electricity Production

Ketep et al. (2013) reported the dominance of *Desulfuromonas acetexigens* in tertiary microbial anode with improved potential in MFC performance. Kaushik and Jadhav (2017) isolated electrogenic bacteria identified as *Pseudomonas fluorescens* from pulp industry wastewater and reported its potential in MFCs using paper and pulp wastewater. Xu et al. (2021) isolated exoelectrogen strain LZ-P1 *Cellulomonas iranensis* from the gut of Giant Panda and was able to produce a power density of 44.05 mW/m^2. Ali et al. (2016) used mixed culture biofilm developed from activated sludge as electrogens for mediator-less MFC treatment of black liquor and molasses. Toczyłowska-Mamińska et al. (2020) developed a syntrophic consortium of *Trichocomaceae* sp., a fungus and two bacterial species namely *Achromobacter insolitus* and *Geobacter sulfurreducens* with potential for bioelectricity generation from the wastewater generated from wood industry.

13.5.3 Substrate Modification in MFCs

Elakkiya and Niju (2021) demonstrated that mixing bagasse wash water and pulping wastewater in 9:1 ratio was able to generate power density of 73 mW/m^2 and current density of 202 mA/m^2 and helped in wastewater management from paper mill based on bagasse. Belleville et al. (2011) described the nitrogen limitation had no negative impact on the electrochemical performance of MFCs working with wastewater. Shankar et al. (2016) used MFC technology for paper industry effluent after electro coagulation and found improvement in electricity generation.

13.5.4 Improvement in Performance of MFC

Ketep et al. (2013) studied the capacity of microbial bioanode (prepared for a week) in a three-electrode MFC having graphite plates as electrodes working with paper wastewater in continuous mode and found the reference electrode placed near to bioanode surface overcome the internal resistance, maintain current distribution and stability in substrate concentration. The polarization of electrode at a low potential improved the microbial anode performance with more electrode respiring bacteria (Torres et al., 2009). The successive scratching/re-inoculation along with the reduction in applied potential resulted in developing biofilms as microbial electrodes on graphite plates helped in efficient microbial fuel cells for paper mill wastewater treatment (Ketep et al., 2013). The control of biofouling of the membrane is possible with back washing and the use of anti-fouling agents (Saratale et al., 2017).

Radha and Kanmani (2017) studied the effect of potassium ferricyanide and potassium permanganate mediators in phosphate buffer solution on the performance of MFC for treating wastewater from the recycling paper industry. The results concluded the potassium permanganate showed high efficiency in power density production with COD removal.

Chen et al. (2020) combined anaerobic moving-bed biofilm reactor and a microbial fuel cell for the enhancement of electricity generation from paper and pulp effluents. Chen et al. (2020) used a moving-bed biofilm reactor having ceramsite as an anode of MFC for the wastewater treatment and found improvement in electricity production and waste removal. Das et al. (2021) reported the significance of membrane crystallisation for product recovery from wastewater.

13.6 Factors Influencing MFC Performance from Paper and Pulp Industry Waste Water

The following factors determine the efficiency of the electricity generation using MFCs from the paper and pulp industry wastewater (Kaushik and Jadhav, 2017).

- Design of the chamber
- Material used for electrodes
- Type of substrate used in paper and pulp industry
- Microorganisms
- Effect of mediator
- pH of wastewater
- Temperature in MFCs

The dual-chamber MFC increased the electrochemical performance of cathode with control of pH, oxygen supply, flow rate and addition of mediators (Flimban et al., 2018). Mathuriya and Sharma (2009a, b) demonstrated dual-chamber MFC with plain carbon paper as anode and cathode working with paper wastewater for treatment of paper-making industry. The dual-chamber MFC with paper wastewater was also studied by Patil et al. (2011).

Huang and Logan (2008) used graphite fibre brush as anode and were able to produce 0.6 A/m^2 in a single chamber MFC having air cathode from fed batch process of wastewater after primary treatment of paper mill. NH_3-treated carbon brush was used as anode by Huang et al. (2009) for single chamber MFC application in continuous mode. The batch mode of MFCs for paper wastewater was also studied by Velasquez-Orta et al. (2011) with carbon cloth used as anode. Patil et al. (2011) used graphite plates as an electrode for wastewater treatment of pulp mill effluent by MFCs. Haavisto et al. (2019) studied various materials for anodes in MFC. The choice of granular activated carbon in a stainless steel cage as an electrode for thermomechanical pulping of wastewater resulted in the average power production of 10–15 mW/m^2. Zhang et al. (2011) reported the inhibition of microbes and reduced MFC performance for a chromium-based COD in paper wastewater of more than 5,200 mg/L.

The inoculation with *Clostridium acetobutylicum* and *C. thermohydrosulfuricum* (Mathuriya and Sharma, 2009b), *Enterobacter aerogenes* (Patil et al., 2011) were reported as suitable inocula for MFC operating with paper wastewater. The electricity production from chemi-thermal mechanical pulping with ferricyanide-based MFC was found higher with more COD removal (Zhang et al., 2011). Kaushik and Jadhav (2017) reported the optimized pH of 7, the temperature of 35°C, inoculum size of 1.5% and humic acid mediator produced high voltage and current from the effluent of paper and pulp industry.

Even though MFC technology focussed on sustainable wastewater treatment, the MFCs in their present form have not been capable to meet the high energy demand (Choudhury et al., 2017). The wastewater from the pulp and paper-making industry rich in organic content can be used for the recovery of other products.

13.7 Co-products Recovery from Wastewater of Pulp and Paper-Making Industry

The single-chamber MFC had a COD removal by fed batch mode from primary clarifier of the paper mill was 29% (Huang and Logan, 2008) increased to 41% in the continuous mode of operation (Huang et al., 2009). Velasquez-Orta et al. (2011) observed a COD removal of 78% from the paper wastewater. Sakdaronnarong et al. (2015) demonstrated the simultaneous production of electricity and methane by combining lignin-rich wastewater and ethanol stillage wastewater using dual-chamber MFC. Rabelo et al. (2018) used the microbial consortium collected from the pulp and paper mill industry wastewater in Brazil for the production of biohydrogen, biogas and organic acids like butyric and acetic acid.

13.8 Other Techniques for Energy Generation from Paper and Pulp Wastewater

Bakraoui et al. (2020) evaluated the annual methane production of 4,137,928 m^3 and electrical energy potential of 510,703 MWh from the effluent of recycled pulp and paper industry with up-flow sludge blanket reactors. The biogas production per day from recycled pulp and paper industry wastewater was 62.5 L with a COD reduction of 80.6%. The lignin recovered from wastewater or the degradation products of lignin namely muconic acid, adipic acid, terephthalic acid, caprolactam, and pyrogallol had converted to resins, lubricants, plastics, and nylon (Haq et al., 2020). The effective pretreatments coupled with energy production technologies enhanced bioenergy from the sludge produced in the paper and pulp industry (Chakraborty et al., 2019). The anaerobic digestion followed by pyrolysis and incineration techniques with the production of biogas-biochar improved energy recovery and also reduced the impact on the disposal of sludge from the pulp and paper-making industry (Mohammadi et al., 2019). Sethupathy and Sivashanmugam (2018) reported the addition of polyhydroxyalkanoate had increased methane production from pulp and paper sludge. Further, the consortium of hydrolytic enzymes assisted poly-3-hydroxybutyrate was found to increase methane production (Sethupathy and Sivashanmugam, 2021).

13.9 Conclusion

The characteristics of spent liquor generated from pulping and bleaching steps in virgin and recycled paper-making industries determine the efficiency of any treatment method of wastewater. The microbial fuel cell is emerging as an attractive and sustainable technology for the degradation of organic matter with bioelectricity production and treatment of wastewater. But had the limitation of fluctuating power density with low yield from pulp and paper mill effluents. The strategies such as improved designs of MFC, selection of ideal exoelectrogens, and integrating with other technologies and constructed wetlands enhanced the MFC performance in terms of electricity and co-products production. The bio-electrochemical system with advancement in research and development has the potential for a sustainable and economic solution for abating environmental pollution and efficient energy production from wastewater of pulp and paper industry.

REFERENCES

Alegria, A.E., Sanchez, S., Quintana, I. 2004. Quinone-enhanced ascorbate reduction of nitric oxide: Role of quinone redox potential. *Free Radical Research* 38(10): 1107–1112. https://doi.org/10.1080/10715760400009852.

Ali, M., Yousaf, S., Anam, M., Bangash, Z., Maleeha, S. 2016. Evaluating the efficiency of a mixed culture biofilm for the treatment of black liquor and molasses in a mediator-less microbial fuel cell. *Environmental Technology* 37(22): 2815–2822. https://doi.org./10.1080/09593330.2016.1166267

Bakraoui, M., Gnaoui, Y.E., Lahboubi, N., Karouach, F., Bari, H.E. 2020. Kinetic study and experimental productions of methane production from UASB reactor treating wastewater from recycled pulp and paper for the continuous test. *Biomass and Bioenergy* 139: 105604. https://doi.org/10.1016/j.biombioe.2020.105604.

Belleville, P., Strong, P. J., Dare, P. H., Gapes, D. J. 2011. Influence of nitrogen limitation on performance of a microbial fuel cell. *Water Science and Technology* 63 (8): 1752–1757. https://doi.org/10.2166/wst.2011.111

Chakraborty, D., Shelvapulle, S., Reddy, K.R., Kulkarni, R.V., Puttaiahgowda, Y.M., Naveen, S., Raghu, A.V. 2019. Integration of biological pre-treatment methods for increased energy recovery from paper and pulp biosludge. *Journal of Microbiological Methods* 160: 93–100. https://doi.org/10.1016/j.mimet.2019.03.015.

Chen, F., Zeng, S., Luo, Z., Ma, J., Zhu, Q., Zhang, S. 2020. A novel MBBR–MFC integrated system for high-strength pulp/paper wastewater treatment and bioelectricity generation. *Separation Science and Technology* 55:14: 2490–2499. https://doi.org/10.1080/01496395.2019.1641519

Choudhury, P., Shankar, U.U., Mahata, N., Tiwari, O., Ray, R., Bandyopadhyay, T., Bhunia, B. 2017. Performance improvement of microbial fuel cells for waste water treatment along with value addition: A review on past achievements and recent perspectives. *Renewable and Sustainable Energy Reviews* 79: 372–389.https://doi.org/10.1016/j.rser.2017.05.098.

Das, P., Dutta, S., Singh, K.K. 2021. Insights into membrane crystallization: A sustainable tool for value added product recovery from effluent streams. *Separation and Purification Technology* 257: 117666. https://doi.org/10.1016/j.seppur.2020.117666.

Elakkiya, E., Niju, S. 2021. Bioelectrochemical treatment of real-field bagasse-based paper mill wastewater in dual-chambered microbial fuel cell. *3 Biotech* 11: 42. https://doi.org/10.1007/s13205-020-02606-6

FAO., UNEP. (2020). The State of the World's Forests 2020. Forests, biodiversity and people. Rome. https://doi.org/10.4060/ca8642en (accessed on 1 October 2021).

Flimban, S., Hassan, S., Rahman, M.M., Oh, S.E. 2018. The effect of Nafion membrane fouling on the power generation of a microbial fuel cell. *International Journal of Hydrogen Energy* 45. https://doi.org/10.1016/j.ijhydene.2018.02.097.

Gavrilescu, M., Teodosiu, C., Gavrilescu, D., Lupu, L. 2008. Strategies and practices for sustainable use of water in industrial papermaking processes. *Engineering in Life Sciences* 8: 99–12.
Haavisto, J., Dessì, P., Chatterjee, P., Honkanen, M., Noori, M.T., Kokko, M., Lakaniemi, A.M., Lens, P.N.L., Puhakka, J.A. 2019. Effects of anode materials on electricity production from xylose and treatability of TMP wastewater in an up-flow microbial fuel cell. *Chemical Engineering Journal* 372: 141–150. https://doi.org/10.1016/j.cej.2019.04.090.
Haq, I., Mazumder, P., Kalamdhad, A.S. 2020. Recent advances in removal of lignin from paper industry wastewater and its industrial applications – A review. *Bioresource Technology* 312: 123636. https://doi.org/10.1016/j.biortech.2020.123636.
Huang, L., Cheng, S, Rezaei, F., Logan, B. 2009. Reducing organic loads in wastewater effluents from paper recycling plants using microbial fuel cells. *Environmental Technology* 30(5): 499–504. https://doi.org/10.1080/09593330902788244.
Huang, L., Logan, B.E. 2008. Electricity generation and treatment of paper recycling wastewater using a microbial fuel cell. *Applied Microbiology and Biotechnology* 80: 349–355. https://doi.org/10.1007/s00253-008-1546-7.
Kassongo, J., Togo, C.A. 2011. Evaluation of full-strength paper mill effluent for electricity generation in mediator-less microbial fuel cells. *African Journal of. Biotechnology* 10: 15564–15570.
Kaushik, A., Jadhav, S.K. 2017. Conversion of waste to electricity in a microbial fuel cell using newly identified bacteria: *Pseudomonas fluorescens. International Journal of Environmental Science and Technology* 14 1771–1780. https://doi.org/10.1007/s13762-017-1260-
Ketep, S.F., Bergel, A., Bertrand, M., Achouak, W., Fourest, E. 2013. Lowering the applied potential during successive scratching/re-inoculation improves the performance of microbial anodes for microbial fuel cells. *Bioresource Technology* 127: 448–455. https://doi.org/10.1016/j.biortech.2012.09.008.
Kumar, V., Thakur, I.S., Shah, M.P., 2020. Bioremediation approaches for pulp and paper industry wastewater treatment: Recent advances and challenges. In: Shah, M.P. (Ed.) *Microbial Bioremediation & Biodegradation*. Springer, Singapore. https://doi.org/10.1007/978-981-15-1812-6_1
Lens, P.N.L., Pol, H.L., Wilderer, P., Asano, T. 2002. *Water Recycling and Resource Recovery in Industry: Analysis, Technologies and Implementation*. IWA Publishing, London, p. 677.
Mandeep, G.K., Shukla, P. 2020. Insights into the resources generation from pulp and paper industry wastes: Challenges, perspectives and innovations. *Bioresource Technology* 297: 122496. https://doi.org/10.1016/j.biortech.2019.122496.
Mathuriya, A. S., Sharma, V. N. 2009a. Bioelectricity production from various wastewaters through microbial fuel cell technology. *Journal of Biochemical Technology* 2(1): 133–137.
Mathuriya, A. S., Sharma, V. N. 2009b. Bioelectricity production from paper industry waste using a microbial fuel cell by *Clostridium* species. *Journal of Biochemical Technology* 1(2): 49–52.
Mishra, D., Kumari, S., Jaiswal, A., Arya, S., Yadav, B.R., Thul, S.T., Kumar, S., Pandey, R., Kumar, R. Pandey, A., 2021. Evaluation of distillery sludge as a soil amendment for improving soil quality and sugarcane (CO-265) yield. *Environmental Technology & Innovation* 23: 101624. https://doi.org/10.1016/j.eti.2021.101624.
Mohammadi, A., Sandberg, M., Venkatesh, G., Eskandari, S., Dalgaard, T., Joseph, S., Granström, K. 2019. Environmental performance of end-of-life handling alternatives for paper-and-pulp-mill sludge: Using digestate as a source of energy or for biochar production. *Energy* 182: 594–605. https://doi.org/10.1016/j.energy.2019.06.065.
Narayan, M., Solanki, P., Rabha, A.K., Srivastava, R.K. 2018. Treatment of pulp and paper industry effluent and electricity generation by constructed wetland coupled with microbial fuel cell (CW-MFC). *Journal of Pharmacognosy and Phytochemistry* 7(6): 493–498.
Patil, V.D., Patil, D.B., Deshmukh, M.B., Pawar, S.H. 2011. Comparative study of bioelectricity generation along with the treatment of different sources of wastewater. *International Journal of Engineering Research and Applications* 2: 162–168.
Pokhrel, D., Viraraghavan, T. 2004. Treatment of pulp and paper mill wastewater – A review. *Science of Total Environment* 333: 37–58. https://doi.org/10.1016/j.scitotenv.2004.05.017.
Quina, M.J., Pinheiro, C.T. 2020. Inorganic waste generated in kraft pulp mills: The transition from landfill to industrial applications. *Applied Sciences* 10(7): 2317. https://doi.org/10.3390/app10072317.
Rabelo, C.AB.S., Soares, L.A., Sakamoto, I.K., Varesche, M.B.A. 2018. Bioconversion of cellulose into hydrogen, biogas and organic acids using microbial consortium from a pulp and paper mill wastewater treatment plant. *Química Nova* 41(2). https://doi.org/10.21577/0100-4042.20170157.

Radha, M., Kanmani, S. 2017. Evaluation of paper recycling industry wastewater treatment and simultaneous bioelectricity generation in a microbial fuel cell. *Desalination and Water Treatment* 62: 161–167.

Sakdaronnarong, C., Ittitanakam, A., Tanubumrungsuk, W., Chaithong, S., Thanosawan, S., Sinbuathong, N., Jeraputra, C. 2015. Potential of lignin as a mediator in combined systems for biomethane and electricity production from ethanol stillage wastewater. *Renewable Energy* 76: 242–248. https://doi.org/10.1016/j.renene.2014.11.009.

Saratale, G.D., Saratale, R.G., Shahid, M.K., Zhen, G., Kumar, G., Shin, H.S., Choi, Y.G., Kim, S.H. 2017. A comprehensive overview on electro-active biofilms, role of exo-electrogens and their microbial niches in microbial fuel cells (MFCs). *Chemosphere* 178: 534–547. https://doi.org/10.1016/j.chemosphere.2017.03.066.

Sethupathy, A., Sivashanmugam, P. 2018. Enhancing biomethane potential of pulp and paper sludge through disperser mediated polyhydroxyalkanoates. *Energy Conversion and Management* 173: 179–186.

Sethupathy, A., Sivashanmugam, P. 2021. Amelioration of methane production efficiency of paper industry waste sludge through hydrolytic enzymes assisted with poly3hydroxybutyrate. *Energy* 214(C): 119083. https://doi/10.1016/j.energy.2020.119083.

Shankar, R., Varma, A.K., Mondal, P., Chand, S. 2016. Simultaneous treatment and energy production from PIW using electro coagulation & microbial fuel cell. *Journal of Environmental Chemical Engineering* 4(4 Part A): 4612–4618. https://doi.org/10.1016/j.jece.2016.10.021.

Shingare, R.P., Thawale, P.R., Raghunathan, K., Mishra, A., Kumar, S., 2019. Constructed wetland for wastewater reuse: Role and efficiency in removing enteric pathogens. *Journal of Environmental Management* 246: 444–461.

Simão, L., Hotza, D., Raupp-Pereira, F., Labrincha, J.A., Montedo, O. 2018. Wastes from pulp and paper mills - A review of generation and recycling alternatives. *Ceramica* 64: 443–453. https://doi.org./10.1590/0366–69132018643712414.

Singh, E., Kumar, A., Mishra, R., You, S., Singh, L., Kumar, S., Kumar, R., 2021. Pyrolysis of waste biomass and plastics for production of biochar and its use for removal of heavy metals from aqueous solution. *Bioresource Technology* 320: 124278.

Singh, P., Srivastava, A., Srivastava, N. 2017. Generation of electricity using non mediated microbial fuel cell by utilizing different types of wastewater. *Journal of Scientific and Industrial Research* 76(8): 512–514.

Smook, G.A. 1992. *Handbook for Pulp & Paper Technologists.* Second edition. Angus Wilde Publications, Bellingham, WA.

Statista. 2021. Paper consumption worldwide from 2020 to 2030 (in million metric tons). URL:https://www.statista.com/statistics/1089078/demand-paper-globally-until-2030/

Toczyłowska-Mamińska, R., Pielech-Przybylska, K., Sekrecka-Belniak, A., Dziekońska-Kubczak, U. 2020. Stimulation of electricity production in microbial fuel cells via regulation of syntrophic consortium development. *Applied Energy* 271: 115184. https://doi.org/10.1016/j.apenergy.2020.115184.

Torres, C.I., Krajmalnik-Brown, R., Parameswaran, P., Marcus, A.K., Wanger, G., Gorby, Y.A., Rittmann, B.E. 2009. Selecting anode-respiring bacteria based on anode potential: Phylogenetic, electrochemical, and microscopic characterization. *Environmental Science and Technology* 43: 9519–9524.

Velasquez-Orta, S.B., Head, I.M., Curtis, T.P., Scott, K. 2011. Factors affecting current production in microbial fuel cells using different industrial wastewaters. *Bioresource Technology* 102(8): 5105–5112. https://doi.org/10.1016/j.biortech.2011.01.059.

Wainaina, S., Awasthi, M.K., Sarsaiya, S., Chen, H., Singh, E., Kumar, A., Ravindran, B., Awasthi, S.K., Liu, T., Duan, Y., Kumar, S., 2020. Resource recovery and circular economy from organic solid waste using aerobic and anaerobic digestion technologies. *Bioresource Technology* 301: 122778. doi.org/10.1016/j.biortech.2020.122778

Wang, W., Zhang, Y., Li, M., Wei, X., Wang, Y., Liu, L., Wang, H., Shen, S. 2020. Operation mechanism of constructed wetland-microbial fuel cells for wastewater treatment and electricity generation: A review. *Bioresource Technology* 314: 123808. https://doi.org/10.1016/j.biortech.2020.123808.

Xu, R., Zhang, K., Xie, S., Liu, P., Yu, Z., Han, H., Zhao, S., Peng, L., Li, X. 2021. Evaluation of electricity production from paper industry wastewater by *Cellulomonas iranensis* LZ-P1 isolated from giant panda. *Journal of Cleaner Production* 278: 123576. https://doi.org/10.1016/j.jclepro.2020.123576.

Zhang, X.X., Zhao, S.H., Li, Y.M., Wan, X.F. 2011. CTMP wastewater treatment by microbial fuel cell process. *China Pulp and Paper* 30: 6–10.

14

Microbial Treatment of Food Processing Wastewater and Recovery of Value-Added Bioactive Compounds: Current Scenario, Challenges, and Future Prospects

Jyoti Kaushal, Madhu Khatri, and Shailendra Kumar Arya
University Institute of Engineering and Technology

CONTENTS

14.1 Introduction

The food processing industry is characteristically involved in the overall transformation of agricultural produce into safe and edible food items which are fit for consumption. The main aim of this industry is to provide food and beverage items after thorough processing of the raw produce so as to get rid of any impure or toxic entities present in the naturally available product. Apart from this, the food processing industry is also focused to increase the shelf life of the consumables by adding certain preservatives and improving the lifespan of the perishable items, i.e. "food". The increasing number of consumers has furthermore increased the importance of this industry so that a maximum amount of consumable

DOI: 10.1201/9781003231738-16

food products can be made available in the market (Earle and Earle, 1997). The food industry comprises several different units within, which depend on the nature of the food being processed. Typically, the industry works towards the processing of grains, fruits, and vegetables, dairy products, meat and meat products, bakeries, and beverages of different kinds.

The food processing industry is a water strenuous industry and the techniques involved in the processing of various kinds of food items generally tend to leave behind wastewater full of undesirable characteristics making it difficult for this effluent to be discarded safely (Roy et al., 2003; Awasthi et al., 2018a; Negi et al., 2020). Food industry being one of the largest industries tends to contribute a maximum amount of wastewater which requires stringent treatments before its disposal. The main characteristics of the wastewater from the food industry include high (chemical oxygen demand) COD and (biological oxygen demand) BOD demand, high content of organic carbon, presence of dissolved and suspended solids, and many other minute impurities which make it very difficult to conjure treatment techniques for its safe disposal (Noukeu et al., 2016). Among the methods applied for the treatment of food industry wastewater, the most commonly used are physical, chemical and biological treatments (Awasthi et al., 2018b; 2020). Out of these treatments, the biological methods have found immense importance because it is a simple, cost-effective, and environment-friendly method along with the ability to provide some useful by-products. Biological treatment of wastewater involves the use of biologically active microorganisms which act upon the waste effluent and utilize this waste as a carbon source for themselves. This process finally delivers a much more eco-friendly effluent that can be easily disposed off without any worry. Moreover, the action of microorganisms on the waste stream generally provides a byproduct such that it can be used for some other purpose, and therefore a well-suited waste management system is generated with the help of small microorganisms. Among the byproducts produces as a result of biological waste degradation the main ones are the production of biofertilizers, methane gas, bio-electricity and bio-fuel which have found great attention owing to the global energy crisis (Tekerlekopoulou et al., 2021).

Various methods of biological treatment of wastewater from the food industry have therefore gained importance since along with the capability of providing a clean effluent stream, the production of valuable by-products shows how beneficial this method can be keeping in mind the future of the world. The treatment of wastewater with active microbes has shown great potential in the cleansing of food industry effluents.

14.2 The Food Processing Industry

The raw food product obtained from the agricultural lands possesses many unwanted elements and entities that need to be removed or modified in order for that food product to be of value in the commercial market. The Food processing industry does this task by acting on the raw produce and transforming it into a more suitable and consumable form. The type of processing that is required for the food products obtained from the agricultural produce depends on the kind of food that is being dealt with. Since food is the main aspect of human life and is essential for our survival, the food processing industry has grown ever since there has been a surge in the population which puts pressure on this industry to provide quality food in large quantities.

The processing of food ranges from simple techniques of grinding of the grains in order to produce flour for bread making to complex purification techniques that are required to provide healthy and convenient food items. Processing of food also helps in improving the shelf life of the raw produce thereby making it more convenient for usage over longer periods of time. The steps involved in the processing of the food are carried out in three stages i.e. primary, secondary and tertiary processing techniques (Knorr and Watzke, 2019).

Primary processing typically involves the initial most basic processing of the raw agricultural produce into a consumable item. In this technique, the raw material like grains, whet kernels etc. are processed by removing their upper husk and providing material for flour making. This is one of the main examples where primary processing is utilized to provide a basic consumable item that can be moulded in any form according to the consumers' needs. Provision of flour for dough making, deboning of meat, extraction of oils, and providing milk, all fall under the primary processing units employed in the food industry. After the primary processing, the secondary food processing is done to produce consumable items utilizing the primarily processed food. Preparation of bread using flour, making milk products like cheese, ghee, etc., production of wine, beer, and other forms of alcohols from processed grains all fall under the secondary processing of food industry. Finally, the tertiary processing helps in the provision of totally refined and processed food which is generally used for manufacturing ready-to-eat food products that have been modified with preservatives to improve their shelf life (Lebovka et al., 2012).

The importance of this industry lies in the fact that it is the main reason that we are able to enjoy different varieties of food items which are free from any toxins and unwanted products. The food processing industry provides a fixed quality of consumables all over the world making it easier to rely on this industry for a healthy lifestyle. The food processing activities mainly rely on a constant and clean supply of water for the processing purpose. Water is utilized in food processing in multiple areas including heating, rinsing, washing, cooling, dissolving, diluting, generation of steam, and various other minute areas where utilization of water is essential. This water is then later on eluted from the industries as wastewater which contains many unnecessary impurities which must be dealt with before it can be safely disposed off.

14.3 Different Types of Consumables Undergoing Industrial Processing

The food processing industry is one of the largest industries in the world since it provides consumables for the ever growing population of the world. There are many different types of food items that are processed in the industry which are discussed further.

- **Processing of Fruits and Vegetables:** Fruits and vegetables constitute one of the main parts of the food processing industry. The processing of the fruits involves the simple steps of procurement of raw produce from agricultural land, separation of wanted leaves, stems, etc. by pruning, removal of excess pesticides, and then making it available for further preservative techniques for commercial usage. Tinned fruits and vegetables are one of the main outcomes of the food processing industry involving fruits and vegetables. The use of preservatives has been exercised extensively in order to improve the shelf life of pre-cut and frozen fruits and vegetables. Apart from canned fruits and vegetables, the processing also involves the manufacturing of fruit juice (very common in the health community), preparation of jams, and also as ingredients in certain cereals and baby food products. The wastewater generally generated from the processing of fruits and vegetables is involved in the cleaning of the produce, thus the presence of pesticides is very common such as wastewaters. The high BOD and COD of the effluents need to be controlled before their discharge (Sumonsiri and Barringer, 2014).
- **Grains:** Grains obtained from agricultural produce undergo various post-harvest treatments and processes in order for them to be consumed and made free of non-edible fibers. The processing typically involves the primary step where the grains are cleaned and sorted and their upper covering (husk/hull) is removed. The grains are further milled and then converted into a form which is considered to be edible for the consumers. In the secondary processing, the above-processed grains are further dried and provided with specific moisture content so that they can be stored for longer time periods (Raack and Risch, 2008). The end result of these processes is the production of flour, daal, and other grains which are used on a daily basis in healthy diets. The usage of water in multiple steps of grain processing diverts the attention toward its safe disposal.
- **Fats and Oils:** The extraction and refinement of different types of oils obtained from a variety of seeds. The processing of oils refers to their refinement so as to produce a consumable product with characteristics such as clear appearance, free from unwanted odor, stability from oxidation, and the ability to fry. The main steps involved in the processing of various fats and oils include lecithin removal, degumming, removal of free fatty acid, bleaching, and removal of odor which finally results in a consumable product. All these steps are water strenuous which indicates that this particular sector of the food industry generates a large amount of wastewater which is a subject of concern (O'brien, 2008).

- **Processing of Milk and Dairy Products:** Several steps are implemented in the procurement of milk and then further utilizing it for the production of milk-based products which include cheese, curd, butter, ice creams etc. Firstly, the milk obtained from the animal source is filtered to remove any solid contaminant present in the freshly squeezed milk. Further, the pasteurization of the milk is done to improve its shelf life and make it free from any microbes that might affect the quality of the milk. This milk is then further processed into products like cheese, whey protein, curd, butter etc. all of which involve a large volume of water (Farizoglu and Uzuner, 2011).
- **Meat, Poultry and Fish Processing:** Multiple steps are involved in the attainment of consumable and healthy fish and meat products from various animal sources. From procurement of different animals and fishes to their final packaging in frozen form, a large amount of water is used in the basic steps that involve cleaning, cutting, washing and further depending upon the type of meat being processed. The use of water in the meat and fish processing industry produces wastewater with high organic content and nutrients from released fats, blood, urine, remnants of meat, slurries as well as the presence of detergents and disinfectants used in the process of cleaning. Thus the water from such facilities has one of the highest COD and BOD since around 5 L of water is used overall for processing 1 kg of meat or fish. The poultry industry uses similar steps for the procurement of raw chicken and also eggs where extensive washing leaves behind wastewater which demands its proper disposal (Aziz et al., 2019).

14.4 Characteristics of Effluents from Food Processing Industries

The extent of treatment required for the wastewater generated from any industry depends on the basic characteristics of that wastewater. The character of wastewater is judged by checking various factors which affect its toxicity and contamination level. These characteristics mainly comprise pH of the water, temperature, biological oxygen demand (BOD), chemical oxygen demand (COD), total dissolved solids (TDS), nitrogen content, phosphorus content, amount of heavy metals present, salinity, and color of the discharged stream. The food processing industry particularly produces wastewater streams that are easily degraded with simple techniques and are mostly free from toxic components. However, the high concentration of BOD, COD, and suspended solids is what mainly affects the treatment system suitable for its effluents (Cristian, 2010). Moreover, the amount of wastewater produced by this industry is very high when compared to the other industries.

All the basic parameters which determine the treatment methods to be implied for food processing wastewater are discussed further (Abdallh et al., 2016).

14.4.1 pH of the Discharge Stream

pH is defined as the amount of hydrogen ions present in a given solution. The pH of the wastewater stream poses a direct influence on the treatment options available for that particular effluent making it a critical component that needs to be monitored before any treatment steps are finalized. The pH of the wastewater stream depends on the processing techniques used in the manufacture of the end products. The processes involving the utilization of weak acids and bases for cleaning, disinfecting, etc. produces varied pH levels of the effluent stream which need to be taken care of immediately. Washing and cleaning various kinds of agricultural products also leaves behind the presence of pesticides and fertilizers that are used in their production which further adds to the varying pH value of the effluents from different types of food processing industries (Noukeu et al., 2016).

14.4.2 Temperature

The wastewater produced from the food processing industry generally comprises different ambient temperatures depending on the type of product being processed. For canning and freezing processed food products, the use of typically chilled water encompasses a stream of waste effluent with very temperatures. However, the processes involving boiling of the raw products to obtain a significant end product require water treatment at a very high temperature which gives off wastewater at high temperatures. This difference in the temperature profiles of the waste effluents affects the process that is used for its treatment. The use of biological wastewater treatment for food industry effluents is mostly affected since the types of bacteria and other microorganisms that are used in the treatment are affected by the temperature of the effluents (Noukeu et al., 2016). Most of the microorganisms with the highest activity at moderate temperatures tend to lose their degrading characteristics at extreme temperatures which hinder with wastewater treatment. It, therefore, is important to keep a check on the effluents temperature in order for the treatment process to work efficiently.

14.4.3 BOD (Biological Oxygen Demand)

BOD refers to the biological oxygen demand which gives an idea about the amount of oxygen that will be used by the microorganisms for the breakdown of organic substances in water. BOD is the most common constituent of the food industry wastewater. The different organic impurities which are present in the food industry wastewater include fecal matter, fats, oils, food particles, grease, detergents, and soaps. The breakdown of this organic matter present in the wastewater streams depends on the ability of microbes which feed upon this matter and convert it into simpler forms and finally into CO_2 and water by utilizing dissolved oxygen for this purpose. The amount of oxygen required by the microbes for this process is referred to as the Biological Oxygen demand. BOD has been used to judge the strength of the wastewater effluents since wastewater high in BOD can result in the depletion of the dissolved oxygen in the receiving systems and cause the killing of the natural habitat of such waters (Noukeu et al., 2016).

14.4.4 COD (Chemical Oxygen Demand)

The chemical oxygen demand is a parameter that is used to measure and determine the amount of total oxygen required by a strong oxidant (chemical compound) to break down organic as well as inorganic matter present in the wastewater. COD helps in representing the overall pollution level of the wastewater effluents and is, therefore, an important factor in its characterization. The higher is the value of COD the more complex it is to treat that particular stream of wastewater.

14.4.5 Total Suspended Solids

Food industry wastewater mostly contains large quantities of suspended solids owing to the release of organic and inorganic food-based wastes like hair and feathers from the meat industry, chunks of fats, feces, etc. which hinders the disposal of the waste stream. The presence of these solids is measured as total suspended solids (TSS) and this term is expressed in mg TSS per liter of wastewater. Primarily, TSS is objectionable because of the hindrance caused by them in wastewater purification by getting stuck in most of the equipment during the primary screening of the effluents. Furthermore, if disposed of without removal of these suspended solids, this wastewater can get trapped within the pores of the soil where it is disposed and hinders the leaching of the water within the soil (Noukeu et al., 2016).

14.4.6 Nitrogen Content

The presence of nitrogen in the food industry wastewater is dependent on the amount of organic matter released from the effluent streams (fecal waste from the meat and poultry industry) and the release of certain detergents and cleaning agents which contain nitrogen. The presence of nitrogen causes eutrophication in natural water where these effluents might be disposed off. Furthermore, nitrogen can get converted to nitrate in presence of certain bacteria which is very harmful to humans causing diseases like methaemoglobinaemia.

TABLE 14.1

Concentration/Range in the Food Processing Industry Waste

Parameter	Concentration/Range	Allowed limits
pH	4–8	7
Temperature (°C)	0–37	20–35 (optimal for most microbes)
Biological oxygen demand (BOD) (mg/L)	700–900	10 (fresh water) 300 (sewage water)
Chemical oxygen demand (COD) (mg/L)	1,336	120
Total suspended solids (mg/L)	300	100
Nitrogen (mg/L)	Variable depending in the food processed	10
Phosphorous (mg/L)	Variable depending in the food processed	0.1
Color and Odor	Milky, grey, maroon/pungent	Normal clear color/odorless

14.4.7 Phosphorous Content

The food industry waste contains bound forms of phosphorous which pollute the wastewater stream originating from these industries. The phosphorous is typically converted to orthophosphates which is the main form of phosphorous found in the food industry wastewaters. Certain detergents used for cleaning purpose and also the presence of phosphorous-based fertilizers further adds to the amount of phosphorous present in the food processing effluents. The presence of phosphorous renders the receiving stream of water unfit for drinking and also causes the problem of eutrophication of surface waters thereby attaining attention for its removal.

14.4.8 Color and Odor

The color and odor of the effluents released from the food processing industry owing to the processing of various food products like meat, which releases blood in the water used for its cleaning, processing of fruits for juice preparation, processing of milk and milk products etc. which hinders with the safe disposal of these effluents into a receiving stream. The presence of color and odor affects the aesthetic appearance of the wastewater and therefore needs to be cleaned and treated from the food industry wastewaters.

The following table gives an idea of the concentration of these different contaminants generally found in the food processing industry wastewater (Table 14.1).

14.5 Treatment of Food Processing Wastewater

After a clear understanding of the basic characteristics of the food processing industry wastewaters, the next step is to evaluate the desirable techniques which can effectively render the effluents free from most of the contaminants. A clear decision on the treatment process enables the selection of the most economic and efficient process for the cleaning of the wastewaters. The first and foremost step in any wastewater treatment process is the physical treatment of the water to get rid of any settling or floating contaminants so as to produce a constant flowing stream. The physical treatment of the wastewater gets rids of the suspended solids, fats, oils etc. which hinder further treatment of these effluents. The basic steps involved in the physical wastewater treatment are discussed below (Tekerlekopoulou et al., 2021).

14.5.1 Physical Treatment

The first and foremost step in the physical treatment of any wastewater effluent is the screening of the effluent. The screening refers to the physical removal of the large suspended solids in the effluents by making them pass through a course screen depending on the size of the solid particles. Large particles from the food processing industry mainly hair, feathers, chunks of meat, dust particles, leaves twigs etc. can be easily removed with this primary method of screening. This is an important step in the treatment process since the presence of these solid particles tends to clog and hinder other equipment used further in the treatment. The size of the screens varies with the size of the suspended particles that are to be removed (Tekerlekopoulou et al., 2021).

After the initial screening, the next basic step in physical treatment is the floatation process which is used to remove any floating liquids like fats, oils, grease etc. which have low densities than that of the water itself. Skimmers are used in the flotation tanks which help in the recovery of oils, fats, etc. from the wastewater and provide a more streamlined effluent for further processing.

Further treatment is done with the help of membrane filters which can separate and concentrate any leftover dissolved and un-dissolved contaminants. This process is carried out under specific pressure and the wastewater is made to pass through the membrane filter with a specific pore size. After these basic treatments, the final settlement of the dissolved solids is done in a settling tank which makes use of the earth's gravity to settle down any dissolved solids having weight greater than that of the effluent water (Sincero and Sincero, 2002).

14.5.2 Chemical Treatment

After the steps of the physical treatment are employed for the basic clarification of the wastewaters from food industry, the next step is to chemically treat the effluent so as to provide a more susceptible stream for further biological processes. The first and foremost step in the chemical treatment plant is to neutralize the pH of the effluents. The addition of weak acids/bases depending upon the pH of the effluent stream is done so as to bring the pH around 7 which is most suitable for performing any further treatment processes involving bacterial degradation since most of the microbes work best at a neutral pH value.

Further, for the treatment of wastewaters, coagulants are used which generally render the effluent stream free of dissolved impurities possessing a particular charge. In general, the coagulants used to possess a positive or negative charge depending on the type of contaminant to be separated out. The desired contaminant gets attracted to a coagulant particle with an opposite charge and settles down with it. Aluminum sulfate and aluminum chloride are the two most commonly used coagulants for wastewater treatment. Removal of phosphorous-containing impurities generally employs this method for their removal.

The chemical treatment further makes use of the ion exchange technique for the removal of certain impurities like calcium and magnesium which impart hardness to the effluent waters. The addition of sodium salts, generally sodium chloride or brine is used for the softening of the water. The exchange between the sodium ions and calcium/magnesium ions takes place and causes the softness of water. Further, certain disinfectants are used in order to supply a more sterile feed for the processes to be used later on for wastewater treatment which makes use of chemical treatment for this purpose (Rao et al., 2012).

14.5.3 Microbial Treatment

The process of microbial treatment of the food industry wastewater comes into play when the organic matter in the effluent stream is to be degraded into simpler and environment-friendly by-products. The basic mechanism of the microbial treatment is that the organic contaminants which are found in large quantities in the food industry wastes are utilized by different microorganisms which feed on this organic matter as their carbon source and convert it into a more suitable by-product after degradation. The degradation mechanism helps in providing energy to the microorganism by utilization of organic matter as a carbon source. The amount of degradable matter present in the wastewater can be analyzed with the help of certain tests that evaluate the amount of organic degradable carbon available in the effluents (Bories, 2000). The most common test is the BOD_5 days test which determines the amount of oxygen uptake over a period of 5 consecutive days by a consortium of seed bacteria confined in a dark container along with the wastewater. In the 5 days period, the organic carbon is utilized by the bacterial consortium and the consequent decrease in the total dissolved oxygen in the wastewater is studied which is the result of bacteria utilizing the organic carbon for energy and respiration (Cecconet et al., 2018).

Since the food processing industry mainly gives away the effluents with typically high organic contents, the use of the microbial treatment of its wastewater has shown strides of advantages. This biological treatment process also paves the way for utilizing the wastewaters for the production of various valuable by-products of great importance thus providing a method to convert waste to value.

The two main strategies employed for biological wastewater treatment with the help of microorganisms is the anaerobic treatment and aerobic treatment. The anaerobic treatment is typically used for the degradation of organic matter in the absence of oxygen. This step considerably reduces the stress on the aerobic microorganisms for further degradation of the organic matter. It is therefore used prior to the aerobic treatment of the wastewater thereby rendering a more easily degradable stream of effluent. One important advantage of using this technique for treating the wastewater is the production of methane gas as a byproduct. The two methods of anaerobic digestion and aerobic digestion are discussed further in detail (Chowdhury et al., 2010).

14.5.3.1 Anaerobic Digestion

The process of anaerobic digestion aims to reduce the stress of wastewaters on the environment by providing a reduction in its hazardous effects on the biosphere. Many different anaerobic microorganisms (those microorganisms which survive without exposure to any oxygen) mutually act together to decompose complex organic matter in the wastewaters into stable and environment-friendly byproducts mainly CO_2 and methane. The microorganisms belonging to four different groups namely fermentative, syntrophic, acetogenic and methanogenic usually cooperate in the fulfillment of this anaerobic digestion process (Lim et al., 2020).

One of the main advantages of employing this method of treatment for the clarification of the wastewater from the food industry is the production of biogas which is a high energy bio-fuel which provides a great alternative to conventional fuels as it is more environment friendly. Biotechnology of biogas production usually refers to the digestion of various types of organic wastes from food industry wastewater mainly from sources like slaughter house, processing of fruits and vegetables and breweries (Pantea et al., 2011).

The process of anaerobic digestion can be anaerobic digestion process can be theoretically divided into four main phases namely hydrolysis, acidogenesis, syntrophic acetogenesis, and methanogenesis (Lv et al., 2010). The overall mechanism of anaerobic digestion is simplified in the figure below (Figure 14.1). Each phase in the sequence requires a set of different microorganisms for the final product to be produced. The primary phase consists of the hydrolysis phase where the organic contaminants in their raw form (carbohydrates, lipids and proteins) are acted upon by the hydrolytic microorganisms. These microorganisms, in presence of water, utilize their extracellular enzymes such as cellulases, pectinases lipases, xylanase, proteases, and amylases for the breakdown of different types of organic wastes and hydrolyze them into simple monomeric and oligomeric compounds including sugars like glucose/xylose etc. from carbohydrates, amino acids from proteins, and long-chain fatty acids/glycerol from lipids. Although the hydrolytic microorganisms are diverse in nature, the main ones belong to the phyla of *Firmicutes* and *Bacteroides*. The hydrolytic bacterias are generally fast growing and are not sensitive to the changes in the environmental conditions and characteristics of the effluent streams.

The products obtained from hydrolysis are further fermented to small chain fatty acids where volatile fatty acids predominate in the products. Production of carbon dioxide, ammonia, hydrogen is also common during this stage of the reaction. Microorganisms belonging to the phyla *Bacteroidetes*, *Firmicutes*, *Chloroflexi*, *Atribacteria*, and *Proteobacteria* are typically involved in the acedogenesis phase of the anaerobic treatment. Acidogenesis can prove to be a rate-limiting step in the anaerobic treatment process since the accumulation of fatty acids can lead to a sharp pH drop hindering with further processes.

Acetogenesis then takes place where the fatty acids are broken down to give acetate, propionate, hydrogen (H_2), and CO_2. These byproducts are then broken down by the action of methanogens for the

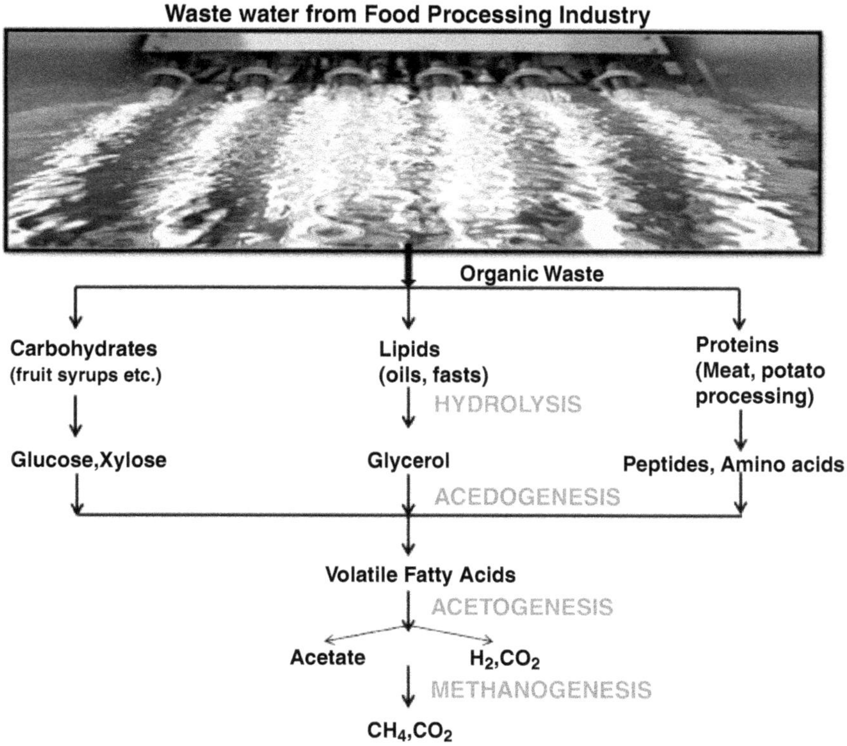

FIGURE 14.1 Basics of the anaerobic treatment process.

specific purpose of biogas production. Methanogens belong to a specialized group of archae which are involved in the process of Methanogenesis. In the anaerobic bioreactor, around two third of the methane production takes place by utilizing acetate and the rest is produced from H_2 and CO_2. In comparison to the other phases, the methanogenesis phase is the slowest since the methanogens grow the slowest and their sensitivity to the environmental changes is very high.

Many of the different microorganisms that are involved in the anaerobic treatment have been isolated and cultured for their biochemical and morphological characterization. After intense studies, the species belonging to the acetogenic and hydrolytic bacteria have been assigned taxonomically to the genus of *Acetivibrio*, *Bacteroides*, *Clostridium*, *Thermotoga* and *Ruminococcus* (in the phylum *Thermotogae*). The acidogens belonging to the Non-hydrolytic family have been assigned to the genus *Bifidobacterium* (in the *Actinobacteria* phylum), the genus of *Lactobacillus*, *Anaerolineaceae* (in the *Chloroflexi* phylum), and also some non-hydrolytic bacteria thermophilic in nature were found to be belonging to the phylum of *Thermotogae*.

Certain commonly occurring hydrogenotrophic methanogens which have been found to be useful in the anaerobic treatment of wastewater include the species belonging to the *Methanoculleus*, *Methanobrevibacter*, *Methanobacterium*, *Methanothermobacter* and *Methanospirillum* genus. On the other hand, acetotrophic methanogens have been only found in the *Methanosaeta* genus which comprises microorganisms which are obligate acetoclastic methanogens and *Methanosarcina* genus belonging to the facultative acetoclastic methanogens.

The genus *Methanosaeta* can only utilize acetate as its substrate and has a very slow growth period. Its affinity for its substrate i.e. acetate is however very high therefore in the presence of the substrate, it grows and propagates very swiftly. On the other hand, *Methanosarcina* species are able to utilize a vast range of substrates which include methanol, H_2, methylamine, CO_2, and methyl sulfides in addition to acetate. Unlike *Methanosaeta*, the species of *Methanosarcina* are able to grow fast whereas they have a low affinity for acetate and can grow well when the concentration of acetate substrate is high (Barua and Dhar, 2017).

TABLE 14.2

Improved Wastewater Characteristics after Anaerobic Treatment

Parameter	Influent Concentration/Range	Concentration after Anaerobic Treatment
pH	4–8	6.9–7.005
Temperature (°C)	0–37	30 (optimal for most microbes)
BOD (mg/L)	700–900	300–400
COD (mg/L)	1,336	400–600
TSS (mg/L)	300	100
Nitrogen (mg/L)	Variable depending on the food processed)	7–8
Phosphorous (mg/L)	(Variable depending on the food processed)	2–3
Color and odor	Milky, grey, maroon/pungent	Light color/improved odor

Owing to the improved technical instrumentation, some new strains belonging to the hydrolytic/acidogenic bacteria e.g. *Clostridium bornimense*, *Herbinix luporum*, *Herbinix hemicellulosilytica*, *Herbivorax saccincola*, *Petrimonas mucosa*, *Proteiniphilum saccharofermentans*, *Fermentimonas caenicola*, and *Proteiniborus indolifex*, have been identifying and cultured successfully. New species from methanogenic archaea, i.e. *Methanosarcina flavescens and Methanobacterium aggregans* have also been reported (Hassa et al., 2018).

The basic improvement in the characteristics of the effluent stream after the anaerobic treatment has been specified in the table below (Table 14.2).

14.5.3.2 Aerobic Treatment for Wastewater

The aerobic treatment of wastewater is a biological technique for the clarification of effluents where organic matter, as well as nitrogen and phosphorous present in the wastewater, are broken down in the presence of oxygen. Aerobic treatment converts the organic biomass in the wastewater to CO_2 and new biomass.

The aerobic treatment process can be used alone as a clarification process for the treatment of raw wastewater but it is generally used after the anaerobic process for further polishing the effluents and providing a clean stream that can be safely disposed off into any receiving stream of water body. This process further removes the BOD and the suspended solids from the wastewaters and also the nutrients like nitrogen and phosphorous. This treatment process is generally more rapid in its degradation process in comparison to the anaerobic treatment and has proved to be very successful for the treatment of the effluents coming out of the food processing industries.

- **The Conventional Technique of Activated Sludge:** The activated sludge treatment process makes use of oxygen-loving microorganisms (aerobes) to breakdown the organic content and also the nutrients (nitrogen, phosphorous) present in the wastewaters into environment-friendly byproducts such that this water can be easily disposed off without any pollution causing a disturbance.

  The main criteria for maintaining the aerobic conditions and to keep the activated sludge suspended in the reaction is to maintain the supply of oxygen at all times. Seed bacteria along with some other microorganisms which work in mutualism for the breakdown of the organic content of the wastewater are suspended and then mixed with the effluents in an aeration tank wherein the presence of the oxygen provided, these microorganisms feed and grow on the organic content and transform it into the water, CO_2, energy and new cell materials. A secondary clarifier is then used for the removal of the suspended activated sludge by making it settle by gravitational force and it can be again recycled back to the main aeration tank for another round of wastewater clarification.

  There are mainly four different groups of microorganisms that are found to be majorly activated during the aerobic treatment of wastewater. These microorganisms are discussed further.

- **Bacteria:** Aerobic bacteria help in the removal of the organic nutrients from the wastewater effluents (Yongkui et al., 2020). Many bacterial species have been identified and testified as important species for the activated sludge process. According to the studies, *Proteobacteria, Acidobacteria, Saccharibacteria, Bacteroidetes, Chloroflexi, Nitrospirae* and *Planctomycetes* have been identified as the most common bacterial species present in the aerobic treatment of wastewater (Liu et al., 2017). Among these species, *Proteobacteria* was found to be the most dominant phylum with total concentration of around 52% in average activated sludge reactors (Vasiliadou et al., 2018). Furthermore, *Chloroflexi* and *Nitrospirae* were found to be the key species in the bacterial phyla for denitrification and nitrite oxidation in activated sludge (Qin et al., 2018). Some other phyla abundant in the activated sludge process were *Chlamydiae, Chloroflexi, Chlorobi, Elusimicrobia, Ignavibacteriae, Spirochaetae, Latescibacteria,* and *Parcubacteria.*
- **Protozoa**: The main aim of the protozoa in the activated sludge process is to digest and remove the dispersed and non-flocculating bacteria and suspended particles. The most common Ciliates are the dominant protozoa in activated sludge, and those which either attach themselves to or crawl over the surface of the sludge are the most common. Without ciliates, highly turbid, low-quality effluents are obtained. The predatory activities of the ciliates upon the dispersed growths of bacteria are responsible for clarification and for the reduction of coliform numbers during the activated-sludge process (Curds, 1982).
- **Metazoa**: Metazoa are multi-cellular organisms and are common in the activated sludge treatment process. These are larger in size when compared to protozoa and are not involved in the removal of organic waste from the wastewater. They are however responsible for the maintenance of the bacterial species since they feed on the bacteria, protozoa, and algae. The presence of metazoa indicates the conditions of the treatment system.

  The three types of metazoan most commonly found in the system for activated sludge treatment are as follows.
  - **Rotifers**: They are involved in the clarification of effluents and are first affected by the amount of toxic load present in the effluents. They can therefore give an idea about the toxicity level of the wastewater stream.
  - **Nematodes**: They feed on the flocculating bacteria, protozoa, fungi, and other nematodes and are responsible for providing a more clear effluent from the industrial outlet.
  - **Tardigrades (Water Bear)**: These are also involved in the consumption of bacteria, protozoa, fungi, and other nematodes and can survive in extreme environmental climate.
- **Algae and Fungi**: Algae and fungi generally represent pH changes and are generally present in the older sludge.

  Apart from the activated sludge process, the aerobic treatment of the wastewater is also done with manipulations of the above strategy in order to enhance the output of the activated sludge process. Such strategies are based on the same basic mechanism of utilizing the aerobic microorganisms for the breakdown of the organic matter and nutrients into simpler forms but with the manipulation that helps enhance the process of providing a much clearer effluent. One such technique used is the moving bed biofilm reactor which involves the growth of the microbial flock on a plastic carrier and that carrier is suspended in an aeration tank where the wastewater is supplied. These reactors provide the benefit of removal of more nitrogen from the effluents by enhancing the nitrification and de-nitrification steps in a voluminous system.

  After the aerobic treatment, the final effluent that is obtained is 80% free of the initial contaminants and such effluent can be safely disposed off into any receiving stream or for irrigation of the agricultural lands.

14.5.3.3 Micro-Algal Based Process

Recently enough, the role of micro-algae in the treatment of food industries wastewaters has been emphasized because of their ability to successfully assimilate significant amounts of nutrients including nitrogen, phosphorous, and also are able to reduce the COD from the wastewaters. Many researchers nowadays propose the cultivation of microalgae using food industry wastewaters as a cost-effective technique for its treatment (Cheng et al., 2013; Di Caprio et al., 2018).

An imperative factor for the cultivation of algal biomass is to provide enough light for the growth of the algae so that they can successfully perform photosynthesis. A variety of pretreatment methods like sedimentation, centrifugation and filtration are suggested to reduce the turbidity and viscosity of the wastewater for the growth of the microalgae. Though the percentage of nutrients like phosphorous and nitrogen can be effectively removed from food processing wastewaters by the micro-algal biomass (*Chlorella* species in dairy industry, *Scenedesmus obliquus* in olive mill effluents), the removal of BOD and COD needs bacterial colonies for their efficient removal. Mixed cultures of bacteria and microalgae are suggested to reduce the cost of cultivation of algae and increase the contaminant removal rates. In such types of cultivation, oxygen is provided to bacteria by the algal biomass through the process of photosynthesis and the carbon dioxide released from the oxidation of organic matter is utilized by the microalgae for their growth.

14.6 Role of Microorganisms in Wastewater Treatment

The removal of organic matter and nutrients from the wastewater depends on the ability of the microorganisms to degrade the contaminants and utilize them for their growth and multiplication. The organic matter is broken down and used for metabolism when ingested by microorganisms. Metabolism refers to the chemical reactions taking place inside an organism which is responsible for the provision of energy to the organism. The metabolism involves enzymatic action on the specific substrate (organic matter in this case), which is converted into a corresponding product to complete the series of chemical reactions involved in the metabolism chain (Davies, 2005).

Two major divisions of metabolism that are involved in the wastewater treatment by the microorganisms are Catabolism and Anabolism. Catabolism is referred to as the energy transfer metabolism where carbon-containing compounds are broken down in order to yield energy. This process involves the utilization of oxygen by the microorganism and is a biological oxidation process. Anabolism on the other hand involves a chain of reactions where small molecules undergo attraction to become larger molecules where the energy produced from the catabolic reactions is utilized for the growth of the microorganism.

The three significant processes that are involved in the metabolism of bacteria are ingestion, respiration, and growth for the division. The ingested carbon from the wastewater organic contaminants of the food processing industry follows catabolic respiration where some of the carbon dioxide produced is lost to the surroundings whereas most of it is used to provide energy for anabolic reactions to take place and helps in growth and new biomass production. In order for the bacteria to utilize the carbon content and other nutrients from the wastewater, they require the secretion of a particular enzyme to digest this stuff. Different bacterial species secrete different enzymes for this purpose depending on the source of carbon contaminant. The presence of a particular substrate only can trigger the enzyme production corresponding to that substrate (Davies, 2005).

Though the requirement of carbon as a substrate is the main criterion, certain amounts of nitrogen and phosphorous are also important for bacterial growth and degradation of effluents to take place. The ratio of C: N: P of 100:17:5 has been reported to be beneficial in bacterial wastewater degradation. The following diagram depicts the three main processes used by bacteria to digest wastewater (Figure 14.2).

14.7 Recovery of Value Added Products after Treatment

The amount of wastewater produced from the food processing industries measures to about millions of tons per year. The need to treat this wastewater has been discussed in the chapter above. It is considered to be beneficial to the environment if this wastewater is put to some use after it has been treated with different techniques. This approach is based on the fact that something useful can always be extracted from wastes (Bories, 2000). Therefore, seeing the amount of wastewater generated from the food processing industries, it has been suggested that after its treatment, this recycled water can be used for the production of value-added products.

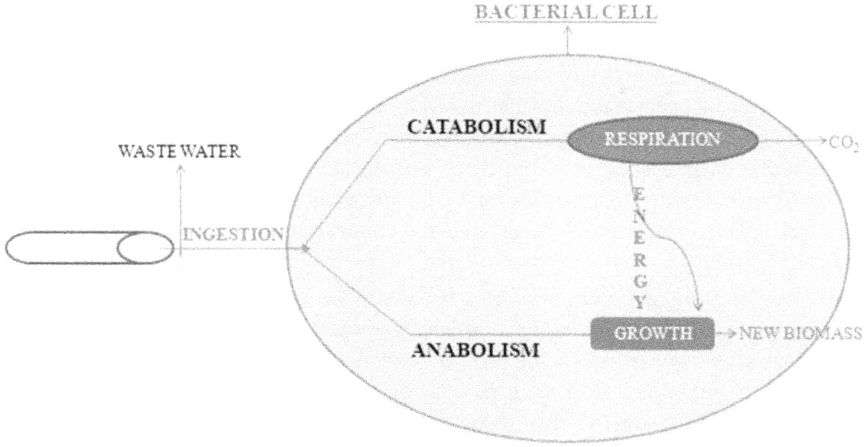

FIGURE 14.2 Bacterial digestion of wastewater through various processes.

The extraction of small amounts of nutrients from the wastes of food processing industries has gained much attention recently. These nutrients generally referred to as bioactive compounds have a beneficial role as additives in the human body and can be used for the treatment of various diseases like cancer. The bio-wastes from food industry wastewaters can be efficiently converted to value-added products instead of their being wasted away by disposal. The role of microbial metabolism in the biotransformation of organic wastes has been discussed above. The resultant products after metabolism of the wastewaters possess many biotechnological applications as pharmaceutical preparations in the form of bioactive compounds. Some of these bioactive compounds are discussed below which are easily extracted from many food processing industries.

- **Phenolics from Olive Mill Wastewaters:** The processing of olive fruit generally leads to the production of wastewaters which contains high amounts of several phenols as bioactive compounds along with some other organics like pectin, and insoluble dietary fibers. The main attraction for the researchers has however been the extraction of different phenolics from the olive mill wastewaters because of their antioxidant properties. The recovery of phenols from olive mill wastewaters has applications in being added into different fortified food preparations (vegetable oils, bakery products, milk products etc.) and also in the cosmetic industry for the preparation of sunscreens. Among all these applications, the use of phenols as additives in food products has proved to be beneficial not only because of its antioxidant properties which help in reducing oil peroxidation during storage and cooking but also because of its antibacterial properties which help in providing a longer shelf life to the perishable items. The extraction is done with the help of ultrasound or microwave-assisted techniques or sometimes pressurized liquid extractions (Tapia-Quirós et al, 2020). Membrane filtration technologies have also been employed for phenolic extraction where micro and nanofiltration operations are performed for the extraction of phenols from wastewaters (Bazzarelli et al., 2016).
- **Pectin from Fruit Processing Industries:** The release of pectin as a bioactive compound in the wastewaters of fruit processing industries is very common. Pectin is a polysaccharide fiber found in the extracts of fruits and vegetables after they have been processed in the industries. Processing of citrus fruits like lemon, oranges, grapefruit, mosambi, etc. results in the generation of wastewaters which can be exploited for the recovery of pectins along with flavonoids and limonene. The use of pectin as a gelling agent is very common in the production of jams and jellies. The extraction of pectins from wastewater is also useful in the preparations of certain pharmaceutical and cosmetic industries where they are used as thickening agents (Waldron, 2009).

- **Carotenoids:** Carotenoids are one of the main food antioxidants which are involved in the establishment of pharmaceutical compositions relating to decrease the risk of certain chronic diseases (Desmarchelier and Borel, 2017). Wastewater from the peach processing industry contains a high amount of carotenoids which can be extracted by ethanol by sedimentation. Carotenoids including β-carotene, zeaxanthin, cryptoxanthin and lutein are easily extracted in large quantities from these wastewaters (Vargas et al., 2016). Carotenoids from tomato processing wastewaters have also been known to possess antioxidant characteristics which can minimize the health risks in cancer patients caused by oxidative damage (Valta et al., 2017).
- **Organic Acids:** Many different organic acids which include lactic acid, citric acid, ascorbic acid, malic acid, fumaric acid, and acetic acid have also been extracted from different food industries' wastewaters. Recovery of tartaric acid from wine industry wastewater has been performed with cation exchange resin for its recovery (Kontogiannopoulos et al., 2017). The transformation of lactic acid into different bioactive compounds like acetaldehyde, acrylic acid, propionic acid etc. has great application in the food, cosmetic, and textile industries. Lactic acid is generally extracted from the wastewaters of starch-rich food products with the help of the action of enzymes like amylase, protease and amyloglucosidase (Ye et al., 2008).
- **Proteins:** The extraction of proteins from the industries processing components with high protein content has also been proved beneficial. The extraction of soy proteins from the wastewaters of soybean processing plants has been exploited for the production of this bioactive compound. The process of separation of foam and also the technique of ultra-filtration along with other membrane filtration techniques have been used for the recovery of proteins from wastewater of soy industry. This extracted protein has benefits in providing high protein diets formulations (Li et al., 2014). Soy protein can also be used in active packaging of food products because of its antioxidant properties.

  The extraction of protein from the sweet potato processing wastewaters (sporamin protein) has been shown to possess antitumor activity and can be exploited in pharmaceutical applications (Jones and Gersdorff, 1931).

14.8 Challenges Faced in Effluent Treatment

The main challenge faced nowadays for the treatment of wastewater from food processing industries is the amount of water produced. Ever since a surge in population, the amount of food generated from different industries has increased exponentially along with the production of millions of tons of wastewater. The disposal of such effluents even after treatment has become a topic of concern because of the lack of options for their disposal. The economical aspect when kept in mind for the treatment of wastewater poses the main challenge in effluent treatment. Many different techniques that are employed in wastewater treatment require heavy machinery which costs fortunes considering the variety of methods applied. The techniques are different for effluents from different processing units which further poses the challenge of carefully recognizing the characteristics of the effluents before starting the treatment. Furthermore, the cost for wastewater treatment is added if the extraction of bioactive and value-added products is to be obtained from the treated water. The appropriate treatment selection also affects the quality of the water supply after treatment. The treatment process should successfully remove all the undesirable components of the wastewater and provide standard grade water either for disposal or reuse.

Apart from the economical concern, the next major concern when talking about wastewater treatment is the amount of energy consumed for this purpose. Huge plants that are set up for food industry wastewater clarification consume an enormous amount of energy each day in the form of electricity which further adds to the burden on the environment. It has therefore become important to first judge the wastewater characteristics and then perform any treatment in a step-wise manner to reduce any kind of cost surge in the whole process.

14.9 Conclusion and Future Perspective

The food processing industry is a water strenuous industry producing millions of tons of wastewater every year. With the increased demand for processed food products for consumption, the rate of wastewater generation is increasing which requires a stringent check before it can be disposed off efficiently without any environmental pollution problem. The food processing industries' wastewater depends on the type of food product being processed and the treatment of such waters is dependent on the basic characteristic of the wastewater. Typically, the amount of organic contaminants is high in such waters which leads to high BOD and COD content in these effluents. A combination of physical, chemical and biological treatment is applied to the wastewaters for their most effective clarification. The most significant among these techniques is the biological treatment which involves the use of aerobic/anaerobic microorganisms for degrading the organic content and certain nutrients present in the wastewater. The microbes are able to effectively utilize the organic contaminants from the wastewater as their source of carbon for metabolism and growth purpose giving out environment-friendly byproducts in return.

The ability to attain bioactive compounds from such wastewaters of food processing industries has recently been exploited to turn waste into value. This approach is still weakly executed and can gain more attention in near future owing to the exponential growth in the amount of wastewater that is produced. The main concern that is of importance is the maintenance of a strict quality standard of these extracted bioactive compounds which needs to be set for their use in various applications. The future holds a great responsibility to lessen the burden we are executing on the environment by combining various techniques together and finally providing a cost-effective and environment-friendly solution to this growing problem of wastewater disposal.

REFERENCES

Abdallh M.N., Abdelhalim W.S., Abdelhalim H.S., 2016, Industrial wastewater treatment of food industry using best techniques. *International Journal of Science and Engineering Invention*, 5(8), 15–28.

Awasthi M.K., Wong J.W., Kumar S., Awasthi S.K., Wang Q., Wang M., Ren X., Zhao J., Chen H., Zhang Z., 2018b. Biodegradation of food waste using microbial cultures producing thermostable α-amylase and cellulase under different pH and temperature. *Bioresource Technology*, 248, 160–170.

Awasthi S.K., Joshi R., Dhar H., Verma S., Awasthi M.K., Varjani S., Sarsaiya S., Zhang Z., Kumar S., 2018a. Improving methane yield and quality via co-digestion of cow dung mixed with food waste. *Bioresource Technology*, 251, 259–263.

Awasthi S.K., Sarsaiya S., Awasthi M.K., Liu T., Zhao J., Kumar S., Zhang Z., 2020. Changes in global trends in food waste composting: Research challenges and opportunities. *Bioresource Technology*, 299, 122555.

Aziz A., Basheer F., Sengar A., Khan S.U., Farooqi I.H. 2019, Biological wastewater treatment (anaerobic-aerobic) technologies for safe discharge of treated slaughterhouse and meat processing wastewater. *Science of the Total Environment*, 686, 681–708.

Barua S., Dhar B.R., 2017, Advances towards understanding and engineering direct interspecies electron transfer in anaerobic digestion. *Bioresource Technology*, 244, 698–707.

Bazzarelli F., Piacentini E., Poerio T., Mazzei R., Cassano A., Giorno L., 2016, Advances in membrane operations for water purification and biophenols recovery/valorization from OMWWs. *Journal of Membrane Science*, 497, 402–409.

Bories A., 2000, A novel bio-physical way for the treatment and volarisation of wastewaters from food industries. In: *Proceedings 1st World Conference on Biomass for Energy and Industry*, Seville, Spain, pp. 5–9.

Cecconet D., Molognoni D., Callegari A., Capodaglio A.G., 2018, Agro-food industry wastewater treatment with microbial fuel cells: Energetic recovery issues. *International Journal of Hydrogen Energy*, 43(1), 500–511.

Cheng H., Tian G., Liu J., 2013, Enhancement of biomass productivity and nutrients removal from pretreated piggery wastewater by mixotrophic cultivation of *Desmodesmus* sp. CHX1. *Desalination and Water Treatment*, 51, 7004–7011.

Chowdhury P., Viraraghavan T., Srinivasan A., 2010, Biological treatment processes for fish processing wastewater–A review. *Bioresource Technology*, 101, 439–449.
Cristian O., 2010, Characteristics of the untreated wastewater produced by food industry. *Analele Universităţii din Oradea, Fascicula: Protecţia Mediului*, 15, 709–714.
Curds C.R., 1982, The ecology and role of protozoa in aerobic sewage treatment processes. *Annual Reviews in Microbiology*, 36(1), 27–28.
Desmarchelier C., Borel P., 2017, Overview of carotenoid bioavailability determinants: From dietary factors to host genetic variations. *Trends in Food Science & Technology*, 69, 270–280.
Di Caprio F., Altimaria P., Iaquaniellob G., Torob L., Pagnanellia F., 2018, *T. obliquus* mixotrophic cultivation in treated and untreated olive mill wastewater. *Chemical Engineering*, 64, 625–630.
Earle M.D., Earle R.L., 1997, Food industry research and development. In: Wallace L.T., Schroder W.R. (Eds.) *Government and the Food Industry: Economic and Political Effects of Conflict and Co-Operation.* Springer, Boston, MA. https://doi.org/10.1007/978-1-4615-6221-4_8.
Farizoglu B., Uzuner S., 2011, The investigation of dairy industry wastewater treatment in a biological high performance membrane system. *Biochemical Engineering Journal*, 57, 46–54.
Hassa J., Maus I., Off S., Puhler A., Scherer P., Klocke M., Schluter A., 2018, Metagenome, metatranscriptome, and metaproteome approaches unraveled compositions and functional relationships of microbial communities residing in biogas plants. *Applied Microbiology and Biotechnology*, 102, 5045–5063.
Jones D.B., Gersdorff C. E. F., 1931, Ipomœin, a globulin from sweet potatoes, *Ipomœa batatas* isolation of a secondary protein derived from ipomœin by enzymic action. *Journal of Biological Chemistry*, 93(1), 119–126.
Knorr D., Watzke H., 2019, Food processing at a crossroad. *Frontiers in Nutrition*, 6, 85.
Kontogiannopoulos K.N., Patsios S.I., Mitrouli S.T., Karabelas A.J., 2017, Tartaric acid and polyphenols recovery from winery waste lees using membrane separation processes. *Journal of Chemical Technology & Biotechnology*, 92(12), 2934–2943.
Lebovka N., Vorobiev E., Chemat F., (Eds.), 2012, *Enhancing Extraction Processes in the Food Industry*. CRC Press, Boca Raton.
Li R., Wu Z., Wang Y., Liu W., 2014, Pilot study of recovery of whey soy proteins from soy whey wastewater using batch foam fractionation. *Journal of Food Engineering*, 142(6), 201–209.
Liu H., Zhu L., Tian X., Yin Y., 2017, Seasonal variation of bacterial community in biological aerated filter for ammonia removal in drinking water treatment. *Water Research*, 123, 668–677.
Lv W., Schanbacher F.L., Yu Z., 2010, Putting microbes to work in sequence: Recent advances in temperature-phased anaerobic digestion processes. *Bioresource Technology*, 101:9409–9414.
Negi, S., Mandpe, A., Hussain, A., Kumar, S., 2020. Collegial effect of maggots larvae and garbage enzyme in rapid composting of food waste with wheat straw or biomass waste. *Journal of Cleaner Production*, 258, 120854.
Noukeu N.A., Gouado I., Priso R.J., Ndongo D., Taffouo V.D., Dibong S.D., Ekodeck G.E., 2016, Characterization of effluent from food processing industries and stillage treatment trial with *Eichhornia crassipes* (Mart.) and *Panicum maximum* (Jacq.). *Water Resources and Industry*. http://dx.doi.org/10.1016/j.wri.2016.07.001.
Pantea E., Romocea T., Ghergheleş, C., 2011, Anaerobic treatment of the wastewater in the food industry. *Analele Universităţii din Oradea, Fascicula: Protecţia Mediului*, 16, 435–440.
Qin H., Ji B., Zhang S., Kong Z., 2018, Study on the bacterial and archaeal community structure and diversity of activated sludge from three wastewater treatment plants. *Marine Pollution Bulletin*, 135, 801–807.
Sahakyan, S. V., Petevotyan, R. A., & Yedoyan, T. V. (2021). Efficient technology for wastewater treatment and desalination: Case study. In *Contemporary Problems of Architecture and Construction* (pp. 385–390). CRC Press.
Sincero A. P., Sincero G. A., 2002, *Physical-Chemical Treatment of Water and Wastewater*. Boca Raton: CRC Press.
Sumonsiri N., Barringer S.A., 2014, Fruits and vegetables–processing technologies and applications. In: *Food Processing: Principles and Applications*, pp. 363–381.
Tapia-Quirós, P., Montenegro-Landívar, M. F., Reig, M., Vecino, X., Alvarino, T., Cortina, J. L., Granados, M., 2020, Olive mill and winery wastes as viable sources of bioactive compounds: A study on polyphenols recovery. *Antioxidants*, 9(11), 1074.
Tekerlekopoulou A.G., Economou C.N., Tatoulis T.I., Akratos C.S., Vayenas D.V., 2021, Wastewater treatment and water reuse in the food industry. In: *The Interaction of Food Industry and Environment*. https://doi.org/10.1016/B978-0-12-816449-5.00008-4.

Valta K., Damala P., Panaretou V., Orli E., Moustakas K., Loizidou M., 2017, Review and assessment of waste and wastewater treatment from fruits and vegetables processing industries in Greece. *Waste and Biomass Valorization*, 8(5), 1629–1648.
Vargas E. F. D., Jablonski A.E., Flores S.H., Rios A.D.O., 2016, Waste from peach (*Prunus persica*) processing used for optimisation of carotenoids ethanolic extraction. *International Journal of Food Science & Technology*, 52(3), 757–762.
Vasiliadou I.A., Molina R., Martinez F., Melero J.A., Stathopoulou P.M., Tsiamis G., 2018, Toxicity assessment of pharmaceutical compounds on mixed culture from activated sludge using respirometric technique: The role of microbial community structure. *Science of the Total Environment*, 630, 809–819.
Waldron K.W., (Ed.), 2009, *Handbook of Waste Management and Co-Product Recovery in Food Processing*. Elsevier, UK.
Ye Z.L., Lu M., Zheng Y., Li Y.H., Cai W.M., 2008, Lactic acid production from dining-hall food waste by *Lactobacillus plantarum* using response surface methodology. *Journal of Chemical Technology & Biotechnology: International Research in Process, Environmental & Clean Technology*, 83(11), 1541–1550.
Yongkui Y., Longfei W, Feng X., Lin Z., Zhi Q., 2020, Activated sludge microbial community and treatment performance of wastewater treatment plants in industrial and municipal zones. *International Journal of Environmental Research and Public Health*, 17, 436.

15

Reclamation of Lead Acid Battery Processing Wastewater through Microbes and Waste Valorization: Progress, Challenges, and Future Prospects

Sankha Chakrabortty and Indranil Saha
KIIT Deemed to be University

Jayato Nayak
Mahindra University

Ramesh Kumar
Hanyang University

CONTENTS

15.1 Introduction

Environmental pollution caused by heavy metal emissions in the atmosphere because of industrial development and population increase is one of the most critical environmental issues faced by many countries in the current times (Singh et al., 2021). A variety of industrial processes, including melting, purification, and extraction of metals, release these metals. Metals are obtained through the discharge of contaminated gases or through industrial processes (Kumar et al., 2021; Chandra and Kumar, 2017; Dutta et al., 2021). These metals are not biodegradable, and their hazardous effects on the body are long-lasting (Shukla et al., 2005). The existence of living things lead is a systemic toxin that affects the vast majority of the body's organs bodily organs after they have been absorbed this element gives a fantastic effect. A tendency to combine with sulfuryl groups (SH). The activity was a success. The activity of enzymes that are dependent

DOI: 10.1201/9781003231738-17

on the SH group is halted or reduced. It has a reducing effect on the subject. This factor has an impact on the calcium's position in the body's biological system, for example, that approximately 90% of it accumulates in the human skeleton Furthermore, it reacts with nucleic acid, which has an impact on protein. Levels as well as the biosynthesis lead is the most common health problem among youngsters. The nerve system controls the body's functions. It also has an effect on the digestive system. It affects the digestive tract and produces colitis as well as loss of appetite, nausea, vomiting, and constipation (Salkhouri and Nouhi, 2011; Malakootian et al., 2015; Malakootian et al., 2011; Malakootian et al., 2009).

Traditional methods for removing heavy metal ions like, lead (Pb) from water include methodologies of chemical precipitation (Malakootian et al., 2018), reverse osmosis (Ferreira et al., 2011), electrochemical treatment techniques (Malakootian et al., 2018), ion exchange (Malakootian et al., 2013), membrane filtration (Malakootian et al., 2018), coagulation (Ferreira et al., 2011), extraction (Goyal et al., 2003), irradiation (Malakootian et al., 2008), and adsorption (Akhtar et al., 2008). Adsorption utilising sorbents is one of the most widely used treatments available, and it is one of the most popular because it produces safe dischargeable effluents when the adsorption process is properly designed. Adsorption is now well acknowledged as a very effective, efficient, and cost-effective equilibrium separation method for water purification applications as well as for water separation.

Adsorbents can be derived from a variety of sources, including mineral, organic, and biological sources. Leading towards the sorption of metallic ions from aqueous solutions, several types of non-biological materials have been investigated, including activated charcoals (Malakootian et al., 2011d), clay minerals (Singh et al., 2012), chelating materials (Malakootian et al., 2016), and chitosan/natural zeolites (Al-Homaidan et al., 2014). Although non-biological sorbents have the potential to eliminate heavy elements from waste effluents, their high monetary involvement, limited adsorption capabilities, and low efficiency prevent them from being widely used in wastewater treatment. In recent years, a variety of methodologies got investigated towards the generation of more affordable and efficient adsorbent preparation from naturally obtainable resources. In biosorption, biomass is used as a sorption medium to remove pollutants from the contaminated water (Kumar, 2018). It entails the detoxification of hazardous compounds rather than the transfer of harmful substances from one source to another through the use of microorganisms and algae. Although less intrusive, such methodology can often be completed on-site, avoiding the requirement of transportation of hazardous waste to the post-treatment facilities section. Because of its great absorption capacity and availability of a very cost-effective raw material source, biosorption is a step forward in the development of a perspective approach. There have been numerous studies conducted on biomaterials to determine their adsorptive attributes, and different types of biomass have demonstrated levels of metal uptake that are high enough to merit further investigation. It was discovered that quite a lot of research were being conducted by employing biologically derived adsorbents from algae, sawdust (Malakootian et al., 2012), sugar beet pulp (Rodrigues et al., 2012), papermill sludges, corncobs, wheat straw (Malakootian et al., 2012), baker's yeast (Sarı and Tuzen, 2009), and fungal biomass (Shen et al., 2013) had been carried out. Microalgae-based sorbents have gotten a lot of attention lately because of their naturally abundant availability and ability to adapt to a wide range of environments, including freshwater and seawater, residential and industrial effluents, salt marshes, and man-made wetlands. Their amazing ability of adsorption and accumulative holding of heavy metals in their vicinity (Sarı and Tuzen, 2009; Chen and Wang, 2006; Gupta and Rastogi, 2008a) can label them as superior quality adsorbents. Due to its straightforward design and lesser initial capital cost involvement in terms of invested money and land, adsorption on microbes or microalgae has the potential to be one of the most effective methodologies with high efficacy for the mitigation and capturing of heavy metals when compared with other technologies used for the same purposes. As an additional point of interest, the growing volume of research articles on the sorptive attributes of microbial biomass on contaminated water indicates overwhelming research concerns in the development of new adsorption systems for acidic wastewater treatment employing such environmentally friendly techniques. The study of marine microalgae has recently focused on them as a potential source of new compounds and biotechnologically relevant genes, notably for wastewater treatment because they account for a significant fraction of primary biomass. Using two marine microalgae, *Nannochloropsis oculata* and *Tetraselmis chuii*, the current study will be done to determine their efficacy as potential biosorbents on lead (Pb) in industrial wastewater effluent. The chloroplasts of microalgae cells that had been exposed to heavy metal ions showed the most significant structural changes, suggesting that they were the principal focus of the metal contamination.

15.2 Lead Acid Battery (LAB) Processing Industry and Its Solid-Liquid Waste

Due to its capacity to offer larger current densities while maintaining reduced maintenance costs, the lead-acid batteries (LAB) it has received a great deal of attention since its founding in 1859. As a result, it is an excellent choice for use as a portable power source in automobiles (Pavlov, 2011).

Laboratory plates are often made of lead dioxide (PbO_2) or lead nitrate ($Pb\ (NO_3)_2$), which are the most commonly utilised materials (Pb). Immediately after a battery's capacity is reduced, deadly lead sulphate is produced on both the positive and negative plates of the battery ($PbSO_4$). Sodium sulphate serves as the electrolyte in this solution. Batteries that have been abandoned, such as waste lead-acid batteries (WLABs), include the most common components, such as lead, lead oxide, and lead sulphate. Because of the lead and lead-containing compounds (oxides and sulphate) that WLABs produce, they generate a significant amount of pollution to the environment. It is possible to reduce or eliminate the amount of these compounds present in soils, water, and individuals, which can help to minimise the toxicity of soils, water, and individuals. Furthermore, waste sulfuric acid (H_2SO_4) from WLABs is detrimental to the environment's soils and water. If these pollutants are not dealt with properly, they can pose a major hazard to both the environment and human health and safety. The recycling and management of wastewater treatment plants are therefore extremely vital and required.

WLABs are different in their configuration, but their essential components are the same regardless of which one you choose. Organic matter (polypropylene, polyethylene, polyvinyl chloride, acryliconitrile-butadiene-styrene plastic (ABS), bakelite, fibres, and other materials), metallic lead (grid plates and connectors), and lead or lead paste are the four main components of an electrolyte (diluted H_2SO_4 solution). Lead and lead-containing compounds are the most prominent constituents of the most prominent components (Li et al., 2011).

15.3 Effluent Compositions Discharged from Lead Acid Battery Industry

The majority of the parameters for on-land irrigation exceeded those established by the Bureau of Indian Standards (BIS 10500: 2012). pH, temperature, and electrical conductivity (EC) all have an impact on effluent ionic concentration, which in turn has an impact on chemical reactions in the aquatic environment, pH, temperature, and electrical conductivity (EC) are all variables that can be controlled (Akpomie and Dawodua, 2015). pH (5.5–9) and EC values that are higher than normal could be produced by the use of different types of salts in the technique (24.167). Among the factors that contribute to the chemical oxygen demand are total dissolved solids (TDS), total suspended solids (TSS), and total solids (TSS), as well as sulphate (43.67 mg/L) and lead (2.30 mg). Total dissolved solids (TDS), total suspended solids (TSS), and total solids (TSS) are some of the factors that contribute to the chemical oxygen demand. When it comes to greater levels of nitrate, phosphate, and sulphate, there are a variety of other elements to consider in addition to hardness and alkalinity. These include the use of various chemicals and salts during the battery manufacturing process, among other things (Ahmed et al., 2012).

15.4 Carcinogenic Effects of Compositions of LAB Wastewater

Wastewater from lead-acid batteries should be considered hazardous waste because it contains sulfuric acid (corrosive) and lead (toxic). The demand for lead-acid batteries is experiencing significant growth. Poisoning of children's environments is a serious threat to their environmental health, and it is also a major cause of occupational diseases. Depending on the quantity and duration of exposure, lead exposure can result in anything from mild neurological symptoms to death. When it affects children or adults, it can have serious consequences for the central nervous system, kidneys, gastrointestinal tract, and haematological system, among other organs and systems. Lead has been demonstrated to have harmful impacts on both men's and women's reproductive systems, according to scientific evidence.

According to the United Nations Environment Program, leap is a "strong neurotoxin" and a "nerve poison," and it poses a major threat to the health and well-being of people of all ages all across the world, especially in developing countries. Acute neurotoxicity is a neurotoxin that affects nearly every organ in the human body. It works by imitating calcium ions to cross the blood-brain barrier and reach the central nervous system. Because of this, it has the potential to cause neurological disorders such as brain damage, mental retardation, neurological system diseases, and encephalopathy (abnormal brain function).

Each year, lead exposure in children costs the United States approximately $43.4 billion. Global figures are not available, however, they would much exceed this figure because exposure levels in poorer countries are significantly higher than the global average. Child residents of impoverished countries who live near battery manufacturing facilities have an average blood lead level that is 13 times higher than the average blood lead level of children in the United States. The number of battery plant workers in low- and middle-income countries is four times higher than the level of surveillance considered high by the Centers for Disease Control and Prevention in the United States. Sulfuric acid (H_2SO_4) is a chemical compound that is widely used in the production of lead-acid batteries. This substance lowers the pH of water and increases the acidity of water when combined. As a result of improper disposal of lead-acid batteries in a solid waste landfill or unlawful burying of batteries, lead and sulfuric acid can seep into soil and groundwater that has been contaminated with sulfuric acid. This has the potential to degrade the quality of the water supply. a tall glass of cool water A wide range of natural and environmental repercussions have occurred in recent years, prompting the government to take action to ensure the appropriate management of this hazardous waste as well as the safety of people and the protection of the environment. It is exceedingly damaging to one's health to be exposed to sulfuric acid. When a fire occurs, carcinogens, corrosives, and reactive compounds are produced, as well as oxidants and toxic gases, and the container itself may catch fire and burn.

15.5 Advancement of Technologies Used for LAB Processing Wastewater Treatment

In metallic materials, lead is by far the most frequent metal to be found; nevertheless, junk can also include internal components of other metals, such as aluminium. Plastics (PP, HDPE, ABS, and PVC) gain robustness as a result of the granulation process, which is a by-product of the process. In order to reach this result, the materials are separated, cleaned, and in some cases reduced in size to a commercially viable size. After being washed in the conveyor, the lead is melted down and reused after being melted down in a kiln.

Because of the high concentration of H_2SO_4, the wastewater generated by washing these materials is extremely acidic and will include significant concentrations of lead and other heavy metals, which must be treated before they can be discharged into the environment. To demonstrate how a typical treatment is carried out, the following reaction is depicted below:

Pb^{2+} is stable in acidic and neutral solutions.

When pH rises, it hydrolyses:

$$Pb^{+2} + OH^- \leftrightarrow Pb(OH)^+$$

At pH 7.8 it begins to precipitate as $Pb(OH)_2$

$$Pb(OH)^+ + OH^- \leftrightarrow Pb(OH)_2$$

However, raising the pH above 12.4 dissolves the precipitate in the form of lead anion because its behaviour is amphoteric:

$$Pb(OH)_2 + OH^- \leftrightarrow HPbO_2^- + H_2O$$

15.5.1 Coagulation and Flocculantion-Based Treatment

A coagulant (PAC type) acting in the high pH range must be used in conjunction with an appropriate polyelectrolyte because of the fine solid nature of the distribution (determined by testing Jar).

15.5.2 Layered Sedimentation

This mixture can then be separated in a layer settling tank at an ascent rate of 4–5 m/h under these conditions. Figure 15.1 shows the floc that results. Prior to dehydration during mechanical drying, it is common to practice to transform extracted sludge into a static thickener. Two distinct phases have been introduced into this method for the simultaneous removal of lead and cadmium, and these are as follows:

15.5.3 Ion Exchange

The concentration of lead in treated wastewater will be near to the legal limits, but an ion exchange column equipped with a chelating resin capable of removing heavy metals, such as Pb and Cd, is frequently included to give a safety buffer. HCl and NaOH are used to regenerate and neutralise the resin, respectively. Even though industrial wastewater accounts for a small percentage of overall outflow, it is the most polluting. — the concentration is approximately 6.6%.

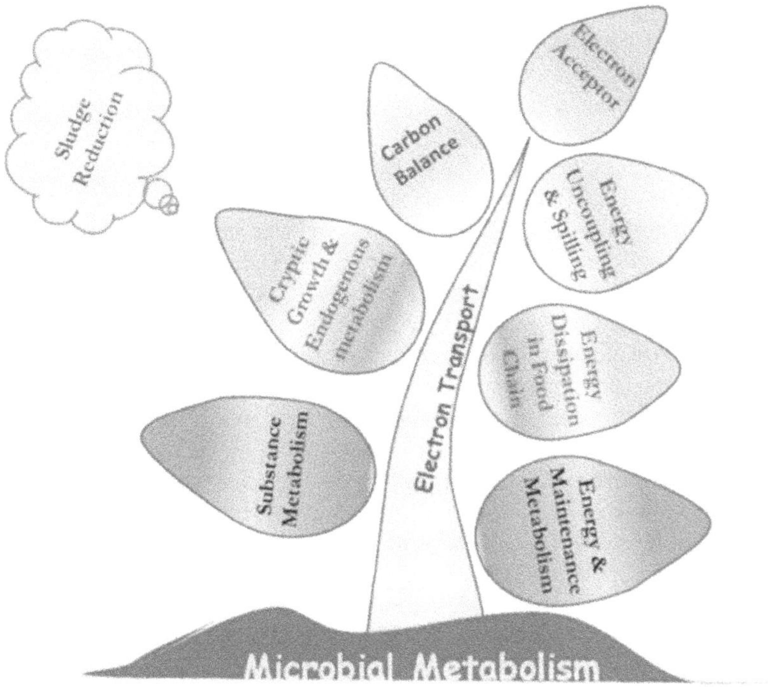

FIGURE 15.1 Metabolism of microbes. [Reproduced with copyright from "Priya AK, Pachaiappan R, Kumar PS, Jalil AA, Vo DV, Rajendran S. The war using microbes: A sustainable approach for wastewater management, *Environ. Pollut.*, 275 (2021) 116598", 61.]

They have become more popular because they have reduced waste creation and high salinity emissions by using evaporator technology that has advanced (such as vacuum equipment, heat pumps, and heat compression systems). Furthermore, when the concentration of is high, it is neutralised with NaOH to form:

$$H_2SO_4 + NaOH \leftrightarrow Na_2SO_4 + H_2O$$

Utilize an evaporator to concentrate the solution until you achieve the desired result (Glauber's salt).

Glauber's salt () is a salt that dissolves when a solution is cooled. The reason for this is due to the entropy effect. Some examples of its multiple market applications include the preparation of pulp (Kraft process), the production of laundry detergents for the wood-processing industry, the production of glass-processing detergents, and pharmaceutical detergents. It was found in 1625 and named after Johann Rudolf Glauber (1604–1670), a Dutch-German scholar and chemist who studied Austrian spring waters and was born in the Netherlands. Alternatively, you can obtain a condensate with a low salinity that can be used in the plant as rinse or cure water (around 100 ppm).

15.5.4 Photocatalytic Process

Though there is a paucity of literature on the use of photocatalysts to degrade lead acid battery production wastewater, innovative composite nanocatalysts have the potential to successfully degrade such contaminated waste effluents. Spinel ferrites (MFe_2O_4, M^{2+}=metal ion) have a better photocatalytic behaviour when it comes to wastewater treatment. Binary composites based on MFe_2O_4 are stable, biocompatible, and low-cost photocatalytic materials. In the photocatalytic destruction of emerging pollutants, MFe_2O_4/metal oxides, MFe_2O_4/carbon-based materials, MFe_2O_4/polymers, MFe_2O_4/metal nanoparticles, and MFe_2O_4/other compounds are effective (Suresh et al., 2021). MFe_2O_4-based binary composites can remove organic and inorganic contaminants. The superparamagnetic properties of binary nanocomposites with easy recovery and reuse offer potential in water purification technology (Kirankumar and Sumathi, 2020). Under UV irradiation, $ZnWO_4$ nanoparticles generated through a solution-based synthetic method can display total mineralisation (100%) at 6,000°C. After five cycles of photocatalytic degradation, there is a high level of stability with a stable structure (Faka et al., 2021). Reduced hydroxylated graphene with low defects has a conductivity of around 1,000 times that of reduced graphene oxide. T/RGOH nanocomposite photocatalysts had a 48.5% greater degradation efficiency for MB than pure TiO_2. T/RGOH nanocomposite photocatalyst had a 19.7% greater degradation efficiency for MB than pure T/RGO photocatalyst (Deng et al., 2021). $Mn_{0.6}Zn_{0.4}Fe_2O_4$@$Zn_{1-x}Mn_xS$, a catalyst made from used batteries, is capable of degrading phenol in the form of $Mn_{0.6}Zn_{0.4}Fe_2O_4$@$Zn_{0.95}Mn_{0.5}S$. Under visible light irradiation, the optimised MZFO@$Zn_{0.95}Mn_{0.05}S$ is particularly effective for photodegradation of phenol, achieving full degradation at 25 mg/L in 120 minutes, with reacting-kinetic constants that are 5.2–23.3 times larger than those of pure $Zn_{0.95}Mn_{0.05}S$ and MZFO, respectively. The process of phenol degradation, as well as the technique and pathways involved in that process, are also proposed. The reusability of MZFO@$Zn_{0.95}Mn_{0.05}S$ is excellent after five cycles, in addition to its strong magnetic separation qualities and reusability. This novel material has cheap prices, is easy to reuse, and has a lot of application potential (Niu et al., 2022). Cu_2O/rGO photocatalyst was created by reducing $Cu(OAC)_2$ and graphene oxide at the same time, leading to the formation of Cu_2O-grafted rGO sheets. Cu_2O/rGO was used as a photocatalyst for the reduction of nitroaromatic compounds to their corresponding ammine in the presence of visible light irradiation with hydrazine monohydrate. The catalyst's recyclability was tested six times and revealed no major differences in reaction conversion (Moghanlou et al., 2021). Under visible-light-emitting diode (LED) irradiation, hollow carbon spheres (HCSs) and titania-modified hollow carbon spheres (TiO_2@HCSs) can perform. The nanocatalyst TiO_2@HCSs was discovered to have diameters of 140–150 nm and a Shell width of 2,035 nm, with activity increasing from 29.1 to 267.9 mol/g/s when the TiO_2 concentration was increased from% 2 to 5%, while activity decreased as the TiO_2 content was increased higher (Boudjemaa et al., 2021).

15.5.5 Adsorption Process

Traditional water treatment methods for removing lead from wastewater include precipitation with hydroxide ions or lime, ion exchange, coagulation, electrochemical process, reverse osmosis, and in some situations, ion fotation, as well as other approaches (Wang and Chen, 2009; Kong et al., 2014). Using these technologies results in secondary wastes such as persistent organic pollutants and volatile organic compounds, both of which are hazardous to the environment. Their high cost is one of these secondary wastes (Salmani et al., 2017; Banerjee et al., 2012). Because of this, researchers have begun to focus their efforts on water adsorption, which has proven to be one of the most effective and cost-effective methods of removing heavy metals from bodies of water (Trans et al., 2010; Uçar et al., 2015; Saka et al., 2011). However, increasing the efficiency of adsorbents may lead to a decrease in the amount of energy and chemicals used. According to current research, Pb^{2+} can be removed from water using easily available, low-cost materials such as agricultural and industrial wastes, residential wastes, algae and fungi, among other things (Saka et al., 2011; Taha et al., 2011; Ibrahim et al., 2012). An agricultural waste product called sugarcane bagasse has been used in multiple experiments to absorb various pollutants, including heavy metals and dyes. There is a waste product known as bagasse (Joseph et al., 2009; Saad et al., 2010). Grass-fed sugar can be obtained from a variety of sources, including sugar mills and juice stores. The wastewater from a battery manufacturing company was treated using bagasse as an adsorbent in order to remove lead (Pb^{2+}). Thermal alteration of the adsorbent resulted in the formation of biochar, which increased the adsorbent's effectiveness. Due to the increase in surface area and porosity caused by heating the adsorbent, it became more effective. It's important to remember that many factors affect the adsorption process such as pH, contact time, adsorbent amount and initial metal concentrations as well as surface area and surface pore size. One of the most critical elements is pH. pH measures the acidity of a solution. It was determined how much adsorbent was used and what the temperature and surface area were, as well as the pH and contact time (Buasri et al., 2012; Nguyen et al., 2013). This study used bagasse biochar as a biosorbent to look at the kinetics of removing lead (Pb^{2+}) from battery production wastewater at the University of Poonam in India. It was published in the journal Environmental Science and Technology that their findings were made public (Poonam, 2018). It was decided to use biochar made from bagasse as a test sample since it could be used to analyse adsorption kinetics and mechanism as well as a variety of other factors (contact time, dose, and pH). The optimal adsorption was discovered at pH 5 after 140 minutes of contact time and 5 g of adsorbent dose were used at room temperature (25.3°C). After that, the pH was adjusted as necessary. With respect to a comparison of a final concentration in the effluent to an initial concentration in the effluent, the maximum removal efficiency was discovered to be 12.741 mg/g, with a removal rate of 75.376% at optimum pH 5. According to the findings of this investigation, a monolayer with uniform Pb^{2+} adsorption was shown to be best for the Langmuir isotherm in this particular case. When it was determined that the process was pseudo-second-order, it was discovered that chemisorption was a significant component in the process. The ultrastructural and functional features of adsorbent biochar were determined using scanning electron microscopy, Energy-dispersive X-ray, and Fourier-transform infrared spectroscopy analysis, which revealed information regarding the organic moiety's ultrastructure and function. Because of this, a porous and rough surface was created, which aided in the adsorption process. The FTIR analysis of the functional groups identified revealed that carboxyl groups are responsible for the binding of Pb^{2+}. The postadsorption elution of metal-loaded bagasse was carried out using 0.1 M HNO_3, with approximately 90% of the bagasse being regenerated after elution.

15.6 Bioremidiation: A Novel Technology Used for Lead Acid Battery Processing Wastewater Treatment

15.6.1 Utilisation of Bacterial Species

According to current research, lacae produce several important ligninolytic enzymes, including lignin peroxidase and manganese peroxidase, which are capable of breaking down and dissolving chromophoreal manufactured colour complexes. Numerous bacterial species, including *Bacillus* and *Pseudomonas*,

Nitrosomonas, Enterobacterial, Nitrobacter, Rhadococcus, and *Streptomonas,* were treated with decolourisation, reduction in body dioxide and COD, and destruction of biological and phenolic compounds (Gulzar et al., 2017). Bacteria and other microbes are the most common animals encountered during the bioaugmentation cycle. Many wastewater treatment systems (Figure 15.1) employ a variety of plant kinds for a variety of purposes, including nutrient removal, metal removal, and pathogenic microbe removal, among others. The study by Shukla et al. (2018) found that lead is a widely diffused and dangerous heavy metal ion that can accumulate in the body, producing various major illnesses and dysfunctions, notably those that disrupt children's learning capacities and their behavioural and mental development (Shukla et al., 2018; Ravenscroft et al., 2018). Biosorption is a straightforward way of eliminating heavy metals from the environment that makes use of a wide range of naturally existing biological materials. Bacteria, fungus, yeast, algae, and agricultural products are the broad categories into which these materials might be divided (Gabr et al., 2008; Luo et al., 2014; Fan et al., 2008; Yu et al., 2008; Gupta and Rastogi, 2008b; He and Chen, 2014; Wan Ngah and Hanafiah, 2008). Their primary characteristics are that they are non-toxic, biodegradable, renewable, and readily available in huge amounts. By making small modifications to these biomaterials, it is possible to improve the selectivity and efficacy of adsorption by the biomaterial.

Inactive (dead) biomass has been found to be more effective at removing heavy metals than active (alive) biomass in a number of experiments. This is because biomass contains no concentration limitations for nutrients or pollutants, but active biomass does. Because biomass does not have concentration limits for pollutants or nutrients, this is the case. Heavy metal pollution can be eliminated from the environment by utilising organic ligands and functional groups degraded from dead biomass (Bairagi et al., 2011; Xu et al., 2017). When dealing with dirty water, recovering microscopic particles of dead biomass is nearly impossible. In terms of large-scale wastewater treatment, only biosorbents with greatly improved adsorption capabilities and selectivity for heavy metals have a bright future ahead of them. The development of the most effective technique for producing excellent biosorbents is therefore crucial. The adsorption capacity and specificity of microorganisms can be improved by adopting modified microorganisms that have undergone a xanthate modification from their original microorganisms. Carbon disulfide interacts with an organic hydroxyl-containing substrate in a caustic environment, forming xanthate as a byproduct as a result of the reaction. This reaction is quite straightforward and just requires a small number of inexpensive chemicals. In addition to its gel-like properties and high insoluble content, xanthate contributes to the formation of a dense and well-coordinated network structure in the adsorbent (Giannouli and Morris, 2003). S groups in xanthate that contain sulfur-containing groups have a strong attraction for heavy metals and high constant values of stability for the heavy metal complexes produced, as previously stated according to the theory of hard and soft acids and bases (Wang et al., 2016).

A possible therapeutic strategy for mercury toxicity is the biological reduction of Pb (II) to metallic lead (Jing and Kjellerup, 2018). This is because Pb(0) is not available for metabolic absorption (Jing and Kjellerup, 2018). Given that metallic Pb particles are composed of Pb metal, they are relatively easy to isolate from their polluted environment. According to previous research (Colantonio and Kim, 2016; Kabutey et al., 2019), mixing cultures can help cut down on the amount of toxic lead (II) discharged into the atmosphere. Theoretical biomass yield on Pb(II) is estimated to be 12.62 g-cells per mole of Pb based on free energy calculations using the thermodynamic model (Bruce, 2020; McCarty, 2007). When the cells are run in an open circuit, the anode chamber of the cells can reduce microbial Pb(II)–Pb(0), according to previous work with MECs (Colantonio and Kim, 2016). According to the results of a follow-up investigation (Jannati, 2019), discovered that using an enriched MEC cathode, it was possible to decrease and remove Pb(II) from synthetic wastewater. When performed under anaerobic conditions, it has been claimed that the use of a microbial consortia originally acquired from the recycling of lead-acid batteries can decrease and remove 90%–99% of increased Pb(II) concentrations (80–1,000 ppm) according to Brink et al. (2017) and Peens et al. (2018) (Brink et al., 2017; Peens et al., 2018). However, even though earlier studies have detailed the reduction of Pb(II) in the literature (Colantonio and Kim, 2016; Brink et al., 2017; Peens et al., 2018), few publications have been published on the microorganisms that reduce Pb(II). With the help of Figure 15.2, it is feasible to summarise and demonstrate the method of Pb(II) removal using a spongy microbe-based biosorbent removal (below).

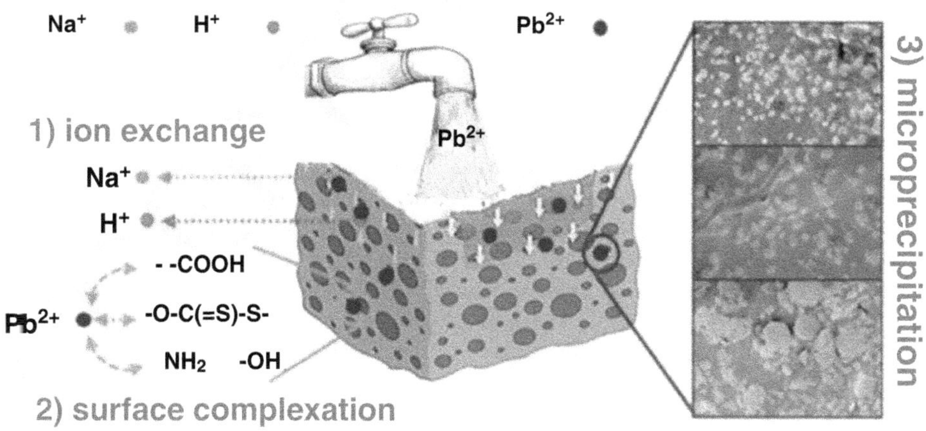

FIGURE 15.2 Mechanism of biosorbent for lead removal. [Reproduced with copyright from "Wang N, Qiu Y, Xiao T, Wang J, Chen Y, Xu X, Kang Z, Fan L, Yu H, Comparative studies on Pb(II) biosorption with three spongy microbebased biosorbents: High performance, selectivity and application, *J. Hazard. Mater.* (2019), https://doi.org/10.1016/j.jhazmat.2019.03.056".]

15.7 Utilisation of Microalgae

Chemical deposition, coagulation, ion exchange, microbial desalination ponds (MDC), and reverse osmosis are some of the common ways used to treat wastewater containing heavy metals. Other methods include aeration and filtration. Because physical and chemical procedures are ineffective in removing low-concentration heavy metals, and because the cost of such approaches is prohibitively high, biological methods for eliminating hazardous metals have become increasingly popular in recent years. Biosorption, which is characterised by the ability of diverse biological components such as algae fungi, bacteria, and yeast to bind metals, has received a significant deal of interest. These compounds are extremely efficient and cost-effective, particularly when used in conjunction with low metal concentrations. It has been demonstrated that the employment of microalgae can improve the adsorption of heavy metals among these biological substances. Algae are being used because of their availability, high metal adsorption capacity, and inexpensive cost. The biosorption effectiveness of the microalga Spirulina in the presence of lead in industrial effluent was investigated. The ability of the algae Spirulina to remove copper from aqueous solutions has been demonstrated. Chromium detoxification. The use of *Ulothrix zonata* algae to remove heavy metals from industrial effluent is becoming more common these days. In a Brazilian study, Rodrigues examined the dry biomass of *Chlorella vulgaris* and Spirulina in terms of their ability to remove heavy metals. In a similar vein, investigate the efficacy of *Chlorella vulgaris* in the removal of heavy metals. According to Aksu and Donmez, the effect of simultaneous adsorption of cadmium (II) and nickel (II) on the dry biomass of common cholera bacteria in Turkey has been investigated.

Heavy metals are commonly treated using a variety of methods including chemical deposition, coagulation, ion exchange, microbial desalination ponds (MDC), and reverse osmosis, among others. Aeration and filtration are two other techniques. The ineffectiveness of physical and chemical techniques in the removal of low-concentration heavy metals, as well as the excessively high cost of such approaches, have led to an increase in the employment of biological methods for the elimination of dangerous metals in recent years. It has piqued the curiosity of many researchers that biosorption, which is defined by the ability of varied biological components such as algae, fungi, bacteria, and yeast to bind metals, is becoming increasingly popular. Particularly when used in conjunction with low metal concentrations, these compounds are exceedingly efficient and cost-effective. It has been established that the use of microalgae can improve the adsorption of heavy metals among these biological substances when they are used in conjunction with other biological substances. In addition to their unique properties, microalgae are a viable alternative feedstock for a variety of bio-refinery applications due to their high oil content.

In addition to having unique features, microalgae may be employed in a wide range of industrial and environmental applications due to their versatility. To increase long-term viability and sustainability, this research investigates ways to widen the scope of merging microalgae with other biotechnological uses. Algae are being used because they are readily available, have a high metal adsorption capability, and are very inexpensive. An investigation into the biosorption efficacy of the microalga Spirulina in the presence of lead in industrial effluent was conducted. The ability of the algae Spirulina to extract copper from aqueous solutions has been established in laboratory experiments. Heavy metal removal from industrial wastewater is becoming more frequent these days, with Ulothrix Zonata algae being used to accomplish this task. Rodrigues et al. (2012) conducted a study in Brazil in which they analysed the dry biomass of *Chlorella vulgaris* and *Spirulina* to determine their potential to remove heavy metals from the environment (Rodrigues et al., 2012). In a similar line, look at the efficacy of *Chlorella vulgaris* in the removal of heavy metals from the environment. Specifically, Aksu and Donmez evaluated the effect of simultaneous adsorption of cadmium II and nickel II on the dry biomass of common cholera bacteria in Turkey, and they found that the bacteria produced more dry biomass.

Using high-rate algae ponds (HRAPs) to treat wastewater is considered to be a cost-effective and environmentally friendly approach to wastewater treatment. The biomass produced by HRAP can also be utilised as animal feed and a feedstock for the production of biofuel, according to the company. Through the use of such ponds, it was observed that *Spirulina platensis* and *Chlorella vulgaris* were capable of creating 99% of the pollutant removal from the whole volume of wastewater (Hoffmann, 1998; Phang et al., 2000). According to Lim et al. (2010), some research has looked into the feasibility of using a blend of microalgae species to clean wastewater (Lim et al., 2010). Mustafa et al. (2012) demonstrated the successful treatment of landfill leachate by a consortium of five microalgal species in the HRAP by a collaboration of five microalgal species (Mustafa et al., 2012). De-Bashan and Bashan (2010) found that by using immobilised microalgae, the removal efficiency of environmental toxicants can be enhanced even further (De-Bashan and Bashan, 2010). According to one study (Chu et al., 2009), colour removal from industrial waste effluents has been reported to be possible by utilising *Chlorella vulgaris* immobilised in alginate beads, which has been shown to be effective (Chu et al., 2009). When immobilised *Chlorella vulgaris* and Scenedesmus obliquus were used in a semi-continuous mode to remove nitrogen and phosphorus from residential wastewater, it was observed that they were quite effective at removing these elements (Ruiz-Marin et al., 2010). When co-immobilised with bacteria, co-immobilised microalgae and bacteria were found to be more successful at removing ammonium, nitrate, and phosphate from wastewater when compared to immobilised microalgae alone (De-Bashan et al., 2002) than immobilised microalgae alone. Because of their great sensitivity to environmental pollutants, microalgae are commonly used in environmental pollution bioassays. *Pseudo kirchneriella subcapitata* (also known as *Pseudo kirchneriella*), *Dunailella tertiolecta* (also known as dunailella), *Isochrysis galbana* (also known as *Isochrysis galbana*), and *Chlorela* sp. are examples of microalgae that have been employed for this purpose, among others. In addition to freshwater habitats, such as microalgae, are also suitable for assessing the availability of nutrients in the environment (nitrogen and phosphorus). The most notable instances of this are the tropical microalgae Ankistrodemsus convolutes, *Chlorella vulgaris*, and *Scenedesmus quadricauda,* which are all found in the tropics and are all known for their medicinal properties.

15.8 Conclusion

It's important to note that both the recycling of lead-acid wastewater and the control of microbial research has proven to be effective initiatives in the past. Sulphur monoxide emissions, which are produced by the combustion of lead and its constituents, are being reduced, with the emphasis currently being placed on recycling lead and its constituents (SO_2). Technical concerns, on the other hand, continue to be a source of contention for scientists. With the integration of existing technical processes with newly developed technologies, lead-acid wastewater recycling will take on a whole new dimension. It is anticipated that things will improve as a result of the implementation of these measures. After their biomass output has been enhanced or they have been combined with other technologies, biocultures will not be practical on a wide scale. A variety of biorefinery applications can benefit from microbial and algal feedstocks due to the specific characteristics of these feedstocks. Their unique characteristics make them valuable in a variety of environmental and industrial settings. Therefore, researchers must instead concentrate on co-cultures as a means of improving the usefulness of microalgae biomass as a bio-refinery feedstock in order to achieve greater efficiency. It is necessary to provide logistical assistance, which may include the establishment of new research facilities, in order to accommodate such a significant shift in research objectives. Systematic study should be conducted to determine the biological potential of a possible feedstock for biotechnological applications. Through the use of techno-economic analysis and life cycle assessment studies, co-processing models should be examined for their economic feasibility as well as their environmental sustainability.

REFERENCES

Ahmed TF, Sushil M, Krishna M, Impact of dye industrial effluent on physicochemical characteristics of Kshipra River, Ujjain City, India, *Int. Res. J. Environ. Sci.*, 1 (2012) 41–45.

Akhtar N, Iqbal M, Zafar SI, Iqbal J, Biosorption characteristics of unicellular green alga *Chlorella sorokiniana* immobilized in loofa sponge for removal of Cr(III), *Environ Sci.*, 20 (2008) 231–239.

Akpomie KG, Dawodua FA, Physicochemical analysis of automobile effluent before and after treatment with an alkalineactivated montmorillonite, *J. Taibah Univ. Sci.*, 9(4) (2015) 465–476.

Al-Homaidan A, Al-Houri J, Al-Hazzani A, Elgaaly G, Moubayed NMS, Biosorption of copper ions from aqueous solutions by *Spirulina platensis* biomass, *Arab J. Chem.*, 7 (2014) 57–62.

Bairagi H, M.M.R. Khan, L. Ray, A.K. Guha, Adsorption profile of lead on *Aspergillus versicolor*: A mechanistic probing, *J. Hazard. Mater.*, 186 (2011) 756–764. https://doi.org/10.1016/j.jhazmat.2010.11.064.

Banerjee K, Ramesh ST, Nidheesh PV, Bharathi KS, A novel agricultural waste adsorbent, watermelon shell for the removal of copper from aqueous solutions, *Iran J. Energy Environ.*, 3 (2012) 143–156.

Boudjemaa A, Nongwe I, Mutuma BK, Matsoso BJ, Bachari K, Coville NJ, TiO_2@hollow carbon spheres: A photocatalyst for hydrogen generation under visible irradiation, *J. Photochem. Photobiol. A Chem.*, 417 (2021) 113355.

Brink HG, Lategan M, Naude K, et al. Microbial precipitation of lead (II) by industrially obtained consortia. In: *International Conference on Energy, Environment and Climate Change (ICEECC 2017)*, Mauritius. 2017.

Bruce E. Rittmann; Perry L. McCarty. *Environmental Biotechnology: Principles and Applications* (McGraw-Hill Education: New York, Chicago, San Francisco, Athens, London, Madrid, Mexico City, Milan, New Delhi, Singapore, Sydney, Toronto, 2001). https://www.accessengineeringlibrary.com/content/book/9781260440591

Buasri A, Nattawut C, Tapang K, Jaroensin S, Panphrom S, Equilibrium and kinetic studies of biosorption of Zn(II) ions from wastewater using modifed corn cob, *APCBEE Procedia*, 3 (2012) 60–64.

Chandra, R., Kumar, V., Phytoextraction of heavy metals by potential native plants and their microscopic observation of root growing on stabilised distillery sludge as a prospective tool for In-situ phytoremediation of industrial waste, *Environ. Sci. Pollut. Res.*, 24 (2017) 2605–2619. https://doi.org/10.1007/s11356-016-8022-1

Chen C, Wang XK, Adsorption of Ni(II) from aqueous solution using oxidized multiwall carbon nanotubes, *Ind. Eng. Chem. Res.*, 45 (2006) 9144–9149.

Chu W-L, See Y-C, Phang S-M, Use of immobilised *Chlorella vulgaris* for the removal of colour from textile dyes. *J. Appl. Phycol.*, 21 (2009) 641.

Colantonio N, Kim Y, Lead(II) removal at the bioanode of microbial electrolysis cells. *ChemistrySelect*, 1 (2016) 5743–5748.

De-Bashan LE, Bashan Y, Immobilized microalgae for removing pollutants: Review of practical aspects. *Bioresour. Technol.*, 101 (2010) 1611–1627.

De-Bashan LE, Moreno M, Hernandez J-P, Bashan Y, Removal of ammonium and phosphorus ions from synthetic wastewater by the microalgae *Chlorella vulgaris* coimmobilized in alginate beads with the microalgae growth-promoting bacterium *Azospirillum brasilense*. *Water Res.*, 36 (2002) 2941–2948.

Deng H, Qin C, Pei K, Wu G, Wang M, Ni H, Ye P, TiO_2/reduced hydroxylated graphene nanocomposite photocatalysts: Improved electron–hole separation and migration. *Mater. Chem. Phys.*, 270 (2021) 124796.

Dutta D, Arya S, Kumar S, Industrial wastewater treatment: Current trends, bottlenecks, and best practices. *Chemosphere*, 285 (2021) 131245. https://doi.org/10.1016/j.chemosphere.2021.131245

Faka V, Tsoumachidou S, M. Moschogiannaki, G. Kiriakidis, I. Poulios, V. Binas, $ZnWO_4$ nanoparticles as efficient photocatalyst for degradation of para-aminobenzoic acid: Impact of annealing temperature on photocatalytic performance, *J. Photochem. Photobiol. A Chem.*, 406 (2021) 113002.

Fan T, Liu Y, Feng B, G. Zeng, C. Yang, M. Zhou, H. Zhou, Z. Tan, X. Wang, Biosorption of cadmium(II), zinc(II) and lead(II) by *Penicillium simplicissimum*: Isotherms, kinetics and thermodynamics, *J. Hazard. Mater.*, 160 (2008) 655–661. https://doi.org/10.1016/j.jhazmat.2008.03.038

Ferreira LS, Rodrigues MS, De Carvalho JC, Lodi A, Finocchio E, Perego P, Converti A, Adsorption of Ni^{2+}, Zn^{2+} and Pb^{2+} onto dry biomass of *Arthrospira* (*Spirulina*) platensis and *Chlorella vulgaris*. I. Single metal systems, *Chem. Eng. J.*, 173 (2011) 326–333.

Gabr RM, Hassan SH, Shoreit AA, Biosorption of lead and nickel by living and non-living cells of *Pseudomonas aeruginosa* ASU 6a, *Int. Biodeter. Biodegr.*, 62 (2008) 195–203. https://doi.org/10.1016/j.ibiod.2008.01.008.

Giannouli, P, Morris, E.R, Cryogelation of xanthan, *Food Hydrocoll.*, 17 (2003) 495–501. https://doi.org/10.1016/S0268-005X(03)00019-5.

Goyal N, Jain SC, Banerjee UC, Comparative studies on the microbial adsorption of heavy metals, *Adv Environ Res.*, 7 (2003) 311–319.

Gulzar T., Huma T, Jalal, F, Iqbal, S, Abrar, S, Kiran, S, 2017. Bioremediation of synthetic and industrial effluents by *Aspergillus niger* isolated from contaminated soil following a sequential strategy. *Molecules*, 22, E2244. https://doi.org/10.3390/molecules22122244.

Gupta VK, Rastogi A, Biosorption of lead from aqueous solutions by green algae *Spirogyra* species: Kinetics and equilibrium studies, *J. Hazard. Mater.*, 152 (2008a) 407–414. https://doi.org/10.1016/j.jhazmat.2007.07.028

Gupta VK, Rastogi A, Sorption and desorption studies of chromium (VI) from nonviable cyanobacterium nostoc muscorum biomass, *J. Hazard. Mater.*, 154 (2008b) 347–354.

He J, Chen JP, A comprehensive review on biosorption of heavy metals by algal biomass: Materials, performances, chemistry, and modeling simulation tools, *Bioresour. Technol.*, 160 (2014) 67–78. https://doi.org/10.1016/j.biortech.2014.01.068.

Hoffmann JP, Wastewater treatment with suspended and nonsuspended algae, *J. Phycol.*, 34 (1998) 757–763.

Ibrahim HS, Ammar NS, Ibrahim M, Removal of Cd(II) and Pb(II) from aqueous solution using dried water hyacinth as a biosorbent. *Spectrochim. Acta A Mol. Biomol. Spectrosc.*, 96 (2012) 413–420.

Jannati M, Bio-Electrochemical Removal of Lead by Exoelectrogens and Phosphate Removal Using Zeolite for Potential Industrial Wastewater Treatment, M.A.Sc. Thesis, McMaster University – Civil Engineering (2019).

Jing R, Kjellerup BV, Biogeochemical cycling of metals impacting by microbialmobilization and immobilization. *J. Environ. Sci.*, 66 (2018) 146–154.

Joseph O, Rouez M, Métivier-Pignon H, Bayard R, Emmaneul E, Gourdon R, Adsorption of heavy metals on to sugar cane bagasse: Improvement of adsorption capacities due to anaerobic degradation of the biosorbent. *Environ. Technol.*, (2009). https://doi.org/10.1080/09593330903139520

Kabutey FT, Antwi P, Ding J, Zhao QL, Quashie FK. Enhanced bioremediation of heavy metals andbioelectricity generation in a macrophyte-integrated cathode sediment microbial fuel cell (mSMFC). *Environ. Sci. Pollut. Res.* 26 (2019) 26829–26843.

Kirankumar VS, Sumathi S, A review on photodegradation of organic pollutants using spinel oxide, *Mater. Today Chem.*, 18 (2020), 100355.

Kong Z, Li X, Tian J, Yang J, Sun S, Comparative study on the adsorption capacity of raw and modifed litchi pericarp for removing Cu(II) from solutions. *J. Environ. Manag.*, 134 (2014) 109–116.

Kumar, V, Mechanism of microbial heavy metal accumulation from polluted environment and bioremediation. In: Sharma D, Saharan BS, (eds), *Microbial Fuel Factories*. Boca Raton: CRC Press (2018).

Kumar, V, Ferreira, LFR, Sonkar, M, Singh, J, Phytoextraction of heavy metals and ultrastructural changes of *Ricinus communis* L. grown on complex organometallic sludge discharged from alcohol distillery. *Environ. Technol. Innov.*, 22 (2021) 101382. https://doi.org/10.1016/j.eti.2021.101382

Li W, Jiang L, Zhan J, Zhang CF, Status and progress of recycling technology of waste lead-acid battery. *China Nonferrous Metall.*, 6 (2011) 53–56 (in Chinese).

Lim S-L, Chu W-L, Phang S-M. Use of *Chlorella vulgaris* for bioremediation of textile wastewater. *Bioresour. Technol.*, 101 (2010) 7314–7322.

Luo S, Li X, Chen L, Chen J, Wan Y, Liu C, Layer-by-layer strategy for adsorption capacity fattening of endophytic bacterial biomass for highly effective removal of heavy metals, *Chem. Eng. J.*, 239 (2014) 312–321. https://doi.org/10.1016/j.cej.2013.11.029.

Malakootian M, Almasi A, Hossaini H, Pb and Co removal from paint industries effluent using wood ash, *Int. J. Environ. Sci. Technol.*, 5 (2008) 217–222.

Malakootian M, Darabi-Fard Z, Amirmahani N, Nasiri A, Evaluation of arsenic, copper, lead, cadmium, and iron concentration in drinking water resources of central and Southern Bardsir plain, Iran, in 2014, *J. Kerman Univ. Med. Sci.*, 22(5) (2015) 542–554.

Malakootian M, Hashemi M, Toolabi A, Nasiri A, Investigation of nickel removal using poly(amidoamine) generation 4 dendrimer (PAMAM G4) from aqueous solutions, *J. Eng. Res.*, 6(2) (2018) 13–23.

Malakootian M, Jaafarzadeh NA, Hossaini H, Efficiency of perlite as a low cost adsorbent applied to removal of Pb and Cd from paint industry effluent, *Desal. Water Treat.*, 26 (2011) 243–249.

Malakootian M, Kalankesh LR, Loloi M, Efficiency of hybrid nanoparticles of TiO_2/SiO_2 in removal of lead from paint industry effluents, *J. Mazandaran Univ. Med. Sci.*, 23 (2013) 244–254.

Malakootian M, Limoni ZK, Malakootian M, The efficiency of lead biosorption from industrial wastewater by micro-alga *Spirulina platensis*, *Int. J. Environ Res.*, 10(3) (2016) 357–366.

Malakootian M, Nouri J, Hossaini H, Removal of heavy metals from paint industry's wastewater using Leca as an available adsorbent, *Int. J. Environ. Sci. Technol.*, 6 (2009) 183–190.

Malakootian M, Tolabi A, Moussavi GH, Kinetic study and bio-absorption isotherm of heavy metals using the algae *Ulothrix zonata* from industrial waste waters, *J. Ilam. Univ. Med. Sci.*, 26 (2012) 37–44.

Malakootian M, Toolabi A, Moussavi SG, Ahmadian M, Equilibrium and kinetic modeling of heavy metals biosorption from three different real industrial waste waters on to *Ulothrix zonata* Algae, *Aust J. Basic Appl. Sci.*, 5 (2011d) 1030–1037.

McCarty PL, Thermodynamic electron equivalents model for bacterial yield prediction: modifications and comparative evaluations, *Biotechnol. Bioeng.*, 97 (2007) 377–388.

Moghanlou AO, Bezaatpour A, Sadr MH, Yosefi M, Salimi F, Cu_2O/rGO as an efficient photocatalyst for transferring of nitro group to amine group under visible light irradiation, *Mater. Sci. Semicond. Process.*, 130 (2021) 105838.

Moussavi Z, Rahmati A, Removing cadmium from aqueous environments using the nano structure γ-alumina, *Water Wastewater J.*, 9 (2011) 4–20.

Mustafa E-M, Phang S-M, Chu W-L, Use of an algal consortium of five algae in the treatment of landfill leachate using the high-rate algal pond system. *J. Appl. Phycol.*, 24 (2012) 953–963.

Nguyen TAH, Ngo HH, Guo WS, Zhang J, Liang S, Yue QY, Li Q, Nguyen TV, Applicability of agricultural waste and byproducts for adsorptive removal of heavy metals from wastewater: Review. *Bioresour Technol.*, 148 (2013) 574–585.

Niu Z, Tao X, Huang H, Qin X, Ren C, Wang Y, Shan B, Liu Y, Green synthesis of magnetically recyclable $Mn_{0.6}Zn_{0.4}Fe_2O_4@Zn_{1-x}Mn_xS$ composites from spent batteries for visible light photocatalytic degradation of phenol, *Chemosphere*, 287 (2022) 132238.

Pavlov D, *Lead-Acid Batteries: Science and Technology*. Oxford, UK: Elsevier, (2011), pp. 29–114.

Peens J, Wu YW, Brink HG, Microbial Pb(II) precipitation: The influence of elevated Pb(II) concentrations. *Chem. Eng. Trans.*, 64 (2018) 583–588.

Phang S, Miah M, Yeoh B, Hashim M, *Spirulina* cultivation in digested sago starch factory wastewater. *J. Appl. Phycol.*, 12 (2000) 395–400.

Poonam, KM, Efficiency of sweet lemon (*Citrus limetta*) biochar adsorbent for removal of chromium from tannery effluent. *Indian J Environ Prot.*, 38 (2018), 246–256

Priya AK, Pachaiappan R, Kumar PS, Jalil AA, Vo DV, Rajendran S, The war using microbes: A sustainable approach for wastewater management, *Environ. Pollut.*, 275 (2021) 116598.

Ravenscroft J, Roy A, Queirolo EI, Mañay N, Martínez G, Peregalli F, Kordas F, Drinking water lead, iron and zinc concentrations as predictors of blood lead levels and urinary lead excretion in school children from Montevideo, Uruguay, *Chemosphere*, 212 (2018) 694–704. https://doi.org/10.1016/j.chemosphere.2018.07.154.

Rodrigues MS, Ferreira LS, de Carvalho JC, Lodi A, Finocchio E, Converti A, Metal biosorption onto dry biomass of *Arthrospira* (*Spirulina*) platensis and *Chlorella vulgaris*: Multi-metal systems, *J. Hazard. Mater.*, 217 (2012) 246–255. https://doi.org/10.1016/j.jhazmat.2012.03.022.

Ruiz-Marin A, Mendoza-Espinosa LG, Stephenson T, Growth and nutrient removal in free and immobilized green algae in batch and semi-continuous cultures treating real wastewater. *Bioresour. Technol.*, 101 (2010) 58–64.

Saad SA, Isa KM, Bahari R, Chemically modifed sugarcane bagasse as a potentially low-cost biosorbent for dye removal. *Desalination*, 264 (2010) 123–128

Saka C, Şahin Ö, Demir H, Kahyaoğlu M (2011) Removal of lead(II) from aqueous solutions using pre-boiled and formaldehyde-treated onion skins as a new adsorbent. *Sep. Sci. Technol.*, 46(3):507–517

Salkhouri NY, Nouhi A, Optimization of the conditions for removing lead by *Saccharomysis cerevisiae* under laboratory conditions, *J. Ghazvin. Univ. Med. Sci.*, 15 (2011) 13–21.

Salmani MH, Abedi M, Mozafari SA, Sadeghian HA, Modifcation of pomegranate waste with iron ions a green composite for removal of Pb from aqueous solution: Equilibrium, thermodynamic and kinetic studies. *AMB Express*, 7 (2017) 225. https://doi. org/10.1186/s13568-017-0520-0

Sarı A, Tuzen M, Equilibrium, thermodynamic and kinetic studies on aluminum biosorption from aqueous solution by brown algae (*Padina pavonica* biomass), *J. Hazard Mater.*, 171 (2009) 973–979.

Shen Q-H, Zhi T-T, Cheng L-H, Xu X-H, Chen H-L, Hexavalent chromium detoxification by nonliving *Chlorella vulgaris* cultivated under tuned conditions, *Chem. Eng. J.*, 228 (2013) 993–1002.

Shukla SR, Pai RS, Adsorption of Cu(II), Ni(II) and Zn(II) on modified jute fibers, *Bioresour. Technol.*, 96 (2005) 1430–1438.

Shukla V, Shukla P, Tiwari A, Lead poisoning, *Indian J. Med. Spec.*, 9 (2018) 146–149. https://doi.org/10.1016/j.injms.2018.04.003.

Singh E, Kumar A, Mishra R, You S, Singh L, Kumar S, Kumar, R, Pyrolysis of waste biomass and plastics for production of biochar and its use for removal of heavy metals from aqueous solution. *Bioresour. Technol.*, 320 (2021) 124278.

Singh SK, Bansal A, Jha MK, Dey A, An integrated approach to remove Cr(VI) using immobilized Chlorella minutissima grown in nutrient rich sewage wastewater, *Bioresour. Technol.*, 104 (2012) 257–265.

Suresh R, Rajendran S, Kumar PS, Vo DV, Cornejo-Ponce L, Recent advancements of spinel ferrite based binary nanocomposite photocatalysts in wastewater treatment, *Chemosphere*, 274 (2021) 129734.

Taha G, Arifen A, El-Nahas S, Removal efciency of potato peels as a new biosorbent material for uptake of Pb(II) Cd(II) and Zn(II) from their aqueous solutions. *J. Solid Waste Technol. Manag.*, 37 (2011) 128–140.

Trans VH, Tran LD, Nguyen TN, Preparation of chitosan/magnetite composite beads and their application for removal of Pb(II) and Ni(II) from aqueous solution. *Mater. Sci. Eng. C*, 30 (2010) 304–310.

Uçar S, Erdem M, Tay T, Karagö S, Removal of lead(II) and nickel(II) ions from aqueous solution using activated carbon prepared from rapeseed oil cake by Na_2CO_3 activation. *Clean Technol. Environ. Policy*, 17 (2015) 747–756. https://doi.org/10.1007/s10098-014-0830-8

Wan Ngah WS, Hanafiah MAKM, Removal of heavy metal ions fromwastewater by chemically modified plant wastes as adsorbents: A review, *Bioresour. Technol.*, 99 (2008) 3935–3948. https://doi.org/10.1016/j.biortech.2007.06.011.

Wang J, Chen C, Biosorbents for heavy metals removal and their future. *Biotechnol. Adv.*, 27 (2009) 195–226.

Wang N, Xu X, H. Li, J. Zhai, L. Yuan, K. Zhang, H. Yu, Preparation and application of a xanthate-modified thiourea chitosan sponge for the removal of Pb(II) from aqueous solutions, *Ind. Eng. Chem. Res.*, 55 (2016) 4960–4968. https://doi.org/10.1021/acs.iecr.6b00694.

Xu X, Li H, Wang Q, Li D, Han X, Yu H, A facile approach for surface alteration of *Pseudomonas putida* I3 by supplying K_2SO_4 into growth medium: Enhanced removal of Pb(II) from aqueous solution, *Bioresour. Technol.*, 232 (2017) 79–86. https://doi.org/10.1016/j.biortech.2017.02.038.

Yu J, Tong M, Sun X, Li B, Enhanced and selective adsorption of Pb2+ and Cu2+ by EDTAD-modified biomass of baker's yeast, *Bioresour. Technol.*, 99 (2008) 2588–2593. https://doi.org/10.1016/j.biortech.2007.04.038.

16

Microbial Approaches for Pharmaceutical Wastewater Recycling and Management for Sustainable Development: Present Status, Challenges, and Opportunities

Prasenjit Chakraborty
The University of Burdwan

Sankha Chakrabortty
Kalinga Institute of Industrial Technology

Jayato Nayak
Kalasalingam Academy of Research and Education

Sujoy Chattaraj
Vignan's Foundation for Science, Technology & Research (Deemed to be University)

CONTENTS

DOI: 10.1201/9781003231738-18

16.1 Introduction

Water treatment and reuse have received a great deal of attention in recent years, necessitating the employment of the most stringent criteria possible in order to meet the expanding global demand for high-quality drinking water. Rivers have been clogged with pollution from industrial water streams, creating serious health risks to those who swim in or ingest river water as a result of the increased pollution burden (Chandra and Kumar, 2015; Kumar et al., 2021a, b; Dutta et al., 2021). When industrial effluents damage the environment, they have an impact not only on the health of workers but also on the health of those who live near chemical synthesis factories and on the health of underground labourer (Asamudo et al., 2005). Pharmaceutical pollutants in aquatic systems are a severe hazard to aquatic life, both in terms of their presence and consequences, as well as their toxicological. Instead of simply treating the symptoms, the recovery of important APIs and prescription medications from diluted streams should be tackled as a priority. It is not uncommon for common generic pharmaceuticals, such as antibiotics, analgesic/anti-inflammatory medications, and antituberculosis therapies, to be used in dosages comparable to pesticides. This means that, for the time being, the full extent and consequences of the existence of these compounds in the environment remain mostly unknown and poorly defined. These compounds have been found in sewage and other waste products, surface waters, groundwater, and drinking water in amounts ranging from parts per trillion to parts per billion, no matter where they have been discovered in the environment. A common misconception about drugs is that they have no negative impact on the environment. As a result of this misconception, it is difficult to undertake an appropriate risk assessment in aquatic environments due to a lack of recognised analytical procedures, reliable monitoring data, and relevant data on the fate and toxicity of pharmaceutical chemicals in aquatic environments, among other factors (Chandra and Kumar, 2017).

The pharmaceutical sector generates raw materials, antibiotics, a wide range of medicines, and cosmetics, all of which result in wastewater that contains compounds that are detrimental to human health and aquatic life, as well as a variety of other substances (Dixit and Parmar, 2013; Chang et al., 2008; Gautam et al., 2007). As a result, the wastewater that emanates from these machines is exceedingly toxic, poisonous, and frequently disagreeable to the human taste. The abrasive molecules were released back into the environment during the wastewater treatment process. Even though the maintenance and housekeeping chores in each plant are equivalent as a result of the variety of pharmaceuticals produced during any given manufacturing process, pharmaceutical businesses do not generate waste streams that are homogeneous (Kavitha et al., 2012; Wei et al., 2012; Madukasi et al., 2010; Chang et al., 2008; Rao et al., 2004). The presence of nonbiodegradable organic substances such as antibiotics, various prescription and non-prescription drugs, beta-lactamides, anti-inflammatories, animal and plant steroids, generative hormones, analgesics, antidepressants, lipid regulators, detergent metabolites, and other such substances in the pharmaceutical industry wastewater that has been left untreated or incompletely treated is a major source of environmental pollution.

Kavitha et al. (2012) analysed the physicochemical attributes of wastewater samples coming out of a pharmaceutical industry where they found the following effluent characteristics: Total suspended solid (TSS) 4,300 ppm, total dissolved solid (TDS) 2,850 ppm, chemical oxygen demand (COD) 7,300 ppm, biological oxygen demand (BOD) 4,200 ppm, chlorides 1,000 ppm, and sulphates 500 ppm. Das et al. (2012) analysed the wastewater samples coming out from various points present in the industry where they observed a wide variation of major characterisation parameters in different ranges like: for sulphates 40–1,500 ppm, TDS 500–1,500 ppm, TSS 20–100 ppm and COD 1,200–1,550 ppm. The pharmaceutical firms use a wide array of interconnected treatment methodologies for the clarification of waste effluent volume produced (Deegan et al., 2011) like neutralisation/pH adjustment, oxidation, sand filtration (Saleem, 2007), activated sludge process, trickling filter, and UV irradiation. Research studies are propagating towards the development of advanced systems using ozonation, Fenton's method, coagulation/flocculation, electrocoagulation (Dixit and Parmar, 2013), Photo-electrocoagulation, peroxi-electrocoagulation, peroxi-photo-electrocoagulation, sedimentation, membrane separation, adsorption, chlorination, distillation, solar photo-Fenton, reverse osmosis, bacterial treatments (Madukasi et al., 2010), fungal treatment (Spina et al., 2012), algal treatment, phytoremediation and methods using membrane bioreactor (MBR) (Chang et al., 2008), anaerobic fixed film reactor (AFFR) (Rao et al., 2004), aerobic sequencing batch reactor (ASBR) (Patil et al., 2013), membrane-aerated biofilm reactor (MABFR) (Wei et al., 2012) etc.

According to the structure of the chemicals (Verlicchi et al., 2012), fewer than 50% of the compounds are removed on average. In the absence of success, wastewater treatment plants (WWTPs) have attempted several physicochemical and biological treatments. Micropollutants are often removed using contemporary oxidation techniques and membrane technologies rather than primary and secondary treatments (Joss et al., 2008). Energy usage, chemical use, and the production of undesirable and harmful by-products (Benotti et al., 2009) are all factors that they must contend with in order to succeed. Researchers have been attempting to develop technologies that are ecologically friendly, commercially viable, and reasonably inexpensive in order to circumvent these difficulties (Wu et al., 2012, Sahu, 2014). In comparison to other wastewater treatment technologies, biological tertiary wastewater treatment is effective, affordable, and environmentally friendly to the environment. Some bacteria have gained notoriety in the field of applied environmental microbiology as a result of their application in the wastewater treatment process (Kumar and Chandra, 2018a). Using fungal and algae-based therapies to combat pharmaceutical contamination has been proposed as a viable solution to the problem of pharmaceutical contamination. This page discusses the research on medicine removal using fungi and algae through bioremediation or biosorption mechanisms, which is being conducted.

Microbial remediation procedures have not received much attention in India, and there aren't many reports on the subject available anywhere else in the world. This review, based on the gaps and lack of expensive work done, emphasises the need to look into the topic further and provides a comprehensive account of the work that has already been done using traditional and other established methodologies.

16.2 Outline of Pharmaceutical Production and Present Status

Indian Pharmaceutical industry plays an important role in the manufacturing and export of medicines worldwide. India holds the third position in pharmaceutical production by volume and the 14th position by value (Cygnus report). As per the ministry of chemical and fertilisers (Government of India) latest report annual turnover of pharmaceuticals for the FY 2019–2020 was 289,998 crore (MoP report, 2021). "Pharma Vision-2020" initiative by the government of India, has taken the country towards a global leader of the pharmaceutical sector. As per the latest data, India is the leading producer of vaccine (~60%) and exporter of generic medicine (>20%) worldwide. Thirty percent of the world's demand for anti-HIV drugs is produced in India. Almost 3,000 drug companies and almost 10,500 manufacturing units are currently established in India (IBEF report, 2021). Among them, medium and large companies are the major contributors to the growth, contributing 75% of the domestic sales and more than 90% of the export. 1.72% of India's GDP is contributed by the pharmaceutical sector (MoP report, 2021). Export in terms of currency from the Indian pharma sector in FY 2019–2020 is rupees 146,260 crores as compared to 132,585 crore rupees in the FY 2018–2019 and 110,195 crore rupees in FY 2017–2018 (DGCIS report). Maharashtra is the leading state in terms of a number of manufacturing units followed by Gujarat, West Bengal, Andhra Pradesh, and Tamil Nadu. Bulk drug production in India falls under 21 major therapeutic classes with Antibiotics constituting more than half of the total production due to their high demand at the domestic level.

The worldwide population growth, the advancement of health and living conditions, and the increase in average longevity of human life have led to a drastic increase in the consumption of pharmaceutical products. As a result of which, in the last few decades pharmaceutical sectors have registered a considerable jump and have contributed significantly to the world economy, especially in developing countries like India. However, as an adverse effect of the increase in the pharmaceutical industry, the untreated or allegedly treated wastewater from this sector has become one of the premier reasons for water pollution.

16.3 Treatment of Pharmaceutical Wastewater

To prevention of water pollution, pharmaceutical industries use different treatment techniques to treat the water for reuse depending on the wastewater characteristics (Deegan et al., 2011). This wastewater quality is solely dependent on the raw material used for manufacturing and the adoption of the manufacturing process (Rana et al., 2017). Mayabhate et al. (1988); Rao et al. (2004); Imran (2005) etc. have identified and tabulated the different parameters of pharmaceutical wastewater and their ranges found from different pharmaceutical product manufacturing units. Disposed wastewater from the pharmaceutical industry mainly contains non-biodegradable organic pollutants such as antibiotics, animal and plant steroids, anti-inflammatories, lipid regulators etc., Besides the organic pollutants, it may also contain heavy metals such as lead, mercury, cadmium, nickel, chromium etc., microbial organisms, strong fluctuation in quantity, high sulfate concentration, biological toxicity, alkali content, high salt etc. (Chen et al., 2008; Sreekanth et al., 2009). Owing to the complex compositions of the pharmaceutical wastewater, all the pharmaceutical industries use a series of water treatment techniques, which includes sand filtration, oxidation, electrocoagulation, flocculation, membrane separation, activated carbon, reverse osmosis, etc. Although much research has been done in recent years on pharmaceutical wastewater treatment, very few focus on the overall scenario of the available treatment processes (Gadipelly et al., 2014).

16.3.1 Conventional Method

Hospital and pharmaceutical industries' effluent are the major sources of wastewater disposal related to pharmaceutical products. Cost-effective and environment-friendly conventional treatment processes are generally employed before discharging to the environment. The conventional treatment method is broadly categorised into two methods *viz.* (i) Physicochemical treatment method and (ii) biological method (US-EPA, 1974). Physicochemical pre-treatment processes prior to biological treatment are considered as physical treatment methods. coagulation and sedimentation, flotation, activated carbon adsorption, advanced oxidation processes, membrane separations etc. are the commonly used physicochemical treatment for the pharmaceutical industry. Among them, coagulation and sedimentations are considered the primary method of physical treatment. Biological treatment process includes Aerobic, anaerobic processes, combination of aerobic and anaerobic processes, composting, vermicomposting etc. (Pal, 2018). Among them, aerobic and anaerobic processes are considered as the most commonly used method for the treatment of pharmaceutical wastewater (Deegan et al., 2011).

16.3.1.1 Physicochemical Treatment

In recent days, the focus of pharmaceutical wastewater treatment has been shifted towards physicochemcical technologies. Starting from pre-treatment to final handling, these processes are executed. Applications of each process are very specific due to the diverse nature of the pharmaceutical wastewater depending upon the type and ranges of pollutants. Physicochemical properties of the pharmaceutical wastewater are mainly driven by suspended solid particles, heavy metals, turbidity, nitrate content, phosphorous content, biodegradability, polarity, adsorbability, absorbability, stability, etc. (Le-Minh et al., 2010; Garrido-Cardenas et al., 2020). Processes like coagulation, flocculation, frothing, precipitation, sedimentation, oxidation etc. are in individual or in combination and are employed as a part of pre or post-treatment of the sewage water. However, the efficiency of these individual processes or combined processes differ significantly based on the source of the effluent water. For example, flocculants like calcium oxide, calcium hydroxide, ferric chloride are conventionally used for the removal of phosphorous depending on the concentration and pH (El-Gohary et al., 2010). Further, a study by Vanerkar et al. (2013) on pharmaceutical effluent water found that the combined process of coagulation using alum, cationic synthetic polyelectrolyte as primary treatment followed by activated sludge process as secondary treatment and Fenton's process as tertiary treatment have the capability to remove the non-biodegradable organics to the permitted level by the pollution control board. Few physicochemical technologies that are frequently used nowadays for the treatment of pharmaceutical wastewater are discussed in this chapter.

16.3.1.2 Coagulation and Precipitation

The process of adding chemicals to the effluent water followed by dispersion of the chemicals by rapid mixing to make the pollutant unstable and precipitable matters is known as coagulation. It is the primary pre-treatment process to remove the oil, grease, and colloids from the water to increase biodegradability (Pal, 2018). Combining coagulation and precipitation processes can reduce the considerable load on the fouling-sensitive membrane for the next stage of the treatment process. Several studies have been reported on the coagulation treatment of pharmaceutical (hospital and domestic) and personal care products (PPCPs) wastewater (Adams et al., 2002; Seitz et al., 2006 etc.). However, there are a limited number of study is available on coagulation techniques for the wastewater from pharmaceutical manufacturing units. Suarez et al. (2009) have performed a detailed study aiming at removing the 13 components from pharmaceutical effluent water through different coagulation techniques. The coagulation technique is also used to neutralise the pH content of the wastewater. However, the generation of secondary pollutants is the main disadvantage of this technique (Feng, 2019). Electrocoagulation, phot-electrocoagulation, peroxi-electrocoagulation, peroxi-photelectrocoagulation, etc. are the few advanced coagulation techniques that have been adopted in the pharmaceuticals industry in recent days.

Precipitation is the next stage of the separation technique of the pollutants after coagulation by means of specific gravity. This process is mostly applicable for large-scale operations, especially for wastewater having high salt and carbon content by adjusting the retention time depending on the water composition (Westerhoff et al., 2005). A considerable quantity of metals and halogen is also separated through this process from the effluent. However, the use of a large number of reagents is the main drawback of this process.

16.3.1.3 Advanced Oxidation Process (AOP)

The treatment process by oxidising the pollutants by forming the free hydroxy radical is known as Advanced Oxidation Processes (AOPs) (Kumar et al. 2021b). Wet air oxidation, ozonisation, ultrasonication, Fentons's treatment etc. are the different categories of AOPs that are applied to pharmaceutical wastewater treatment. The main objective of the AOPs is to generate reactive oxygen species (ROS) to produce reclaimed effluents (Pignatello et al., 2006). Generated hydroxyl radicals destroy the organic pollutants by hydrogen abstraction, combination or addition of radicals, and electron transfer (Deng and Zhao, 2015). AOPs are generally employed as a tertiary treatment method after physicochemical process as a primary and biological process as a secondary technique in the pharmaceutical industry and they show significant efficiency (Andreozzi et al., 2005, 2006; Helmig et al., 2007). However, employment of AOPs is not always possible at an industrial scale due to the high cost of chemical agents and energy sources (Legrini et al., 1993). Implementation of Solar irradiation can reduce the capital cost of AOPs substantially.

16.3.1.4 Adsorption

Persistent pharmaceutically active compounds (PhACs) from the effluent water are captured by means of physisorption or chemisorption in this process. High surface area and well-developed pore structure of the adsorbent like activated carbon reduces the pollutant content by adsorption (Gadipelly et al., 2014). Mestre et al. (2007) have studied the adsorption of ibuprofen by activated carbon from pharmaceutical wastewater. Adsorption property regeneration for re-use is one of the most advantages of this technology. A recent study by De Ridder et al. (2012) has demonstrated the advantageous use of zeolite as an adsorbent over conventional activated carbon for the removal of micropollutants from pharmaceutical wastewater. Owing to its higher surface area, adsorption by nanoparticles and nanotubes has also gained attention for the removal of pharmaceutical pollutants from effluent water (Ghauch et al., 2009). Deryło-Marczewska and Marczewski (2002) found that apart from the surface area of the adsorbents, the adsorption process efficiency also depends on the pH, temperature, and agitation speed operational conditions of the effluent water.

16.3.1.5 Membrane Separation

High separation efficiency membrane technology is one of the advanced and efficient wastewater treatment technology nowadays that are being used in the pharmaceutical industry. However, the filtration efficiency of this process is solely dependent on the pore size (Martina Hamingerova and Beckmann, 2010). Removal of heavy metals, organic pollutants, and reducing COD values from sewage water are the major advantages of the membrane technology (Lee et al., 2008; Koczka, 2009). Membrane technology is also beneficial for the treatment of the bottom product of rectification to concentrate non-volatile components along with the decrease of BOD values without scaling and fouling of membrane (Tóth et al., 2011). However, in spite of several major advantages, accumulation of non-degraded pollutants (fouling) and deposition of pollutants, causing it to plug (scaling) are the main drawbacks of membrane technology.

In case of arbitrary water composition where conventional membrane technology fails, nanofiltration (NF) and ultra-filtration (UF) are introduced. MF treatment is specially used for the removal of microorganisms in wastewater, whereas UF is effective for the removal of macromolecules and colloids in wastewater. However, the MF membrane holds more endocrine-disrupting compounds (EDCs) and PhACs as compared to UF depending on the membrane pore size (Kim et al., 2008). Few recent studies on membrane separation technology concluded that nanofiltration combined with forward osmosis is very effective for the treatment of complex pharmaceutical wastewater (Kumar and Pal, 2014).

16.4 Biological Treatment of Pharmaceutical Wastewater

It has been discovered that conventional chemical treatment methods are ineffective when dealing with the complex pollution load of pharmaceutical industrial wastewater effluents. In fact, conventional chemical treatment methods may even contribute to the formation of a more difficult-to-treat and more polluting waste product in the form of a complex byproduct. Purification methods such as chemical and physico-chemical purification rely on expensive chemicals and treatment units that are difficult to operate at the unit level, making them uneconomical (Amin et al., 2013). Mayabhate et al. (1988) reported that pharmacological therapy using inorganic coagulants to treat wastewater (salts of Fe and Al) was less effective in a pharmaceutical factory. Phenols that have been chlorinated are more hazardous than their non-chlorinated counterparts. Phenol toxicity rises with increasing lipophilicity of chlorophenol and chlorination degree (Lee et al., 1994). Exogenous chemicals, also known as xenobiotics, are chemical molecules that have been synthesised artificially and are resistant to degradation in the environment. For wastewater treatment, microorganisms and plants are employed in biological processes (Kumar et al., 2020, 2021c; Kumar and Chandra, 2018b). Microbes decompose or turn trash into something useful. It is vital to guarantee that the final product is not more poisonous and damaging than the original component, regardless of whether the compound was destroyed or converted. The treatment of wastewater via biological methods involves the employment of bacteria and plants. Microbes decompose the trash or convert it into another form. It is critical to ensure that the end result is not more harmful and troublesome than the original chemical. The product's biodegradation is influenced by various factors, including:

1. stereochemistry,
2. concentration,
3. toxicity,
4. microbial strain efficiency,
5. degradation conditions,
6. the presence and concentration of other chemicals, and
7. retention time.

Microorganisms are essential for the breakdown of xenobiotics and the maintenance of steady-state chemical concentrations in the atmosphere (Misal et al., 2011; Kumar, 2018). When hazardous residues undergo microbial change, they become less stable and less toxic. In this procedure, biodegradation assisted Raj and Anjaneyulu (2005) in achieving a maximum COD reduction of 86.6% in pharmaceutical industry effluent, according to Raj and Anjaneyulu (2005). Das et al. (2012) discovered that the bioremediation procedure considerably reduced the quantities of COD, TSS, TDS, and sulphate in the water samples.

The use of fungal and bacterial consortia of microorganisms in the pharmaceutical wastewater treatment process has been demonstrated to be beneficial in other studies (Rosen et al., 1998). The treatment of pharmaceutical industry waste biologically can be accomplished through the use of aeration and anaerobic treatment (Deegan et al., 2011). A number of reactors with high-rate have been developed for use in large-scale biological treatment applications, including wastewater treatment. The membrane aerated biofilm reactor (MABR) technology for pharmaceutical wastewater treatment was shown to have a COD removal efficiency of 90% and an ammonium nitrogen removal efficiency of 98%, respectively. When combined with an immobilised biomass reactor (IBR), solar photo-Fenton, and PIWW were utilised to treat the wastewater, and the result was a 95% reduction in the amount of DOC. In this treatment method, a solar photo-Fenton reduction of 33% and an immobilised biomass reactor reduction of 62% were both utilised to reduce the amount of phosphorus in the water. The findings of this study demonstrate that biological processes can improve therapeutic effects. A biological approach was employed to treat PIWW by Mayabhate and colleagues in 1988. In terms of therapeutic efficacy, a comparison was done between the physicochemical and biological processes used in the study. When it came to COD reduction, chemical processes were found to be unsuccessful and to require high doses, whereas biological processes resulted in an 86% to a 91% reduction in COD levels and a 50% drop in the phenol concentrations in the wastewater samples. This shows that biological processes may be a preferable alternative for wastewater treatment in the future. Other bioremediation components have also been investigated in depth.

16.4.1 Anaerobic Treatment

In the past, anaerobic technology was used to treat wastewater from a variety of industries like paper and pulp manufacturing, distilleries and tanneries, textile manufacturing, and food processing. Anaerobic sludge blanket (UASB), fluidised bed reactor (FBR), anaerobic contact reactor (ACR), and anaerobic fixed film reactor (AFFR) are some of the methods used to treat water from diverse industries (AFFR). They are subject to several operational constraints despite their broad use, even though they are widely available (Rao et al., 2004). Microorganisms assist in the degradation of industrial wastewater's high organic content into methane and CO_2 through the production of methane and CO_2 in the anaerobic treatment. Anaerobic treatment of wastewater has several interesting advantages, including the production of very little sludge and a low energy requirement, as well as the production of biogas, which can be used to generate electricity during the treatment process. Because of its numerous benefits, anaerobic treatment of wastewater is becoming extremely prevalent. (Nandy et al., 2002). Efficacy of the treatment can be influenced by a number of factors, including the source of the inoculum and the manner in which the feed has been pretreated. Because of the low pH and slow growth, the hydraulic retention time (HRT) is significantly longer than it would be in a normal environment. This issue was handled with the development of a high-rate setup for treating industrial wastewater at shorter HRTs (Patel and Madamwar, 2000). It was found that using anaerobic biological treatment, Enright et al. (2005) could remove between 60% and 70% of the COD from pharmaceutical wastewater.

16.4.1.1 UASB Reactors (Upflow Anaerobic Sludge Blanket)

According to Tóth and colleagues (2011), the usage of upflow anaerobic sludge blanket (UASB) reactors for the treatment of industrial wastewater is prevalent. Using this reactor, it has proven possible to achieve high-rate anaerobic wastewater treatment. UASB will not be successful unless there are active and competent granules present. Because of their self-immobilisation, aggregated organisms in granule form are adept at retaining their position in the reactor (Akunna and Clark, 2000). In addition to its independence from mixing mechanically and the sludge biomass reusing, the UASB reactor's ability to endure high loading rates is one of its most significant advantages. A UASB reactor is an excellent choice for industrial wastewater treatment in psychrophile circumstances. Some investigators have employed hybrid UASB reactors to treat pharmaceutical wastewater, and this has been shown to be effective (Oktem et al., 2007).

16.4.1.2 Anaerobic Fixed Film Reactor (AFFR)

It is equipped with a biofilm support structure that allows biomass to be attached to the anaerobic fixed film reactor. The simplicity of design, the absence of mechanical mixing, the improved stability, and the ability to withstand toxic shock loads are all advantages of this reactor. This kind of reactor may swiftly recuperate after being subjected to prolonged starvation (Rajeshwari et al., 2000). Packing materials such as foam made of polyurethane, red drain clay, rubber of waste tyre, corrugated plastic, polyacryl nitrile acrylamide, nylon fibre, coconut coir, and charcoal can be utilised as sustenance media for industrial wastewater treatment in this reactor, improving the reactor performance (Acharya et al., 2008).

16.4.1.3 Fixed-Film Anaerobic Reactor (AFFR)

For biomass attachment, the anaerobic fixed film reactor incorporates a biofilm support structure. The advantages of this reactor are ease of construction, the absence of mechanical mixing, improved stability, and the ability to absorb toxic shock loads. After a period of fasting, this sort of reactor can quickly recover (Rajeshwari et al., 2000). In the AFFR, Glass bead, foam of polyurethane, red drain clay, polyacryl nitrile acryl amide, rubber of waste tyre, corrugated plastic, splintered glass, nylon fibre, coconut coir, and charcoal may be employed as packing material to improve the performance of reactor (Acharya et al., 2008). To achieve significant organic load removal, this technology guarantees excellent contact with anaerobic biomass in suspended form. Rao et al. (2004) investigated the use of an anaerobic fixed film reactor (AFFR) to treat wastewater from a bulk drug manufacturing facility with high suspended solids, concluding that the AFFR could be used effectively to treat wastewater from bulk drug manufacturing facilities with high COD (60%–70% removal), TDS, TSS, and BOD (80%–90% removal). COD elimination effectiveness was discovered to be 45%.

16.4.2 Aerobic Treatment

In the wastewater treatment industry, activated sludge and aerobic sequencing batch reactor (ASBR) technology have been the most widely used aerobic technologies thus far. In conjunction with a primary settling tank, an aerobic tank as well as two storage tanks are implemented. Aerobic biological techniques have been reported to be used in the treatment of pharmaceutical effluent (Lapara et al., 2001). It was discovered that the aerobic treatment of pharmaceutical waste in a biological reactor was feasible in some circumstances. The researchers constructed and ran a batch-type integrated aerobic biological treatment system for pharmaceutical wastewater treatment on a laboratory scale, and the system was then tested and evaluated. Fresh air was pumped into the reactor from the bottom of a cylindrical open tank, which served as the reactor. During this investigation, the hydraulic retention time (HRT) was investigated and determined to be 15 days in length. For the purpose of determining the effectiveness of the plant, they collected daily samples of treated water and tested them for the following parameters: COD, TSS, pH, conductivity, and total dissolved solids (TDS). With a 15-day HRT, it was possible to remove around 75% of the COD from the body. The pH, BOD, COD, TSS, TDS, and colour of the treated wastewater were all found to be within permissible ranges, demonstrating the plant's efficiency and validating the plant's operation. They discovered that air injection had the greatest influence on COD removal from wastewater. The ability of this treatment plant to remove COD and TSS from industrial wastewater makes it a feasible alternative for industrial wastewater treatment. Such a facility, when used for industrial wastewater treatment, is environmentally friendly, minimises chemical use, is cost-effective, and is simple to manage and maintain. An SBR was used to treat the mixed pharmaceutical industrial wastewater, and significant reductions in BOD (approximately 85%), carbon dioxide (approximately 87%), suspended solids (approximately 98%), ammonia (approximately 96%), and phosphate (approximately 96%) were obtained (approximately 5%). It has been demonstrated that by employing an activated sludge system, Vanerkar et al. (2013) were able to reduce COD, BOD, and suspended particles in the wastewater of the herbal pharmaceutical sector by 69%–91%, respectively.

16.4.3 Hybrid Anaerobic/Aerobic Systems

While a single anaerobic scheme can reduce COD in pharmaceutical effluent effectively, the results can vary substantially. Different wastewater parameters, such as salinity, refractory chemicals and toxicity are scarcely biodegradable under anaerobic circumstances, which could explain the variance in anaerobic treatment efficacy. As a result, combining anaerobic and aerobic processes to achieve more thorough oxidation and steady treatment performance has been discovered to be a practical and promising approach. In general, a consecutive aerobic-anaerobic process comprises an anaerobic stage that removes the majority of organic contaminants, followed by an aerobic stage that polishes the anaerobic effluent by removing organic residues and reducing ammonia by nitrification. According to the literature, a prominent combination for pharmaceutical wastewater treatment is sequential UASB + aerobic continuous stirred tank reactor (CSTR). According to Sponza and Demirden (2007), with an initial sulfamerazine concentration of 90 mg/L, the nine treatability of sulfamerazine in a sequential UASB and CSTR achieved COD removal effectiveness of 97%. A few years later, they discovered that COD derived from rapidly degradable organics had no effect on the anaerobic biodegradation process, however, COD derived from slowly degradable organics and inert microbial products had a considerable negative impact on the performance of anaerobic ABR reactors.

16.4.4 Fungal Treatment

Many fungal strains perform a significant role in the process of industrial wastewater treatment, but their long-life cycles and high spore production limit their effectiveness in this process. Spina and colleagues examined the efficacy of fungal treatment to activated sludge for the pharmaceutical industry effluent treatment in their study (2012). According to the findings of this investigation, the fungal strain *Bjerkandera adusta* MUT 2,295 reduced COD by 91% when compared to activated sludge, which reduced COD by 78% in the same study. An increasing number of fungal species have been demonstrated to reduce the amount of carbon dioxide (CO_2) present in diluted industrial effluent, including *Aspergillus niger, Aspergillus fumigatus,* and *Aspergillus niveus.* An imperative role is played in the industrial effluent treatment by a cluster of fungi recognised as Ascomycetes, which includes species such as *Penicillium decumbens* and *Penicillium lignorum* (Mohammad et al., 2006).

16.4.5 Treatment through Bacteria

It has also been demonstrated that bacterial culture can be used for the bioremediation of industrial wastewater. It has been discovered that bacteria such as *Pseudomonas, Enterobius, Streptobius, Aeromonas, Acinetobactor,* and *Klebsiella* can reduce COD by as much as 44% when exposed to nitrate (Ghosh et al., 2004). The researchers discovered 15 rhizosphere bacteria within 30 days, which lowered 76% of the colour, as well as 85%–86% of the BOD and COD. The usage of E. coli cells and a methanogenic consortium was employed in the process of eliminating hazardous pollutants. Pentachlorophenols and phenols used as replacements are much more dangerous, carcinogenic, and mutagenic than the phenols and pentachlorophenols used as originals. In order for the bacterial population to be able to decompose organic pollutants, the toxic organic molecules must be totally mineralised (Tewari and Malviya, 2002). There is some evidence that bacteria belonging to the genera *Arthrobacter, Comamonas,* and *Rhodococcus* can break down phenolic and complex chemical compounds. A number of *Clostridium* species (fermenting bacteria) have been shown to break down résorcinol. It is believed that Pseudomonas bacteria, namely those prevalent in rhizosphere soil, have a special ability to break down phenolic chemicals. Duffner et al. (2000) proposed the thermophilic Bacillus thermoglucosidasius A7, which degrades phenol at 65°C via the meta-cleavage pathway, as a potential phenol degrader. Phenol and catechol biodegradation kinetics were investigated using the bacteria *Pseudomonas putida* MTCC 1194. It was discovered that as soon as the well-acclimatised *P. putida* culture was developed, the original 100 mg/L of phenol concentration and 500 mg/L catechol concentration were entirely destroyed. It has been discovered that the bacteria *P. fluorescence* is capable of degrading phenol and chlorophenols as well. According to Das and colleagues, bioremediation can be employed to regulate the properties of pharmaceutical wastewater (2012). After samples were collected from nine different locations in the industry, sulphates were reduced to 6–65.8 mg/L from 44 to 1,527 mg/L, total suspended solids (TSS) were reduced from 24–84 mg/L to 12–56 mg/L, COD was reduced from 1,257.9 to 1,542.9 mg/L to 11 and total dissolved solids (TDS) were reduced from 484–1,452 mg/L to 68–540 mg/L, *Rhodobactor spheroids,* which are phototrophic bacteria, were employed to eliminate COD from pharmaceutical wastewater by the team of researchers. Some researchers were able to extract 62% of COD from PIWW at 30°C and 38% at 60°C using a mixed bacterial culture (Lapara et al., 2001).

16.4.6 Phytoremediation

It is possible to remove metal ions from industrial effluent using phytoremediation, which is a low-cost, revolutionary method that is still in the experimental stage (Chandra and Kumar, 2018; Kumar et al., 2021d). Heavy metals such as cadmium and lead are incapable of being absorbed by microorganisms. Plants that have been genetically changed, such as transgenic plants, have the ability to accumulate these toxins in their tissues and organs (Amin et al., 2013). Aquatic plants have the ability to lower toxic metals, BOD, and total solids in wastewater. Billore et al. (2001) reported that *Typha latipholia* and *Phragmitis karka* were employed to purify industrial wastewater in their study. As a result of this treatment, the concentrations of COD, BOD, total solids, and phosphorus were lowered. The peroxidases found in tomato hairy root cultures have been shown to be effective in phytoremediating phenol-contaminated industrial effluent (Gonza'lez et al., 2006).

16.4.7 Membrane Bioreactor (MBR)

The MBR has been employed in high-scale effluent treatment, wastewater from industries, home wastewater, and other types of effluent treatment. The MBR provides a number of advantages, including Complete elimination of suspended particles, small plant size, and other factors high rate of deterioration, operational flexibility, and low rate of deterioration generation of sludge, disinfection and odour regulator, long-term retention time of microorganisms, and treatment of stubborn microorganisms as well as organic and inorganic contaminants that are hazardous. Chang et al. (2008) were successful in achieving pharmaceutical wastewater. COD levels were measured after treatment with the membrane bioreactor. 95% removal of BOD and 99% removal of BOD correspondingly. Some researchers employed Zee Weed MBR technology for sewage treatment for pharmaceuticals.

16.5 Sustainable Development of Microbial Pharmaceutical Wastewater Treatment: Challenges and Scope

To carry out sustainable initiatives, one must first examine the natural processes that have sustained life on Earth since the dawn of time (Indigo Development, 2004). Carbon, nitrogen, sulphur, phosphorus, and other biogeochemical cycles indicate that, as their names suggest, they work in cycles with little interaction with the environment. Waste from one cycle component is repurposed as a resource in another. If sustainable development is to be realised, industries must imitate these natural closed cycles.

Since the industry is a point source of pollution, contaminants produced during the method of manufacturing are more familiar to manage. Pollution-specific treatment facilities can be installed. The providence of these recalcitrant organic substances in various treatment schemes has been greatly aided by various treatment methods for pharmaceuticals in water and wastewater found in the literature. The behaviour of these compounds in these systems would allow for a better understanding of the fate and risk of pharmaceuticals in the environment, but the extensive difference in removal efficacies through treatment methods, and even between distinct investigations for similar substances, is preventing this trend.

In order to improve treatment efficiencies, identify degradation compounds, and determine the cost and feasibility of full-scale operations, more data is needed in this area. Most of these physicochemical processes removed most of colloidal organic material and suspended components, but refractory compounds stayed in the liquid effluent. These disruptive organic composites are mineralised using biological processes and encounter the necessities for wastewater release to sewage. Such a combination of methods will be competent and have the possibility to be used on a large scale.

Pharmaceutical wastewaters have typically been managed using biological treatment technologies. Activated sludge, membrane batch reactors, sequence batch reactors, anaerobic film reactors, anaerobic sludge reactors, and anaerobic filters are examples of aerobic and anaerobic processes. The features of the wastewater are crucial in deciding which biological treatments to use. Bioremediation procedures are technological tools which make nearly zero waste by-products, have the probabilities to be less expensive, and, when used in conjunction with various advanced oxidation processes and physicochemical processes provide greater pharmaceutical industrial effluent treatment and thus aid in the sustainable use of accessible water resources. It has the potential to become one of the most effective technologies for cleaning up and protecting our environment.

16.5.1 Processes of Waste Recovery

Pretreatment and recovery of various useful byproducts that find their way into waste streams, like acids, solvents, heavy metals, and numerous significant APIs, is an imperative waste control approach for the pharmaceutical industry. The fermentation broth comprises a lot of solvent and mycelia in the fermentation plants. Pharmaceutical product recovery can lower or even remove waste dumping costs in the primary unit method, as well as necessities of raw water in the secondary unit process, rapidly compensating waste-treatment operative prices and refining the economy of the process. The recovered waste stream can be used elsewhere in the process, and the water can be used for cooling towers, boiler feed, and other procedures, decreasing the consumption of raw water and operational expenses dramatically. Overall, pharmaceuticals have molecular weights greater than 250 Da, and may improve using efficient membrane technologies if the product is present alone in the stream. Reverse osmosis, nanofiltration, and ultrafiltration can all provide significant financial benefits. For high operating efficacy and safety of the product, nanofiltration can be employed to recuperate more than 80% of the complicated waste stream with high-worth feedwater.

16.6 Recommendations and Future Prospects

The pharmaceutical industry is subject to stringent regulations from food and drug officials around the world, and if a pharmaceutical company wants to continue operating, it must meet acceptable water quality criteria for use, disposal, or reuse elsewhere in the plant. In order to meet regulatory requirements, such as putting limitations on the amount of certain waste contaminants present or even assuring that they do not exist, large amounts of ultrapure water must be utilised. Water that is deposited into waste streams, whether municipal or otherwise, may be subject to volume restrictions. Manufacturing facilities' improper usage and disposal of pharmaceuticals is the most common reason pharmaceuticals end up in the environment. Medications discovered in the environment are generally made up of effluent from chemical synthesis and fermentation processes, which can be found in a wide variety of production sites and laboratories. Waste is generated during the manufacturing, filtration, and cleanup/washing operations, and this waste is not insignificant. The majority of wastewater research is done on a small scale in laboratories using synthetic media, even though wastewaters are complex matrices that differ from WWT to WWTP and may vary by geographic region. The intricacies of processes, their dependence on physical, chemical, and biological components as well as the mechanisms involved in order to ensure high pharmaceutical removal efficiency, necessitates additional research, which is why only a small number of scientists are working on real-world applications (Shi et al., 2015). A large number of papers have been published in recent decades on the issue of pharmacological substances and endocrine disruptive chemicals. In the vast majority of cases, wastewater from chemical and fermentation processes is the primary focus of wastewater treatment technology. A number of compounds have been treated with hybrid approaches when single-stage treatment has failed to completely eliminate the toxin. In the vast majority of circumstances, hybrid techniques allow for the removal of the contaminant almost completely or within safe discharge limits to be achieved. A pretreatment stage consisting of advanced oxidation methods is most frequently used to remove chemicals that are recalcitrant or resistant, some of which are nonbiodegradable, from the final product. Biological treatment systems are better able to handle rubbish because of the enhanced biodegradability of the waste. There are two types of biological treatment procedures now available: aerobic treatment and anaerobic treatment. Aerobic treatment is the more common of the two. Both of these technologies have limitations, and a membrane bioreactor is a viable alternative. Additionally, anaerobic reactors are becoming increasingly popular since the process's byproduct, namely biogas, may be utilised in agriculture to create fertiliser from treated sludge, which reduces the need for fossil fuels (Shi et al., 2017).

The use of microorganisms derived from saline settings could provide a solution to the salinity problem that plagues pharmaceutical wastewater treatment and is preventing biological treatment from performing correctly. When comparing microorganisms isolated from saline environments such as oceans or salt lakes to common microbial populations found in traditional biological wastewater treatment plants, it is observed that the latter are more resistant to high salinity. Being able to remain working under high osmotic pressure, or actively synthesising organic compounds within their cell to balance the osmotic potential of their surrounding environment, both of which contribute to this unique property, helps to the survival of bacteria (Shi et al., 2015). To aid in the cleanup of organic pollutants derived from industrial salt effluent, the heterotrophs in this community make use of organic carbon as a fuel source. It is now being investigated whether wastewater treatment may be improved by using halo-tolerant/halophilic bacteria that have been isolated from saltwater environments. Several studies have demonstrated that they are capable of degrading a wide range of refractory organics, including polycyclic aromatic hydrocarbons (PAHs), phenol, and azo dyes (Arulazhagan and Vasudevan, 2011; Xu et al., 2016; Jiang et al., 2015). Earlier this year, Shi et al. (2015) published the results of the first laboratory-scale study on the use of intertidal wetland sediment (IWS) as an inoculation source for the treatment of actual pharmaceutical wastewater using anaerobic sludge bioreactor systems. Among those who have contributed to this work are Shi et al. According to the findings of the researchers, the IWS reactor reduced COD by 71.4%, which is more than twice as much as a conventional digested sludge reactor, which reduced COD by 32.3%, on average. In another study, Ng et al. (1989) discovered that using marine microorganisms to treat pharmaceutical wastewater, they were able to outperform the standard seeded biomass reactor in terms of organic removal. They did this by operating a unique anaerobic bio-entrapped membrane bioreactor (AnBEMR) and discovering that the bio-augmented reactor outperformed the standard seeded biomass reactor in terms of organic removal. Rather than "removal" technologies, emphasis should be given on "recovery" technologies. In order to recover useful reagents, garbage, and solvents that may be used, several scientists have experimented with various recovery procedures. It is necessary to conduct a complete study of the system's characteristics from both a personal and a global aspect in order to fully comprehend its advantages and disadvantages and, as a result, make an informed conclusion on the system's overall economic viability. More emphasis should be focused on the recovery and repurposing of pharmaceutical wastewaters in general (Gadipelly et al., 2014).

16.7 Conclusion

A variety of concerns, such as the future of medicine and the treatment of urban wastewater are prompting the development of new wastewater treatment methods and technologies. Examples include the discussion of how fungus and algae might remove pharmaceuticals from aquatic matrices by bioremediation and biosorption, as detailed in this study. Because of their rapid growth and ability to remove medicines as well as nutrients from wastewater, algae have a major advantage over fungus in the treatment of residential wastewater. This is critical because tertiary therapies are frequently necessary. A useful bioresource for the development of biofuels and high-value byproducts due to its high carbon content, biomass is an excellent source of carbon. Comparing bioremediation to conventional approaches, the results of the study showed that bioremediation and biosorption were more environmentally friendly and cost-effective alternatives for drug elimination than biosorption. A new understanding of genetic engineering is necessary in order to select and amplify the most successful algal or fungal strains for the removal of drugs from the environment. A wide spectrum of organisms stands to profit from metabolic genes produced from bacteria, according to recent research. Increasing the ability of genetically modified mushrooms and algae to degrade a variety of compounds will almost certainly have a significant impact on the future of this field, as it will allow these microorganisms to effectively remove pharmaceutical waste from wastewaters while also allowing their continued use, taking into account legal restrictions as well as market and cost considerations.

REFERENCES

Acharya BK, Mohana S, Madamwar D (2008) Anaerobic treatment of distillery spent wash: a study on up flow anaerobic fixed film bioreactor. *Bioresour Technol* 99:4621–4626.

Adams C, Wang Y, Loftin K, Meyer, M (2002) Removal of antibiotics from surface and distilled water in conventional water treatment processes. *J Environ Eng* 128(3):253–260.

Akunna JC, Clark M (2000) Performance of a granular-bed anaerobic baffled reactor (GRABBR) treating whisky distillery wastewater. *Bioresour Technol* 74(3):257–261.

Amin A, Naik ATR, Azhar M, Nayak H (2013) Bioremediation of different wastewaters—a review. *Cont J Fish Aquat Sci* 7(2):7–17.

Andreozzi R, Canterino M, Giudice R, Marotta R, Pinto G, Pollio A (2006) Lincomycin solar photo degradation, algal toxicity and removal from wastewaters by means of ozonation. *Water Res* 40(3):630–638.

Andreozzi R, Canterino M, Marotta R, Paxeus N (2005) Antibiotic removal from wastewaters: the ozonation of amoxicillin. *J Hazard Mater* 122(3):243–250.

Arulazhagan P, Vasudevan N (2011) Biodegradation of polycyclic aromatic hydrocarbons by a halotolerant bacterial strain *Ochrobactrum* sp. VA1. *Mar Poll Bull* 62:388–394.

Asamudo NU, Daba AS, Ezeronye OU (2005) Bioremediation of textile effluent using *Phanerochaete chrysosporium*. *Afr J Biotechnol* 4(13):1548–1553.

Benotti M, Trenholm RA, Vanderford BJ, Holady HC, Stanford BD, Snyder SA (2009) Pharmaceuticals and endocrine disrupting compounds in U.S. drinking water. *Environ Sci Technol* 43:597–603, http://dx.doi.org/10.1021/es801845a.

Billore SK, Singh N, Ram HK, Sharma JK, Singh VP, Nelson RM, Dass P (2001) Treatment of a molasses based distillery effluent in a constructed wetland in central India. *Water Sci Technol* 44(11–12):441–448.

Chandra R, Kumar V (2015) Biotransformation and biodegradation of organophosphates and organohalides. In: Chandra R (Ed.) *Environmental Waste Management*. Boca Raton: CRC Press. http://dx.doi.org/10.1201/b19243-17.

Chandra R, Kumar V (2017) Detection of *Bacillus* and *Stenotrophomonas* species growing in an organic acid and endocrine-disrupting chemicals rich environment of distillery spent wash and its phytotoxicity. *Environ Monit Assess* 189, 26. http://dx.doi.org/10.1007/s10661-016-5746-9.

Chandra R, Kumar V (2018) Phytoremediation: a green sustainable technology for industrial waste management. In: Chandra R, Dubey N, Kumar V (Eds.) *Phytoremediation of Environmental Pollutants*. Boca Raton: CRC Press, https://doi.org/10.1201/9781315161549-1.

Chang C, Chang JS, Vigneswaran S, Kandasamy J (2008) Pharmaceutical wastewater treatment by membrane bioreactor process-a case study in southern Taiwan. *Desalination* 234:393–401.

Chen Z, Ren N, Wang A, Zhang ZP, Shi Y (2008) A novel application of TPADMBR system to the pilot treatment of chemical synthesis-based pharmaceutical wastewater. *Water Res* 42:3385–3392.

Cygnus Report. Cygnus Business Consulting and Research is a services organization focusing on analysis and research of Economies, Industries and Companies.

Das MP, Bashwant M, Kumar K, Das J (2012) Control of pharmaceutical effluent parameters through bioremediation. *J Chem Pharm Res* 4(2):1061–1065.

Deegan AM, Shaik B, Nolan K, Urell K, Oelgemo¨ller M, Tobin J, Morrissey A (2011) Treatment options for wastewater effluents from pharmaceutical companies. *Int J Environ Sci Technol* 8(3):649–666.

Deng Y, Zhao R (2015) Advanced oxidation processes (AOPs) in wastewater treatment. *Curr Pollut Rep* 1(3):167–176.

Deryło-Marczewska, A, Marczewski, AW (2002) Effect of adsorbate structure on adsorption from solutions. *Appl Surf Sci* 196:264–272.

De Ridder DJ, Verberk JQJC, Heijman SG, Amy GL, Van Dijk JC (2012) Zeolites for nitrosamine and pharmaceutical removal from demineralised and surface water: mechanisms and efficacy. Separation and *Purification Technology* 89:71–77.

Dixit D, Parmar N (2013) Treatment of pharmaceutical wastewater by electro-coagulation and natural coagulation process: review. *VSRD Int J Technol NonTechnol Res* 4(5):79–88.

Duffner FM, Kirchner U, Bauer MP, Muller R (2000) Phenol/cresol degradation by the thermophillic *Bacillus thermoglucosidasius* A7: cloning and sequence analysis of five genes involved in the pathway. *Gene* 256:215–221.

Dutta D, Arya S, Kumar S (2021) Industrial wastewater treatment: current trends, bottlenecks, and best practices. *Chemosphere* 285, 131245.

El-Gohary F, Tewfik, A, Mahmoud, U (2010) Comparative study between chemical coagulation/precipitation (C/P) versus coagulation/dissolved air flotation (C/DAF) for pre-treatment of personal care products (PCPs) wastewater. *Desalination* 252:106–112.

Enright A, McHugh S, Collins G, O'Flaherty V (2005). Low-temperature anaerobic biological treatment of solvent containing pharmaceutical wastewater. *Water Res*, 39(19):4587–4596.

Feng H (2019). Research progress in pharmaceutical wastewater treatment technology. In: *ICAEER 2019. E3S Web of Conferences* 118, 04019.

Gadipelly C, Pérez-González A, Yadav GD, Ortiz I, Ibáñez R, Rathod VK, Marathe KV (2014) Pharmaceutical industry wastewater: review of the technologies for water treatment and reuse. *Ind Eng Chem Res* 53:11571–11592.

Garrido-Cardenas JA, Esteban-García B, Agüera A, Sánchez-Pérez JA, Manzano-Agugliaro F (2020) Wastewater treatment by advanced oxidation process and their worldwide research trends. *Int J Environ Res Public Health* 17: 170.

Gautam AK, Kumar S, Sabumon PC, 2007. Preliminary study of physico-chemical treatment options for hospital wastewater. *J Environ Manage* 83:298–306.

Ghauch A, Tuqan, A., Assi, HA (2009) Antibiotic removal from water: elimination of amoxicillin and ampicillin by microscale and nanoscale iron particles. *Environ Pollut* 157:1626–1635.

Ghosh M, Verma SC, Mengoni A, Tripathi AK (2004) Enrichment and identification of bacteria capable of reducing chemical oxygen demand of anaerobically treated spent wash. *J Appl Microbiol* 6:241–278.

Gonza'lez PS, Capozucca CE, Tigier HA, Milrad SR, Agostini E (2006) Phytoremediation of phenol from wastewater, by peroxidases of tomato hairy root cultures. *Enzyme Microbial Technol* 39(4):647–653.

Helmig E, Fettig J, Cordone L (2007). API removal from pharmaceutical manufacturing wastewater – results of process development, pilot-testing, and scale-up wef session, Rhode Island.

IBEF report (2021). www.ibef.org.

Imran H, (2005) Wastewater monitoring of pharmaceutical industry: treatment and reuse options. Electron J Environ Agric Food Chem, 4(4):994–1004.

Indigo Development. (January 2004) Indigo Planing Scheme Amendment C14 Report of Panel.

Jiang Y, Yang K, Wang H, Shang Y, Yang X (2015) Characteristics of phenol degradation in saline condition of a halophilic strain JS3 isolated from industrial activated sludge. *Mar Poll Bull* 99:230–234.

Joss A, Siegrist H, Ternes TA (2008) Are we about to upgrade wastewater treatment for removing organic micropollutants? *Water Sci Technol* 57:251–255, http://dx.doi.org/10.2166/wst.2008.825.

Kavitha RV, Murthy VK, Makam R, Asith KA (2012) PhysicoChemical analysis of effluents from pharmaceutical industry and its efficiency study. *Int J Eng Res Appl* 2(2):103–110.

Kim JH, Park PK, Lee CH, Kwon HH (2008) Surface modification of nanofiltration membranes to improve the removal of organic micro-pollutants (EDCs and PhACs) in drinking water treatment: graft polymerization and cross-linking followed by functional group substitution. *J Membr Sci* 321(2):190–198.

Koczka K (2009) Environmental conscious design and industrial application of separation processes, BME, Budapest, PhD Dissertation.

Kumar R, Pal P (2014) Turning hazardous waste into value added products: production and characterization of struvite from ammoniacal waste with new approaches. *J Cleaner Prod* 43:59–70.

Kumar V (2018) Mechanism of microbial heavy metal accumulation from polluted environment and bioremediation. In: Sharma D, Saharan BS (Eds.) *Microbial Fuel Factories*. Boca Raton: CRC Press.

Kumar V, Chandra R (2018a) Characterisation of manganese peroxidase and laccase producing bacteria capable for degradation of sucrose glutamic acid-Maillard reaction products at different nutritional and environmental conditions. *World J Microbiol Biotechnol* 34:32.

Kumar V, Chandra R (2018b) Bacterial assisted phytoremediation of industrial waste pollutants and eco-restoration. In: Chandra R, Dubey NK, Kumar V (Eds.) *Phytoremediation of Environmental Pollutants*. Boca Raton: CRC Press.

Kumar V, Ferreira LFR, Sonkar M, Singh J (2021d) Phytoextraction of heavy metals and ultrastructural changes of *Ricinus communis* L. grown on complex organometallic sludge discharged from alcohol distillery. *Environ Technol Innov* 22:101382. http://dx.doi.org/10.1016/j.eti.2021.101382.

Kumar V, Kaushal A, Singh K, Shah MP (2021c) Phytoaugmentation technology for phytoremediation of environmental pollutants: opportunities, challenges and future prospects. In: Kumar V, Saxena G, Shah MP (Eds.) *Bioremediation for Environmental Sustainability: Approaches to Tackle Pollution for Cleaner and Greener Society*. Elsevier. https://doi.org/10.1016/B978-0-12-820318-7.00016-2.

Kumar V, Shahi SK, Ferreira LFR, Bilal M, Biswas JK, Bulgariu L (2021a) Detection and characterization of refractory organic and inorganic pollutants discharged in biomethanated distillery effluent and their phytotoxicity, cytotoxicity, and genotoxicity assessment using *Phaseolus aureus* L. and *Allium cepa* L. *Environ Res* 201:111551. http://dx.doi.org/10.1016/j.envres.2021.111551.

Kumar V, Singh K, Shah MP (2021b) Advanced oxidation processes for complex wastewater treatment. In: Shah MP (Ed.) *Advance Oxidation Process for Industrial Effluent Treatment*. Elsevier. https://doi.org/10.1016/B978-0-12-821011-6.00001-3.

Kumar V, Thakur IS, Singh AK, Shah MP (2020) Application of metagenomics in remediation of contaminated sites and environmental restoration. In: Shah M, Rodriguez-Couto S, Sengor SS (Eds.) *Emerging Technologies in Environmental Bioremediation*. Elsevier. http://dx.doi.org/10.1016/B978-0-12-819860-5.00008-0.

Lapara TM, Nakatsu CH, Pantea LM, Alleman JE (2001) Aerobic biological treatment of a pharmaceutical wastewater: effect of temperature on COD removal and bacterial community development. *Water Res* 35(18):4417–4425.

Lee CM, Lu CJ, Chuang MS (1994) Effects of immobilized cells on the biodegradation of chlorinated phenols. *Water Sci Technol* 30:87–90

Lee J, Lee BC, Ra JS, Cho J, Kim IS, Chang NI, Kim KH, Kim SD (2008) Comparison of the removal efficiency of endocrine disrupting compounds in pilot scale sewage treatment processes. *Chemosphere*, 71(8):1582–1592.

Le-Minh N, Khan SJ, Drewes JE, Stuetz RM (2010).Fate of antibiotics during municipal water recycling treatment processes. *Water Research* 44(15):4295–4323.

Legrini O, Oliveros E, Braun AM (1993) Photochemical processes for water treatment. Chemical *Reviews*, 93(2), 671–698.

Madukasi EI, Dai X, He C, Zhou J (2010) Potentials of phototrophic bacteria in treating pharmaceutical wastewater. *Int J Environ Sci Technol* 7(1):165–174.

Martina Hamingerova LB, Beckmann M (2010) Membrane technologies for water and wastewater treatment on the european and indian market. Technical Report, Fraunhofer Center for International Management and Knowledge Economy, 37pp.

Mayabhate SP, Gupta SK, Joshi SG (1988) Biological treatment of pharmaceutical wastewater. *Water Air Soil Pollut* 38(1–2):189–197.

Mestre AS, Pires J, Nogueira JMF, Carvalho AP (2007) Activated carbons for the adsorption of ibuprofen. *Carbon* 45:1979–1988.

Misal SA, Lingojwar DP, Shinde RM, Gawai KR (2011) Purification and characterization of azoreductase from alkaliphilic strains *Bacillus badius*. *Process Biochem* 46(6):264–269.

Mohammad P, Azarmidokht H, Fatollah M, Mahboubeh B (2006) Application of response surface methodology for optimization of important parameters in decolorizing treated distillery wastewater using *Aspergillus fumigatus* UB2.60. *Int Biodeterior Biodegrad* 57:195–199.

MoP Report, 2021. Annual Report. Government of India, Ministry of Chemicals & fertilizers (Department of Pharmaceuticals). p. 151.

Nandy T, Shastry S, Kaul SN (2002) Wastewater management in a cane molasses distillery involving bio resource recovery. *J Environ Manage* 65:25–38.

Ng WJ, Yap MGS, Sivadas M (1989) Biological treatment of a pharmaceutical wastewater. *Biol Wastes* 29(4):299–311

Oktem YA, Ince O, Sallis P, Donnelly T, Ince BK (2007) Anaerobic treatment of a chemical synthesis-based pharmaceutical wastewater in a hybrid upflow anaerobic sludge blanket reactor. *Bioresour Technol* 99(5):1089–1096.

Pal P (2018) Treatment and disposal of pharmaceutical wastewater: toward the sustainable strategy. *Sep Purif Rev* 47:3, 179–198. http://dx.doi.org/10.1080/15422119.2017.1354888.

Patel HD, Madamwar D (2000) Biomethanation of low pH petrochemical wastewater using up-flow fixed film anaerobic bioreactors. *World J Microbiol Biotechnol* 16:69–75.

Patil PG, Kulkarni GS, Kore SV, Kore VS (2013) Aerobic sequencing batch reactor for wastewater treatment: a review. *Int J Eng Res Technol* 2(10):534–550.

Pignatello JJ, Oliveros E, MacKay A (2006) Advanced oxidation processes for organic contaminant destruction based on the fenton reaction and related chemistry. *Crit Rev Environ Sci Technol* 36: 1–84.

Raj DSS, Anjaneyulu Y (2005) Evaluation of biokinetic parameters for pharmaceutical wastewaters using aerobic oxidation integrated with chemical treatment. *Process Biochem* 40(1):165–175.

Rajeshwari KV, Balakrishnan M, Kansal A, Lata K, Kishore VVN (2000) State-of art of anaerobic digestion technology for industrial wastewater treatment. *Renew Sust Energ Rev* 4:135–156.

Rana RS, Singh P, Kandari, V, Singh R, Dobhal R, Gupta S (2017). A review on characterization and bioremediation of pharmaceutical industries' wastewater: an Indian perspective. *Appl Water Sci* 7:1–12.

Rao AG, Naidu GV, Prasad KK, Rao NC, Mohan SV, Jetty A, Sharma PN (2004) Anaerobic treatment of wastewater with high suspended solids from a bulk drug industry using fixed film reactor (AFFR). *Bioresour Technol* 93:241–247.

Rosen M, Welander T, Lofqvist A, Holmgren J (1998) Development of a new process for treatment of a pharmaceutical wastewater. *Water Sci Technol* 37(9):251–258.

Sahu O (2014) Reduction of organic and inorganic pollutant from wastewater by algae. *Int Lett Nat Sci* 8:1–8. http://dx.doi.org/10.18052/www.scipress.com/ILNS.13.1.

Saleem M (2007) Pharmaceutical wastewater treatment: a physicochemical study. *J Res (Sci)* 18(2):125–134.

Seitz W, Jiang J, Weber WH, Lloyd BJ, Maier M, Maier D (2006) Removal of iodinated X-ray contrast media during drinking water treatment. *Environ Chem* 3(1):35–39.

Shi X, Leong KY, Ng HY (2017) Anaerobic treatment of pharmaceutical wastewater: a critical review. *Bioresour Technol*. http://dx.doi.org/10.1016/j.biortech.2017.08.150a

Shi X, Ng KK, Li X, Ng HY (2015) Investigation of intertidal wetland sediment as a novel inoculation source for anaerobic saline wastewater treatment. *Environ Sci Technol* 49:6231–6239.

Spina F, Anastasi A, Prigione V, Tigini V, Varese GC (2012) Biological treatment of industrial wastewaters: a fungal approach. *Chem Eng Trans* 27:175–180.

Sponza DT, Demirden P (2007) Treatability of sulfamerazine in sequential upflow anaerobic sludge blanket reactor (UASB)/completely stirred tank reactor (CSTR) processes. Separation and Purification Technology 56(1):108–117.

Sreekanth D, Sivaramakrishna D, Himabindu V, Anjaneyulu Y (2009) Thermophilic treatment of bulk drug pharmaceutical industrial wastewaters by using hybrid up flow anaerobic sludge blanket reactor. *Bioresour Technol* 100(9):2534–2539.

Suarez S, Lema JM, Omil F (2009) Pre-treatment of hospital wastewater by coagulation–flocculation and flotation. *Bioresour Technol* 100:2138–2146.

Tewari L, Malviya P (2002) Biodegradation of catechol by fluorescent *Psuedomonas* for sustainable environment. *J Sci Ind Res* 61:70–74.

Tóth AJ, Gergely F, Mizsey P (2011) Physicochemical treatment of pharmaceutical process wastewater: distillation and membrane processes. *Chem Eng* 55(2):59–67.

US-EPA (1974) "Report to Congress Disposal of Hazardous Wastes", US-EPA Publication in the Solid Waste Management Series.

Vanerkar AP, Satyanarayan S, Dharmadhikari DM (2013) Full scale treatment of herbal pharmaceutical industry wastewater. *Int J Chem Phys Sci* 2:52–62.

Verlicchi P, Al Aukidy M, Zambello, E (2012) Occurrence of pharmaceutical compounds in urban wastewater: removal, mass load and environmental risk after a secondary treatment—a review. *Sci Total Environ* 429:123–155, http://dx.doi.org/10.1016/j.scitotenv.2012.04.028.

Wei X, Li B, Zhao S, Wang L, Zhang H, Li C, Wang S (2012) Mixed pharmaceutical wastewater treatment by integrated membraneaerated biofilm reactor (MABR) system—a pilot-scale study. *Bioresour Technol* 122:189–195.

Westerhoff P, Yoon Y, Snyder S, Wert E (2005). Endocrine disruptor, pharmaceutical, and personal care product fate during simulated drinking water treatment processes. *Environ Sci Technol* 39(17):6649–6663.

Wu S, Zhang L, Chen J (2012) Paracetamol in the environment and its degradation by microorganisms. *Appl Microbiol Biotechnol* 96:875–884, http://dx.doi.org/10.1007/s00253-012-4414-4.

Xu F, Mou Z, Geng J, Zhang X, Li C (2016) Azo dye decolorization by a halotolerant exoelectrogenic decolorizer isolated from marine sediment. *Chemosphere* 158:30–36.

17

Biotechnological Approaches for Microbial Treatment of Textile Wastewater and Resource Recovery: Opportunities, Challenges, and Future Perspectives

Himanshu Tiwari and R.S. Singh
Indian Institute of Technology (BHU)

CONTENTS

17.1 Introduction

The textile industry is one of the leading overgrown sectors and proliferates as a result of rapid industrialization. Across the globe, as the population surges, the demand for textile allied products increases. The global textile dyes market surges with an 8.13% of compound annual growth rate (CAGR) during the forecast period of 2016–2023, which is as per the study by Market research future (MRFR) (Samsami *et al.*, 2020). The production of synthetic dyes is about one million tons across the globe, which annually emits 280,000 tons of textile effluents (Roy *et al.*, 2020). Throughout the product manufacturing process, water consumes abundantly. Various processing units emit wastewaters polluted with organic waste such as dyes and pigments. Textile effluents have mostly synthetic dyes, salts, heavy metals, organic surfactants such as alkyl ether phosphates, carboxylates, sulphonates, and solvents namely 1, 1, 1-trichloroethane, l-chlorinated hydrocarbons, and Dichloromethane are also present (Behera *et al.*, 2021). Therefore, the direct discharge of these pollutants may lead serious threat to our environment. The disposal of textile effluents not only leads to eutrophication, and diminishes sunlight penetration into the aquatic body but also enhances the COD, BOD, pH, temperature, and turbidity.

Meanwhile, various physical and chemical techniques viz. coagulation-flocculation, photocatalysis, membrane-based separation techniques, adsorption, ozonation, Fenton oxidation are also effective; but the generation of secondary pollutant, high expenses, and use of toxic chemicals restrict their wide range of applications (Mustafa *et al.*, 2021). Bioremediation has proved to be the cost-effective, environmentally friendly sustainable, most popular, and desirable technique for textile allied wastewater effluents treatment. Microbial species naturally develop into new strains, readily mineralized recalcitrant, carcinogenic organic dye effluents into less toxic products and ultimately into simpler compounds such as CO_2 and H_2O.

DOI: 10.1201/9781003231738-19

The enzyme acts as a biocatalyst, effectively utilized in free or immobilized form during a typical bioprocess. Immobilization of enzymes imparts operational features including long-term stability, no loss in their activity, easy recovery, and reusability, which improves process performance significantly and makes the overall process cost-effective. However, immobilization of enzymes conserves their tertiary structure and active sites, the thermal stability of enzymes improves, and so its process utility further proliferates (Kashefi et al., 2019b). Enzymatic degradation techniques are also efficiently used for textile wastewater treatment. Laccases, reductases, and oxygenases, also known as bacterial enzymes, have been utilized successfully for azo bond cleavage under anaerobic conditions, also after that, the aromatic amines are biologically transformed under aerobic conditions. The degradation mechanism follows anaerobic/aerobic or sometimes sequential methods.

Nowadays, nanoparticle-based treatment techniques have become much more important in biological processes. A wide variety of nanomaterials have been adopted in the bioremediation of textile dyes. The size range of nanoparticles is 10–100 nm, which is somehow quite reasonably similar to the size range of biological molecules viz. proteins, lipids, and DNA. A variety of enzymes are merely immobilized into the Nano support materials and enhance the process performance. Furthermore, enzymes immobilize nanomaterials such as graphene oxide (GO) and carbon nanotubes. GO is considered the suitable support material for enzyme immobilization, providing numerous large developed surface areas. Enzymes and GO-based nanomaterials can interact with each other either in the form of covalent bonding through cross-linkers or weak van der Waals forces, electrostatic interaction, π–π stacking, and hydrophobic interactions (Kashefi et al. 2019a). On the other hand, GO exhibits good sorption characteristics, having hydroxyl groups on its surface, therefore, responsible for dye sorption, the immobilized enzyme could lead the dye degradation process simultaneously (Ariaeenejad, Motamedi and Hosseini Salekdeh, 2021). Thus, the synergistic nanobiocatalytic system explores an advanced textile pollutant mitigation technique, which superficially resembles the bioremediation technique.

This chapter discusses the existing biotechnological approaches based on microbes that are especially useful for textile wastewater treatment. Furthermore, new aspects of Nanobiotechnology will discuss.

17.2 Textile Wastewater Treatment Technology

There is no doubt that the textile sector is one of the leading creators of wastewater. The wastewater consists of a complex mixture of heavy metals, detergents, inorganic salts, dispersants, surfactants, alkalis, dioxin, volatile organic compounds (VOCs), pentachlorophenol, reducing agents, and binders (Kishor *et al.*, 2021). Synthetic dyes such as acidic, basic, reactive, vat, azo, direct, and sulfide have been used. To tackle textile effluents and maintain environmental sustainability, various physicochemical and biological methods have traditionally been used. Sometimes, a combination of these methods adopts for enhancing the overall process performance.

17.2.1 Physico-Chemical Treatment Methods

Adsorption, membrane-based separation techniques, and ion exchange are the most popular physical treatment techniques, whereas electrochemical, coagulation/flocculation, and oxidation techniques are chemical methods. Adsorption, a surface-based phenomenon, solid adsorbent (such as activated carbon, charcoal, silica gel, alumina, and zeolite) attract adsorbate molecules/ions (from polluted water) to its surface. Carbon-based adsorbents such as activated carbon, charcoal, and biochar have traditionally been used, even though activated carbon has lots of importance at the industrial level. This method allows the mass transfer of pollutants from one phase into another and does not mineralize the pollutant. Membrane-based separation technique (membrane filtration), size exclusion method, selectively permeate solute molecule through its pores. Microfiltration (MF), ultrafiltration (UF), nanofiltration (NF), and reverse osmosis (RO) are extensively in use for textile dye color removal, salinity, and COD reduction. The ion exchange process has applications in heavy metal removal, water purification, and textile effluents treatment. Ion exchange resins have their functional groups charged on them, charged dyes also have a unique charge, the ionic attraction between resin and charge dye molecule makes a strong bond, deposits dye on the surface of the resin, and color removal occurs.

Chemical techniques (exemption for electrochemical technology) are relatively more expensive than physical and biological methods. The dye removal via this technique requires proper equipment as well as a large number of chemicals and electrical energy, the secondary waste generates from it is toxic and creates waste disposal problems, restricting its wide range of process applicability. Electrocoagulation (EC) tank consists of two metal electrodes, a direct current source, and in-situ formed coagulant particles. Electrolytic reduction and oxidation take place, anode made up of metal acts as a catalyst as well as a coagulant reagent. Overall, the reactor generates odorless and colorless water. Other advantages are no additional chemical requirement, no secondary waste generation, economically viable, and environmentally friendly. Electro-Fenton (EF), integration of electrolysis and Fenton's reaction, removes organic wastes through the process of combination of coagulation and oxidation. Advanced oxidation processes (AOPs) such as photocatalysis, ozonation, plasma, Photo-Fenton, ultrasound, and UV-based methods are specially used for wastewater treatment. In another process, called coagulation-flocculation, chemical coagulants such as aluminum sulfate solution ($Al_2(SO_4)_3.18H_2O$) inject under vigorous mixing conditions. Neutralization of oppositely charged dispersed particles happens, flocculants thereby add, a conglomeration of fine particles forms a larger particle. The sedimentation process allows the removal of these particles.

17.2.2 Biological Treatment Methods

As a matter of custom, the biological process overwhelmingly gains widespread application for the treatment of industrial textile wastewater. Bacteria, fungi, algae, and yeast play a vital role during a typical biodegradation process of organic pollutants. Because bacterial culture is grown (cultured) easily, has high specific growth rates, and possesses enough potential for the mineralization of recalcitrant organic compounds, and so is abundantly used for dye degradation (Yang *et al.*, 2014). Bacterial degradation of azo dyes occurs significantly under anaerobic conditions; azoreductase enzyme breaks the azo bond. In most cases, microbial consortia consider to enhance the degradation rate, two or more bacterial species degrade the intermediate/metabolite compounds generated by others, so pollutant mineralization facilitates. *Pseudomonas, Rhodococcus, Lysinibacillus, Klebsiella, Aeromonas sp., Arthrobacter, Shigella* and *Bacillus* are bacterial strains responsible for azo dye degradation (Samsami *et al.*, 2020). *Chlorella Vulgaris* and *Spirulina Platensis* are the algae that show a better degradation performance (more than 90% degradation) for Reactive Black 5, Disperse Red 1, and Indigo Blue dye, respectively (Ishchi et al., 2020; Robledo-Padilla *et al.*, 2020). Whereas, fungus species *Achaetomium Strumarium, Cylindrocephalum Aurelium* RY06, and *Trametes Versicolor* also perform significantly towards the degradation of Acid Red 88 (99% removal), Mordant Orange-1 (85% removal), and Methyl Red (91% removal) (Mostafa et al., 2019; Bankole et al., 2018; Dayi et al., 2019). Various well-known yeast species belong to the genera of *Saccharomyces, Ascomycetous, Candida, Debaryomyces, Trichosporon, Kluyveromyces,* and *Rhodotorula* show significant dye degradation potential (Martorell *et al.*, 2018).

17.2.3 Enzymatic Treatment

Enzymes, biomacromolecular proteins (Ding *et al.*, 2015) act as a biocatalyst, accelerate a typical biological process to several folds. It also plays a vital role in almost all metabolic processes, catalyze all the individual intermediate metabolic pathways. As usual, they are specific in nature like the chemical catalyst, i.e., one specific enzyme can catalyze a unique biological reaction. Likewise, microbial degradation, enzymes now become increasingly viable biocatalysts for the treatment of textile wastewater at the industrial level. It, too, exists in liquid form, has enormous efficiency towards dye degradation. The chief quantity of interest during an enzymatic reaction is 'enzymatic activity'; as the pH or temperature reaches an extreme range, denaturation seriously lowers down the enzyme activity, and this is due to the loss of its active sites, ultimately deactivation of biocatalyst occurs. Poor stability, difficulty in recovery, and reusability may hamper its process performance at pilot plant wastewater treatment. These process difficulties have been suppressed by the immobilization of the enzyme. Enzyme immobilization preserves the enzyme, getting attached to insoluble and inert solid support material.

Immobilization, however, improves its activity, imparts structural rigidity (stability), redresses inactivation of the enzyme, and minimizes microbial contamination; compared to its free enzyme counterpart. Some common outcome of enzyme immobilization is: (i) mass transfer limitations; (ii) steric hindrance; and (iii) conformational change in the structure of an enzyme, therefore, may impose the limitation in enzyme performance (Ding *et al.*, 2015; Longoria *et al.*, 2010). In general, organic (natural and synthetic polymers, e.g., cellulose, chitin, chitosan, and collagen) and inorganic (silica, kaolin, glass beads, celite, metals, and metal oxides) substances use as the support materials for enzyme immobilization (Jun *et al.*, 2019). Adsorption, encapsulation, cross-linking, entrapment, and covalent bonding; are the methods used for enzyme immobilization on the typical solid support. Quantitatively, immobilization efficiency (IE) and relative activity (RA) are the principal measurements of the biocatalytic performance of immobilized enzymes. They are defined below by governing equations (17.1) and (17.2) (Jun *et al.*, 2019).

$$\text{Immobilization efficiency}(\text{IE}) = \frac{\text{Enzyme immobilized}}{\text{Enzyme loaded}} \times 100\% \tag{17.1}$$

$$\text{Relative activity}(\text{RA}) = \frac{\text{Specific activity of immobilized enzyme}\left(\dfrac{\text{U}}{\text{mg}}\ \text{protein}\right)}{\text{Specific activity of free enzymes}\left(\dfrac{\text{U}}{\text{mg}}\ \text{protein}\right)} \times 100\% \tag{17.2}$$

Microorganisms, carrying out the degradation process, are responsible for the secretion of the extra- and intracellular enzymes. Enzymes including azoreductase, peroxidase, laccase, polyphenol oxidases, NADH-DCIP reductase, manganese peroxidases, veratryl alcohol oxidase, and lignin peroxidases have already shown their great potentiality at the industrial level for textile wastewater remediation.

17.3 Hybrid Biotechnological Approaches for Textile Wastewater Treatment

There are numerous physicochemical and biological techniques have been discovered for the remediation of recalcitrant textile waste effluents. However, none of them appears to be a single, universally applicable method that will always give significant degradation performance. It realistically paves a way toward the emergence of the combination of suitable methods to mitigate the limitations of a single technique. Hybrid techniques always adhere to the advantages of two separate methodologies. The ultimate aim is to attain the maximum possible extent of pollutant mineralization; without secondary sludge inculcation with minimum consumption of energy; so that the entire process will be economically viable and environmentally friendly.

17.3.1 Microbial Fuel Cell (MFC) Technology

Microbial fuel cell (MFC) is an invasive, sustainable bioelectrochemical technique that simultaneously degrades organic wastes and converts chemical energy into electrical energy. It integrates the chief advantages of biocatalyst (microbes) activity and abiotic electrochemical reactions (Rodrigues de Almeida et al., 2019). Along with textile waste degradation, can be successfully utilized for the treatment of domestic as well as industrial wastewater treatment. Microbial species, carrying the strong potential for targeted waste mineralization, facilitate the degradation of refractory organic compounds in waste and generate electricity simultaneously. Thus, the recovery of renewable energy (bioenergy) from waste paves the way toward higher power generation in the future. Recently, it has gotten widespread application due to its environmentally friendly behavior and low cost.

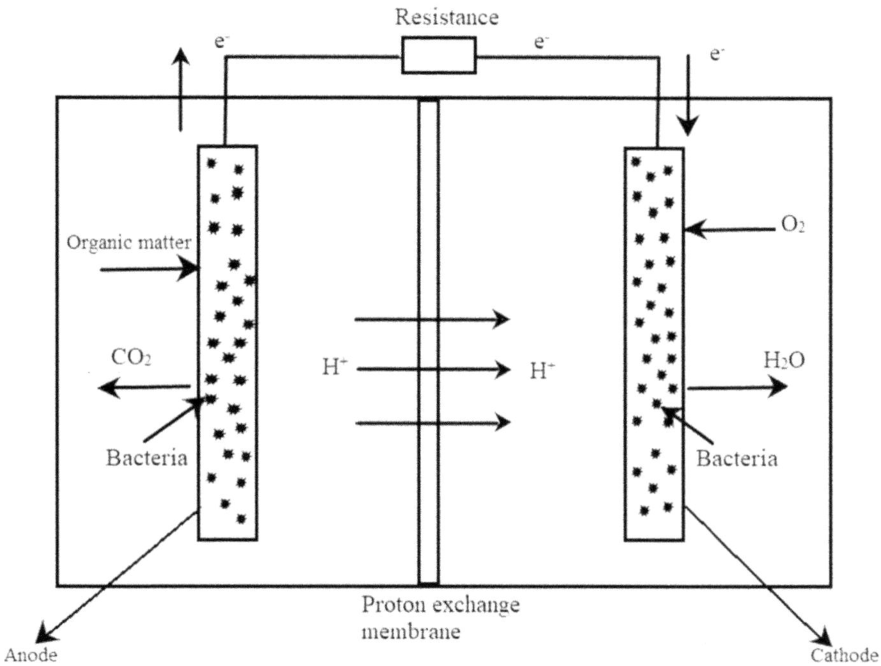

FIGURE 17.1 A typical microbial fuel cell reactor.

In an MFC, oxidation of organic compounds occurs in the anode, and proton generation takes place, on the other hand, reduction of oxygen (O_2) to water (H_2O) takes place at the cathode, as shown in Figure 17.1. The membrane ensures the separation of anode and cathode compartments (Khan et al. 2020). Meanwhile, the bacteria responsible for dye degradation and the electrochemically active bacteria (EAB) go towards competing for dye degradation and bioelectricity generation from electrons simultaneously (Li et al., 2021). External resistance attaches between the cathode and anode compartment for the collection of generated bioelectricity. MFCs at the industrial level exhibits limitation in terms of low power production and high cost of membrane and cell electrodes. A variety of microbial cultures have been reported in the literature for dye degradation, as shown in Table 17.1.

TABLE 17.1
MFC Configuration and Dye Degradation by Various Microbial Cultures

S.N.	MFC Configuration	Microbial Culture	Dye	References
1.	Single chambered, Nafion membrane, bianodic	*Dysgonomonas* and *Pseudomonas* consortium	Procion Red MX-5B (PRMX)	Rodrigues de Almeida et al. (2019)
2.	Two double anode stacked MFC	Microbial consortium	Reactive brilliant red X-3B	Cao et al. (2021)
3.	Two membranes less baffled MFC	Microbial consortium	New Coccine	Oon et al. (2021)
4.	Aerobic biocathode two-chamber	Microbial consortium	Congo red	Hou et al. (2019)
5.	Single chamber	Anaerobic sludge (Mixed culture)	Acid Black 10B	Yeruva et al. (2018)
6.	Single chamber	*Enterobacter cancerogenus*	Reactive Blue 160 Reactive Green 19	Chen et al., (2016)

17.3.2 Membrane Bioreactor

Membrane bioreactor is popular and overwhelmingly used especially for dye-contaminated wastewater remediation. It integrates the chief advantages of bioremediation and membrane-based advanced separation technology and degrades the recalcitrant textile wastewater. Microfiltration, Ultrafiltration, Nanofiltration, and Reverse osmosis are the major applications of membrane-based separation technology. The pore size of the selected membrane directly depends on the selective membrane process, which has to do with a typical separation process. Microbial species, enzymes, and plant/animal cells act as the biocatalyst. Biocatalysts restrict the physical space of the membrane throughout the reactor system. Enzymes and microorganisms have immobilized to the membrane surface through the covalent bonds, and sometimes they are allowed to be entrapped within the pores. "Membrane fouling" is one of the undesirable phenomena, significantly diminishing the flow rate and simultaneously enhancing the

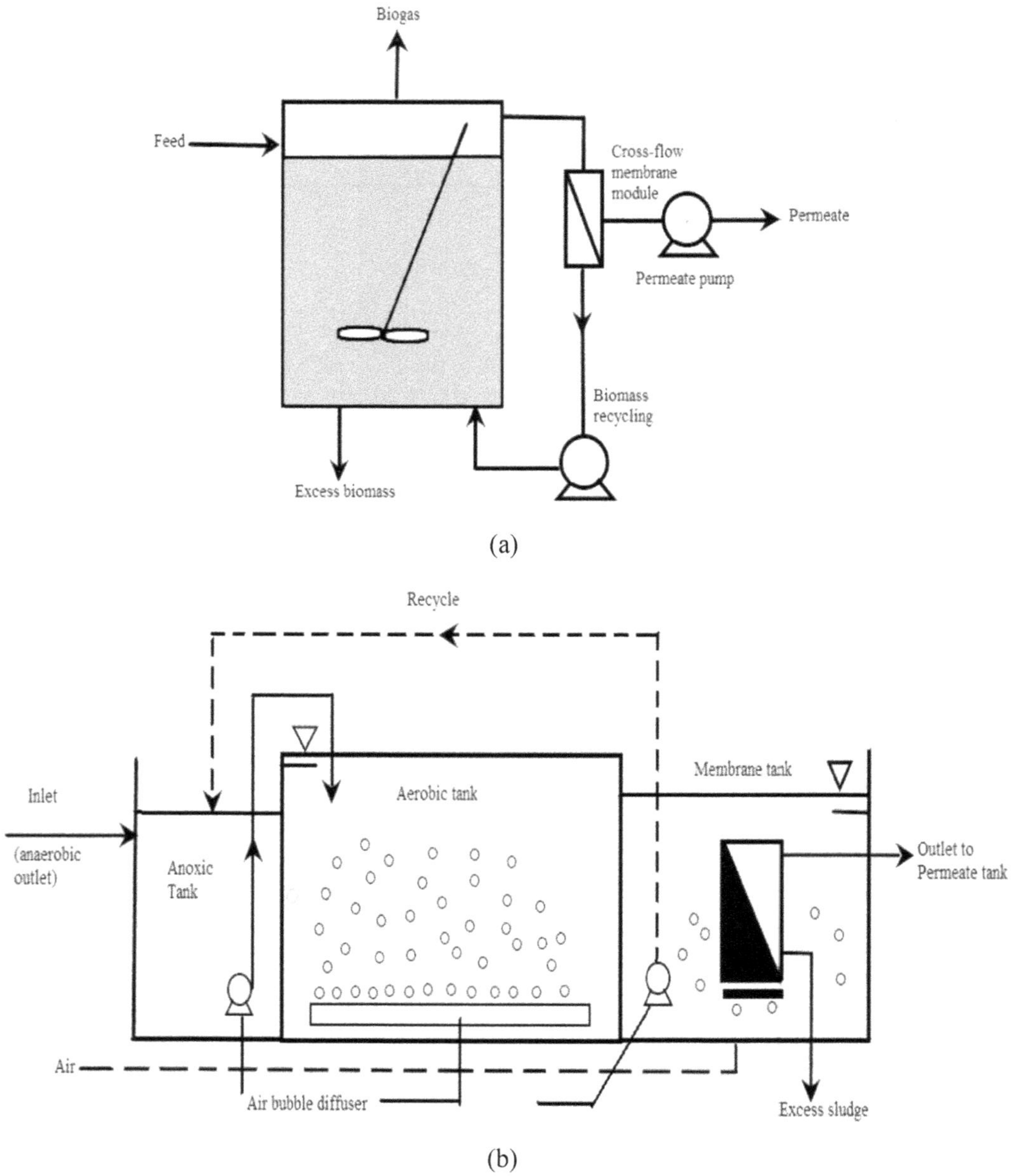

FIGURE 17.2 Membrane bioreactor (a) external loop recycle reactor, (b) submerged membrane reactor (Galinha et al., 2018.)

TABLE 17.2
MBR Configuration for Textile Dye Wastewater Remediation

S.N.	MBR Configuration	Microbial Culture	Dye	References
1.	Anoxic/oxic electro-membrane bioreactor (A/O-EMBR),	Janibacter and *Lactococcus* genus	Remazol brilliant violet-5R	Ravadelli et al. (2021)
2.	The membrane bioreactor inoculates with an enriched anaerobic methane oxidation (AOM) culture.	*Methanomethylovorans* are dominant archaea, and *Moranbacteria* is the leading bacteria	Methyl orange (MO)	Bai et al. (2020)
3.	Bioaugmentation in a membrane bioreactor (MBR)	*Aeromonas hydrophila* LZ-MG14	Malachite green	Ji et al. (2020)
4.	Sequencing anaerobic moving bed biofilm reactor (AnMBBR) and powdered activated carbon – aerobic membrane bioreactor (PAC-AeMBR)	*Alphaproteobacteria*	Real textile wastewater	Kozak, Cirik and Basak (2021)
5.	Anoxic & oxic biofilm-membrane bioreactors	*Proteobacteria* (44% abundance)	Textile wastewater	Zhou et al. (2021)
6.	Aerobic membrane bioreactor	IHK22 consortium	Reactive Blue Bezactiv S-GLD 150 (BB150)	Khouni, Louhichi and Ghrabi (2020)

trans-membrane pressure (TMP), therefore, imposing the limitation on process applicability, restricting the life span of the membrane, and quite often requiring periodic membrane cleaning (Jegatheesan et al., 2016). A typical membrane bioreactor setup has shown in Figure 17.2. Various organic, inorganic molecules and biomass are present in the wastewater that accumulates over the membrane surface during a typical separation process and leads to membrane fouling. Table 17.2 summarises the process application of MBR technology for textile wastewater treatment.

17.4 Nanobiotechnogy

Nanobiotechnology is a new evolving era of science that deals especially with nanostructured materials and nanotechnology tools, their biological and physicochemical properties, along applications among various fields including medical, agriculture, and life science. In most cases, it could be considered for the integration of biotechnology and nanotechnology, having the dimensions of materials within 1–100nm (Anwar, Zwar and de Souza, 2020). These nanomaterials at the molecular level interact uniquely with biological systems due to their size similarity. Nanomaterials have desirable characteristic features: (i) high surface area to volume ratio; (ii) high adsorption capacity; (iii) low intraparticle diffusion resistance, and (iv) expeditiously come into adsorption equilibrium, which makes them a perfect candidate for being the adsorption process has to be carried out (Mehta *et al.*, 2021). Nowadays, consistent research efforts are carried out to eliminate traditional wastewater treatment methods by incorporating nanobiotechnology. This technology doesn't require any large infrastructures, and could efficiently treat the wastewater along with other treatment methods such as adsorption, membrane technology, photocatalysis, and coagulations (Sahu *et al.*, 2019). Nanotubes, nanostructured catalytic membranes, nano-catalysts, and nano-adsorbent are used extensively for wastewater remediation. Nanomaterials of various types comprising carbonaceous nanomaterials, dendrimers, metal-containing nanoparticles, and zeolites can also be successfully utilized (Sharma *et al.*, 2017). Nanomaterials have been synthesized as the perfect supporting materials for enzyme immobilization and thereby enhancing the biocatalyst efficiency. Nanomaterials are classified into inorganic, organic, and carbon-based nanomaterials, as illustrated in Figure 17.3.

Furthermore, the major types of inorganic nanomaterials are Metal-based nanoparticles (NPs), transition metal chalcogenide NPs, and metal-oxide/hydroxide NPs. Cu, Au, Co, Pt, Pd, Ru, Re, Pb, Cd, Ag,

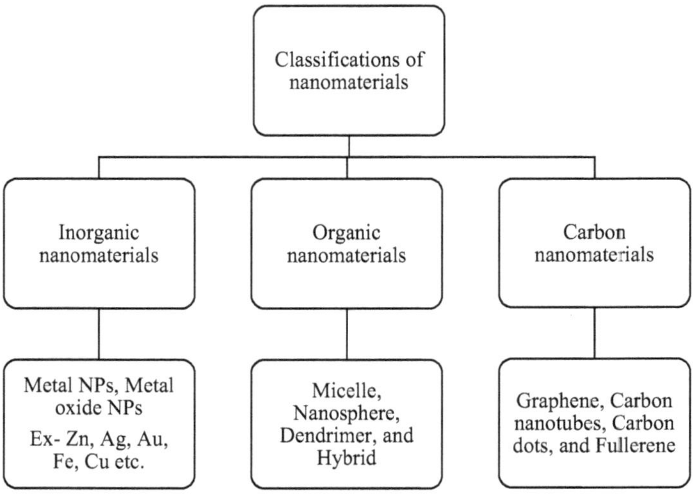

FIGURE 17.3 Classifications of nanomaterials: inorganic nanomaterials, organic nanomaterials, and carbon nanomaterials.

Zn, and Ni are the popular metals that have been employed for metal-based nanoparticles synthesis, metal oxides such as Fe_3O_4, $MnFe_2O_4$, WO_2, CdS, TiO_2, NiO, and ZnO are the well-known metal oxide NPs. Metal NPs, either in the appearance of colloidal solutions or solid particles, are generally produced by techniques including hydrothermal synthesis, bio-assisted synthesis, and microwave-assisted synthesis. Carbon and hydrogen-based nanoparticles are underlining in the category of organic nanomaterials; examples are liposomes, ferritin, compact polymer nanoparticles (NPs) (nanospheres, nanocapsules), dendrimers, micelles, and hybrid NPs. Their non-toxic and biodegradable nature makes them environmentally friendly. Carbon nanomaterials, also called new generation nanomaterials, are graphene, carbon nanotubes (single wall and multi-wall carbon nanotubes), carbon black, carbon nanofibers, carbon dots, nanodiamond, and fullerene; having high specific surface area, high sensitivity to cost ratio, and are non-toxic in nature (Joshi *et al.*, 2019).

Enzymes have been immobilized successfully on carbon-based nanomaterials for investigating the dye biodegradation performance. Graphene and graphene oxide have both high structural strength and high specific surface area, so they overwhelmingly prefer as the support material for the immobilization of biocatalysts. Graphene oxide and enzyme could interact through covalent bonding (cross-linking), physical adsorption, and non-covalent bonding such as electrostatic interaction, weak van der Waals forces, hydrophobic interactions, and π–π stacking. Recently, nanobiocatalysts have become very popular in the remediation of various industrial wastewater effluents such as textile, hospital, and municipal wastewater. Graphene and other metal-based nanoparticles have been subjecting to some modifications before the textile wastewater treatment.

Kiran *et al*(2020) fabricated graphene oxide functionalized $MnFe_2O_4$ nanoparticles, 10 nm in size, using sol-gel auto-combustion. Lignin Peroxidase (LiP), secreted from *Pseudomonas fluorescens* LiP-RL5, was immobilized on nanoparticle support material for investigating the degradation study of various textile dyes. Nanobiocatalyst had been reported as thermostable and hyperactive. LiP enzymatic activity had retained 50% after nine successive runs, and it remained stable after attaining the temperature of 50°C with 14 hours of half-life duration. Experimental analysis revealed that methylene blue dye showed a maximum of 88% decolorization, bromophenol blue exhibited 82% decolorization. LiP showed its effectiveness with up to 88% dye degradation during 1 hour of incubation at a temperature of 30°C. In another study, Kashefi et al (2019a) used covalently immobilized laccase enzyme on graphene oxide nanosheets as a nanobiocatalyst through the facile and large-scale method. Laccase (EC 1.10.3.2), extracted from genetically modified white-rot fungi *Aspergillus*, also gave significant results in terms of enzyme loading (156.5 mg/g), concentration (0.9 mg/mL), and immobilization yield (64.6%). The targeted anionic dyes were Direct Red 23 (DR23) and Acid Blue 92 (AB92). Nanobiocatalyst dose of 300.0

mg/L exhibited 88.7% and 48.7% bioconversions of DR23 and AB92, respectively. The reusability of nanobiocatalyst was more than 75% even after completing six cycles of dye decolorization.

Thus nanobiotechnology will seem to be one of the green technology in the future for different types of wastewater remediation. Excellent mechanical strength and reusability are one of desirable catalytic characteristics. It not only enhances the enzymatic activity but also withstands it under a mild reaction environment. Therefore, the nanobiocatalyst and allied nanobiotechnology need further proliferation in textile wastewater reclamation.

17.5 Conclusions

As the textile demand expeditiously surges with population, wastewater effluents from textile production and associated synthetic waste treatment become one of challenging tasks, need an innovative, reliable, economically viable, and environmentally friendly technique. The biological methods have been used for textile effluents treatment and significantly degrade the recalcitrant, carcinogenic, and mutagenic organic compounds. Enzymatic treatment believes to give promising results but suffers from activity reduction and fluctuation in pH and temperature of the reaction. Nanobiotechnology merely enhances the biodegradation rate through the incorporation of immobilization of biocatalyst over supported nanomaterials. The enzymatic treatment limitations have almost been circumventing by the immobilization technique. The separation of nanoparticles from treated effluents leads the difficulty and results in losing it. Nanobiotechnology, akin to "Green Technology", needs an exploration towards industrial-scale wastewater treatment.

REFERENCES

Anwar, U. B., Zwar, I. P. and de Souza, A. O. (2020) 'Biomolecules produced by extremophiles microorganisms and recent discoveries', *New and Future Developments in Microbial Biotechnology and Bioengineering*, pp. 247–270. doi: 10.1016/B978-0-444-64301-8.00012-3.

Ariaeenejad, S., Motamedi, E. and Hosseini Salekdeh, G. (2021) 'Application of the immobilized enzyme on magnetic graphene oxide nano-carrier as a versatile bi-functional tool for efficient removal of dye from water', *Bioresource Technology*, 319, p. 124228. doi: 10.1016/J.BIORTECH.2020.124228.

Bai, Y. N. et al. (2020) 'High-rate anaerobic decolorization of methyl orange from synthetic azo dye wastewater in a methane-based hollow fiber membrane bioreactor', *Journal of Hazardous Materials*, 388(October). doi: 10.1016/j.jhazmat.2019.121753.

Bankole, P. O., Adekunle, A. A. and Govindwar, S. P. (2018) 'Enhanced decolorization and biodegradation of acid red 88 dye by newly isolated fungus, *Achaetomium strumarium*', *Journal of Environmental Chemical Engineering*, 6(2), pp. 1589–1600. doi: 10.1016/J.JECE.2018.01.069.

Behera, M. et al. (2021) 'A review on the treatment of textile industry waste effluents towards the development of efficient mitigation strategy: An integrated system design approach', *Journal of Environmental Chemical Engineering*, 9(4), p. 105277. doi: 10.1016/J.JECE.2021.105277.

Cao, X et al. (2021) 'Limitation of voltage reversal in the degradation of azo dye by a stacked double-anode microbial fuel cell and characterization of the microbial community structure', *Science of the Total Environment*, 754, p. 142454. doi: 10.1016/J.SCITOTENV.2020.142454.

Chen, C. Y. et al. (2016) 'Analysis of bacterial diversity and efficiency of continuous removal of Victoria Blue R from wastewater by using packed-bed bioreactor', *Chemosphere*, pp. 17–24. doi: 10.1016/j.chemosphere.2015.11.061.

Dayi, B., Duishemambet Kyzy, A. and Akdogan, H. A. (2019) 'Characterization of recuperating talent of white-rot fungi cells to dye-contaminated soil/water', *Chinese Journal of Chemical Engineering*, 27(3), pp. 634–638. doi: 10.1016/J.CJCHE.2018.05.004.

Ding, S. et al. (2015) 'Increasing the activity of immobilized enzymes with nanoparticle conjugation', *Current Opinion in Biotechnology*, 34, pp. 242–250. doi: 10.1016/J.COPBIO.2015.04.005.

Galinha, C. F., Sanches, S. and Crespo, J. G. (2018) 'Membrane bioreactors', *Fundamental Modelling of Membrane Systems*, 6, pp. 209–249. doi: 10.1016/B978-0-12-813483-2.00006-X

Hou, B. et al. (2019) 'Performance of microbial fuel cells based on the operational parameters of biocathode during simultaneous Congo red decolorization and electricity generation', *Bioelectrochemistry*, 128, pp. 291–297. doi: 10.1016/J.BIOELECHEM.2019.04.019.

Ishchi, T. and Sibi, G. (2020) 'Azo dye degradation by Chlorella vulgaris: optimization and kinetics', *International Journal of Biological and Chemical Sciences,* 14(1), pp. 1–7.

Jegatheesan, V. et al. (2016) 'Treatment of textile wastewater with membrane bioreactor: A critical review', *Bioresource Technology*, 204, pp. 202–212. doi: 10.1016/J.BIORTECH.2016.01.006.

Ji, J. et al. (2020) 'Bioaugmentation of membrane bioreactor with *Aeromonas hydrophila* LZ-MG14 for enhanced malachite green and hexavalent chromium removal in textile wastewater', *International Biodeterioration and Biodegradation*, 150, p. 104939. doi: 10.1016/j.ibiod.2020.104939.

Joshi, S. J. et al. (2019) Laccases in nano-biotechnology: Recent trends and advanced applications. In: Joshi, S. J. and Geetha S. J. (Eds.), *Recent Advances in Biotechnology*. Nova Science Publishers, Hauppauge, NY, pp. 39–62. ISBN: 978-1-53615-860-1

Jun, L. Y. et al. (2019) 'An overview of immobilized enzyme technologies for dye and phenolic removal from wastewater', *Journal of Environmental Chemical Engineering*, 7(2), p. 102961. doi: 10.1016/J.JECE.2019.102961.

Kashefi, S., Borghei, S. M. and Mahmoodi, N. M. (2019a) 'Covalently immobilized laccase onto graphene oxide nanosheets: Preparation, characterization, and biodegradation of azo dyes in colored wastewater', *Journal of Molecular Liquids*, 276, pp. 153–162. doi: 10.1016/J.MOLLIQ.2018.11.156.

Kashefi, S., Borghei, S. M. and Mahmoodi, N. M. (2019b) 'Superparamagnetic enzyme-graphene oxide magnetic nanocomposite as an environmentally friendly biocatalyst: Synthesis and biodegradation of dye using response surface methodology', *Microchemical Journal*, 145, pp. 547–558. doi: 10.1016/J.MICROC.2018.11.023.

Khan, R., Patel, V. and Khan, Z. (2020) 'Bioremediation of dyes from textile and dye manufacturing industry effluent', In *Abatement of Environmental Pollutants: Trends and Strategies*, pp. 107–125. doi: 10.1016/B978-0-12-818095-2.00005-9.

Khouni, I., Louhichi, G. and Ghrabi, A. (2020) 'Assessing the performances of an aerobic membrane bioreactor for textile wastewater treatment: Influence of dye mass loading rate and biomass concentration', *Process Safety and Environmental Protection*, 135, pp. 364–382. doi: 10.1016/j.psep.2020.01.011.

Kiran et al. (2020) 'Fabrication of thermostable and reusable nanobiocatalyst for dye decolourization by immobilization of lignin peroxidase on graphene oxide functionalized $MnFe_2O_4$ superparamagnetic nanoparticles', *Bioresource Technology*, 317, p. 124020. doi: 10.1016/J.BIORTECH.2020.124020.

Kishor, R. et al. (2021) 'Ecotoxicological and health concerns of persistent coloring pollutants of textile industry wastewater and treatment approaches for environmental safety', *Journal of Environmental Chemical Engineering*, 9(2), p. 105012. doi: 10.1016/J.JECE.2020.105012.

Kozak, M., Cirik, K. and Basak, S. (2021) 'Treatment of textile wastewater using combined anaerobic moving bed biofilm reactor and powdered activated carbon-aerobic membrane reactor', *Journal of Environmental Chemical Engineering*, 9(4), p. 105596. doi: 10.1016/J.JECE.2021.105596.

Li, T. et al. (2021) 'Biological detoxification and decolorization enhancement of azo dye by introducing natural electron mediators in MFCs', *Journal of Hazardous Materials*, 416, p. 125864. doi: 10.1016/J.JHAZMAT.2021.125864.

Longoria, A. et al. (2010) 'Enzyme technology of peroxidases: Immobilization, chemical and genetic modification', In *Biocatalysis Based on Heme Peroxidases*, pp. 209–243, Springer, Berlin, Heidelberg. doi: 10.1007/978-3-642-12627-7_9.

Martorell, M. M. et al. (2018) 'Optimization and mechanisms for biodecoloration of a mixture of dyes by Trichosporon akiyoshidainum HP 2023', *Environmental Technology*, 39(24), pp. 3169–3180. doi: 10.1080/09593330.2017.1375024.

Mehta, M. et al. (2021) 'Degradation of synthetic dyes using nanoparticles: A mini-review', In *Environmental Science and Pollution Research 2021*, pp. 1–13. doi: 10.1007/S11356-021-15470-5.

Mostafa, A. A. F. et al. (2019) 'Decolorization and biotransformation pathway of textile dye by *Cylindrocephalum aurelium*', *Bioprocess and Biosystems Engineering*, 42(9), pp. 1483–1494. doi: 10.1007/S00449-019-02144-3.

Mustafa, G. et al. (2021) 'Biodegradation and discoloration of disperse blue-284 textile dye by Klebsiella pneumoniae GM-04 bacterial isolate', *Journal of King Saud University - Science*, 33(4), p. 101442. doi: 10.1016/J.JKSUS.2021.101442.

Oon, Y. S. et al. (2021) 'Innovative baffled microbial fuel cells for azo dye degradation: Interactive mechanisms of electron transport and degradation pathway', *Journal of Cleaner Production*, 295, p. 126366. doi: 10.1016/J.JCLEPRO.2021.126366.

Ravadelli, M. et al. (2021) 'Anoxic/oxic membrane bioreactor assisted by electrocoagulation for the treatment of azo-dye containing wastewater', *Journal of Environmental Chemical Engineering*, 9(4), p. 105286. doi: 10.1016/j.jece.2021.105286.

Robledo-Padilla, F. et al. (no date) 'Evaluation and predictive modeling of removal Condition for bioadsorption of Indigo Blue dye by *Spirulina platensis*', *Microorganisms*, 8(1), p. 82. doi: 10.3390/microorganisms8010082.

Rodrigues de Almeida, E. J. et al. (2019) 'Azo dyes degradation and mutagenicity evaluation with a combination of microbiological and oxidative discoloration treatments', *Ecotoxicology and Environmental Safety*. doi: 10.1016/j.ecoenv.2019.109484.

Roy, D. C. et al. (2020) 'Bioremediation of malachite green dye by two bacterial strains isolated from textile effluents', *Current Research in Microbial Sciences*, 1, pp. 37–43. doi: 10.1016/J.CRMICR.2020.06.001.

Sahu, J. N. et al. (2019) 'Applications of nano-biotechnology for sustainable water purification', *Industrial Applications of Nanomaterials*, pp. 313–340. doi: 10.1016/B978-0-12-815749-7.00011-6.

Samsami, S. et al. (2020) 'Recent advances in the treatment of dye-containing wastewater from textile industries: Overview and perspectives', *Process Safety and Environmental Protection*, 143, pp. 138–163. doi: 10.1016/j.psep.2020.05.034.

Sharma, N. et al. (2017) 'Bio nanotechnological intervention: A sustainable alternative to treat dye bearing waste waters', *Indian Journal of Pharmaceutical and Biological Research*, 5(1), pp. 17–24. doi: 10.30750/IJPBR.5.1.3.

Yang, H. Y. et al. (2014) 'Degradation of recalcitrant aliphatic and aromatic hydrocarbons by a dioxin-degrader *Rhodococcus* sp. strain p52', *Environmental Science and Pollution Research*, 21(18), pp. 11086–11093. doi: 10.1007/S11356-014-3027-0.

Yeruva, D. K. et al. (2018) 'Spatial variation of electrode position in bioelectrochemical treatment system: Design consideration for azo dye remediation', *Bioresource Technology*, 256, pp. 374–383. doi: 10.1016/J.BIORTECH.2018.02.030.

Zhou, L. et al. (2021) 'Core nitrogen cycle of biofoulant in full-scale anoxic & oxic biofilm-membrane bioreactors treating textile wastewater', *Bioresource Technology*, 325, p. 124667. doi: 10.1016/j.biortech.2021.124667.

18

Strategic Re-use and Recycling of Grey Water through Treatment Systems for Resource Recovery

Divita Gautam and Praveen Dahiya
Amity University Uttar Pradesh (AUUP)

CONTENTS

18.1 Introduction

Out of four necessities (food, shelter, air, and water) essential for the survival of living organisms, the most important is water. Due to Industrialization, urbanisation, and increase in population, there is a rising demand for water, which intends to be wiping out many parts of the world to be hit by an acute shortage of water due to over-exploitation of water resources. By the year 2025, people living under the threshold of water stress would increase from 800 million to 3 billion people across the globe (Qureshi and Hanjra, 2010). Although India, already constituting 16% of the overall world's population, is now feared to be left with 4% of the freshwater resources throughout the world. Being a tropical country, enhanced levels of global warming and changing weather patterns have strained out most of the water resources present in the country. Based on the data collected by the Central Ground Water Board in 2017, 256 districts out of 700, have been regarded as 'over-exploited' or 'critical' due to the degradation in the water table.

Urbanisation and population growth have over-paced in contrast with the infrastructural development and regulation, which has intended an alarming situation to command urban waste-water management. Every year, several populations have been moving to cities and suburbs making a notable mark in establishing India with more peri-urban than rural areas till 2030. This has majorly impacted and forced a direction for the planners to make progressive and sustainable plans to provide all the possible and basic utility services, water being the priority. Consecutively, while the demand and supply increase the return flows, i.e., the wastewater, also peaks which is usually not wholly re-used. As the government is not paced with the infrastructural development for wastewater and sewage treatment, the polluted water is being discharged into waterways. This may result in deterioration of natural water quality and high toxins levels running in the environmental water above the permissible levels (Kumar et al., 2021; Chandra and Kumar, 2015).

DOI: 10.1201/9781003231738-20

The level of deterioration is such that the grey water cannot be used for cultivation and agricultural purposes. Hence, efficient and precise treatment strategies of grey water have become a focus for sustainable development.

The major issue faced by the cities of India along with the wide-spreading Indian economy is that the wastewater procured from the cities is usually a mixture of wastewater (domestic, industrial) which makes recycling and re-use of greywater, a hefty and difficult task. Due to technological innovation, capacity and knowledge, several water re-usable resources have come into play which are divided into three basic categories:

I. White water-normally referred to as the drinking portable water
II. Grey water-referred to as the water extracted from the household except for toilet water
III. Black water-referring to the water extracted from the toilets.

Grey water is generated in a huge amount in day-to-day human life. This water can be used for agricultural irrigation, plants and garden keeping, cleaning floors, washing off vehicles etc. The grey water usually constitutes about 65% of total household water. This grey water is distinctively different from the black water (generated in household wastewater) in the concentration of organic and pathogenic contaminants (Samayamanthula et al., 2019). The wastewater is usually constituting blackwater and grey water. Hence, to tone down the contaminants (biological, chemical), several treatments are used for recovery and re-use depending upon the degree of contaminated wastewater procured. Through various innovations and advanced techniques, grey water treatment for non-potable re-use has great recoupment in decreasing the pollution load and increasing the availability of water resources in urban households.

Grey water treatment and utilising it to get along with everyday chores would have a huge impact on overall potential water resources availability. Efficient and sustainable use of grey water management holds a promising improvement in natural water/fresh water and less damage to the environment as the wastewater would not be discharged into the waterways preventing the health hazards and life threats of several water residing bodies, laying out a sustainable plan for the conservation of water on which future generations could rely on. Though grey water composition varies from one household to another, but the microbial levels are somewhat similar in general. However, physical, and chemical composition highly differs due to several factors like lifestyle, health status, water usage patterns, presence of children, number of members and age: all of these contribute equally (Abedin and Rakib, 2013; Katukiza et al., 2014).

18.2 Sources of Grey Water

On average, approximately 115 L/person/day water is used in a daily household. Out of this, 85% of water can be re-used and recycled. Grey water is a mixture of various chemicals and wastes which depends on the source, duration of the activity, the activity type, and the types of products used. The insoluble chemicals are hazardous for the environment and are the real reason for contamination and pollution index of the grey water. Figure 18.1 represents different sources of grey water obtained from daily household activities.

18.3 Composition and Characteristics of Grey Water

Grey water composition largely varies and depends on the lifestyle, choices, and types of detergents and chemicals used in the household in cleaning, bathing, washing, and especially the waste oil quality and grease as well as organic wastewater procured from the kitchen sinks.

It usually comprises organic matter which is biodegradable and is present in large quantities. This organic matter consists of phosphates and their derivatives, nitrates and their derivatives, Xenobiotic organic compounds (XOCs), and biological microbes like *Salmonella,* faecal coliform, protozoans, and helminths (parasitic worms present in the intestine of vertebrates) as reported by Fatta-Kassinos

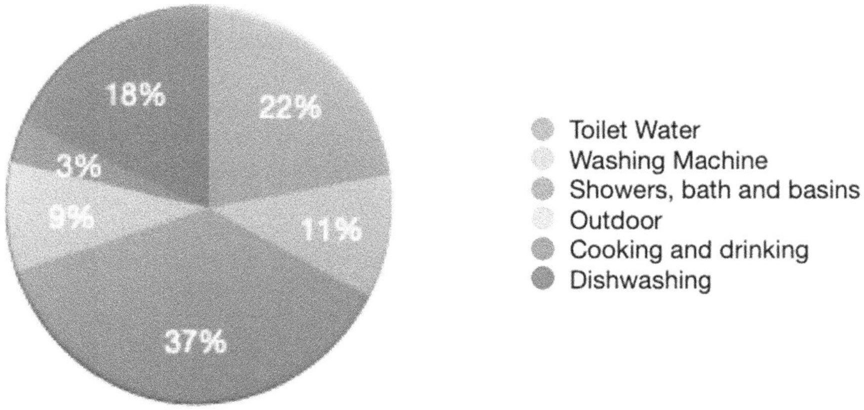

FIGURE 18.1 Depiction of grey water obtained from different sources in daily household.

et al. (2011). Besides all the above, noticeable quantities of lead, cadmium, copper, mercury, nickel, and chromium (Aonghusa and Gray, 2002) are reported in grey water because of the various industries like pharmaceuticals, pigments, aerosols, cosmetics as well as health products. All these contaminants show that grey water is of complex nature and therefore, it needs proper treatment strategies for its safe re-use. Table 18.1 includes grey water sources like laundry, bathroom, and kitchen and their important characteristics.

18.4 Physical, Chemical and Biological Characteristics of Grey Water

Physical characteristics are determined by the constituents which are associated with determining the physical appearance of the grey water. These constituents include turbidity, electrical conductivity, suspended solids, etc. Conventionally, the temperature range of grey water varies between 18°C and 35°C. The higher temperatures are because of the use of warm water for personal health and hygiene and for washing and cooking activities. These high temperatures can be favourable for microbial and pathogenic attacks and growth. It is undesirable and may also cause precipitation of inorganic salts like $CaCO_3$ that can be easily procured from soap, detergents, and washing chemicals. The grey water extracted from the kitchen and basins as well as laundry contains high levels of suspended solids (i.e., Total Suspended Solids (TSS) caused due to the washing of several items like shoes, vegetables, tubers, fruits, clothes, sand, and even clay, etc (Ciabatti et al., 2009). The electrical conductivity of grey water is observed to be higher when compared to groundwater resources and water from scarce areas. Leaching in grey water may be the result from the presence of poor or old plumbing materials in grey water, concluding to an increase in conductivity. Turbidity, on the other hand, is also comparatively high due to the availability of several detergents and surfactants (Edwin et al., 2014). The turbidity increases depending on lifestyle choices, usage patterns, and increased levels of suspended particles or TSS.

TABLE 18.1

Table Showing Various Grey Water Sources and Their Characteristics

Water Sources	Characteristics
Laundry	Laundry grey water usually comprises all sorts of soaps and detergents. This is highly chemically polluted containing surfactants, phosphates, nitrogen, ammonia, and sodium in large amounts. Due to high chemical content, physically it is turbid and contains lint.
Bathroom	Contains soaps, hair dyes, shampoos, cleaning chemicals, toothpastes and mouthwashes etc. being physically turbid in appearance and having hair in variable quantity.
Kitchen	Contains high levels of fats, grease, and food particles. Also contains chemicals and detergents used for dishwashing.

The grey water source is highly crucial in understanding the chemical composition. The detergents used in laundry consist of alkaline materials like $CaCO_3$, which leads to an increase in the pH of the grey water. The pH and alkalinity range observed in grey water is around 5–9. Another critical constituent that plays a prime role in deciding the chemical behaviour of grey water is surfactants. Surfactants are compounds which are responsible for forming micelles and hence contribute as a major ingredient in washing and cleaning chemicals used in daily household activities. Depending upon charges, the surfactants can differ in two ways-the cationic and anionic. Cationic surfactants are available in salt-based formulation which majorly contributes to ammonium concentrations in the grey water. Apart from ammonium, cationic surfactants also contribute to producing nitrates and phosphates residues in grey water. Although mostly used washing, cleaning and laundry products are anionic in nature (Eriksson et al., 2002). The wastewater procured from kitchen is the primary source of Nitrogen contamination whereas for phosphate contamination it comes from the washing detergents i.e., from laundry wastewater. The kitchen wastewater also contributes to sodium contamination which is present in cooking items and preservatives in enormous amounts. The soaps containing sodium also contribute to the overall contamination in grey water (Eriksson et al., 2010).

Various parameters which can chemically determine the quality of wastewater include Biological Oxygen Demand (BOD) and Chemical Oxygen Demand (COD). In general, the COD of grey water overpowers the BOD levels. This ascendency of COD over BOD is because of the availability of XOCs which are present in pharmaceuticals and chemicals like bleaches, softeners, surfactants, builders, and beauty products. XOCs are usually intractable to treat through basic treatment methods and hence tend to pile up in animals and plants, making them a great risk to the environment (Rathore et al., 2022). Monocyclic aromatics, phthalates, brominated flame retardants, and polycyclic aromatic hydrocarbons etc. are hazardous substances which are present in grey water in appreciable amounts. Chemical and physical parameters of grey water characteristics are presented in Table 18.2 as per the data collected by the Indian government (FAO, 2010; Kumar et al., 2005)

A thorough understanding of the biology of grey water is crucial before emphasising on any of the treatment methods. Biologically, several micro-organisms like protozoa, bacteria and helminths show their presence evidently in grey water composition. These microorganisms are resultant of improper food handling techniques and usage of contaminated food. In grey water, the most common enteric pathogenic bacteria include *Campylobacter* and *Salmonella*. Due to the laundry discharge water, faecal contamination is commonly found in grey water. Faecal contamination is highly related to ill personal hygienic habits. Therefore, *E. coli* and *coliform* bacteria are the common indicators for biological contamination in grey water. Other excreta-related pathogens like *Cryptosporidium*, *Staphylococcus aureus*,

TABLE 18.2

Table Showing Chemical and Physical Characteristics of Grey Water

Parameters	Values [Ranging from Min-Max (Average)]
pH	5.90–8.34 (7.4)
EC (μS/cm)	489–550
Turbidity (NTU)	20.6–38.7
K (mg/L)	8.3–15.2
Na (mg/L)	43.8–48.1
B (mg/L)	1.3-1.5
Cl^- (mg/L)	7.4–12.9
TSS (mg/L)	53.80–788.00 (337.2)
BOD_5 (mg/L)	17.10–290.00 (244.2)
COD (mg/L)	43.90–733.00 (705.4)
TN (mg/L)	17.00–28.82 (17.8)
TP (mg/L)	0.01–3.84
TC (CFU/100 mL)	3.74×10^4–3.8×10^4
FC (MPN/100 mL)	5.0E+01–1.2E+02

Legionella, Salmonella typhi, Salmonella sp. and *Pseudomonas* show significant presence in grey water. *Coliform* bacteria, *Enterococci*, spores of anaerobic bacteria as well as bacteriophages are the common biological indicators (Ghaitidak and Yadav, 2013; Ramprasad et al., 2017; Dahiya et al., 2021).

Coliforms are heterogeneous groups that can easily originate from any environmental condition. Thermostable coliform bacteria are highly efficient in not only showing their presence in wilted or degraded plants but also most significantly in sewage pollution. The coliform is hard to keep up with due to its regrowth abilities which depend on favourable conditions. Enterococci are a great supplement to coliform bacteria as they are far more tolerant even in hard environmental conditions. Enterococci are especially present in faeces in various densities. They serve as indicators of various enteric viruses, especially in sludge and seawater. Bacteriophages are viruses harmless to human cells but can infect bacterial cells. As enteric viruses are immune to various harsh conditions of the environment, the bacteriophages are used as an indicator to detect and analyse the presence of specific bacteria and viruses. In India, the following parameters are usually observed on average in extracted grey water from usual Indian households (Tilve, 2014; Vakil et al., 2014).

18.5 Parameters Affecting Characteristics of Grey Water

The composition of grey water differs due to various factors such as the basis of grey water, the sources and also the types of lifestyle of the user and the products used. Some important parameters affecting the characteristics include:

- Type and quality of water supplied to the household – can be piped water or groundwater
- Locality and the source of water
- Type of distribution net/filter of grey water
- Type of distribution net/filter for drinking water
- Level of occupancy
- Discharged amount quantity
- Presence of substance amount
- Activities and lifestyle of household
- Types of chemical or non-chemical products used
- Geographical location

Not only these factors but greywater exhibits a significant variation in composition and properties. The variation is due to the disparity in lifestyle radiation in water consumption, and the type and amount of discharge substances.

18.6 Treatment of Grey Water

The grey water treatment strategies provide sustainable short as well as long-term implementations for the improvement of quality of life. Re-using grey water after treatment not only decreases freshwater requirement but will also help in reducing the accelerating demand for water supply to a great extent. Also, the problem of running wastewater into sewers and in on-site treatment systems would be reduced aiming to benefit the community (Flowers, 2004). The treatment ensures an ecological balance by various benefits including reduced fish water renewal from aquifers and rivers, reduced use of energy, lowered chemical pollution, reclamation of nutrients back to nature, less impact on ground and surface water improving their quality, and no or low water bills. Grey water treatment methodology ranges from simple low-grade cost devices, that easily divert this water to the toilets or other household outdoor landscaping, to complex treatment processes which include bioreactors, filters, disinfection, pumps, and sedimentation tanks. Treatment systems usually reduce the level of contamination in grey water before final disposal or re-use. These systems use a conventional sequence for treatment, i.e., pre-treatment, primary, secondary, and tertiary. The systems either acquire the biological means or the physio-chemical means of treatment.

Biological treatment uses various combinations of sunlight and oxygen manipulation, and selective microbial culture. Few of the biological methods include trickling filters, coating biological contractors, waste stabilisation ponds, activated sludge systems etc. On the other hand, physiochemical method means physical as well as chemical methods like adsorption, filtration, and reverse osmosis can be used to treat grey water efficiently (Asano, 1998). Membrane bioreactors, up flow anaerobic sludge blankets, trickling filters, constructed wetlands, filtration and rotating biological contractors are a few widely preferred systems used for grey water treatment.

18.6.1 Primary Treatment

The primary treatment is the first step of treating grey water which includes the removal of the objects which floats or sink into the water, extracting about 30% of the carbonaceous BOD from the domestic and household sewage water procured. Large amounts of grease and oil, grit (sand, gravel, large food waste particles) and suspended solids result in improper treatment, which further makes it unfit for reuse making disposal the only resort. Hence the main aim of primary treatment is to eradicate all sorts of settable suspended solids, oil and grease, coarse solids, and any part of organic matter. All possible heavy metal-associated particles are removed in this stage, however, only dissolved particles and colloidal particles remain in the system. The screens used in the primary treatment are mainly made of long-spaced and narrow metal bars. They are successful in blocking any floating debris that is present in the grey water like wood, fat and oils, or any food-related particles which can clog on the pipes and drains (Prathapar et al., 2005; Ciabatti et al., 2009). The mechanism used in primary treatment is usually categorised as screening, sedimentation, flotation, and filtration.

1. **Screening:** It stands for scanning the particles present in grey water that needs to be processed for reuse. The grey water procured from household comes from kitchen, laundry and bathroom having various sorts of surfactants, in differential types of contamination, that need to be retreated. For laundry, various surfactants and nutrients like nitrogen, phosphorus and sulphates are easily found and these need to be treated for safe reuse. The bad bidets, sinks and showers contain more microbial contaminants when compared to physical or chemical contaminants. Kitchens sink mainly contain lots of oil and grease, food-related particles, and sodium contamination that usually comes from cooking food. Grey water needs to be screened properly and the source of such water needs to be ascertained to figure out the appropriate method of treatment that can be used to reduce maximum amount of BOD, COD, microbial contamination, and presence of toxic or heavy metals.
2. **Sedimentation**: It is a simple technology used for recycling of water by removing larger solids found in the greywater. This method helps in the reduction of several parameters like turbidity, temperature and increase or decrease in total solids. Through the process of sedimentation, all the settleable solids are accumulated at the bottom of the device clarifying the liquid.
3. **Flotation**: It is a physical process in which light components of grey water like oil, fat and grease can be extracted from the surface of the water. The entrapment of grease is a simple method and small-scale treatment. It is a low-cost alternative to septic tanks or sedimentation. Flotation process is highly important when the treatment of kitchen grey water is to be done.
4. **Filtration**: Filtration is a type of primary treatment utilized to remove solid contaminants from entering the irrigation sites or in any further treatment steps. The particles of a certain size are retained in the filter media or screens and only small size particles are filtered out. In the filtration process, several types of systems vary with respect to the filter used, size and material. The filter type can differ from simple mesh-like structure to a complicated one, i.e., activated charcoal or ceramic that are usually available in the markets. Another physical barrier which is usually used in kitchen households is the drain screens. Screens in drains are installed in showers, basins, and kitchen sinks, to prevent any large particles like hair or food residue to

clog the drainage, pipes, or filters as the screen can easily stop litter to rush down the sewage, the cleaning process becomes a lot easier. The drain screens should be anticorrosive and easily detached to clean all the solid particles procured. Also, for the grey water from laundry, a mesh of larger sizes is used to prevent any clogs in pipes, and infiltration trenches.

18.6.2 Secondary Treatment

The next or the second stage consists of digesting the organic parts of waste with the assistance of bacteria. With the help of the correct fusion of sunlight, oxygen, and bacterial activity, the waste can be treated. This stage is successful in removing the floating as well as settable solid particles reducing the BOD concentrations by 90%. The main objective of this stage is the reformation and reduction of organic matter, nutrients as well as pathogenic load present in the grey water. Several sorts of filter media are utilized for treating grey water to get rid of COD and BOD at the maximum possibility using aerobic and anaerobic digestion. The filter media collects and decomposes the growth of organic matter (pathogens and microorganisms).

1. **Aerobic Digestion**: Degradation of organic matter by aerobic digestion of microorganisms, i.e., in the presence of oxygen, results in the production of water, carbon-dioxide, and mineral products. This is a faster process when compared with anaerobic digestion. The common systems in this process include planted sand filters and tricking filters.
2. **Anaerobic Digestion**: Degradation of organic matter present in grey water in the absence of oxygen produces methane/hydrogen sulphide and carbon di-oxide. Few surfactants like LAS are non-biodegradable and hence don't break down in aerobic conditions. It is a slower process as compared to aerobic digestion (Boyjoo et al., 2013). The common systems include septic tanks, anaerobic filters, constructed wetlands, and anaerobic baffled reactors.

18.6.3 Tertiary Treatment

Tertiary treatment techniques are highly advanced steps of all treatment techniques. In the tertiary treatment of grey water, the water is maximal treated afresh by removing nutrients like nitrogen, phosphorus, suspended solids, and all the BOD possibly left.

18.6.3.1 Treatment Systems

1. **Sedimentation Ponds**: They are used as faecal sludge treatment. The grey water is loaded into the pond, after which the solid settles and accumulates at the bottom of the pond, and hence the liquid clarifies and flows out of the pond. The ponds used in sedimentation are designed to have high retention time. The major advantages of sedimentation ponds are that they are sturdy in construction, have simple operation, have good sedimentation and stabilisation capabilities and properties. The only major drawback is that two parallel ponds are required for continuous operation which means high land requirement.
2. **Septic Tanks:** These are commonly used small-scale decentralised treatment plant. Septic tanks are underground sedimentation tanks that consist of 2–3 compartments which help in settlement of sludge which is stabilised by anaerobic digestion. The dissolved or suspended particles that are available in the grey water before the treatment leaves the tank untreated. The septic tanks are usually used for treating domestic grey water. The settled sludge is pumped out periodically to maintain a stabilised environment. Septic tanks are successful in decreasing the chemical oxygen demand by about 25%–50%. The major advantages include their underground nature and requiring less space, being user-friendly, cost-efficient, and carrying out simple operations. The disadvantages include their low efficiency in the treatment of grey water, foul smell emission as it depends on anaerobic digestion process.

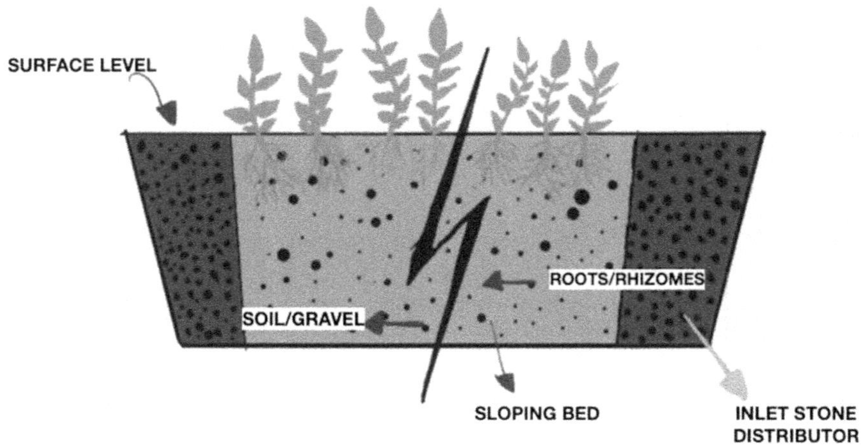

FIGURE 18.2 Constructed wetlands for greywater treatment.

3. **Imhoff Tanks:** These tanks are utilized to treat domestic wastewater having a flow over 3 m^3/d. They are successful in separating the influent from sludge with the help of settling chambers being just above the digestion chambers. They also contain baffle walls which prevent the flowing of foul sludge particles so that they don't get mixed with the effluent. This also helps in the removal of COD by about 25%–50%. Since these are underground and require little space, this is a durable system giving out a clear distinction of the fermentation and sedimentation processes. It also gives out an odourless effluent. The major disadvantages include regular desludging, low treatment efficiency, and complicated process when compared to septic tanks.
4. **Constructed Wetlands:** They are artificial wetlands which utilise ecological technology to mimic conditions of a natural wetland. Figure 18.2 represents the design for constructed wetlands for grey water treatment. The technology is designed to adapt the soil texture, flora, and fauna as well as the microorganisms that helps in the removal of unnecessary pollutants. The wetlands are usually classified into three main categories: surface flow, subsurface flow, and floating treatment wetlands. The most widely used constructed wetlands is the subsurface flow and they come in two main technologies, horizontal flow constructed wetlands and vertical flow constructed wetlands (Halalsheh et al., 2008). Both the technologies are a combination of chemical, physical, and biological processes, and the treatment efficacy relies on different factors such as the availability of electron acceptors and the loading rate. The constructed wetland bed contains coarse sand and plants like *Phragmites karka* and *Canna* sp. *Phragmites karka* is an important wetland and medicinally potent plant used for its various applications (Soni and Dahiya, 2015; Varshney et al., 2020). Constructed wetland technologies have high potential in eliminating suspended solids, BOD_5, and various heavy metals such as iron, zinc, lead etc. Although the effluent has a very low amount of BOD_5 and TSS levels meeting the regulatory limits, constructed wetlands are not very well sufficient in eliminating the microbiological agents like helminths eggs proliferation and presence of *E. coli* (Travis et al., 2010). For this purpose, the effluent is further passed through specific biological treatment systems to get a contamination-free effluent.
5. **Rotating Biological Contractors:** These are fixed bed contractors that are exhibited on a horizontal shaft. These biological contractors consist of rotating disks that are partially submerged and cause rotation for the wastewater to flow right through them (Figure 18.3). The rotating biological contractors use the aerobic digestion of microorganisms to digest the microbial waste available in grey water (Pathan et al., 2011). The microbes acting upon the grey water are constantly exposed to air causing aeration as well as assimilation which helps in faster and efficient degradation of dissolved nutrients and organic pollutants.

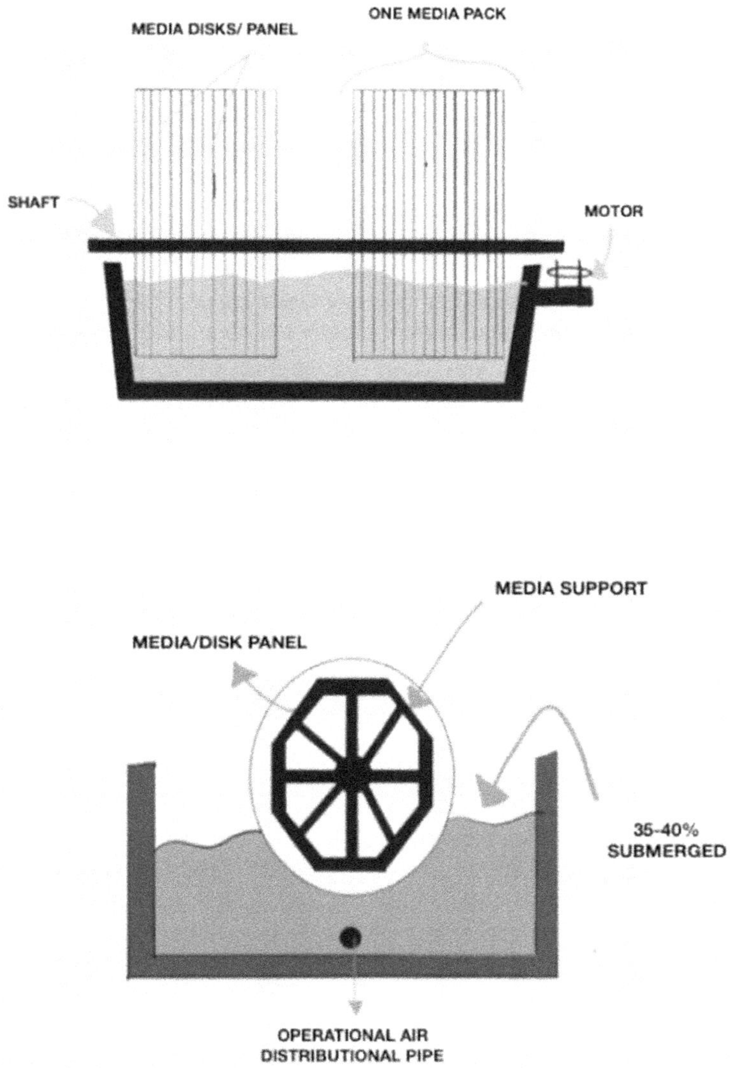

FIGURE 18.3 Figure represents the frontal and the side view of instrumentation of the rotating biological contractors.

Various indicator microbes and pathogenic cultures including heterotrophic bacteria, coliforms, *Staphylococcus aureus, Staphylococcus* sp., *Pseudomonas aeruginosa*, etc. were reported to be removed using RBC with high removal efficiency in the range of 88%–99.9% (Friedler et al., 2011). The RBCs are efficient in removing a significant amount of microbial contamination including indicator bacteria and pathogens (Gilboa and Friedler, 2008). A higher (99%) microbial removal efficiency was observed leading to the removal of coliforms, *Clostridium sp., Pseudomonas aeruginosa, Staphylococcus aureus* from grey water by RBC (Gilboa and Friedler, 2008). Various studies have concluded that RBCs are efficient in eliminating COD, BOD, and pH and reduce microbial loads creating a pollution-free effluent discharge.

6. **Sequencing Batch Reactor:** A sequencing batch reactor (SBR) is a type of activated sludge process that is used for the treatment of wastewater. All three processes-physical, chemical, and biological takes place inside the reactor tank and the treatment occurs in batches. The batches are sequenced according to the treatment stages - primary, secondary, and tertiary and successfully perform secondary clarification, biological treatment, and equalisation in a single tank (Scheumann and Kraume, 2009; Hernandez et al., 2010). The major drawback of SBR is

the retention time as their limiting factor. The effluent qualities of various sources are affected due to the high retention time offered in the SBR.

7. **Membrane Bioreactor:** Membrane bioreactor is a selectively permeable process used in integration with the biological process to treat grey water. This ultimate combination of ultrafiltration, microfiltration, and biological filtration helps in achieving the goal of pollution-free effluent. The membrane reactors are very space friendly, i.e., compact in size, and hence serve as a significant solution to be used in urbanised localities (Merz et al., 2007). Membrane bioreactors are highly efficient in lowering the COD contamination level by 80%–95% and a high level of 80% and 85% contamination free from nitrogen as well as phosphorus respectively. Membrane bioreactors are also great in removing faecal contamination using 0.1 µm of nominal pore size to achieve contamination free effluent (Jong et al., 2010).
8. **Upflow Anaerobic Sludge Blanket:** These are the most widely used treatment systems to treat different categories of wastewater. Under the UASB, the anaerobic process of digestion is used to retain high concentration of suspended biomass and efficiently produce more settleable sludge than produced in any other treatment method. The UASB is a compact system hence needs low land requirements (Lettinga et al., 1980). The major advantages of UASB are that low construction and operating cost is required, require low energy consumption, 65%–75% of COD and BOD removal efficiencies, low sludge production, and good dewatering characteristics of excess sludge produced (Henze et al., 2008). The major drawback to UASB is its requirement of a post-treatment stage and the system needs long time intervals for a new and fresh start-up. UASB used for grey water treatment showed a significant decrease in the levels of TSS, BOD5, oil and grease, nitrates, and COD as reported by (Abdel-Shafy et al., 2015).
9. **Nature-Based Systems for Treatment:** Recently, natural treatments of grey water have also been popularly adapted and used. Natural methods also serve as an alternate to conventional treatment methods. Vertical flow constructed wetlands (VFCW), green roofs/living walls, vertical flow filters (VFF), horizontal flow constructed wetlands (HFCW), urban green spaces, and constructed soil filters (CSF) are some commonly used natural treatment system for grey water treatment. All these systems consist of all three processes - physical, chemical, and biological altogether. Usually, these methods are utilized to treat light to medium grey water to produce high and good quality of effluent with relatively low operational and maintenance cost (European Commission, 2015). NBS combines physical processes like a use of sand, rocks, and gravel in combination with the biological process such as anaerobic and aerobic degradation of grey water with the help of microorganisms that can be found within the systems like wetlands etc.

   The nature-based systems famously known as NBS are referred to as the procedures to manage sustainability, provide protection to the environment, addressing the challenges focused on societal issues, and emphasize on biodiversity benefits as well as the wellbeing of human race. NBS concentrates on saving energy through various ways like reducing flooding risks, reducing cooling loads, and encourages to promote economic growth in a sustainable way (Rizzo et al., 2018; Alexandri and Jones, 2008). The constructed wetlands were the very first NBS used to treat grey water. The vertical and horizontal subsurface constructed wetlands are subjected to be the two most common types of constructed wetlands. Both types are highly efficient in providing the optimal level of total suspended solids as well as removal of organic matter. These are also efficient in restriction of wastewater contact. Constructed wetlands are a great asset for environmental issues as they have higher biodiversity levels as compared to any other possible wetland configurations. It is not only an environmentally friendly treatment method but these wetlands are also low in terms of operating and maintenance cost i.e., cost efficiency and are highly recommended in low-income or middle-income countries. The only disadvantage of constructed wetlands is that they require large areas, making it difficult for urbanised cities to install personal wetlands.

   Another used method in the Urban Green infrastructure is green roofs or green walls. Both green walls and green boards are the most important practices in the pacification of ecology and integrating a sustainable and global future (Alexandri and Jones, 2008; Francis and

Lorimer, 2011). These are vegetated surfaces, either roofs or walls (basically cracks) made up of a modular system consisting of a waterproof membrane, an insulated layer, and the vegetated layer of plants critically placed at the growing substrate. Green walls are vertical systems in which the planters are attached to the cracks up in the walls themselves. The vegetation here is rooted across the whole vertical surface of the structure. Just like the reed beds, the green beds act similar and use up media like sand, growstone, vermiculite, coco coir, expanded clay, Phyto foam and perlite as well. The major benefit of green beds over constructed wetlands is that they have low space requirements, and the plants provide various climatic benefits to the environment as well (Manso and Castro-Gomes, 2015; Teotonio et al., 2018).

18.8 Re-use of Grey Water

Grey water re-use is our immediate need. With global warming rising at an alarming speed with water table depletion, it has become a necessity to save our resources and control the overutilization of water resources and introduce re-usability methods to lead a sustainable life in future years. Various nations across the globe adopted different treatment methods as mentioned above, leading to maximum utilization of grey water which in turn controls the wastage and counters the shortage of water (Ottosson and Stenstrom, 2003). Re-cycled grey water can be easily utilized for non-potable activities like landscape irrigation, agricultural irrigation, irrigation of golf courses, and use in public parks. Toilet flushing, use of grey water in paper mills as well as in carpet dyer industry, controlling dust, mixing of concrete, and creating artificial lakes are few of the industrial uses of recycled grey water (Dolnicar and Saunders, 2006). Although currently, it is subjected to be mainly used at non-potable levels nevertheless various ongoing studies by scientists are progressing in making the recycling of grey water to turn into a portable consumption method (Nolde, 1999). This practice would further result in controlling water scarcity in tropical countries and making a progressive remark for the development of the planet globally. Currently, due to over-exploitation of current major resources, planet earth's carbon–footprint has amplified to another level, significantly causing extreme natural weather-related calamities like draught, wildfires, tropical storms. Re-use and recycling strategies would be a great asset for the environment due to the reduction of water load which helps in leading a sustainable lifestyle (Fane and Reardon, 1999).

By efficient use of recycling technologies for grey water, it can be used for non-potable use and purposes. By let alone, the re-use of grey water for toilet flushing intends in reducing almost 10%–20% of water demands (Alhumoud and Madzikanda, 2010). Moreover, recycled grey water usage in garden or landscape irrigation and toilet flushing could lower the water consumption demand in domestic use by at least 50%. However, in tropical countries, recycled grey water is more inclined to be used in agricultural irrigation (Parkinson, 2008; Ottosson and Stenstrom, 2003). In a matter of fact, it has been reported that crop-rotation and yielding practices more often results in degradation of soil quality. Using re-cycled grey water this issue can be easily countered as treated grey water tends to contain a great amount of nutrients and salts that can be highly essential in nurturing plant life. The reuse of greywater through landscape irrigation or agricultural irrigation works as an absolute low-cost treatment for reusability as well as lowering the biological contamination of grey water by countering disinfection of *Salmonella* sp., helps in eliminating turbidity, and controlling the faecal contamination of coliforms in Grey water (Kariuki et al., 2011). Also, depending upon the strength of grey water, light/low strength and dark/high strength, different treatment technologies as well as different treatment methods are adopted and practiced maximizing the usage of re-cycled grey water in a non-hazardous and ethical way (Jefferson et al., 2001). Re-usability of recycled grey water is a very questionable topic as mishandling can lead to consequences like illness of humans as well as animals and, drag hazards to the environment as well. Hence, to practice re-use, various countries have put up several guidelines depending upon the type and use of grey water procured from their households and industries (Dixon et al., 1999). These regulations must be practiced vigorously to aim and acquire maximum as well as appropriate usage of recycled grey water. Depending upon the type of usage, the rules and guidelines have been subjected accordingly. Basic guidelines for usage are as follows (WHO, 1989):

1. For fruiting trees, fodder crops and ornamental usage following terms can be adopted.
    a. Before irrigation, the type of soil as well as the quality of recycled water should be tested.
    b. To avoid grey water leaching into surface water bodies or groundwater bodies, the system flow rate on grit should be analysed closely.
    c. Any direct contact should be completely eluded, and hence the primary grey water/dark grey water reuse should only be incorporated with underground irrigational techniques. Direct contact can lead to serious bacterial transmission and infection to humans as well as animals.
    d. The very first step of reusability starts at amending lifestyle and household practices which includes restricting the release of harsh materials to grey water like heavy detergents, or any substance containing bleaching agents or boron materials.
    e. Any sort of washing of clothes (soiled/bloodstained etc.), or diapers should not take place. In draining sinks.
    f. Grey water is highly suitable for matured flora due to its high tolerance in several factors including pH, and salinity levels.
    g. Bougainvillea, Bermuda Grass, Oaks, Juniper, Citrus, Nuts, Date Palms, Rose, Rosemary, Olives, Cottonwood, and Honey suckle are a few plants that have a great affinity to grey water and proliferated well.
    h. Do not practice grey water storage as it can result in increase of total coliform count which would be highly infectious to humans as well as animals.
2. For vegetables consumed uncooked.
    a. Secondary treatment of grey water is essentially needed for vegetables to especially minimalize the number of coliforms and other microorganisms.
    b. The vegetable crop should not be in direct contact with the treated grey water.
    c. Do not practice grey water storage as it can result in increase of total coliform count which would be highly infectious to humans as well as animals and it might develop unusual and unpleasant odour.
    d. Grey water is highly suitable for matured flora due to their high tolerance in several factors including pH, and salinity levels.
3. For car washing and toilet flushing
    a. Secondary treatment in addition to disinfection is the only treatment for the suggested method for usage.
    b. The treated grey water should not be stored for more than 24–48 hours.

The recycled water is highly perishable because of the microbial contamination levels; therefore, it is very important to keep a check on the levels of recycled grey water for future re-use. Some suggested regulations are as follows:

1. An extensive management plan for grey water re-use and discharge needs to be adopted by the significant department for industrial, household, commercial as well as public buildings.
2. The significant department should study and analyse the appropriate for grey water treatment systems taking account of the quality, quantity of water and issues and impact on environment.
3. The system should include:
    a. Holding tank/watertight settling tank for grit removal.
    b. Domestic flow rate not above 500 L per day.
    c. Maximal allowance is four people per household.
    d. Designed in such a way that the effluent should be of the required quality.
    e. Proper maintenance and inspection by a trained professional to prevent any issues.

4. Regular checks as often as every week should be performed on the effluent to check BOD, COD, TDS, TSS as well as coliform count.
5. A clear indication like an alarm system, both visual and audio, should be placed to denote any failures in the treatment system.

18.9 Conclusion

The above detailed study of the water shows the characteristics, treatment technologies, and methods that depend on the perception of users and their needs. It has been evident that the characteristics and volumetric generation of grey water highly depend upon the lifestyle, choices, and types of settlement. For reuse strategies, different techniques have been reviewed which differ highly on user consumption and the type of influent received. It is quite evident that despite all the technological treatments available, none of them provides a contamination free effluent or ensures total pollutant removal. Hence, these systems are integrated to fit in the user perceptions to provide ample and maximum amount of treatment giving out pollution free effluent. For household-level treatments, several schemes have been applied by individual societies of people working and establishing various techniques for maximum benefit. Also, various natural methods and treatments can be used in eliminating the physical, chemical as well as microbial contamination offering recycling and effective reuse of grey water.

One of the major visions for reusing grey water after treatment is to reduce the black water footprint and limiting the freshwater demands leading to a sustainable livelihood. The recycled grey water after lowering the microbial load and controlling the physical and chemical contamination, can easily be used for non-potable purposes like landscaping, agricultural irrigation, and public parks maintenance. It can also be used for cleaning and washing floors as well as vehicles in daily household. The above-mentioned and currently available technologies focus only on the removal of harsh contaminants from grey water and are not performing a full treatment of it. Hence, strict rules and regulations for reuse and recycling should be followed. All the criteria of the technological systems should be paid full attention to successfully acquire a desired effluent. It is noteworthy that the contribution of technologies to make sustainable water management may vary in regional preferences and local circumstances. Since the information about life cycle and total energy consumption by various technologies is limited, several methods of combining two different technologies to make an efficient one should be inculcated and practiced enabling productive grey water treatment for reuse.

REFERENCES

Abdel-Shafy HI, Al-Sulaiman AM, Mansour MS. Anaerobic/aerobic treatment of greywater via UASB and MBR for unrestricted reuse. *Water Science and Technology*. 2015; 71:630–637.

Abedin SB, Rakib ZB. Generation and quality analysis of greywater at Dhaka City. *Environmental Research, Engineering and Management*. 2013; 64:29–41.

Alexandri E, Jones P. Temperature decreases in an urban canyon due to green walls and green roofs in diverse climates. *Building and Environment*. 2008; 43:480–493.

Alhumoud JM, Madzikanda D. Public perceptions on water reuse options: the case of Sulaibiya wastewater treatment plant in Kuwait. *International Business and Economics Research Journal*. 2010; 9(1). https://doi.org/10.19030/iber.v9i1.515

Aonghusa CN, Gray NF. Laundry detergents as a source of heavy metals in Irish domestic wastewater. *Journal of Environmental Science and Health. Part A, Toxic/Hazardous Substances & Environmental Engineering*. 2002; 37:1–6.

Asano T. *Wastewater Reclamation, Recycling, and Reuse: An Introduction to Wastewater Reclamation and Reuse*. Technomic Publishing, Pennsylvania. 1998; p. 40.

Boyjoo Y, Pareek VK, Ang M. A review of greywater characteristics and treatment processes. *Water Science and Technology*. 2013; 67(7):1403–1424.

Chandra R, Kumar V. Biotransformation and biodegradation of organophosphates and organohalides. In: Chandra R (ed.): *Environmental Waste Management*. Boca Raton: CRC Press, 2015. https://doi.org/10.1201/b19243-17.

Ciabatti I, Cesaro F, Faralli L, Fatarella E, Tognotti F. Demonstration of a treatment system for purification and reuse of laundry wastewater. *Desalination*. 2009; 245:451–459.

Dahiya P, Varshney A, Mohan S. Molecular biology techniques for the identification of microbial community in wastewater reactors. In: Shah M, Couto S (eds) Wastewater Treatment Reactors: Microbial community structure, Elsevier, (2021), pp. 387-405. ISBN: 9780128239919 https://doi.org/10.1016/B978-0-12-823991-9.00005-8.

Dixon A, Butler D, Fewkes A, Robison M. Measurement and modelling of quality changes in stored untreated greywater. *Urban Water*. 1999; 1(4):293–306.

Dolnicar S, Saunders C. Recycled water for consumer markets-a marketing research review and agenda. *Desalination*, 2006; 187:203–214.

Edwin GA, Gopalsamy P, Muthu N. Characterization of domestic grey water from point source to determine the potential for urban residential reuse: a short review. *Applied Water Science*. 2014; 4:39–49.

Eriksson E, Auffarth K, Henze M, Ledin G. Characteristics of grey wastewater. *Urban Water*. 2002; 4:85–104.

Eriksson E, Srigirisetty S, Eilersen AM. Organic matter, and heavy metals in grey-water sludge. *Water SA*. 2010; 36:139–142.

European Commission. *Towards an EU Research and Innovation Policy Agenda for Nature-Based Solutions and Re-Naturing Cities*, RTD Publications, Brussels. 2015.

Fane S, Reardon C. *Wastewater Reuse*. University of Technology Sydney, Sydney, 1999.

FAO. *Global Water Statistics of the Land and Water*. Division of the Food and Agricultural Organisation, United Nations. 2010.

Fatta-Kassinos D, Kalavrouziotis IK, Koukoulakis PH, Vasquez MI. The risks associated with wastewater reuse and xenobiotics in the agroecological environment. *Science of the Total Environment*. 2011; 409:3555–3563.

Flowers B. *Domestic Water Conservation: Greywater, Rainwater and Other Innovations*. Canadian Standards Council, Ottawa, 2004.

Francis RA, Lorimer J. Urban reconciliation ecology: the potential of living roofs and walls. *Journal of Environmental Management*. 2011; 92:1429–1437.

Friedler E, Yardeni A, Gilboa Y, Alfiya Y. Disinfection of greywater effluent and regrowth potential of selected bacteria. *Water Science and Technology*. 2011; 63:931–940.

Ghaitidak DM, Yadav KD. Characteristics and treatment of greywater-a review, *Environmental Science and Pollution Research*. 2013; 2795–2809.

Gilboa Y, Friedler E, UV disinfection of RBC-treated light greywater effluent: kinetics, survival and regrowth of selected microorganisms. *Water Research*. 2008; 42(4–5):1043–1050.

Halalsheh M, Dalahmeh S, Sayed M, Suleiman W, Shareef M, Mansour M, Safi M. Grey water characteristics and treatment options for rural areas in Jordan. *Bioresource Technology*. 2008; 99:6635–6641.

Henze M, van Loosdrecht CM, Ekama GA, Brdjanovic D. *Biological Wastewater Treatment*. IWA Publishing, London, 2008.

Hernandez Leal L, Temmink H, Zeeman G, Buisman C. Comparison of three systems for biological greywater treatment. *Water*. 2010; 2:155–169.

Jakobi G, Lohr A. *Detergents and Textile Washing*. VCH, Weinheim. 1987.

Jefferson B, Judd S, Diaper C. Treatment methods for greywater. In: Lens P, Zeeman G & Lettinga G (eds.): *Decentralised Sanitation and Reuse: Concepts, Systems and Implementation*. IWA Publishing, London, UK. 2001; pp. 334–353.

Jong J, Lee J, Kim J, Hyun K, Hwang T, Park J, Choung Y. The study of pathogenic microbial communities in grey water using membrane bioreactor. *Desalination*. 2010; 250:568–572.

Kariuki FW, Kotut K, Ngángá VG. The potential of a low-cost technology for the grey water treatment. *The Open Environmental Engineering Journal*. 2011; (4): 32–39.

Katukiza AY, Ronteltap M, Niwagaba CB, Kansiime F, Lens PNL. Grey water characterisation and pollutant loads in an urban slum. *International Journal of Environmental Science and Technology*. 2014; 12:423–436.

Kumar, V, Shahi, SK, Ferreira, LFR, Bilal, M, Biswas, JK, Bulgariu, L. Detection and characterization of refractory organic and inorganic pollutants discharged in biomethanated distillery effluent and their phytotoxicity, cytotoxicity, and genotoxicity assessment using *Phaseolus aureus* L. and *Allium cepa* L. *Environmental Research*. 2021; 201, 111551. https://doi.org/10.1016/j.envres.2021.111551

Kumar R, Singh RD, Sharma KD. Water resources of India. *Current Science*. 2005; 89(5):1518–1527.

Lettinga G, van Velsen AFM, Hobma SW, de Zeeuw W, Klapwijc A. Use of upflow sludge blanket (USB) reactor concept for biological wastewater treatment, especially for anaerobic treatment. *Biotechnology Bioengineering*. 1980; 22:699–734.

Manso M, Castro-Gomes J. Green wall systems: a review of their characteristics. *Renewable Sustainable Energy Review* 2015; 41:863–871.

Merz C, Scheumann R, El Hamouri B, Kraume M. Membrane bioreactor technology for the treatment of greywater from a sports and leisure club. *Desalination*, 2007; 215:37–43.

Nolde E. Greywater reuse systems for toilet flushing in multi-storey buildings-over ten years' experience in Berlin. *Urban Water*. 1999; 1:275– 284.

Ottosson J, Stenstrom TA. Faecal contamination of greywater microbial risks. *Water Research*. 2003; 37(3):645–655.

Parkinson J. Waste not: The facts about indirect potable reuse. *Humanist*. 2008; 68:4–6.

Pathan AA, Mahar RB, Ansari K. Preliminary study of greywater treatment through rotating biological contactor Mehran Univ Res. *Journal of Engineering Technology*. 2011; 30:531–538.

Prathapar SA, Jamrah A, Ahmed M, Al Adawi S, Al Sidairi S, Al Harassi A. Overcoming constraints in treated greywater reuse in Oman. *Desalination*. 2005; 186:177–186.

Qureshi ME, Hanjra MA. Global water crisis and future security in an era of climate change. *Food Policy*. 2010; 35:365–377.

Ramprasad C, Smith CS, Memon FA, Philip L. Removal of chemical and microbial contaminants from grey water using a novel constructed wetland: GROW, *Ecological Engineering*. 2017; 106:55–65.

Rathore S, Varshney A, Mohan S, Dahiya P. An innovative approach of bioremediation in enzymatic degradation of xenobiotics. Biotechnology & Genetic Engineering Reviews. 2022; 38(1):1-32.

Rizzo A, Bresciani R, Masi F, Boano F, Revelli R, Ridolfi L. Flood reduction as an ecosystem service of constructed wetlands for combined sewer overflow. *Journal of Hydrology*. 2018; 560:150–199.

Samayamanthula DR, Sabarathinam C, Bhandary H. Treatment and effective utilization of greywater. *Applied Water Science*. 2019; 9(90).

Scheumann R, Kraume M. Influence of hydraulic retention time on the operation of a submerged membrane sequencing batch reactor (SM-SBR) for the treatment of greywater. *Desalination*. 2009; 246:444–451.

Soni A, Dahiya P. Screening of phytochemicals and antimicrobial potential of extracts of Vetiver zizanoides and *Phragmites karka* against clinical isolates. *International Journal of Applied Pharmaceutics*. 2015; 7(1): 22–24.

Teotonic I, Silva CM, Cruz CO. Eco-solutions for urban environments regeneration: the economic value of green roofs. *Journal of Cleaner Production*. 2018; 121–135.

Tilve MM. Seasonal variation on greywater quality for a real-life system, *International Journal of Research in Engineering and Technology*. 2014; 3:768–770.

Travis MJ, Wiel-Shafran A, Weisbrod N, Adar E, Gross A. Greywater reuse for irrigation: effect on soil properties. *Science of the Total Environment*. 2010; 408(12):2501–2508.

Vakil KA, Sharma MK, Absar AB, Sarkar S. Characterisation of grey water in an Indian middle-class household and investigation of physicochemical treatment using electrocoagulation. *Separation and Purification technology*. 2014; 130:160–166.

Varshney A, Mohan S, Singh N, Dahiya P. An insight into ethnobotanical study and research of wetland plant phragmites. In: Daniels JA (ed.): *Advances in Environmental Research*, Nova Sciences, Hauppauge, NY. 2020; 199–226.

WHO. *Health Guidelines for the Use of Wastewater in Agricultural and Aquaculture*. WHO, Geneva. 1989. Technical report series No. 778.

19

A Sustainable Approach of Biodiesel Production and Water Treatment Using Oleaginous Microorganisms

Heilyn Pérez Soler and Jasiel García
CINVESTAV

Myriam A. Amezcua-Allieri
Mexican Institute of Petroleum

CONTENTS

19.1 Oleaginous Organisms for Biodiesel Production

Oleaginous organisms currently represent an important contribution to the investigation of new energy alternatives that can be obtained through the metabolic transformation of structurally simple compounds. This group of microorganisms is mainly composed of bacteria, fungi, yeasts, and algae, which have been isolated from bodies of water with hydrocarbons, saline environments, and contaminated soils. Oleaginous microorganisms play an important role in biodiesel production because they can accumulate significant quantities of lipids, mostly in the form of triacylglycerols (TAGs), and have fatty acid profiles similar to those of vegetable oils traditionally used as feedstock for biodiesel. However, the high cost of sugar-based growth substrate is an economic issue to be addressed for biofuel production (Lin et al., 2013). In order to achieve sustainable and economical production of lipids by oleaginous microorganisms, a cheap and abundantly available growth substrate is required, such as lignocellulosic biomass that can serve as an ideal source of sugar for culturing oleaginous microorganisms (Kumar et al., 2017). In Section 19.4 of this chapter, the characteristics that oleaginous organisms possess for the use of lignocellulosic biomass are described.

DOI: 10.1201/9781003231738-21

Recently Vasaki et al. (2021) studied biofuels (biodiesel) and clean environment (reduce wastewater) by producing biodiesel through oleaginous biomasses grown on pulp and paper industrial wastewater. This is an updated relevant study because deals with wastewater and biodiesel production in a biorefinery scheme. Results reported by the authors showed that the maximum yield of 0.48 (g-biodiesel/g-lipid) was obtained under the 2.5 g of lipid dosage with 0.02 g/mL of catalyst concentration by constant stirring at 70°C. It was found that *Yarrowia lipolytica, Metschnikowia pulcherrima* and *Lipomyces starkeyi* are fast-growing microorganisms and their main advantages are that they do not demand large cultivation areas and are easy to scale. These two characteristics are important in a biorefinery scheme. In addition to producing biodiesel, the process does not generate wastewater, whose treatment generates associated costs.

Another way to improve maximum yield and lower costs is to modify organisms using genetic tools that involve not just recombinant strains, but also the study of the fatty acid metabolic pathway.

19.1.1 Genetic Modifications for the Production of Biodiesel

Some of the techniques involve the construction of recombinant strains, either by inserting the exogenous genetic material of an oleaginous organism into another that does not have these characteristics, this can be carried out due to the difficulty that some oleaginous present in terms of isolation, access and cultivation *in vitro*. Generally, the construction of a recombinant one is based on the enzymes and primordial genes of the fatty acid ester pathway.

The activity of the enzyme Acyl CoA carboxylase (ACC) is of utmost importance in the synthesis. The overexpression of this in non-oleaginous organisms has shown a higher performance in the production of triglycerides. ACC of *Mucor rouxii* over-expressed in models such as *E. coli*, *Hansenula polymorpha*, which has reported a 40% increase in production in *Hansenula polymorpha* and having a production of 100 times more in *E. coli*. Also, the over-expression of a citrate lyase (ACL), can mean an important activity in oleaginous organisms and recombinant lignocellulolytic since it converts pyruvate into oxaloacetate and acetyl CoA which can result in an active source that provides substrate CoA (Cho and Park, 2018).

Research and modifications have also revolved around the expression of malate dehydrogenase enzymes, which, due to their activity, catalyze the passage from malic acid to pyruvate, generating the reduction of a molecule of NADP to NADPH+that works as reducing power for enzymes as ACC, ACL, and FAS carry out their activity.

Finally, in these systems, it has been taken into account to silence those genes involved in the ß oxidation of TAGs. In order to reduce this process, in *E. coli* cells an increase in the accumulation and secretion of fatty acids has been verified when the *fadD* genes are turned off, which codes for an acyl CoA synthetase, and *fadE*, which gives its corresponding acyl CoA dehydrogenase (Hui et al., 2010).

The coordinated enzymatic complex of the *Saccharomyces cerevisiae* FAS exhibits two domains, one alpha and one beta. The alpha domain is made up of four functional domains that are identified as follows. Acyl group carrier protein (ACP), ketoacyl reductase (KS), ketoacyl synthase (KS), phosphopantethenyl transferase (PPT) and two structural ones (SDα). Each of these domains is organized and arranged as ACP, SD1α, KR, KS, SD2α, PPT.

The enzymes involved present a catalytic site for the recognition, catalysis, condensation, reduction and condensation of the fatty acid chain. For example, in ACP, the amino acid serine 180 (Ser 180) plays an important role in the activity of the enzyme, as well as the catalytic triad of cysteine (Cys 1305), histidine (His142 and 1582) of acetoacyl reductase (Leibundgut et al., 2007).

The architecture of the beta (ß) subunit is made up of four structural domains (SDß) and four of activities such as malonyl transasylase (MPT), acetyltransferase (AT), dehydratase (DH), each one accommodated in the subunit of the following way: SDIß, AT, ER, SD2ß, SD3ß, DH, SD4ß and MPT. For the MPT and AT sites, the presence of an active site made up of a series of amino acids such as Arg 117, His 201, Gln 11, Ser 92 that are necessary for the recognition and catalysis of the substrate is reported. In DH the catalytic region is constituted by Asp 1559, His 1564 (Lomakin et al., 2007).

The structural sites in the FAS complex present functions of approach turn and transport of the substrate and elongation chains, on the other hand, the sites of action will support activities of transformation of acetyl CoA into fatty acids such as triacylglycerides.

Modifications can be generated in the structures of these enzymes, specifically mutations or changes in the amino acids of the active site, this allows us to observe if the activity of these macro-molecules increases, decreases, or is the same at the last moment. Studies on molecular dynamics can provide favorable data on the behavior of structural sites. Simulating changes in structure, analyzing, and possibly modifying the distances between amino acids can provide data on the speed and efficiency of the FAS complex in the biosynthesis of fatty acids such as TAGs.

19.2 Type of Oleaginous Microorganisms for Wastewater Treatment

The treatment of wastewater using oleaginous microorganisms is a process with significant advantages: on the one hand, the cost of operation, this technology can be developed in a bioreactor, which requires less investment in equipment, and on the other hand, the short period in which this process can treat wastewater and obtain valuable bioproducts. Table 19.1 shows examples of wastewater treatment using oleaginous microorganisms.

TABLE 19.1

Wastewater Treatment Using Oleaginous Microorganisms

Microorganism	Type of Wastewater	Biomass (g/L)	Lipid Content (%)	COD/ Contaminants Removal (%)	Nutrients and Contaminants Removal (%)	Reference
Oleaginous yeast						
Candida tropicalis	Sago processing wastewater	0.021/h	49.0	84.0	100 (NO_3) 98 (NH_4^+) 85 (PO_4)	Thangavelu et al. (2020)
Trichosporon cutaneum (ACCC 20271)	Cellulosic ethanol fermentation wastewater	N/A (maximum lipid concentration obtained was 2.61 g/L)		55.05	N/A	Wang et al. (2017)
Candida tropicalis ATCC 750	Olive mill wastewater (OMW)	25.0	11.0	68.2	39.1 (TPh) 58.7 (TOC) 47.5 (TN)	Dias et al. (2021)
Trichosporon dermatis *Cryptococcus albidus*	Wastewater after butanol fermentation.	9.76 7.59	15.69 10.25	N/A	N/A	Li et al. (2018)
Oleaginous yeast and microalgae						
Co-culturing microalgae-yeast (CCMY)	Secondary sedimentation tank of a wastewater plant	N/A (biomass productivity was 0.06 g/L*d)	55	N/A	N/A	Suastes-Rivas et al. (2020)
Oleaginous microalgae						
Scenedesmus sp. (GN 171)	Anaerobically digested effluents	3.2	34.0	N/A	92.0 (TN) 100.0 (NH_3-N) 74.0(NO_3-N) 98.0 (TP)	Yang et al. (2017)
Chlorella sp.	Secondary effluent from palm oil mill	0.72	43.0	65.0	99.0 (TN) 64.0 (TP)	Cheirsilp et al. (2017a)

(Continued)

TABLE 19.1 (*Continued*)

Wastewater Treatment Using Oleaginous Microorganisms

Microorganism	Type of Wastewater	Biomass (g/L)	Lipid Content (%)	COD/ Contaminants Removal (%)	Nutrients and Contaminants Removal (%)	Reference
Nannochloropsis sp.	Secondary effluent from palm oil mill	1.30	N/A	>50.0	>90.0 (TN) >90.0(TP)	Cheirsilp et al. (2017b)
Oleaginous bacteria						
Bacillus cereus	Palm oil mill effluent (POME)	8.09	18.04	79.35	75.29 (TOC) 69.34 (TPC) 41.76 (TN) 36.92 (NH_3-N)	Karim et al. (2019)
Serratia sp. ISTVKR1	Wastewater	N/A	N/A	88.93	N/A	Gupta et al. (2017)
Rhodococcus opacus (BM985)	Synthetic dairy wastewater	3.7	68.0	88.0	N/A	Mandal et al. (2019)
Oleaginous fungi						
Mucor circinelloides (MTCC1297) *Trichoderma reesei* (NCIM992)	Wastewater from equalization tank	0.60 0.68	22.11 9.82	88.72 86.75	N/A	Bhanja et al. (2014)
Mucor rouxii (DSM1191)	Potato processing wastewater	13.5	N/A	90.0	93.0 (TSN) 81.0 (TSP)	Muniraj et al. (2015)
Aspergillus awamori (MTCC 11639)	Corncob waste liquor (CWL)	10.2	N/A	68.0	N/A	Subhash and Mohan (2015)

N/A, not available; TN, total nitrogen; TOC, total organic carbon; TP, total phosphorus; TPC, total phenolic content; TPh, total phenols; TPP, total polyphenols; TSN, total soluble nitrogen; TSP, total soluble phosphorus.

19.2.1 Oleaginous Yeast

One of the most widely used microorganisms in biological wastewater treatment is oleaginous yeast. The high organic load of these waters provides the carbon and nitrogen sources necessary for their optimal growth and development. Oleaginous yeasts possess the ability to accumulate a lipid content greater than 20% of their cell mass (Das et al., 2021). Although the number of lipids may be different within these microorganisms, it has been perceived that the fatty acids produced are similar to those produced by plants (Beopoulos & Nicaud, 2012). The principal yeast genera used to remediate wastewater are *Trichosporon* such as *Trichosporon fermentans* (Yu et al., 2018), *Cryptococcus* such as *Cryptococcus albidus* (Li et al. 2018), and *Rhodotorula glutinis* (Zhang et al., 2018).

19.2.2 Oleaginous Microalgae

Microalgae are considered critical oleaginous microorganisms for biofuel production. Although they rarely use specific substrates in fermentative processes, they possess many advantages that make them potential candidates for wastewater treatment. Microalgae can grow and develop using inorganic nitrogen and phosphorus in wastewater under extreme pH conditions at fluctuating temperatures (Razzak et al., 2013). They can fix CO_2 and remediate these waters by producing oxygen and removing xenobiotic pollutants. In addition, they can mitigate CO_2 from flue gas (Cheirsilp et al., 2017a, b). Therefore, a promising economically viable and environmentally friendly biorefinery concept for sustainable and renewable energy production would combine wastewater treatment with algae cultivation for biofuel production (Yoshida et al., 2012).

19.2.3 Oleaginous Bacteria and Fungi

Bacteria and fungi have less application in wastewater treatment than their analogs, microalgae, and yeasts. However, these microorganisms can generate value-added products that microalgae and yeasts rarely produce (Huang et al., 2017), so they have unique applications in the industry. Specifically, some bacterial species possess a less complex genome and high metabolism, which favours an accumulation of different types of lipids superior to other oleaginous microorganisms (Karim et al., 2019). Bacteria and fungi of note in wastewater treatment are *Serratia* sp. (Gupta et al., 2017), *Rhodococcus opacus* (Mandal et al., 2019), *Mucor rouxii,* and *Aspergillus flavus* (Muniraj et al., 2015).

19.3 Treatment of Wastewaters by Oleaginous Microorganisms

Proper management and treatment of wastewater is a challenge today. Different methods have been developed to restore and maintain the biological, physical, and chemical properties of water, highlighting biological processes (Vijatov et al., 2020). Oleaginous microorganisms can treat different types of wastewater, the main ones being: municipal wastewater (sludge), fermentation wastewater, and food industry wastewater. These bioprocesses are described below.

19.3.1 Treatment of Municipal Wastewater (Sludge)

Municipal wastewater contains the nutrients and carbon source necessary for the growth and development of oleaginous microorganisms. Its low caste and availability make it an alternative raw material for biofuel production.

Activated sludge or biosolids, by-products of wastewater treatment, have also been used to cultivate oleaginous microorganisms. These microorganisms are capable of using them as substrates for lipid production However, under conditions where the available nutrients are not optimal, the maximum amount of lipids they can accumulate is not obtained (Chen et al., 2021).

Several studies in recent years have used oleaginous microorganisms to treat this type of wastewater and its residues. Patel et al. (2017) employed the yeast *Rhodosporidium kratochvilovae* to remediate wastewater from the pulp and paper industry. The results obtained showed decreased organic load COD 94.22% BOD_5 77.36%, and phenolic compounds 99.60%. The maximum cell dry weight was 13.87 g/L after 144 hours of culture. Selvakumar and Sivashanmugam (2018) studied the potential of *Naganishia liquefaciens* yeast for obtaining lipids using activated sludge as substrate. Chaudhary et al. (2018) demonstrated removing phosphate and ammonia ions by the microalgae *Chlorella vulgaris* and *Scenedesmus obliquus* in municipal wastewater.

19.3.2 Treatment of Food Industry Wastewater

Large amounts of wastewater are generated in the food industry. Generally, these waters possess high fermentability (Huang et al., 2017). So it is possible to employ oleaginous microorganisms for their treatment. Zeng et al. (2018) used a nutrient-free food waste hydrolysate as a culture medium for oleaginous yeasts and microalgae. The results showed that it is not feasible to use a pure culture when a single residue is used because of the low lipid content obtained. The mixed culture of *Chlorella vulgaris* and *Rhodosporidum toruloides* improved this bioprocess and led to higher removal of organic matter in a shorter cultivation time. Tan et al. (2021) showed that in biological processes of food wastewater with microalgae, a large amount of biomass is obtained. The study was conducted at a pilot scale in a photobioreactor, and the microorganism of interest was *Chlorella pyrenoidosa.* The microalgae proliferated in different periods reaching a growth rate of 1.59/day, and dissolved oxygen and biological contamination were monitored for optimal development. The highest percentages of TP and TN removal were 67.6% and 59.7%, respectively. The above studies show that oleaginous microorganisms represent an environmentally friendly alternative for the remediation of food wastewater and can add value to this type of waste.

19.3.3 Treatment of Fermentation Wastewater

The fermentative processes used in different sectors of industry generate wastewater. These waters contain nutrients and organic substances that can be harmful to the aquatic environment if not properly treated. A pioneering study that demonstrated the potential of *Trichosporon dermatis* in the reduction of COD in butanol fermentation wastewater was conducted by Peng et al. (2013). The highest percentage of removal of this parameter was obtained at 120 hours of the fermentative process with a value of 68%. A similar reduction rate, but in only 48 hours, was achieved in the research conducted by Xiong et al. (2015) using butanol acetone and ethanol acetone fermentation wastewater as substrate. The *Trichosporon cutaneum* strain produced 14.9% lipids with a biomass yield of 4.9/L. Zhang et al. (2018) developed a method to produce lipids by culturing *Rhodotorula glutinis* in ethanol wastewater. The nutrients available in these waters were conducive to optimal yeast growth, obtaining a maximum biomass yield of 11.31 g/L. The results of TN (70.52%) and TP (67.46%) removal showed the capacity of this microorganism for the biological treatment of these fermentation waters.

19.4 Use of Lignocellulosic Biomass by Oleaginous Microorganisms

Oleaginous microorganisms present innovative characteristics that can contribute to obtaining potentially applicable energies. These groups have the ability to synthesize different enzymes capable of generating a degradation of recalcitrant compounds that will be converted into valuable products. The degradation of various carbon sources promises the use of products that were already destined as waste (stubble) or that are associated with a mixture (wastewater). Such is the case of plant biomass, an inexhaustible source of carbon found throughout the plant kingdom, structurally made up of cellulose hompolymers, hemicellulose heteropolymers, and lignin. Some genera of ascomycete fungi such as *Aspergillus, Trichoderma, Mortierella, Fusarium*; yeasts such as *Rhodotorula, Cryptococcus,* and bacteria such as *Bacillus*, can express enzymatic cocktails made up of endo, exo glucanases, and xylanases, which when in contact with plant fibers, are capable of hydrolyzing these polymers into reducing sugars (glucose, mannose, xylose, etc.), which later can be metabolized to obtain energy and in the case of oleaginous plants will also be used in the biogenesis of TAGs as a reserve source (Gutiérrez-Rojas et al., 2015).

The presence and abundance of lignocellulosic material is closely related to the expression and overexpression of proteins for the degradation of plant biomass (Figure 19.1). The biological hydrolysis of hemicellulose causes the release of hexoses and pentoses such as D-xylose; D-galactose; D-manose and L-arabinose, which area described in different organisms as inducers of the expression of lignocellulolytic enzymes (Kowalczyk et al., 2015).

Some fungi of the genus *Aspergillus,* D-xylose monomers, for example, enter the cell through the plasma membrane (transporters), once there, they interact with the XlnR protein which triggers the transcription of different genes of the xylanase system (*xlnA, xlnB, xlnC, xlnD* and *axeA*), some endoglucanases (*eglA, eglB*) and cellobiohydrolases.

On the other hand, a substrate rich in cellulose and poor in hemicellulose, generates a disadvantage, since it can turn off the xylanolytic protein expression system, which is caused by the entry of glucose. Once the exoza is in the intracellular fraction, it is converted to glucose six phosphate (G-6-P) and turns on the repression system, activating the transcription of the complex of genes creA, B, C, D that are translated into its corresponding proteins *CreA, B, C, D.* Already present these protein products, CreB and C form a complex that desubiquitin to the CreA protein which has been described as a repressor protein of the transcription and translation of the *xlnR* gen, for which the

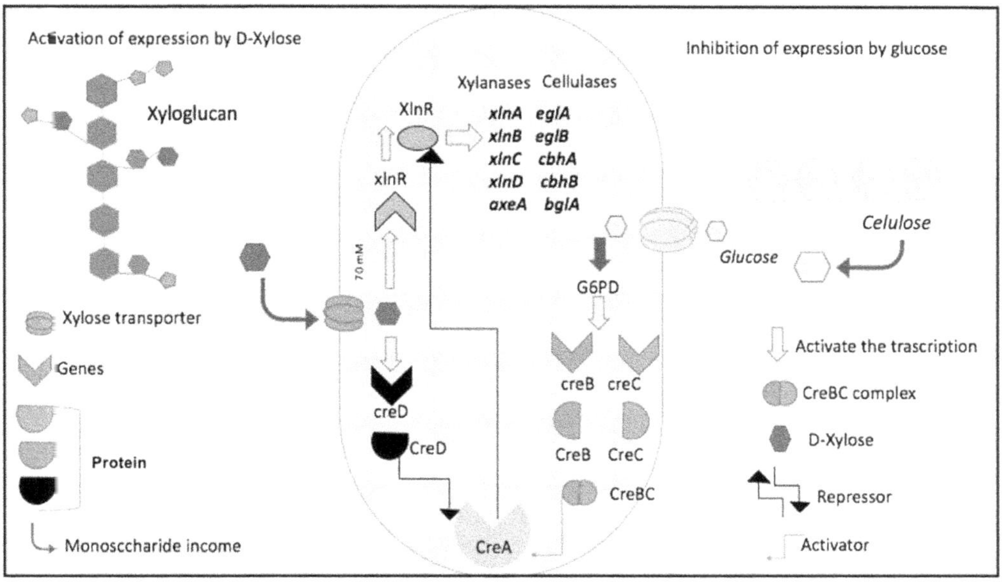

FIGURE 19.1 Expression of lignocellulolytic enzymes (left) by the presence of hemicellulose and repression of expression by glucose and glucose-6-phosphate (right). Elaborated by O. Jasiel Quintero G.

microorganism stops producing xylanases and cellulases. The activation of the system is carried out when the absence of G-6-P and the presence of 70 mM of D-xylose are detected. These two phenomena activate CreD, which is responsible for ubiquitinating CreA, generating a post-translational modification that may be related to the inactivation of this protein (Figure 19.1) (Gutiérrez-Rojas et al., 2015). Therefore, the microorganisms that take advantage of the plant biomass must generate a regulated hydrolysis of the substrate, which will keep the expression systems active and the supply of reducing sugars.

19.5 Reducing Sugars as a Source for Fatty Acid Production

The genomic, transcriptomic, and proteomic analyzes constitute a significant contribution of information, on the pathways, genes, and proteins, which is activated in the presence of various monosaccharides involved in the manufacture of TAGs. The conversion of glucose to pyruvic acid and the oxidation of the latter gives rise to the precursors for the generation of TAGs. Likewise, the pentose phosphate route promises the production and bioavailability of NADPH as one of the main reducing agents for the biosynthesis of TAGs as a substrate in the manufacture of biodiesel. The intracellular generation of fatty acids is mediated by the activation of a coordinated enzyme system that can carry out the final production of various fatty acid esters. The availability of lignocellulose monomers and their metabolism represents a source of raw material for the generation of Acetyl-CoA as a pioneer in the synthesis of TAGs. Through various works, two enzymatic systems for lipid biosynthesis, FAS and FAS II have been described. The former is expressed in yeast and FAS II in bacteria and plants. They have also been documented as two very well-conserved systems that perform in the same way in all biological systems. Their characterization of genes and enzymes has been extensively studied in models such as *E. coli*.

Fatty acid biosynthesis is sustained by the expression of various genes such as *fabD, fabH, fabB, fabF, fabG, fabA, fabZ, fabI, fabK,* and *fabL* which are translated into their respective enzymes such as acetyl-CoA carboxylase: acyl -CoA: ACP acyl carrier protein (FabH): ß-ketoacyl-ACP synthase III (FabG): ß-ketoacyl-ACP reductase (FabA, Z): ß-hydroxyacyl-ACP dehydratase-isomerase and ß-hydroxyacyl-ACP dehydratase (FabI, KL) and enoyl ACP-reductase (FabB, F) (Hui et al., 2010).

The initiation of biosynthesis is caused by an acetyl CoA carboxylase which is responsible for carbixylating acetyl CoA to obtain a malonate residue and is then transported by a FabH transasylase producing malonyl-ACP.

As part of the chain elongation, the first step consists of the condensation of the malonyl-ACP molecule in the presence of acetyl-CoA and a FabH. Later, the elongation is carried out by the FabB-FabF synthases, leading the molecule to the formation of TAGs of 16–18 carbon atoms.

Finally, the chain is subjected to reduction by the reductase activity of FabG, followed by dehydration by FabA-FabZ, finally the intermediate trans-2-enoyl-ACP undergoes reduction by the reductase FabI-FabK-FabL. As an important part of this synthesis pathway, the regulation of expression generated by the FadR repressor has been reported, which in the absence of long-chain acyl-CoAs (12–18 carbon atoms) remains in its binding site to DNA disabling the transcription of genes involved in ß-oxidation and leaving active the expression of genes for the synthesis of TAGs. It is very important to consider the bioavailability and abundance of saturated and unsaturated extracellular fatty acids since their catabolism involves the presence of long-chain acyl-CoAs, which, when detected, trigger the repression of fatty acid biosynthesis genes (Faife-Pérez et al., 2012).

19.6 Oleaginous Organisms and Some Genetic Modifications

Biotechnological and genetic engineering tools have focused on studying and modifying various characteristics, physiological, metabolic, and genetic, involved in alternative processes, to obtain materials with a positive impact on society and the environment. Oleaginous organisms are a viable option in the manufacture of biodiesel, due to their particularity of synthesizing and accumulating more than 20% of their biomass in the form of fatty acids, these molecules can serve as raw material in the production of this biofuel (Tapia et al., 2012). Previous research works have reported various strains that have proven to be suitable candidates for the storage of TAGs in controlled cultures, that is, rich in a carbon source and deficient in nitrogen.

Different biological models have been proposed and studied, where the efficiency of bacteria such as *E. coli* and yeasts such as *Saccharomyces cerevisiae* in the production of fatty acids with a length of 16–18 carbon atoms is demonstrated. In particular, *S. cerevisiae* is a very interesting case because it can synthesize fatty acid esters in the presence of an acyltransferase, which catalyzes the reaction from substrates such as alcohols and fatty acyl coA esters (Cho and Park, 2018).

Likewise, lignocellulosic organisms such as *Trichoderma reesei*, *Aspergillus niger*, *Aspergillus versicolor*, *Fusarium verticillioides*, among others, could be suitable candidates because they have a diverse group of carbohydrate-dependent enzyme families such as Glycosyl hydrolase (GH), carbohydrate esterase (CE), polysaccharide lyase (PL) and accessory activity (AA) that degrades plant biomass, allowing it to have its own supply of carbon for synthesis of acetyl-CoA (Zhang et al., 2019). Due to this characteristic, it would be convenient to insert the plasmid from oleaginous organisms such as *Yarrowia lipolytica* and *Mucor rouxii*, *Lipomyces starkeyi* in lignocellulolytic species in order to produce and accumulate fatty acids. These alternatives can reflect a great significant saving in the production process, due to the decrease in the supply of raw material (Tapia et al., 2012).

19.7 Conclusions

Oleaginous organisms currently represent an important contribution to the research of new energy alternatives obtained through the metabolic transformation of structurally simple compounds, even for biodiesel production and wastewater treatment in a biorefinery concept.

Oleaginous microorganisms currently represent an important contribution to the research of new energy alternatives obtained through the metabolic transformation of structurally simple compounds. The versatility and biological diversity of oleaginous organisms make it possible to effectively clean wastewater in a short period due to these being fast-growing microorganisms that are easy to scale up.

In addition, it is important to consider that oleaginous microorganisms play an important role in biodiesel production because they can accumulate significant quantities of lipids, and have fatty acid profiles similar to those of vegetable oils traditionally used as feedstock for biodiesel. Therefore, the use of biological tools as energy alternatives promises a new source of available and abundant resources that do not imply the generation of recalcitrant waste that can generate environmental pollution processes. In particular, the use of oleaginous microorganisms in a biorefinery approach let the integration of conversion process with high-end equipment facilities for the generation of biodiesel and value recovery through wastewater treatment.

REFERENCES

Beopoulos, A., & Nicaud, J. M. 2012. Yeast: A new oil producer? *Oléagineux, Corps gras, Lipides* 19(1): 22–28.

Bhanja, A., Minde, G., Magdum, S., & Kalyanraman, V. 2014. Comparative studies of oleaginous fungal strains (*Mucor circinelloides* and *Trichoderma reesei*) for effective wastewater treatment and bio-oil production. *Biotechnology Research International* 2014: 479370.

Chaudhary, R., Dikshit, A. K., & Tong, Y. W. 2018. Carbon-dioxide biofixation and phycoremediation of municipal wastewater using *Chlorella vulgaris* and *Scenedesmus obliquus*. *Environmental Science and Pollution Research* 25(21): 20399–20406.

Cheirsilp, B., Thawechai, T., & Prasertsan, P. 2017a. Immobilized oleaginous microalgae for production of lipid and phytoremediation of secondary effluent from palm oil mill in fluidized bed photobioreactor. *Bioresource Technology* 241: 787–794

Cheirsilp, B., Tippayut, J., Romprom, P., & Prasertsan, P. 2017b. Phytoremediation of secondary effluent from palm oil mill by using oleaginous microalgae for integrated lipid production and pollutant removal. *Waste and Biomass Valorization* 8(8): 2889–2897.

Chen, J., Zhang, X., & Tyagi, R. D. 2021. Impact of nitrogen on the industrial feasibility of biodiesel production from lipid accumulated in oleaginous yeast with wastewater sludge and crude glycerol. *Energy* 217: 119343.

Cho, H. U., & Park, J. M. 2018. Biodiesel production by various oleaginous microorganisms from organic wastes. *Bioresource Technology* 256: 502–508.

Das, M., Yash, K., & Khan, M. A. 2021. Oleaginous yeast as potential lipid producing organism: A review. *Plant Archives* 21(1): 1830–1833.

Dias, B., Lopes, M., Ramôa, R., Pereira, A. S., & Belo, I. (2021). *Candida tropicalis* as a promising oleaginous yeast for olive mill wastewater bioconversion. *Energies* 14(3): 64.

Faife-Pérez, E., Otero-Rambla, M. A., & Alvarez-Delgado, A. 2012. Producción de biodiesel a partir de microorganismos oleaginosos. Una fuente de energía renovable. Parte I. Levaduras y bacterias. *ICIDCA. Sobre los derivados de la Caña de Azúcar* 46(1): 22–32.

Gupta, A., Kumar, M., & Thakur, I. S. (2017). Analysis and optimization of process parameters for production of polyhydroxyalkanoates along with wastewater treatment by *Serratia* sp. ISTVKR1. *Bioresource Technology* 242: 55–59.

Gutiérrez-Rojas, I., Moreno-Sarmiento, N., & Montoya, D. 2015. Mecanismos y regulación de la hidrólisis enzimática de celulosa en hongos filamentosos: Casos clásicos y nuevos modelos. *Revista Iberoamericana de Micología* 32(1): 1–12.

Huang, C., Luo, M. T., Chen, X. F., Xiong, L., Li, X. M., & Chen, X. D. 2017. Recent advances and industrial viewpoint for biological treatment of wastewaters by oleaginous microorganisms. *Bioresource Technology* 232: 398–407.

Hui, L., Wan, C., Hai-Tao, D., Xue-Jiao, C., Qi-Fa, Z., & Yu-Hua, Z. 2010. Direct microbial conversion of wheat straw into lipid by a cellulolytic fungus of *Aspergillus oryzae* A-4 in solid-state fermentation. *Bioresource Technology* 101(19): 7556–7562.

Karim, A., Islam, M. A., Yousuf, A., Khan, M. M. R., & Faizal, C. K. M. 2019. Microbial lipid accumulation through bioremediation of palm oil mill wastewater by *Bacillus cereus*. *ACS Sustainable Chemistry & Engineering* 7(17): 14500–14508.

Kowalczyk, J. E., Gruben, B. S., Battaglia, E., Wiebenga, A., Majoor, E., & de Vries, R. P. 2015. Genetic interaction of *Aspergillus nidulans* galR, xlnR and araR in regulating D-galactose and L-arabinose release and catabolism gene expression. *PLoS One* 10(11): e0143200.

Kumar, D., Singh, B., Korstad, J. 2017. Utilization of lignocellulosic biomass by oleaginous yeast and bacteria for production of biodiesel and renewable diesel. *Renewable and Sustainable Energy Reviews* 73: 654–671.
Leibundgut, M., Jenni, S., Frick, C., & Ban, N. 2007. Structural basis for substrate delivery by acyl carrier protein in the yeast fatty acid synthase. *Science* 316(5822): 288–290.
Li, X., Song, C., Xiong, L., Chen, X., Huang, C., Lin, X., Chen, X., & Ma, L. 2018. Microbial conversion of wastewater from butanol fermentation to microbial oil and biomass by oleaginous yeasts. *Environmental Progress & Sustainable Energy* 37(3): 1220–1226.
Lin, H., Wang, Q., Shen, Q., Zhan, J., & Zhao, Y. 2013. Genetic engineering of microorganisms for biodiesel production. *Bioengineered* 4(5): 292–304.
Lomakin, I. B., Xiong, Y., & Steitz, T. A. 2007. The crystal structure of yeast fatty acid synthase, a cellular machine with eight active sites working together. *Cell* 129(2): 319–332.
Mandal, B., Prabhu, A., Pakshirajan, K., & Dasu, V. V. 2019. Construction and parameters modulation of a novel variant *Rhodococcus opacus* BM985 to achieve enhanced triacylglycerol-a biodiesel precursor, using synthetic dairy wastewater. *Process Biochemistry* 84: 9–21.
Muniraj, I. K., Xiao, L., Liu, H., & Zhan, X. 2015. Utilisation of potato processing wastewater for microbial lipids and γ-linolenic acid production by oleaginous fungi. *Journal of the Science of Food and Agriculture* 95(15): 3084–3090.
Patel, A., Arora, N., Pruthi, V., & Pruthi, P. A. 2017. Biological treatment of pulp and paper industry effluent by oleaginous yeast integrated with production of biodiesel as sustainable transportation fuel. *Journal of Cleaner Production* 142: 2858–2864.
Peng, W. F., Huang, C., Chen, X. F., Xiong, L., Chen, Y., & Ma, L. L. 2013. Microbial conversion of wastewater from butanol fermentation to microbial oil by oleaginous yeast *Trichosporon dermatis*. *Renewable Energy* 55: 31–34.
Razzak, S. A., Hossain, M. M., Lucky, R. A., Bassi, A. S., & De Lasa, H. 2013. Integrated CO_2 capture, wastewater treatment and biofuel production by microalgae culturing—a review. *Renewable and Sustainable Energy Reviews* 27: 622–653.
Selvakumar, P., & Sivashanmugam, P. 2018. Study on lipid accumulation in novel oleaginous yeast *Naganishia liquefaciens* NITTS2 utilizing pre-digested municipal waste activated sludge: A low-cost feedstock for biodiesel production. *Applied Biochemistry and Biotechnology* 186(3): 731–749.
Suastes-Rivas, J. K., Hernández-Altamirano, R., Mena-Cervantes, V. Y., Valdez-Ojeda, R., Toledano-Thompson, T., Tovar-Gálvez, L. R., López-Adrián, S. & Chairez, I. 2020. Efficient production of fatty acid methyl esters by a wastewater-isolated microalgae-yeast co-culture. *Environmental Science and Pollution Research* 27(23): 28490–28499.
Subhash, G. V., & Mohan, S. V. 2015. Sustainable biodiesel production through bioconversion of lignocellulosic wastewater by oleaginous fungi. *Biomass Conversion and Biorefinery* 5(2): 215–226.
Tan, X. B., Wan, X. P., Yang, L. B., Wang, X., Meng, J., Jiang, M. J., & Pi, H. J. 2021. Nutrients recycling and biomass production from *Chlorella pyrenoidosa* culture using anaerobic food processing wastewater in a pilot-scale tubular photobioreactor. *Chemosphere* 270: 129459.
Tapia, E., Anschau, A., Coradini, A. L., Franco, T. T., & Deckmann, A. C. 2012. Optimization of lipid production by the oleaginous yeast *Lipomyces starkeyi* by random mutagenesis coupled to cerulenin screening. *AMB Express* 2(1): 1–8.
Thangavelu, K., Sundararaju, P., Srinivasan, N., Muniraj, I., & Uthandi, S. 2020. Simultaneous lipid production for biodiesel feedstock and decontamination of sago processing wastewater using *Candida tropicalis* ASY2. *Biotechnology for Biofuels* 13(1): 1–14.
Vasaki, EM; Karri, RR; Ravindran, G.; Paramasivan, B. 2021. Predictive capability evaluation and optimization of sustainable biodiesel production from oleaginous biomass grown on pulp and paper industrial wastewater. *Renewable Energy* 168: 204–215.
Vijatov, T., Dražić, G., & Jovanović, F. 2020. Environmental aspects of biological wastewater treatment by different methods and microorganisms. *Sustainable Forestry: Collection* (81–82): 133–147.
Wang, J., Hu, M., Zhang, H., & Bao, J. 2017. Converting chemical oxygen demand (COD) of cellulosic ethanol fermentation wastewater into microbial lipid by oleaginous yeast *Trichosporon cutaneum*. *Applied Biochemistry and Biotechnology* 182(3): 1121–1130.
Xiong, L., Huang, C., Li, X. M., Chen, X. F., Wang, B., Wang, C., Zeng, X.A., & Chen, X. D. 2015. Acetone-butanol-ethanol (ABE) fermentation wastewater treatment by oleaginous yeast *Trichosporon cutaneum*. *Applied Biochemistry and Biotechnology* 176(2): 563–571

Yang, S., Xu, J., Wang, Z. M., Bao, L. J., & Zeng, E. Y. 2017. Cultivation of oleaginous microalgae for removal of nutrients and heavy metals from biogas digestates. *Journal of Cleaner Production* 164: 793–803.

Yoshida, M., Tanabe, Y., Yonezawa, N., & Watanabe, M. M. 2012. Energy innovation potential of oleaginous microalgae. *Biofuels* 3(6): 761–781.

Yu, D., Wang, X., Fan, X., Ren, H., Hu, S., Wang, L., Shi, Y., Liu, Na., & Qiao, N. 2018. Refined soybean oil wastewater treatment and its utilization for lipid production by the oleaginous yeast *Trichosporon fermentans*. *Biotechnology for Biofuels* 11(1): 1–12.

Zeng, Y., Xie, T., Li, P., Jian, B., Li, X., Xie, Y., & Zhang, Y. 2018. Enhanced lipid production and nutrient utilization of food waste hydrolysate by mixed culture of oleaginous yeast *Rhodosporidium toruloides* and oleaginous microalgae *Chlorella vulgaris*. *Renewable Energy* 126: 915–923.

Zhang, F., Bunterngsook, B., Li, J. X., Zhao, X. Q., Champreda, V., Liu, C. G., & Bai, F. W. 2019. Regulation and production of lignocellulolytic enzymes from *Trichoderma reesei* for biofuels production. *Advances in Bioenergy* 4: 79–119.

Zhang, X., Liu, M., Zhang, X., & Tan, T. 2018. Microbial lipid production and organic matters removal from cellulosic ethanol wastewater through coupling oleaginous yeasts and activated sludge biological method. *Bioresource Technology* 267: 395–400.

20

Biological Treatment and Value-Added Products Recovery from Wastewaters Discharged from Food Processing Industries: A Review

Spriha Raven and Archana Tiwari
Amity University

Lipi Rina
SHUATS

Aditya Francis
ISARA-Lyon

Sashita Bindu Ekka
Indira Gandhi National Tribal University

CONTENTS

DOI: 10.1201/9781003231738-22

20.1 Introduction

Wastewater from food processing industries is characterized by moderate to high BOD, high COD, high dissolved and suspended solids, very high nutrient content such as phosphorous, nitrogen, heavy oil, and grease. This effluent is amicable to non-chemical biological treatment comprising of anaerobic treatment like hybrid upflow anaerobic sludge blanket reactor (HUASBR) followed by aerobic treatment. The HUASBR generates biogas for use in electricity generator which meets the total captive requirement of the Effluent Treatment Facility (Porwal et al., 2015; Scherhag and Ackermann, 2021; Suelson and Lowery, 1992; US EPA, 2004; World Bank Group, 2007). Aerobic biological treatment, either in the form of high-rate trickling filters or activated sludge plants generally suffice for this industry. However, due to the intermittent nature of wastewater discharge or due to higher shock loadings, advanced biological treatments such as moving bed bioreactor (MBBR), membrane bioreactor (MBR) are suggested instead of the conventional activated sludge process (ASP) (Salama et al., 2017; Scherhag and Ackermann, 2021). The treated effluent can be recycled at par with food grade water by employing advanced treatment i.e., Ultrafiltration (UF) and Reverse Osmosis (R.O.) Alternatively, Membrane Bioreactor (MBR) can be used in place of MBBR which can yield industrial quality water, and by additional R.O. food grade water could be generated (Muhamad et al., 2021). Whenever food is processed, there will remain an inherent generation of wastewater. The quantity of processing wastewater that is produced and its general quality have both economic and environmental problems regarding its treatment and disposal. When considering the discharging of food processing wastewaters, the degree of pre-treatment required is determined by the specified discharge limitations defined in the municipal ordinances for discharging to sewers. This chapter discusses the application of different food processing wastewater and its treatment, composition, the role of microbes in remediation, and perspectives on nutrient recovery from dairy wastewater (Muhamad et al., 2021; Porwal et al., 2015). Water is an essential part of our life. Food industries such as dairy, beverage, meat, and others are generating effluents that contain high biochemical oxygen demand (BOD) and high chemical oxygen demand (COD), high nutrient contents. These needed to get remediated to obtain high-quality water for the industrial and domestic use. Wastewater generated from food industry and agricultural activities is a major reason of environmental pollution. Food processing wastewater is also the most difficult and expensive waste to handle as food processing wastewater carries an enormous amount of nutrients, organic carbon, nitrogenous organics, inorganics, suspended, and dissolved solids. Food processing wastewater has high BOD and COD. There are various treatment technologies being employed to use for food processing wastewater that is not unusual among wastewater treatment options and it includes various array of biological and physical-chemical procedures. The involvement can be oxidative as well as anaerobic processes. It involves various methods such as flotation, coagulation, sedimentation, filtration, adsorption, membranes, primary settling, secondary activated sludge, anaerobic digestion, and others. In addition, treated wastewater is being utilized for various agricultural land. However, due to excessive loss of nutrients present in it can damage the quality of the food being grown in the soil. Therefore, it can be beneficial as well as harmful to some extent. This chapter covers about characteristics and treatment of food processing wastewater, high-value products from microbes, recovery of value-added bioactive compounds from microbial treatment food processing wastewater, and others (Scherhag and Ackermann, 2021).

20.2 Characteristics and Treatment of Food Processing Wastewater

There is economic growth in the food industry growth and that makes food more accessible. Wastewater discharged from the food processing industries often needs to be graded to meet the standards of the regulatory authority. Various treatments are being implemented for these wastewaters to reduce environmental pollution to the ecosystem. Characterization of wastewater assists in advancing different treatment procedures among are biological techniques, advanced oxidation process (AOP), and more lately adsorption. The biological treatment and AOP go through different examinations in the past few years and have preferable ranging from low-performance cost to no waste products but not as well organized as adsorption has low-performance cost and high productivity. This chapter focuses on the characterization as well as the treatment and recovery of food processing of the wastewater (Asgharnejad et al., 2021). Water is an essential source of life, and it is basically used for various purposes such as industrial, environmental, recreation, and others. Several methods are being used for treating wastewater and those include physical, chemical, and biological processes like coagulation/flocculation process. However, the latest method is adsorption using activated carbon (Kolev, 2017). This flow chart depicts about treated water quality in carbonated drink wastewater treatment that involves the following components as certain temperature, Water, pH, COD, BOD, dissolved oxygen (DO), and suspended solids (SS) (Figure 20.1).

20.3 Meat Processing Wastewater

Meat processing wastewater establishes one of the significant concerns in the upsurge over environmental pollution because of haste growing animal industry that has prompted the advancement of new and complex wastewater treatment cycles of the agro-industries. Wastewater effluents created at butcher house typically contain undeniable degrees of organic matter (Al-Mutairi et al., 2004). The principal benefactors of organics load to these effluents are faecal, paunch, fat and lard, blood, undigested food, feathers bones, urine, suspended material, loose meat, excrement, soluble proteins, chemical oxygen demand biological oxygen demand, and suspended solids (Jian and Zhang, 1999). The meat processing wastewater produces enormous volumes of wastewater in numerous nations, which needs extensive treatment if it is released to the environment which is to be renewable (Johns, 1995). There are a few strategies for the treatment of meat processing wastewater, with most concentrating on biological cycles like activated sludge, stabilization ponds, and aerobic reactors (Davarnejad and Nasiri, 2017). Albeit these frameworks are effective and economic, they additionally generally require a long retaining time, have an enormous footprint, have a high energy interest because of aeration, and produce high volumes of sludge (Bayar et al., 2014; Daneshvar et al., 2007; Davarnejad and Nasiri, 2017).

20.3.1 Microbial Treatment

The process of meat involves huge potable water, and it transforms out into high-strength COD and nutrient supplement wastewater. The meat processing wastewater is dealt with principally by screens, and sedimentarily by pond frameworks. Although the pond framework eliminates a considerable measure

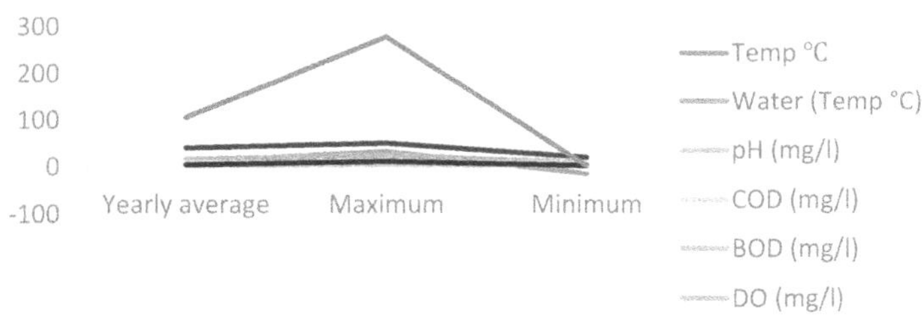

FIGURE 20.1 Treated water quality in carbonated drink wastewater treatment.

of organic carbon, nutrient evacuation is restricted. The organically treated wastewater which constitutes ammonia nitrogen and nitrate nitrogen causes unfavourable effects when released into the environment. Also, the gradually stricter principles for effluent release worldwide have made the production of advanced wastewater treatment technology obligatory (Environment Canada, 2000, 2012; US EPA, 2004; World Bank Group, 2007).

Biochemical oxygen demand is pivotal to comprehend because of the organic nature of a considerable lot of the waste materials produced into wastewater streams by the meat processing industry. BOD can be pretty much as high as 10–20 kg or 8,000 mg/L per metric ton (kg/t) of slaughtered animals and suspended solids levels could be 800 mg/L and higher. Now and again, a hostile smell might arise. The measures of wastewater produced, and the contamination load rely upon the sort of meat being processed.

COD is just a proportion of the amount of the quality of dissolved oxygen burned through during the chemical oxidation of wastewater, which could be like BOD. Wastewater for its high measure of organic material and it gives COD and suspended solids elimination.

20.3.2 Composition of Meat Wastewater

The worldwide meat production of pork, beef, and poultry meat was doubled over the most recent 30 years (Bouwman et al., 2013; Mekonnen and Hoekstra, 2012) has projected a steady consistent multiplying of meat production until 2050. Meat processing can be sorted into primary and secondary treatments.

20.3.2.1 Primary Treatment

The efficacy of the meat processing wastewater treatment relies to a great extent upon the primary treatment. The main principle motivation behind the primary treatment is the evacuation of suspended solids and fats, oil, and greases from slaughter wastewater. Screening and save all and dissolved air floatation (DAF) are broadly utilized for these purposes. Screening is capturing the gross solid particles typically by a 0.5–2 mm sieve. An assortment of screen types is utilized by the meat industries, for example, shaker, stationary, and rotating screens. The screen material might be a wedge wire or a woven wire contingent upon these purposes' application. Excrement is best eliminated by a shaker screen, and revolving wedge wire trommel screen eliminates the leftover squanders. Gross solids are recovered from the screening measures are moved to the rendering department for processing without postponement (Figure 20.2).

The effective removal of gross solids by screen, the effluent is passed through a settling tank. Settling is utilized on the principle of the natural forces that can drift the fat and settle out the solids. The bulk of the excess fat and settleable solids is eliminated and the re-established fat could be utilized as tallow. The tank can be a basic catch pit for the evacuation of fat and solids. The conducts of the settling tanks rely upon the proficient velocity, purging of short-circuiting, an adequate settling path, minimum turbulence persuaded by scrapers, and the overflow rate optimum of around $0.8\,m^3\,(m^2d)^{-1}$. A very well-designed tank and managed settling tank can eliminate 85% of the effluent fat. Nevertheless, higher detainment time and higher temperature can decrease the proficiency of a settling tank.

DAF works on a similar standard to a settling tank. The fine air bubbles are shaped when dissolved air in the solution under pressure is delivered with the approaching effluent in the floatation tank. Saturated air is framed with recycled effluent from the DAF. From an appropriately well-designed and managed DAF, more than 90% fat and 70% suspended solid elimination can be anticipated.

20.3.2.2 Secondary Technique

Most of the organics from the wastewater from the meat industry emanate from blood, oil and so on that are insoluble and gradually biodegradable. Because of the great percentage of organic materials, biological strategies for purification are ideal. The secondary technique incorporates biological treatment which is additionally partitioned into an anaerobic method or aerobic method.

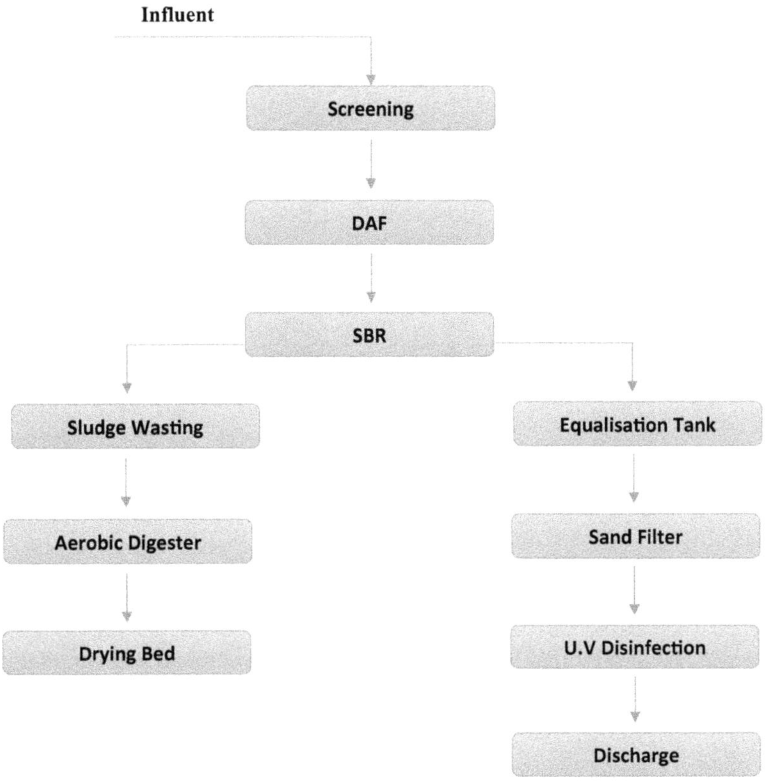

FIGURE 20.2 Schematic diagram of a meat processing wastewater treatment system.

20.3.2.2.1 Anaerobic Method

Anaerobic wastewater treatment is an interaction where microbes digest bio-solids without oxygen. Primary treated abattoir wastewater contains a high concentration of organic carbon (Manjunath et al., 2000) which requires high air circulation, and sludge disposal on the expenses of wastewater is dealt with utilizing an anaerobic treatment. It is fixed-bed biological reactor with filtration chambers. At the point when the SWW goes through the filtration chambers, particles are restricted inside then, organic material is eliminated by the dynamic biomass appended to the filter surface. AFs are utilized as secondary treatment because of their high solid evacuation and biogas recuperation rates. An AF is planned as an anaerobic digestion segment loaded with various kinds of media (Mittal, 2004). The exhibition of up-flow anaerobic filters (UAFs) has been inspected under thermophilic and mesophilic conditions for SWW treatment (Gannoun et al., 2009, 2013) (Figure 20.3).

20.3.2.2.2 Aerobic Methods

Aerobic wastewater treatment is an interaction where microorganisms use oxygen to debase organic matter and pollutants associated with the different production systems and furthermore it is utilized for decontamination and expulsion of nutrients in the wake of utilizing anaerobic procedures (Chernicharo, 2006). It can be a conventional method with active sludge, aerobic filter, and aerobic lagoons. For wastewater from the slaughterhouse can be applied biological methodology with aerobic lagoons.

20.3.2.2.2.1 Activated Sludge Process (ASP) ASP an aerobic treatment strategy that carries the effluent into contact with air and free-floating flocs of microorganisms including bacteria and protozoa. The ASP cycle has been broadly applied in various industries as a regularly known savvy technique for the treatment of SWW. The motivation behind the ASP cycle is to eliminate soluble and insoluble organics

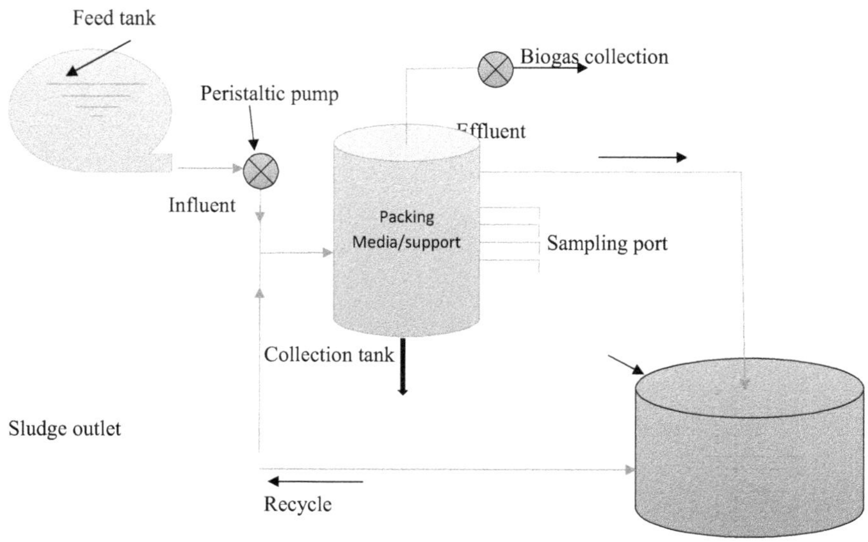

FIGURE 20.3 Schematic diagram of a typical anaerobic filter system.

from the wastewater and to alter this material into a flocculent microbial adsorption that is then settle down into a clarifier. Two discrete mechanisms are applied in AS, adsorption, and oxidation of the organic matter (Al-Mutairi, 2009; Bull et al., 1982) (Figure 20.4).

20.3.3 Role of Microbes in Remediation

The principal emanation of microorganisms to atmospheric air was from the mechanical sewage treatment devices of the wastewater treatment plant (WWTP). The presence of enteric bacteria, particularly *Enterobacteriaceae* replicates the degree of air contamination with bioaerosols from sewage and is a significant factor in examining the quality of the air around WWTPs. Despite's bacteria being the prevailing organisms in WWTPs, fungi, protists (Foissner, 2016), and microalgae (Caprio et al., 2019) likewise play vital parts in pollutant evacuation. It is revealed that microalgae can proficiently eliminate contaminants initiated from pharmaceuticals (Tolboom et al., 2019) and personal care products.

Microalgae demonstrate higher efficiency in nutrient evacuation than different microorganisms (ammonia, nitrate, phosphate, urea, and trace elements just like COD and BOD. Several species *Chlorococcum, Chlorella (C. sorokiniana, Scenedesmus, S. acutus, S. quadricauda, S. rubescens), C. minutissima),*

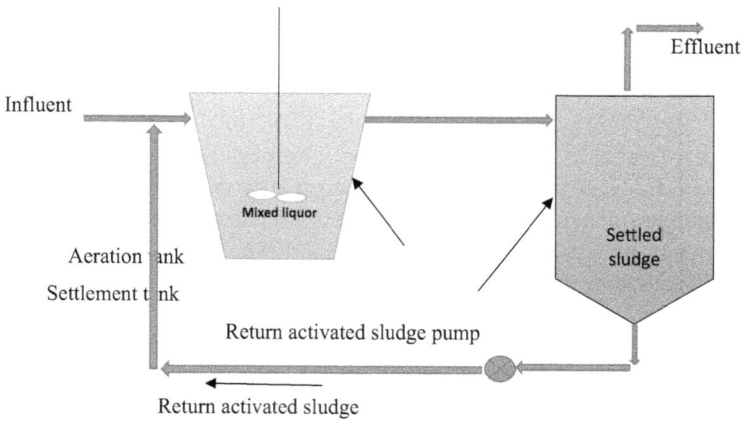

FIGURE 20.4 Schematic diagram of an activated sludge process.

Auxenochlorella protothecoides, Oscillatoria, Nostoc, and *Botryococcus braunii* have been featured as the most effective in municipal WW treatment. *Galderia sulphuraria, Chlorella sorokiniana,* and *Chlamydomonas acidophila* segregated from an extreme environment, were also effectively utilized in WW treatment (Wollmann et al., 2019).

Apparently indifferent wastewaters, potential advancement in the area of microalgal cultivation combined with wastewater treatment in open and shut frameworks has prompted an improvement in algal biomass production (Salama et al., 2017). It surveys the microalgae-mediated treatment of various wastewaters including municipal, piggery/swine, municipal, anaerobic wastewater, and industrial and feature the wastewater attributes reasonable for microalgae cultivation. Various pre-treatment techniques, (for example, autoclaving, filtration, dilution, autoclaving, and UV application) are required for wastewater before its utilization for microalgae cultivation.

20.4 Beverage Wastewater Processing

Beverage industry is one of the well-known in the food industry with a great market value of about billions worldwide and is growing day by day (Sandhar et al., 2013). However, these beverage industries also face challenges related to their effluent treatment. Effluent or the wastewater discharged from a Beverage manufacturing industry consists of the following chemicals:

Glucose, sucrose, lactose, fructose, maltose, fruit juice concentrates, coloring agents, artificial sweeteners, preservatives, mineral salts, and flavoring agents. And, therefore these effluents discharging from the beverage industries to any water sources may affect the quality of the water and degrade it in many ways, also it increases the chemical oxygen demand as well as biological oxygen demand in the water sources by increasing in organic water content. (Ghosh and Henry, 1981). Therefore, it is necessary to treat the effluent or the water discharge containing organic matter in high concentrations before being discharged into the water sources to protect the environment and control water pollution (Bulawayo City Council, 1980; EMA Act Statutory Instrument 6, 2007).

20.4.1 Microbial Treatment

Microbial Treatment is an important and useful constituent of several wastewater pants which treats wastewater from the industrial or municipal sectors containing soluble organic materials (Mittal, 2011).

A huge amount of fresh water is being used by the Beverage industry for different processes, which ultimately generates a considerable amount of wastewater during these processes such as- production of drinks, bottle washing, the washing of the plant, and the washing of floors and the working area. Almost 3–4 L of fresh water is used to prepare 1 L of soft drink (Gumbo et al., 2003). Microbial or biological treatment is the most used treatment for beverage wastewater. Wastewater of higher strength can be treated through an anaerobic process and that with the low strength can be together with the wastes from the anaerobic treatment through the aerobic process.

An overall Microbial treatment process includes screening, neutralization, anaerobic and aerobic treatment, sludge separation process (which also includes sedimentation or dissolved air flotation), and finally sludge disposal. Microbial Treatment is an important and useful constituent of several wastewater pants which treat wastewater from the industrial or municipal sectors containing soluble organic materials (Mittal, 2011). A huge amount of fresh water is being used by the Beverage industry for different processes, which ultimately generates a considerable amount of wastewater during these processes such as the production of drinks, bottle washing, washing of the plant, and washing of floors and the working area. Almost 3–4 L of fresh water is used to prepare 1 L of soft drink (Gumbo et al., 2003). Microbial or biological treatment is the most used treatment for beverage wastewater. Wastewater of higher strength can be treated through an anaerobic process and that with low strength can be together with the wastes from the anaerobic treatment through the aerobic process.

An overall Microbial treatment process includes screening, neutralization, anaerobic and aerobic treatment, sludge separation process (which also includes sedimentation or dissolved air flotation), and finally sludge disposal (Muhamad et al., 2021).

20.4.1.1 Anaerobic Wastewater Treatment

Anaerobic wastewater treatment is useful for both wastewater treatment and sludge digestion. Anaerobic treatment of wastewater is an effective method that is well known for treating wastes containing organic contents. This process is more effective than aerobic process so as to treat the wastes with higher organic content ranging above or equal to 5 g COD/L. The anaerobic process includes facultative microorganisms and anaerobic microorganisms that help in the conversion of organic matters/contents into several gases like carbon dioxide and methane, and this conversion is carried out in the absence of oxygen. Anaerobic treatment is done in two steps: (i) fermentation of acid and, (ii) formation of methane.

20.4.1.2 Anaerobic Digestion Treatment of Wastewater

The use of Anaerobic digestion is increasing day by day in industrial sectors. This process mainly involves the fermentation of organic matter which is degraded to produce biogas in the absence of oxygen by using certain microbes. This process is helpful in the treatment of industrial as well as domestic wastewaters (Muhamad et al., 2021). This process involves several microbiological stages that become the outline of the pathway to organic matter degradation. It requires environmental & nutritional factors of the microorganisms to be fulfilled to produce the biogas (Muhamad et al., 2021) (Figure 20.5).

20.4.1.3 Aerobic Wastewater Treatment

Due to the high concentration of organic content found in beverage wastewater, Wastewater from the beverage industry is treated biologically by aerobic treatment process. Even if the wastewater does not contain high organic concentration still it can be treated by aerobic treatment process because this process is quite an easy process. Removal of BOD and COD can be carried out using several aerobic suspensions or a treatment done by using fixed film growth treatment.

20.4.1.3.1 Suspended Aerobic Growth Treatment Process Aerobic Suspended Growth Treatment Process

The suspended aerobic growth treatment process includes an activated sludge process, sequencing batch reactors and aerated lagoons. The reaction time between the microorganism and the organic concentration needs to be higher than the wastewaters produced from domestic areas (Scherhag and Ackermann, 2021).

20.4.1.3.2 Attached (Fixed Film) Growth Treatment Process

The trickling filter and rotating biological contractor are the major parts of the Attached fixed film growth treatment process. This process is carried out by fixing microorganisms into an inert biofilm. Removal of BOD and COD is carried out by supplying air so as to oxidize organic wastes.

Trickling filter has the following materials which help in the treatment process: rock, gravel, sand, redwood, slag, and plastics of different ranges and synthetic materials. The process of organic waste degradation is carried out with the help of these materials when flows along with the attached biofilm.

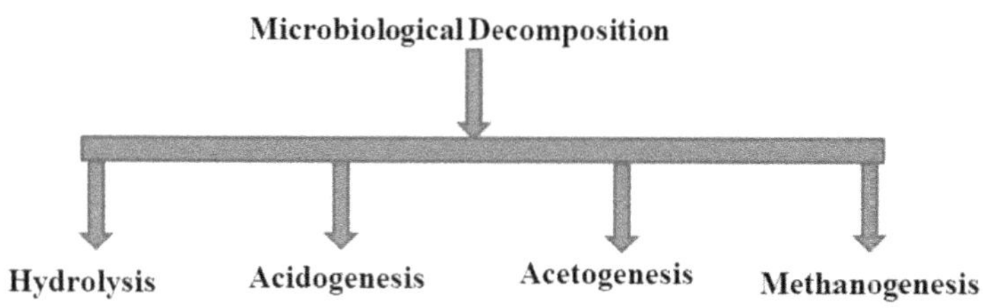

FIGURE 20.5 Different stages of microbiological decomposition.

20.4.2 Beverage Wastewater Composition

Wastewater from beverage industries is the result of different processes taking place in the beverage industries such as bottle washing, product filling process, heating or cooling process, cleaning working areas, washing of floors, manufacturing of drinks, and cleaning of pipe networks.

Biodegradable wastes from industries that contain organic contents are produced from wastes that come from different activities like manufacturing and bottling processes of wines, beverages, and several drinks in different beverage industries. The increase in organic concentration in the wastewater from beverage industries is due to the utilization of raw materials like sugars, phosphates, and agricultural products (Wilkie, 2002).

The concentration of the beverage industrial wastewater parameters is usually high. Following are the parameters.

1. Electrical Conductivity (EC)
2. Chemical Oxygen Demand (COD)
3. Biological Oxygen Demand (BOD)
4. Total Suspended Solids (TSS)
5. Total Dissolved Solids (TDS)
6. Total Kjeldahl Nitrogen (TKN)
7. Total Nitrogen (TN)
8. Total Phosphorus (TP), and
9. pH

Distilleries, wineries, and breweries produce alcoholic beverages. The major raw material used in the beverages produced by these industries is barley, corn, wheat, potatoes, sugar beets, rice, molasses, and sugar canes (Blonskaja and Mulbry, 2002). Wine is produced from the total or fractional fermentation process done using fresh grapes. Distillery & Winery is also known to produce residual unfermented carbohydrates.

Beer brewing includes the blending of sugar-based raw materials such as barley, maize sorghum, wort which is fermented along with the yeast using a huge quantity of fresh (Gray, 2002) (Figure 20.6).

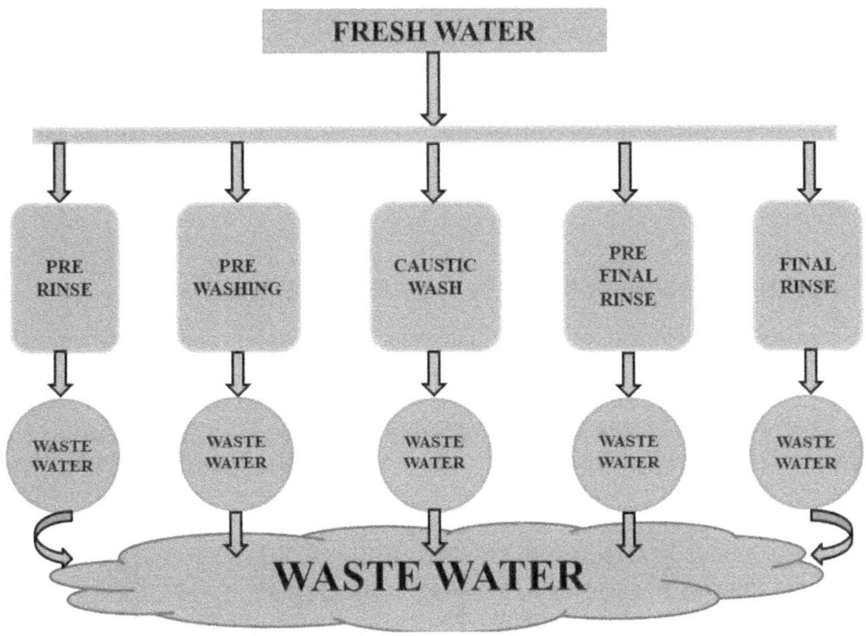

FIGURE 20.6 Diagram showing bottle washing plant/unit.

20.4.3 Role of Microbes in Remediation

The role of Microbes in the remediation of beverage wastewater involves the engagement of some of the wide ranges of microbes such as bacteria, protozoa, viruses, algae, fungi, and helminths. Their roles are as follows:

20.4.3.1 Bacteria

Bacteria play an important role in wastewater remediation through the conversion of organic matter content into less complex compounds. Important bacterias which are involved in wastewater treatment are *Achromobacter, Arthrobacter, Citromonas, Alcaligenes, Flavobacterium, Zoogloe, Acinetobacter,* and *Pseudomonas.* Bacteria are responsible for the stabilization of effluent waste and are known to form floc particles. Floc particles are the clusters of Bacteria which helps in the breakdown of the waste (Paillard et al., 2005).

20.4.3.2 Protozoa

Protozoans are ten times bigger than Bacteria and they prey on pathogenic bacteria which makes them advantageous in wastewater. Example: *Microstoma, Aspidiscacostata, Carchesium polypinum* (Amaral et al., 2004; Caccio et al., 2003).

20.4.3.3 Fungi

Fungi are other examples of microorganisms which are involved in wastewater treatment. They are multicellular organisms and find their role in activated sludge process. Fungi are capable to oxidize ammonia into nitrite and nitrate. Example of fungus is *Zoogloea sp.* and *Sphaerotilus natans* (Le Chevallier and Au, 2004; Painter, 1970). Some of the fungi like *Aspergillus, Penicillium, Fusarium* acts like a host and help in the removal of carbon and nutrients sources in wastewater (Akpor et al., 2013). Some of the fungi can break down organic matter present in the sludge processing system. Studies have shown that some types of fungi having filaments are very useful in wastewater treatment as they secrete a kind of enzyme that is helpful in the degradation of substrates (Molla et al., 2004).

20.4.3.4 Algae

Algae are helpful microorganisms in the biological purification of wastewater. Algae has gained much importance clue to its ability in biological wastewater treatment over the years (Lloyd and Frederick, 2000). Algae is helpful in accumulating heavy metals, Pesticides, nutrients, inorganic & organic toxic substances from the wastewater and hence can purify the wastewater. Example: *Euglena sp., Oscillatoria sp., Chlamydomonas sp.* etc.

20.4.4 Perspective in Nutrient Recovery from Beverage Wastewater

Microalgae is an alternate source of wastewater treatment and is helpful in recovering nutrients from the wastewater (Craggs et al., 2013; Olguín, 2012). Nutrient recovery from wastewater has economic benefits which makes it a real profitable process of technique. Microalgae production for the treatment of wastewater can be widely seen in different effluents from different effluents from different sectors largely from the beverages industry sewage & manure etc. (Craggs et al., 2013; Hernández et al., 2013; Olguín, 2012). The overall process of Beverage wastewater treatment is based upon the composition of the effluent which requires an adequate design & operation. Microalgae production on such wastewater effluents from beverage industry helps to recover of N (Nitrogen) & P (Phosphorus) found in these valuable biomass esculents.

Upto 1 kg of dry microalgae biomass can be produced using per m^3 of sewage and 10 kg of microalgaelgae biomass can be produced as per m^3 of manure. Therefore, microalgae are found to be the most suitable components to be used in nutrient recovery technology (Acién et al., 2016). Higher the productivity of the biomass more will be the nutrient recovery capacity.

Main reason for using Microalgae for recovering nutrients from the wastewater is that it also minimizes the emissions of greenhouse gases and is helpful in saving energy (Craggs et al., 2013; Olguín, 2012).

However, while using the microalgae you need to keep in mind that there is a community of microalgae and bacteria. This community, either of microalgae or bacteria is the naturally occurring final substrate for the composition of wastewater, environmental conditions, rector design, and operational conditions in the reactor to carry out the treatment process. The bacterium in the culture helps in the oxidation of organic matter content and converts it into inorganic compounds and uses all the oxygen in this first step while the microalgae on the other hand with the help of light uptakes the inorganic nutrients/compounds released by the bacteria to produce biomass. And releases oxygen in the second step so as to be utilized by bacteria for the oxidation process (Scherhag and Ackermann, 2021). According to this process/technique, a neutral equilibrium between microalgae & bacteria is established in the reactor. Thus, the consortium determines the production of oxygen, consumption of CO_2, nitrogen and phosphorus fixation etc. during this process (García et al., 2018; Matamoros et al., 2016; Mittal, 2004; Pal et al., 2020; Siracusa et al., 2020) (Figure 20.7).

20.5 Dairy Processing Wastewater

20.5.1 Microbial Treatment

With the quick industrialisation watched in the last century and the developing rate of the production of milk, dairy processing is most often known to be the biggest industrial food wastewater birthplace, chiefly in Europe (Kolev Slavov, 2017). The milk-processing enterprises that discharge untreated/partially treated wastewater cause major environmental problems due to the high pollutant load of dairy wastewater (Porwal et al., 2015). Seeding inoculation of microbes for degrading waste materials in streams, rivers, and treatment tanks is becoming increasingly popular in many countries since it is cost-effective and simple to implement (Porwal et al., 2015). According to the research from (Porwal et al., 2015), it was found out that microbial treatment is proved to be beneficial for dairy wastewater treatment in various dimensions. For example, the study found out that after an aeration period of 48h, their DS_{14} (mixed culture) was found to be the most effective in lowering certain water physiochemical characteristics, also the DSI_3 (bacterial isolate) came to the second position in reduction of EC, TSS, TDS, TS, COD, BOD, chlorides, and sulphates. However, when compared to DSI_2 and DS_{13}, DS_{11} (yeast isolate) was found to be more successful in reducing turbidity and O&G. In comparison to the other three cultures, DSI_2 (bacterial isolate) demonstrated the least decline in all selected water parameters (Porwal et al., 2015). Below is a process flow chart is presented for the process behind the treatment of dairy effluent by utilising the isolates of microbes.

Another study in which microbial treatment of dairy product wastewater was shown to be effective comes from (Mazzucotelli et al., 2014). In order to improve aerobic treatment of milk processing wastewater, 11 fat/protein degrading microorganisms of the genera *Bacillus, Serratia, Lactococcus, Enterococcus, Stenotrophomonas, Klebsiella,* and *Escherichia* were evaluated in the Plackett-Burman design as possible degrading bacteria. Of these, four strains such as *Lactococcus garvieae, Bacillus thuringiensis, Escherichia coli* and *Stenotrophomonas* sp were selected as potential members of the microbial consortium. The Plackett-Burman design has proven to be a useful tool for selecting a bacterial mix that will improve aerobic treatment of dairy effluents. The values of the percentage of total degradation (TD) of the selected consortium were higher at 24 and 48 hours at ideal pH and temperature (6.5°C and 32°C, respectively) than any other combination of variables.

20.5.2 Composition of Dairy Wastewater

Normally the effluents of milk processing have higher temperature and large variations in pH, TSS, biological oxygen demand (BOD), COD, total nitrogen (TN), total phosphorous (TP) and fat, oil, and grease (FOG) (Kolev Slavov, 2017). Dairy wastewater is typically white in color (while whey is yellowish-green) also, the annual temperature of dairy wastewater is between 17°C and 25°C, which is

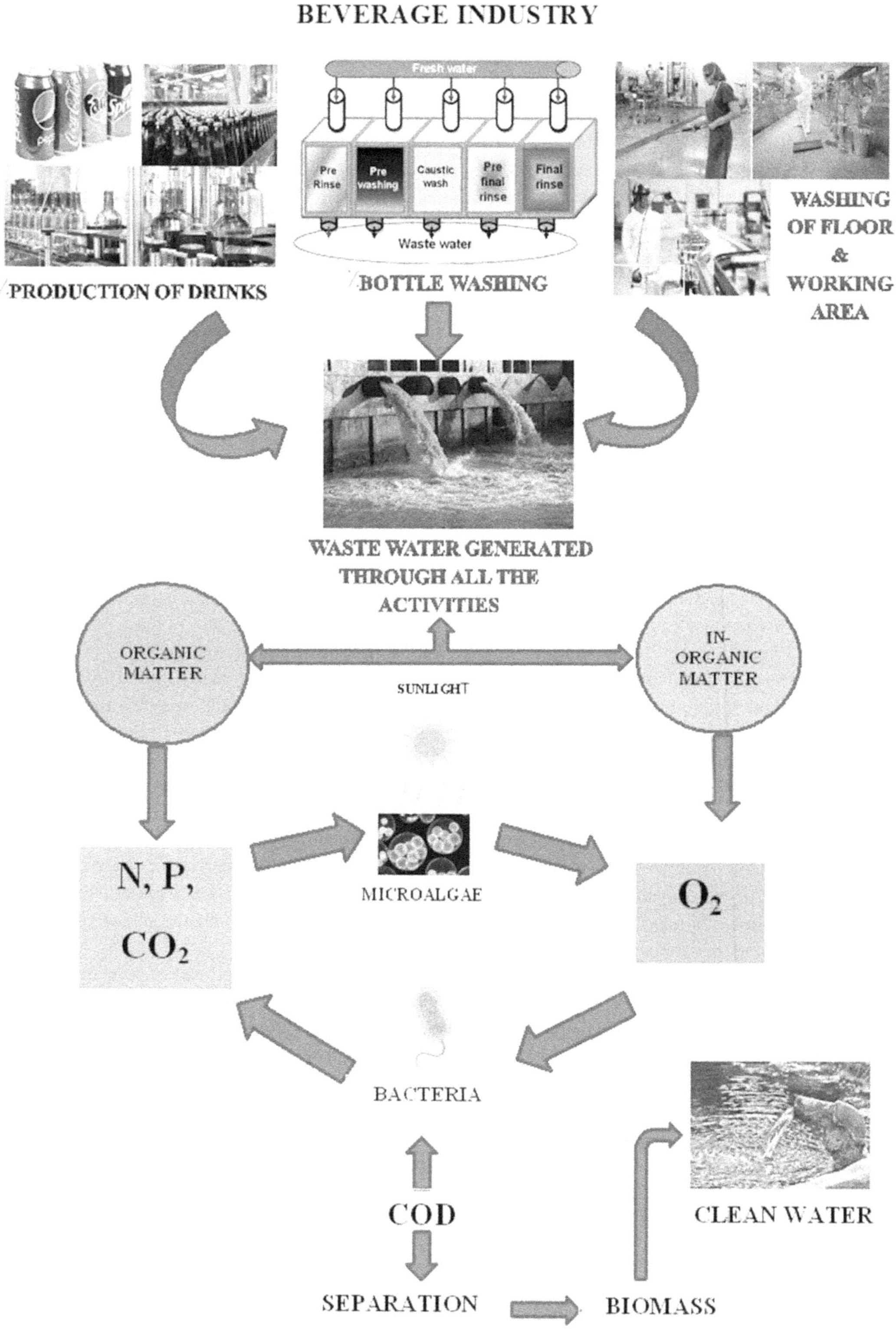

FIGURE 20.7 Diagram showing the use of microalgae/bacteria consortia for nutrient recovery from beverage industry.

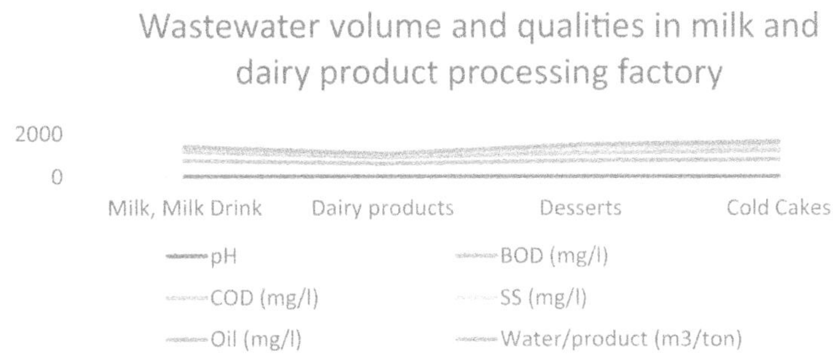

FIGURE 20.8 Wastewater volume and qualities in milk and dairy product processing factory.

typically warmer than municipal wastewater (10°C–20°C) resulting in the biological degradation at a much faster rate compared to sewage treatment plants and has an unpleasant odor and turbidity (Kolev Slavov, 2017). Also, the pH of the effluents from milk and butter factories with active reaction is near neutral (pH=6.8–7.4), in the manufacturing of cheese, sweet whey is barely acidic, with pH=5.9–6.6, while mineral acid coagulation gives an acidic whey with pH=4.3–4.6 (Kolev Slavov, 2017). The FOG concentration of dairy wastewater withinside the manufacturing of high-fats products is 0.2–0.4 g/L. Dairy wastewater is characterized by excessive BOD and COD values differing between 0.1 and 100 g/L (Kolev Slavov, 2017). According to the research, the wastewaters produced because of a technological method from the dairy industry are characterized by high levels of organic contaminants, lipids, nitrates, ammonium, and phosphates. Also, the pH value varies according to the time. For example, during the day it can be strongly acidic and during the week it can be strongly alkaline.

This graph depicts wastewater volume and quality in milk and dairy product processing factories (Figure 20.8).

20.5.1 Role of Microbes in Remediation

Microbes play a crucial role in the conversion and modification of toxic pollutants in order to obtain energy and biomass during the process (Abatenh et al., 2017). The bioremediation process is a microbiologically carefully planned activity that is used to break down contaminants or convert them into less toxic or non-toxic elemental and composite forms (Abatenh et al., 2017). Bioremediation can be divided into the following categories based on location, principles, benefits, and limitations. The wastewater from the dairy industries typically contains high convergences of impurities and, since the volume created is likewise high, the complete toxin load is exceptionally huge (Mazzucotelli et al., 2014). Below is the (Table 20.1) the microbes used in the remediation of dairy wastewater and the results associated with them have been analysed from different works of literature.

20.5.1.1 Perspectives in Nutrient Recovery from Dairy Wastewater

The dairy industry generates a significant amount of nutrient-rich wastewater that must be purified before being released into the environment (Kurup et al., 2019). There are several methods that are used in the recovery of nutrients from dairy wastewater. According to studies by (Braun et al., 2021; Bulawayo City Council, 1980; Kurup et al., 2019), an inexpensive food grade sodium lignosulfonate (nalignosulfonate) was used to recover proteins and lipids from dairy wastewater and reduce BOD. First, a raw dairy wastewater sample was taken from a commercial manufacturing plant in North Victoria that is used to make baby food and cheese. Below in (Table 20.2) the values are presented in which the experiment for nutrient recovery from dairy wastewater was conducted. The experiment was also carried out at a higher temperature and different pH levels to investigate its effect on the efficiency of recovery (Table 20.3).

There is also great potential for nutrient recovery from dairy effluents through the use of freshwater algae. According to research by (Wilkie, 2002), a combination of conventional farming systems with an

TABLE 20.1

Microbes Used in the Remediation of Dairy Wastewater and the Results Associated with Them

Microbes	Results	Sources
Pseudomonas aeruginosa, Bacillus subtilis, Lactobacillus delbrueckii, Staphylococcus aureus and Enterococcus hirae	BOD removal %=78.7%, TSS removal %=99.3%	Al-Wasify et al. (2017)
Bacillus subtilis and pseudomonas aeruginosa	COD removal with Pseudomonas aeruginosa=84.2%	Gulhane and Shome (2019)
Candida albicans, Candida krusei, Lactobacillus casei and Lactobacillus plantarum	Lactobacillus casei, Lactobacillus plantarum and Candida albicans (COD reduction=48%)	Keffala et al. (2017)
Pseudomonas fluorescens, Pseudomonas aeruginosa, Bacillus cereus, Bacillus subtilis, Enterobacter, Streptococcus faecalis, Escherichia coli	Lactobacillus casei, Lactobacillus plantarum and Candida albicans (COD reduction=48%)	Porwal et al. (2015)

BOD, biological oxygen demand; COD, chemical oxygen demand; EC, electrical conductivity; TDS, total dissolved solids; TS, total solids; TSS, total suspended solids.

TABLE 20.2

Values Used in the Experiment for the Nutrient Recovery in Dairy Wastewater

pH	Additional Na-lignosulphonate Concentrations % (w/v)	Temperature °C	Reference
1.0–3.5	0.002, 0.004, 0.008, 0.016 & 0.032	22	Kurup et al. (2019)
1.0, 2.0 and 3.5	-	40	Kurup et al. (2019)

TABLE 20.3

Results of the Conducted Experiment for Nutrient Recovery in Dairy Wastewater

pH	Additional Na-lignosulphonate Concentrations % (w/v)	Temperature °C	Turbidity Removal %	Recovered Lipids and Proteins (Mainly Caseins %	BOD Reduction %	Reference
3.5	0.016	22 and 40	98	-		Kurup et al. (2019); Botis (2015)
3.5	0.016	22	-	96 and 46	73	Kurup et al. (2019); Botis (2015)

algae treatment system can lead to more productive crop production and nutrient management on farms, resulting in a more sustainable dairy farm with less land use. Several other technological advances are currently being made for the recovery of nutrients from wastewater. According to (Paucar and Sato, 2021), microbial fuel cells (MFC) have reached formidable importance due to their versatility in their applications in wastewater treatment, power generation, removal of toxic pollutants, environmental monitoring sensors and more. However, progress is needed to create more effective MFC innovations that are relevant to the large area, as most studies have been carried out on a research scale and, exceptionally, few considerations have used a full-scale wastewater treatment plants. Next, the efficacy of MFC for the recovery of nutrients from dairy wastewater and the functional environment is presented

20.6 High Value Products from Microbes Grown on Wastewater

High-value products are being produced from various microbes resulting in high nutritional quality for fish, poultry, eggs, fruits, and beverage, soybean products, starch products, edible oil, and others. There are various technologies for the recovery such as Ion exchange, organic solvent extraction, SFE, MAE, UAE, PLE, EAE, adsorption, membrane processes, foam fractionation, precipitation, and solvent

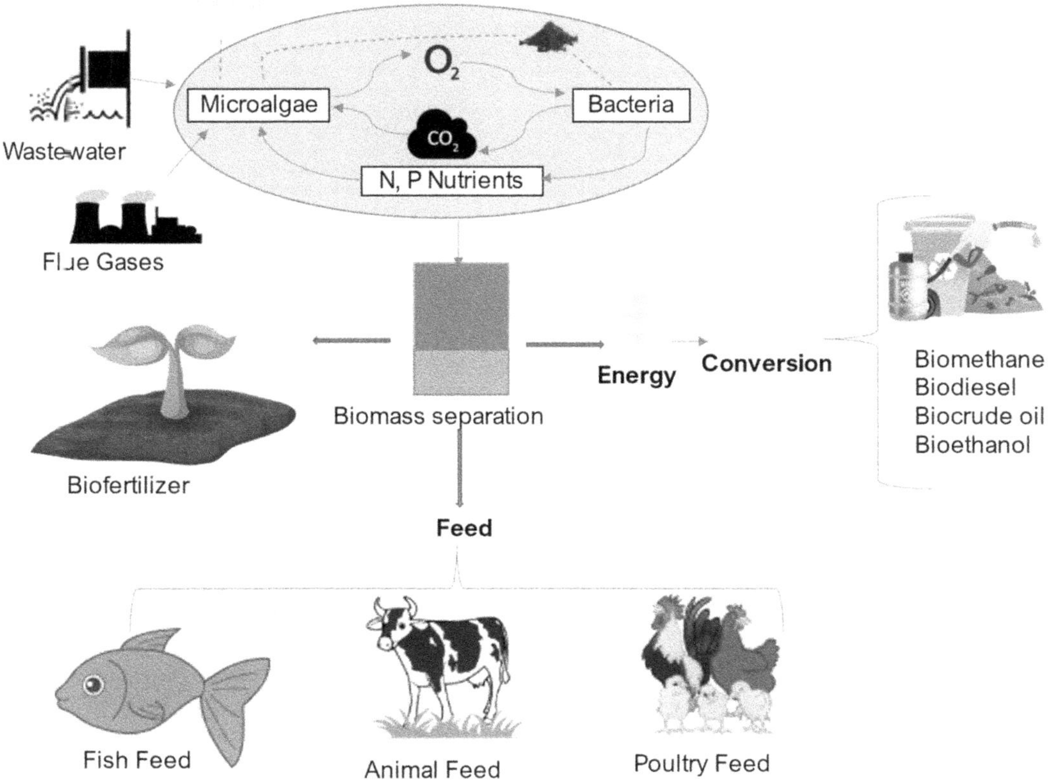

FIGURE 20.9 High-value products from the microbes grown on wastewater.

extraction. These technologies appear as a good alternative for nutrient recovery. There are few chemical technologies such as precipitation, solvent extraction, and other extraction technologies that play an important role in protein and enzyme recovery. There are few physical technologies such as membrane technologies, adsorption, and foam fractionation that are useful in the enrichment of bioproducts such as protein recovery from food (Akpor and Muchie, 2010; Di Caprio et al., 2020; Environment Canada, 2000, 2012; Flugaur, 2003; Kundu et al., 2013; Mazzucotelli et al., 2014) (Figure 20.9).

Some of the high value-added products produced from wastewater through the help of Microbes are:

20.6.1 Bio-Hydrogen

Production of Bio-hydrogen through third-generation feedstock i.e., microalgae is the most known sustainable and bio-refinery step towards balancing the energy crisis (Kumar et al., 2020). Hydrogen is the cleanest and most well-founded authentic and dependable alternative energy resource for non-renewable fossil fuels; it has a high calorific value of 122 kJ/g and produces water molecules as the product after the combustion.

20.6.2 Bioplastic

Increased consumption and production of plastics lead to environmental pollution. (Bhatia et al., 2021). Commonly used normal plastics are made up of petrol-based oils that cannot be decomposed by microbes. And, when these plastics are disposed into the environment, it causes severe environmental impacts (Geyer et al., 2017). Therefore, bioplastics are the only alternative as it disintegrates/degrades/decompose easily. Bioplastic is a biodegradable material produced with the help of biological resources such as bacteria, microalgae, yeast, and transgenic plants (Park et al., 2020). Such biologically derived plastics are of several types.

a. Polylactic acid-based plastics
b. Starch-based plastics
c. Carbohydrate based plastics
d. Protein-based plastics, etc.

20.6.3 Exopolysaccharides

Exopolysaccharides are high-value products obtained from algal secretions with outstanding and extremely valuable & important therapeutic properties (Sathiyanarayanan et al., 2016). These products are known to have properties like immunomodulatory, anticoagulant, anti-mutagenic, radio-protective, anti-bacterial, anti-cancer, anti-inflammatory bioactivities, and anti-ulcer (Bhatia et al., 2021). The chief producers of Exopolysaccharides are Cyanobacteria and Red Algae.

20.6.4 Bio-Fertilizer

Bio-fertilizer: Bio-fertilizers are complex substances that include live and latent micro-organisms that colonizes at the roots of the plants and helps in the fixation of atmospheric nitrogen, helps in nutrient fixation, help in phosphate solubilization, and help to make use of soil fertility to enhance the healthy crop production. Microbes/Micro-organisms present in the soil helps in nutrient balance and are widely used for soil conditioning as they are helpful in recovering excess nutrients from the wastewater during cultivation and growth of algae or microbes in the wastewater and make it clean and available for further other uses (Han et al., 2019). Cyanobacteria is the best example of a microalgae that has a high water-holding capacity, helps in atmospheric nitrogen fixation, is capable to withstand extreme environmental conditions and has a short generation period (Singh et al., 2016) (Figure 20.10).

20.6.5 Biodiesel

Biodiesel can be defined as the Fatty acid Methyl esters that are produced through the transesterification of oils along with the alcohols in the presence of catalysts (Bhatia et al., 2021). The major activities in the production of biodiesel involves the following steps:

a. Cultivation of Microalgae
b. Drying of Algal biomass
c. Oil extraction from Algal biomass
d. Transesterification of extracted oil into Fatty acid Methyl esters (Biodiesel)

20.7 Recovery of Value-Added Bioactive Compounds from Microbial Treatment Food Processing Wastewater

Natural bioactive compounds are being examined, for the treatment and prevention of different human illnesses. The nutrients in wastewater are getting less consideration in food industry. The relinquishment of such renewable wastewater prompts genuine ecological dangers as well as results in a loss of nutrient resources and helpless financial matters. Subsequently, there is a solid inspiration to foster the techniques and advancements to recuperation of nutrient supplements and wastewater reuse frameworks (Asgharnejad et al., 2021; Salama et al., 2017; Xie et al., 2016).

The food waste can be converted into beneficial value-added bioactive compounds viz., food supplements, phytochemicals, bioactive compounds, livestock feed, colourants, food supplements, biopigments and colourants, emulsifiers, proteins, edible and essential oils, biopreservatives, biofertilizers, polyphenols,

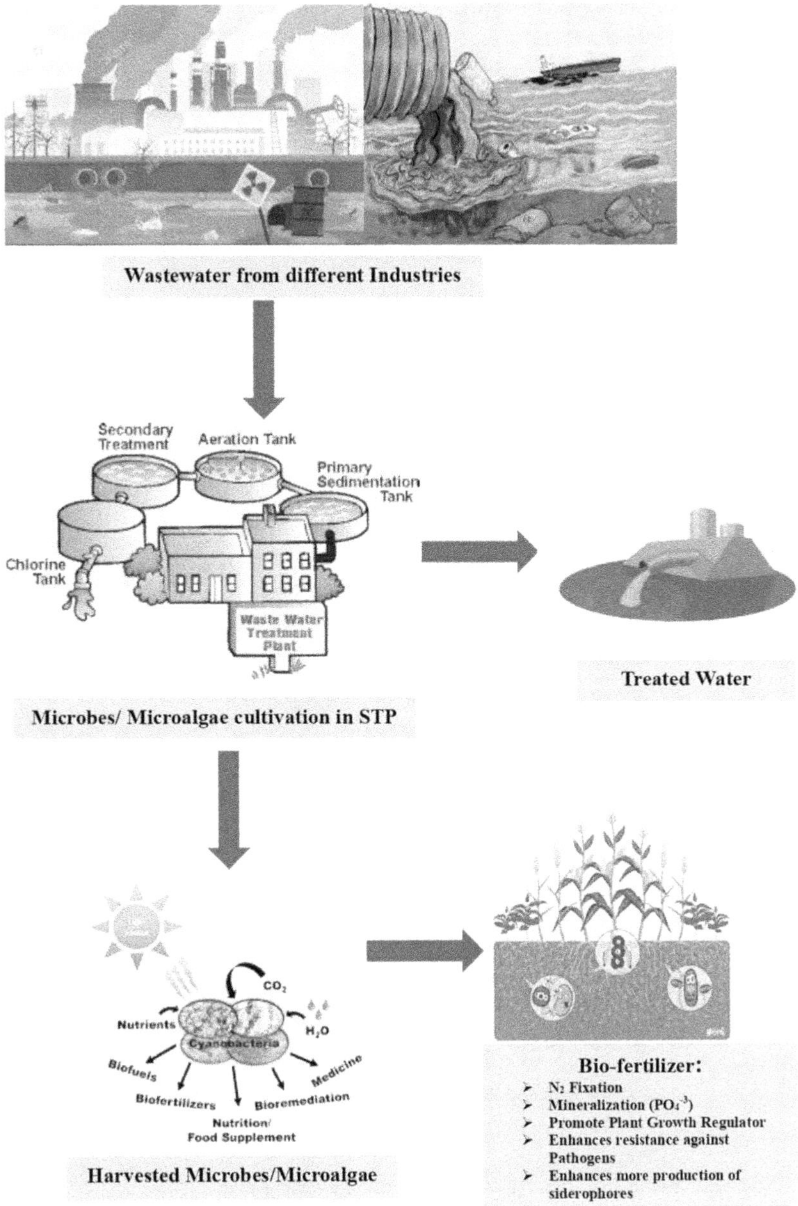

FIGURE 20.10 Production of bio-fertilizer using microbes/microalgae in wastewater.

biofuels, single-cell protein, dietary fiber, proteins, polysaccharides, pigments, and flavonoids (Barbera and Gurnari, 2018; Braun et al., 2021; Cassini et al., 2010). The bioactive compounds signify an interesting prospect for wastewater processing, a portion of these might give sorts of food additives, antioxidants, antimicrobial agents and others may be additionally changed into additional chemicals, sophisticated natural, macromolecules or biofuels for various applications (Federici et al., 2010; Scherhag and Ackermann, 2021), and surprisingly other nutritious food may be created with these nutrient supplement (De et al., 2007; Goula and Lazarides, 2015). Among different nutritions (*e.g.* nitrogen, phosphorous, and carbon) and sources, food processing wastewater is relatively safe and appropriate for microalgae development because of its high natural substance and low poisonousness (Li et al., 2019; Paucar and Sato, 2021).

Furthermore, the composition of these wastes relies upon the source or sort of waste. Around half of the waste is lignocellulosic in nature created from food processing industries (Sadh et al., 2018).

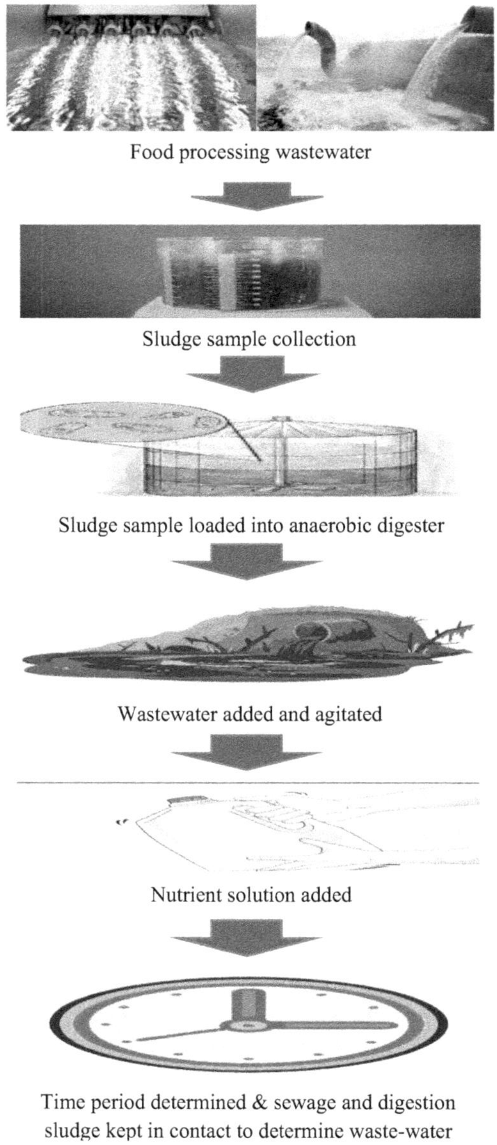

FIGURE 20.11 Microbial treatment for food processing wastewater (Wang et al. 2020.)

Sooner rather than later, the chance of totally or partially replacing chemical preservatives utilized today with natural additives derived from the by-products of the food industry will be expanded. Not least, as far as premium, is the chance of incorporating the feed formula of domesticated animals for creating meat and milk with phenols and other related compounds with the reason for the improvement of the derivative products (milk, fresh meat, mik products and meat) with antioxidant and antimicrobial substances (Veneziani et al., 2017) (Figure 20.11).

20.8 Conclusion

The release of the food processing wastewater into the aquatic system poses a great risk to the health and survival of the pelagic and benthic organisms owing to its rich nutrient composition. The solution to the problem is the envisaging cost-effective remediation strategies to quench the different nutrients present in the wastewater and microbes can be of great value in this regard. The microbial cells metabolize the nutrients for their growth and produce biomass which can be employed as great sources of valuable products like biofuels and bioplastics, biofertilizers etc and thus generate wealth from waste.

REFERENCES

Abatenh, E., Gizaw, B., Tsegaye, Z., & Wassie, M. 2017. The role of microorganisms in bioremediation-A review. *Open Journal of Environmental Biology*, 2(1), 038–046.

Acién, F. G., Gómez-Serrano, C., Morales-Amaral, M. D. M., Fernández-Sevilla, J. M., & Molina-Grima, E. 2016. Wastewater treatment using microalgae: How realistic a contribution might it be to significant urban wastewater treatment. *Applied Microbiology and Biotechnology*, 100 (21), 9013–9022.

Akpor, O. B., Adelani-Akande, T., & Aderiye, B. I. 2013. The effect of temperature on nutrient removal from wastewater by selected fungal species. *International Journal of Current Microbiology and Applied Sciences*, 2(9): 328–340.

Akpor, O. B., & Muchie, M. 2010. Remediation of heavy metals in drinking water and wastewater treatment systems: Processes and application. *International Journal of the Physical Sciences*, 5(12): 1807–1817.

Al-Mutairi, N. Z. 2009. Aerobic selectors in slaughterhouse activated sludge systems: A preliminary investigation. *Bioresource Technology*, 100(1), 50–58.

Al-Mutairi, N. Z, Hamoda, M. F., & Al-Ghusain, I. 2004. Coagulant selection and sludge conditioning in a slaughterhouse wastewater treatment plant. *Bioresource Technology*, 95 (2), 115–119.

Al-Wasify, R. S., Ali, M. N., & Hamed, S. R. 2017. Biodegradation of dairy wastewater using bacterial and fungal local isolates. *Water Science and Technology*, 76(11), 3094–3100.

AmaralA. L., Motta M. D., Pons M. N., Vivier H., Roche N., Mota M., & Ferreira, E. C. 2004. Survey of protozoa and metazoan populations in wastewater treatment plants by image analysis and discriminate analysis. *Environmetrics*, 15: 381–390.

Asgharnejad, H., Khorshidi Nazloo, E., Madani Larijani, M., Hajinajaf, N., & Rashidi, H. 2021. Comprehensive review of water management and wastewater treatment in food processing industries in the framework of water-food-environment nexus. *Comprehensive Reviews in Food Science and Food Safety*, 20(5), 4779–4815.

Barbera, M., & Gurnari, G. 2018. Water reuse in the food industry: Quality of original wastewater before treatments. In: M. Barbera & G. Gurnari (Eds.), *Wastewater Treatment and Reuse in the Food Industry* (pp. 1–16). Springer Briefs in Molecular Science. Cham, Switzerland: Springer.

Bayar, S., Yildiz, Y. S., Yilmaz, A., & Koparal, A. S. 2014. The effect of initial pH on treatment of poultry slaughterhouse wastewater by electrocoagulation method. *Desalination and Water Treatment*, 52(16–18), 3047–3053.

Bhatia, S. K., Otari, S. V., Jeon, J.-M., Gurav, R., Choi, Y.-K., Bhatia, R. K., & Yang, Y. H. 2021. Biowaste-to-bioplastic (polyhydroxyalkanoates): Conversion technologies, strategies, challenges, and perspective. *Bioresource Technology*, 326, 124733. doi: 10.1016/j.biortech.2021.124733

Blonskaja. A. C., & Mulbry, W. W. 2002. Recovery of dairy manure nutrients by benthic freshwater algae. *Bioresource Technology*, 84(1), 81–91.

Botis, M. 2015. Purification of the wastewater from meat industry. *Journal on Processing and Energy in Agriculture*, 19(1), 21–23

Bouwman. L., Goldewijk, K. K., Van Der Hoek, K. W., Beusen, A. H., Van Vuuren, D. P., Willems, J., ... & Stehfest, E. 2013. Exploring global changes in nitrogen and phosphorus cycles in agriculture induced by livestock production over the 1900–2050 period. *Proceedings of the National Academy of Sciences*, 110(52), 20882–20887.

Braun, R., Padsala, R., Malmir, T., Mohammadi, S., & Eicker, U. 2021. Using 3D City GML for the modeling of the food waste and wastewater generation—A case study for the city of Montréal. *Frontiers in Big Data*, 4, 662011.

Bulawayo City Council. 1980. Sewerage, drainage and Water By-laws, Statutory Instrument 390, Part 1.

Bull, M. A., Sterritt, R. M., & Lester, J. N. 1982. The treatment of wastewaters from the meat industry: A review. *Environmental Technology*, 3(1–11), 117–126.

Caccio S.M., Degiocome M., Aulicino F., & Edoardo P. (2003), *Giardia* cysts in wastewater treatment plants in Italy. *Application Environmental Microbiology*, 69, 3393–3398.

Cassini, A. S., Tessaro, I. C., Marczak, L. D. F., & Pertile, C. 2010. Ultrafiltration of wastewater from isolated soy protein production: A comparison of three UF membranes. *Journal of Cleaner Production*, 18 (3), 260– 265.

Chernicharo, C. D. 2006. Post-treatment options for the anaerobic treatment of domestic wastewater. *Reviews in Environmental Science and Bio/Technology*, 5(1), 73–92.

Craggs, R. J., Lundquist, T. J., & Benemann, J. R. 2013. Wastewater treatment and algal biofuel production. In *Algae for Biofuels and Energy* (pp. 153–163). Springer, Dordrecht.

Daneshvar, N., Khataee, A. R., Ghadim, A. A., & Rasoulifard, M. H. 2007. Decolorization of CI Acid Yellow 23 solution by electrocoagulation process: Investigation of operational parameters and evaluation of specific electrical energy consumption (SEEC). *Journal of Hazardous Materials*, 148(3), 566–572.

Davarnejad, R., & Nasiri, S. 2017. Slaughterhouse wastewater treatment using an advanced oxidation process: Optimization study. *Environmental Pollution*, 223, 1–10.

De, L. A., Macciola, V., Lembo, G., Aretini, A., & Nag, A. 2007. Studies on oxidative stabilisation of lard by natural antioxidants recovered from olive-oil mill wastewater. *Food Chemistry*, 99(3), 998–1004.

Di Caprio, F., Altimari, P., & Pagnanelli, F. 2020. New strategies enhancing feasibility of microalgal cultivations. In *Studies in Surface Science and Catalysis* (Vol. 179, pp. 287–316). Elsevier. doi:10.1016/B978-0-444-64337-7.00016-1.

EMA Act Statutory Instrument 6. 2007. 4thSchedule, Part II Table 1.

Environment Canada. 2000. Framework and Recommendations Concerning Effluent Quality of Wastewater Disposed by Federal Institutions. Final Report. Federal Committee on Environmental Management Systems (FCEMS)/Wastewater Working Group (accessed 10.03.15.)

EnvironmentCanada. 2012. Wastewater Systems Effluent Regulations. Fisheries Act. SOR/2012–139 (accessed 10.03.15.)

Federici, F., Fava, F., Kalogerakis, N., & Mantzavinos, D. 2010. Valorisation of agro-industrial by-products, effluents and waste: Concept, opportunities and the case of olive mill wastewaters. *Journal of Chemical Technology & Biotechnology Biotechnology*, 84(6), 895–900.

Flugaur, N. J. 2003. Wastewater effluent treatments and control technologies in the beef processing industry.

Foissner, W. 2016. Protists as bioindicators in activated sludge: Identification, ecology and future needs. *European Journal of Protistology*, 55, 75–94.

Gannoun, H., Bouallagui, H., Okbi, A., Sayadi, S., & Hamdi, M. 2009. Mesophilic and thermophilic anaerobic digestion of biologically pretreated abattoir wastewaters in an upflow anaerobic filter. *Journal of Hazardous Materials*, 170(1), 263–271.

Gannoun, H., Khelifi, E., Omri, I., Jabari, L., Fardeau, M. L., Bouallagui, H., … & Hamdi, M. 2013. Microbial monitoring by molecular tools of an upflow anaerobic filter treating abattoir wastewaters. *Bioresource Technology*, 142, 269–277.

García, D., Posadas, E., Blanco, S., Acién, G., García-Encina, P., Bolado, S., & Muñoz, R. 2018. Evaluation of the dynamics of microalgae population structure and process performance during piggery wastewater treatment in algal-bacterial photobioreactors. *Bioresource Technology*, 248, 120–126.

Geyer, R., Jambeck, J. R., &Law, K. L. 2017. Production, use, and fate of all plastics ever made. *Science Advances*, 3, e1700782. doi: 10.1126/sciadv.1700782

Ghosh, S., & Henry, M. P. 1981. *Stabilization and Gasification of Soft-Drink Manufacturing Waste by Conventional and Two-Phase Anaerobic Digestion* (No. CONF-810577-1). Inst. of Gas Tech., Chicago, IL.

Goula, A. M., & Lazarides, H. N. 2015. Integrated processes can turn industrial food waste into valuable food by-products and/or ingredients: The cases of olive mill and pomegranate wastes. *Journal of Food Engineering*, 167, 45–50.

Gray F.N. (2002), *Water Technology: An Introduction for Environmental Scientists and Engineers* (pp. 35–80). Butterworth-Heinemann, Oxford.

Gulhane, V., & Shome, S. D. 2019. Treatment Efficiency Enhancement of Dairy Effluent by Bioaugmentation Using Bacterial Species. *SSRN Electronic Journal*. Published.

Gumbo, B., Mlilo, S., Broome, J., & Lumbroso, D. 2003. Industrial water demand management and cleaner production potential: A case of three industries in Bulawayo, Zimbabwe. *Physics and Chemistry of the Earth, Parts A/B/C*, 28(20–27), 797–804.

Han, P., Lu, Q., Fan, L., & Zhou, W. 2019. A review on the use of microalgae for sustainable aquaculture. *Applied Sciences*, 9, 2377. doi: 10.3390/app9112377

Hernández, D., Riaño, B., Coca, M., & García-González, M. C. 2013. Treatment of agro-industrial wastewater using microalgae–bacteria consortium combined with anaerobic digestion of the produced biomass. *Bioresource Technology*, 135, 598–603.

Jian, T., & Zhang, X. 1999. Bioprocessing for slaughterhouse wastewater and its computerized control and supervisory system. *Resources, Conservation and Recycling*, 27(1–2), 145–149.

Johns, M. R. 1995. Developments in wastewater treatment in the meat processing industry: A review. *Bioresource Technology*, 54(3), 203–216.

Keffala, C., Zouhir, F., Abdallah, B. H., & Kammoun, S. 2017. Use of bacteria and yeast strains for dairy wastewater treatment. *International Journal of Research in Engineering and Technology*, 6(4), 108–113.

KolevSlavov, A. 2017. Dairy wastewaters – general characteristics and treatment possibilities – A review. *Food Technology and Biotechnology*, 55(1), 14.

Kundu, P., Debsarkar, A., & Mukherjee, S. 2013. Treatment of slaughter house wastewater in a sequencing batch reactor: Performance evaluation and biodegradation kinetics. *BioMed Research International*, 2013, 134872.

Kumar, G., Mathimani, T., Sivaramakrishnan, R., Shanmugam, S., Bhatia, S. K., & Pugazhendhi, A. 2020. Application of molecular techniques in biohydrogen production as a clean fuel. *Science of the Total Environment*, 722, 137795. doi: 10.1016/j.scitotenv.2020.137795

Kurup, G. G., Adhikari, B., & Zisu, B. 2019. Recovery of proteins and lipids from dairy wastewater using food grade sodium lignosulphonate. *Water Resources and Industry*, 22, 100114.

LeChevallier M.W., & Au K. 2004. Inactivation (disinfection) processes. In: *Water and Treatment and Pathogen Control* (pp. 41–65). IWA Publishing, London.

Li, S., Zhao, S., Yan, S., Qui, Y., Song, C., Li, Y., & Kitamura, Y. 2019. Food processing wastewater purification by microalgae cultivation associated with high value-added compounds production. *Chinese Journal of Chemical Engineer*. 27(12), 2845–2856.

Lloyd B.J., & Frederick G.L. 2000. Parasite removal by waste stabilization pond systems and the relationship between concentrations in sewage and prevalence in the community, *Water Science and Technology*, 42(10): 375–386.

Manjunath, N. T., Mehrotra, I., & Mathur, R. P. 2000. Treatment of wastewater from slaughterhouse by DAF-UASB system. *Water Research*, 34(6), 1930–1936.

Matamoros, V., Uggetti, E., García, J., & Bayona, J. M. 2016. Assessment of the mechanisms involved in the removal of emerging contaminants by microalgae from wastewater: A laboratory scale study. *Journal of Hazardous Materials*, 301, 197–205.

Mazzucotelli, C. A., Durruty, I., Kotlar, C. E., Moreira, M. D. R., Ponce, A. G., & Roura, S. I. 2014. Development of a microbial consortium for dairy wastewater treatment. *Biotechnology and Bioprocess Engineering*, 19(2), 221–230.

Mekonnen, M. M., & Hoekstra, A. Y. 2012. A global assessment of the water footprint of farm animal products. *Ecosystems*, 15(3), 401–415.

Mittal, A. 2011. Biological wastewater treatment. *Water Today*, 1, 32–44.

Mittal, G.S. 2004. Characterization of the effluent wastewater from abattoirs for slaughterhouse waste - A review. *Bioresource Technology*, 83(1), 13–26

Molla, A.H., Fakhru'l-Razi, A., & Alam, M.Z. 2004. Evaluation of solid-state bioconversion of domestic wastewater sludge as a promising environmental friendly disposal technique. *Water Research*, 38(19), 4143–4152.

Muhamad Ng, S. N., Idrus, S., Ahsan, A., Tuan Mohd Marzuki, T. N., & Mahat, S. B. 2021. Treatment of wastewater from a food and beverage industry using conventional wastewater treatment integrated with membrane bioreactor system: A pilot-scale case study. *Membranes*, 11(6), 456.

Muñoz, R., Guieysse, B., Muñoz, R., & Guieysse, B. 2006. Algal-bacterial processes for the treatment of hazardous contaminants: A review. *Water Research*, 40, 2799–2815.

Olguín, E. J. 2012. Dual purpose microalgae-bacteria-based systems that treat wastewater and produce biodiesel and chemical products within a biorefinery. *Biotechnology Advances*, 30, 1031–1046.

Paillard D., Dubois V., Thiebaut R., Nathier F., Hogland E., Caumette P., & Quentine C. 2005, Occurrence of *Listeria* spp. In effluents of French urban wastewater treatment plants. *Applied and Environmental Microbiology*, 71(11), 7562–7566.

Painter H.A. 1970, A review of literature on inorganic nitrogen metabolism in microorganisms. *Water Research*, 3, 241–250.

Pal, A. K., Singh, J., Soni, R., Tripathi, P., Kamle, M., Tripathi, V., & Kumar, P. 2020. The role of microorganism in bioremediation for sustainable environment management. In *Bioremediation of Pollutants* (pp. 227–249). Elsevier, Amsterdam.

Park, Y.-L., Bhatia, S. K., Gurav, R., Choi, T.-R., Kim, H. J., Song, H.-S., ... & Yang, Y. H. 2020. Fructose based hyper production of poly-3-hydroxybutyrate from *Halomonas* sp. YLGW01 and impact of carbon sources on bacteria morphologies. *International Journal of Biological Macromolecules*, 154, 929–936.

Paucar, N. E., & Sato, C. 2021. Microbial fuel cell for energy production, nutrient removal and recovery from wastewater: A review. *Processes*, 9(8), 1318.

Porwal, H., Mane, A., & Velhal, S. 2015. Biodegradation of dairy effluent by using microbial isolates obtained from activated sludge. *Water Resources and Industry*, 9, 1–15.

Sadh P.K., Kumar. S.K., Chawla, P., & Duhan, J. S. 2018. Fermentation: A boon for production of bioactive compounds by processing of food industries wastes (by-products). *Molecules*, 23(10), 2560. doi: 10.3390/molecules23102560

Salama, E. S., Kurade, M. B., Abou-Shanab, R. A., El-Dalatony, M. M., Yang, I. S., Min, B., & Jeon, B. H. 2017. Recent progress in microalgal biomass production coupled with wastewater treatment for biofuel generation. *Renewable and Sustainable Energy Reviews*, 79, 1189–1211.

Sandhar, S. K., Nim, D., & Agrawal, S. 2013. Consumption pattern of soft drinks and fruit juices: A comparative study. *Shri Vaishnav Institue of Mangement e-Journal of Applied Mangement*, 1, 39–49.

Sathiyanarayanan, G., Bhatia, S. K., Kim, H. J., Kim, J.-H., Jeon, J.-M., Kim, Y.-G., ... & Yang, Y. H. 2016. Metal removal and reduction potential of an exopolysaccharide produced by Arctic psychrotrophic bacterium *Pseudomonas* sp. PAMC 28620. *RSC Advances*, 6, 96870–96881. doi: 10.1039/C6RA17450G

Scherhag, P., & Ackermann, JU. 2021. Removal of sugars in wastewater from food production through heterotrophic growth of *Galdieria sulphuraria. Engineering in Life Sciences*, 21(3–4), 233–241.

Singh, J. S., Kumar, A., Rai, A. N., & Singh, D. P. 2016. Cyanobacteria: A precious bio-resource in agriculture, ecosystem, and environmental sustainability. *Frontiers in Microbiology*, 7, 529. doi: 10.3389/fmicb.2016.00529

Siracusa, G., Yuan, Q., Chicca, I., Bardi, A., Spennati, F., Becarelli, S., ... & Di Gregorio, S. 2020. Mycoremediation of old and intermediate landfill leachates with an ascomycete fungal isolate, *Lambertella* sp. *Water*, 12(3), 800.

Suelson, K.A. and Lowery, K.W. 1992. Designing a sequencing batch reactor system for the treatment of high strength meat processing wastewater. In *Proceedings of the 47th Purdue Industrial Waste Conference* (pp. 765–774). Lewis Publishers, Inc., Chelsea, MI.

Tolboom, S. N., Carrillo-Nieves, D., de Jesús Rostro-Alanis, M., de la Cruz Quiroz, R., Barceló, D., Iqbal, H. M., & Parra-Saldivar, R. 2019. Algal-based removal strategies for hazardous contaminants from the environment–a review. *Science of the Total Environment*, 665, 358–366.

USEPA. 2004. Effluent limitations guidelines and new source performance standards for the meat and poultry products point source category. *Federal Register*, 69(173), 54475–54555.

Veneziani, G., Novelli, E., Esposto, S., Taticchi, A., & Servili, M. 2017. Applications of recovered bioactive compounds in food products. In *Olive Mill Waste* (pp. 231–253). Academic Press. doi: 10.1016/B978-0-12-805314-0.00011-X

Wang, Y., Chen, X., Huang, J., Chen, L., Liu, Y., Huang, J., Wu, T., & Li, Y. (2020). Research on anaerobic treatment of food processing waste-water based on watershed water protection. *Journal of Physics: Conference Series*, 1549, 022044. doi: 10.1088/1742-6596/1549/2/022044.

Wilkie, A. 2002. Recovery of dairy manure nutrients by benthic freshwater algae. *Bioresource Technology*, 84(1), 81–91. doi: 10.1016/s0960-8524(02)00003-2

Wollmann, F., Dietze, S., Ackermann, J. U., Bley, T., Walther, T., Steingroewer, J., & Krujatz, F. 2019. Microalgae wastewater treatment: Biological and technological approaches. *Engineering in Life Sciences*, 19(12), 860–871.

World Bank Group. 2007. Environmental, Health and Safety (EHS) Guidelines for Meat Processing. General EHS Guidelines: Environmental Wastewater and Ambient Water Quality.

Xie, M., Shon, H. K., Gray, S. R., & Elimelech, M. 2016. Membrane-based processes for wastewater nutrient recovery: Technology, challenges, and future direction. *Water Research*, 89, 210–221.

21

Integrated Wastewater Treatment and Biofuel Production Using Microalgae

Mohit Aggarwal and Neelancherry Remya
Indian Institute of Technology Bhubaneswar

CONTENTS

21.1 Introduction

Finding clean drinking water and a sustainable energy supply are the two fundamental challenges the world faces right now. Water is one of our planet's most valuable natural resources. However, presently, the water quality, especially in developing countries, has been deteriorating to alarming levels by various human activities. The pollution is mainly caused due to the unchecked dumping of municipal, agricultural, and industrial wastewater into the water bodies. These wastewaters contain a massive amount of organic and inorganic contaminants, such as microplastics, xenobiotics, heavy metals, and high concentrations of nitrates, phosphates, and carbon compounds (Chandra and Kumar, 2017a). This puts a tremendous strain on the food chain and consequently on human life. At the same time, as fossil fuel resources are depleting, the need for renewable energy is growing. The idea of combining biological waste treatment with bioenergy production is gaining popularity. Microalgae are the most promising candidates to deal with both energy and wastewater challenges. Due to their robust metabolisms and ability to quickly adapt to environmental stresses, microalgae can survive in various ecosystems on the planet, freshwater, brackish water, seawater, and even wastewater (Aggarwal and Remya, 2021).

DOI: 10.1201/9781003231738-23

Microalgae can grow quickly, treat wastewater, and produce biofuels from their biomass at a lower cost than other current technologies. Microalgae can thrive in a variety of trophic environments. Microalgae, for example, can grow photoautotrophically to trap sunlight and CO_2 from the air, assisting in CO_2 sequestration while producing valuable products. Microalgae can grow heterotrophically, too, absorbing nutrients from wastewater and producing biofuel. While other microalgal strains can grow mixotrophically, capturing CO_2, treating wastewater, and producing valuable products all at the same time (Wollmann et al., 2019). However, to employ microalgae in wastewater treatment and generate biomass for biofuel production, it's vital to choose suitable wastewater, robust microalgal species, pretreatment methods, and appropriate operating conditions.

21.2 Application of Microalgae for Wastewater Treatment

21.2.1 Municipal Wastewater

In developing countries, municipal wastewater is discharged directly into the water bodies without any pretreatment. This is due to mismanagement of conventional water and wastewater treatment plants and virtually non-existent environmental regulations. Municipal wastewater has a high concentration of organic and inorganic materials and hence, is considered a suitable nutrient medium for several microorganisms such as bacteria and viruses. The organic fraction in municipal wastewater comprises amino acids, fats, carbohydrates, and volatile acids. The inorganic part consists of ammonium salts, calcium, magnesium, potassium, sodium, chlorine, bicarbonates, sulfur, phosphate, and heavy metals (Abdel-Raouf et al., 2012). The municipal wastewater is subjected to tertiary treatment to remove these nutrients. However, the conventional tertiary treatment is much more costly than the primary treatment. Hence, this approach is unsustainable. Cultivating microalgae in municipal wastewater treatment is an environmentally and economically sustainable solution due to the ability of microalgae to consume nutrients such as phosphorus and nitrogen from the wastewater for their growth. The biomass thus harvested can be used to generate biofuels further, reducing the total operation cost of the wastewater treatment plant. Microalgae cultivation in the municipal wastewater treatment systems for nutrient removal has been quite extensively studied. *For example, chlorella* sp. and *Scenedesmus* sp. have been successfully reported to remove carbon, ammonia, nitrate, and phosphate from the municipal wastewater with the removal rates in the region of 80%–100% in many cases (Pittman et al., 2011).

21.2.2 Agricultural Wastewater

Agriculture is another major wastewater producer, especially in developing countries. The majority of agricultural wastewater is generated in animal farming. As a result, the wastewater contains high levels of phosphorus and nitrogen. These excess nutrients leach to the nearby water bodies and cause eutrophication deteriorating the water quality. Piggery wastewater is generally treated with anaerobic bacteria. But the nutrient removal capacity of microalgae is comparatively higher than these bacteria (Bohutskyi et al., 2016).

21.2.3 Industrial Wastewater

The conventional way of treating industrial wastewater includes ion exchange and precipitation. Since they contain lower levels of phosphorus and nitrogen and higher levels of heavy metals and toxic elements such as oils, organic acids, polycyclic aromatic compounds, phenols, and nitrogenous compounds, and which are difficult to remove, it is crucial to choose particular strains that are tolerant to heavy metals and can absorb metals to be able to treat the industrial wastewater successfully (Chandra and Kumar, 2017b). So far, very few strains have been explored that can remove metals from these wastewaters. Depending on the industry, the composition of industrial wastewater can significantly vary as every sector produces its characteristic set of pollutants. Several microalgae species such as *Chlorella* sp., *Arthrospira* sp., *Nannochloropsis* sp., and *Scenedesmus* sp. have been reported to treat industrial wastewater from the brewery, olive mill, pharmaceutical, tannery, and electronic device factories (Wollmann et al., 2019).

21.3 Use of Algal Biomass for Biofuels

Microalgal biomass has the potential to produce multiple forms of biofuels through different routes. Microalgal biomass can be transformed into biochar, bio-oil, and biogas through thermochemical conversion; biodiesel through lipid transesterification; bioethanol, biohydrogen through the fermentation of carbohydrates; biochar by thermal decomposition. A microalgal cell contains 25%–50% lipid, depending upon the species. The presence of specific stressors such as nitrogen limitation in the growth medium improves lipid productivity. This amount can be further enhanced by modifying the acetyl CoA pathway, which is responsible for fatty acid production in microalgae (Goh et al., 2019). The common fatty acids present in the lipid have a chain length that ranges from C*16* to C*18,* which is similar to that of higher plants. These fatty acids can be both saturated and unsaturated. The chain length and the amount of saturation determine the characteristics of the lipid. The quality of lipids also depends on the operational parameters of the extraction process.

Biochar is carbon-rich charcoal made of thermochemical decomposition (pyrolysis, hydrothermal liquefaction, torrefaction) of microalgal biomass in an oxygen-free atmosphere and high temperature. Among the thermal conversion process, pyrolysis and torrefaction are more prominent technologies for biochar production.

Biogas majorly consists of CH_4 and CO_2. One of the most significant advantages of biogas is that it can be generated from any microalgae species irrespective of its low lipid or high ash content.

Microalgae-based biofuels production offers several advantages such as (i) the ability to utilize non-arable land to cultivate microalgae, which does not threaten the land requirement for food crops (ii) higher photosynthesis activity as compared to terrestrial plants, which results in high biomass productivity as well as offer higher CO_2 sequestration; (iii) Addition of external nutrients is not required; and (iv) ability to remove the pollutants from wastewater. Several techniques are currently deployed to convert the algal biomass into biofuel. The nature of biofuel depends on the algal species, conversion process, and the operating parameters of the process.

21.3.1 Transesterification

The lipids present inside the microalgal cells can be converted to fatty acid methyl ester (FAME), the primary components of biodiesel, by direct transesterification with a catalyst or via in-situ transesterification of the microalgal lipids. The lipid present in microalgae is more viscous than the conventional diesel, due to which it cannot be directly used in internal combustion engines. Transesterification reduces the microalgal oil's viscosity and improves its flow characteristics (Wicker et al., 2021). In transesterification, the first step is breaking down large, branched Triglyceride (TAG) molecules into smaller Diglyceride (DAG) molecules using alcohol. Next, the DAG further breaks down to release straight-chain glycerol molecules whose size is comparable to the fatty acids present in the diesel fuel. Alcohols like propanol, butanol, ethanol, and methanol are used in the reaction as a catalyst. Methanol is the most preferred option among these alcohols as it is cheap and easy to recover. The transesterification of 1 mole of TAG requires 3 moles of alcohol which produces 1 mole of glycerol and 3 moles of fatty esters (de Jesus et al., 2020).

21.4 Anaerobic Digestion

Certain microalga such as cyanobacteria cannot accumulate high levels of lipids. For such microalgae, anaerobic digestion is more suitable. It converts all the three macromolecules (protein, carbohydrate, and lipid) into volatile fatty acids (VFAs), which can later be converted to biodiesel. VFAs majorly consist of propionate, butyrate, and acetate. They can also be converted to CH_4 and CO_2, which are significant components of biogas. The VFAs are more stable than biogas. Hence, they are safer and easier to store. Biogas production is also an attractive option when the spent microalgal biomass is used as a feedstock after lipid extraction. It is produced when the biomass is subjected to anaerobic digestion. Anaerobic digestion happens in four steps: hydrolysis, acidogenesis, acetogenesis, and methanogenesis. Firstly, the macromolecules are hydrolyzed to monomers, which are then converted to VFAs. Acidogenic bacteria also form additional products such as alcohols, CO_2, and H_2. Some of the propionate, acetate, alcohols, and butyrate are further converted to acetate via the acetogenic pathway. VFAs are produced as intermediates during acidogenesis and acetogenesis stages, which are later converted to biogas in the methanogenesis stage as a final product of the anaerobic digestion. In addition to biodiesel production, VFAs can also generate electricity and remove phosphorus and nitrogen from the wastewater (Cho et al., 2017; Waqas et al., 2018).

21.4.1 Fermentation

Fermentation is another anaerobic process in which microorganisms such as bacteria or yeast breaks down the carbohydrate molecules in the biomass into ethanol. The process takes about 1–3 days to complete at 20°C–32°C. To convert the cellulosic and hemicellulosic components of the algal biomass into ethanol, it is first hydrolyzed to glucose using enzymes or acids. The most common bacteria and yeast used for fermentation are *Bacillus subtlis* and *Saccharomyces cerevisiae* (Zanchetta et al., 2021).

21.4.2 Torrefaction

Torrefaction involves heating the biomass between 200°C and 350°C in an inert environment and under atmospheric pressure. Recently, torrefaction can be carried out at elevated pressures and in the presence of different gases such as O_2, CO_2, and even flue gases to reduce the total operating cost. The heating rates are usually kept low (<50°C/min). The biomass is then held at the maximum temperature for few minutes, generally less than 1 hour. torrefaction generates three products: biochar, biogas, and bio-oil. Biogas includes H_2, CO_2, CO, and CH_4. The bio-oil contains water, lipids, and organic compounds. The ratio depends upon the operating conditions. Among the three products, biochar is the main product and generally accounts for 70% of the raw biomass and 90% of the total energy contained in the raw biomass. One drawback of torrefaction is that it generates high ash content (Bach and Skreiberg, 2016). The biochar needs to be compressed or pelletized to be used as fuel. But since the resultant biochar is more porous, brittle, and less dense than the feedstock, it requires more pelletizing pressure and temperature to be made into a durable pellet. The biochar also loses some of its binding ability due to the removal of hemicellulose and lignin in the biomass, which acts as natural binding agents. As a result, the binder requirement is also increased (Gan et al., 2018). Overall, the increased need for binder, temperature, pelleting pressure to produce commercial-grade pellets from biochar makes the post-treatment of torrefaction products costly and challenging.

Recently, wet torrefaction has gained prominence for obtaining biochar as it can convert algal biomass even with high moisture content (Bach et al., 2017). In this process, the biomass is treated in hot compressed water at a temperature range of 180°C–260°C. Before the heating, initial pressure is applied to the system to ensure water remains in the liquid phase throughout the process. After the maximum temperature is reached, it is held at that temperature for a specific time (called holding time). Afterward, the system is quickly cooled to room temperature. Finally, the pressure is released from the system. Under this process, a solid substance, called hydrochar, is the main product that can account for up to 88% of the feedstock. CO_2 is the major gaseous component, with H_2, CH_4, and CO being the minor components. Although the amount of solid fuel or hydrochar increases in the wet torrefaction process, the HHV is comparable to biochar (Bach and Skreiberg, 2016).

21.4.3 Pyrolysis

Pyrolysis involves heating the biomass in an oxygen-free atmosphere (usually N_2 or CO_2) at 300°C–800°C. At such temperatures, the different macromolecules in biomass are decomposed into several compounds that are collected as solid fractions (biochar), gas (pygas), and liquid (bio-oil). The pygas contain CO, CO_2, light hydrocarbons (C_1–C_4), and hydrogen. The solid fraction or biochar has C, H, and little O (less than 10 wt%). The bio-oil contains multiple oxygenated compounds, fine char particles, and some moisture. The moisture content of the bio-oil is in the range of 15%–40%. Sometimes, the bio-oil automatically separates into aqueous and organic fractions. Bio-oil is produced more in the intermediate temperatures (500°C–600°C), at extremely low residence time, generally a few seconds, and with high heat transfer and heating rates. Under these conditions, the bio-oil yield can easily cross 70%. All these products have a high HHV (Yang et al., 2019). The biochar produced by pyrolysis is characterized by having higher surface area and hence are good bio-adsorbents and, therefore, can be used to remove pollutants from the wastewater. If added to the soil, it can improve the soil quality. However, the torrefied biochar has a higher heating value as compared to pyrolytic biochar. In addition to higher heating values, the torrefied biochar has a higher carbon number, improved grindability, lower moisture content, and ash content. This makes the torrefied biochar more suited to be used as a fuel.

Biochar can be used as a biofertilizer, carbon sequester, or as absorbent for wastewater treatment. Recent studies suggest that it can be used as a coal substitute for electricity generation because of its comparable HHV to coal. Biochar production is favored when pyrolysis is carried out at low temperatures (400°C) for an extended period which can last from an hour to days. This type of pyrolysis is called slow pyrolysis or carbonization.

Microwave-assisted pyrolysis has recently gained prominence due to its fast and selective heating properties. The microwaves directly heat the biomass by penetrating the wall of the container. In this process, no heat is absorbed by the container. This is much more energy-efficient as in conventional pyrolysis, a chunk of heat is first absorbed by the container walls, wasting a lot of energy. Microwave heating is also found to have a higher biochar yield than conventional heating with the same operating parameters.

21.4.5 Hydrothermal Processing

Hydrothermal processing is also a popular thermochemical conversion method to convert wet algal biomass into biofuels. At low temperatures (170°C–250°C), hydrothermal carbonization occurs, and biochar is formed. At moderate temperatures (250°C–370°C), hydrothermal liquefaction (HTL) occurs and produces bio-oil. On further increasing the temperature to 800°C, hydrothermal gasification occurs, giving rise to energy-rich gases (Cheng et al., 2017). In HTL, all the components present in an algal cell: carbohydrates, lipids, and proteins, can contribute to bio-oil production. This results in HTL generating more bio-oil than the actual lipid content in the algal cell (Yu et al., 2011). Bio-oil developed from HTL has higher HHV, lower oxygen, and lower moisture content than the pyrolytic bio-oil. The use of specific homogeneous and heterogeneous catalysts can remove unwanted heteroatoms from the bio-oil (Duan et al., 2013). In the temperature range of 180°C–370°C and 5–21 MPa, multiple reactions such as dehydration, hydrolysis, decarboxylation, deamination, repolymerization, and Maillard reactions occur. These reactions convert all three macromolecules into solid biochar, water-soluble molecules, bio-oil, and non-condensable gases. The water present in the biomass provides H^+ and OH^- ions, which act as a catalyst for these reactions. At temperatures below 250°C, the lipid is hydrolyzed into free fatty acids of varying lengths. When the temperature increases, the algal cell wall breaks, and some carbohydrates and proteins undergo several changes such as decarboxylation, repolymerization, deoxygenation, and deamination. The heteroatom-containing organic compounds are added to the bio-crude. It is also worth noting that the temperature required to produce bio-oil is relatively lower for high-lipid-containing algal cells than for high-carbohydrate or high-protein cells (Li et al., 2017).

21.4.5 Gasification

Gasification includes the algal biomass that is partially oxidized into syngas at elevated temperatures (800°C–900°C). Gasification is an excellent option to extract energy from low lipid-containing algal biomass. Algal biomass can produce two forms of gases that can be used as biofuels: biogas and biohydrogen. Biogas comprises CH_4 and CO_2 and small amounts of other gases such as CO, H_2S, NH_3, and N_2. Bio-hydrogen is a much cleaner and more sustainable source of fuel. It contains much fewer contaminants, almost zero carbon, and has a higher HHV than biogas. Bio-hydrogen can be used in two ways: by direct combustion or by using in fuel cells. One of the most significant advantages of using biohydrogen as a fuel is that water is generated as a byproduct. This property makes biohydrogen one of the most attractive fuel options of the future (Salakkam et al., 2021).

If the algal biomass has high moisture, it can be subjected to hydrothermal gasification, converting wet biomass into gaseous fuels. In this process, the biomass is heated at high temperatures (300°C–700°C) in a water medium at elevated pressure. The water present in the biomass can act as a reactant as well as a reaction medium. The long-chain molecules in the feedstock are broken down by hydrolysis.

Due to high temperature, the reaction happens faster and more efficiently than fermentation and anaerobic digestion. The main products of hydrothermal gasification are H_2, CO_2, and CH_4 (Jayaraman et al., 2021).

21.5 Economic Prospective

Despite the potential of microalgae to treat wastewater and produce biofuels, it is still uncertain if algal biofuels can be produced economically and sustainably. Two criteria widely used to study the economic and environmental viability of biofuel are the life cycle assessment (LCA) and techno-economic analysis (TEA). The lifecycle includes studying the lifecycle energy analysis, lifecycle analysis of CO_2 emission, and lifecycle analysis of cost. Certain parameters such as Net Energy Ratio (NER) and global warming potential (GWP) are employed in LCA to help the decision-makers in determining the system's sustainability (Mälkki and Alanne, 2017).

21.5.1 Life Cycle Assessment

A study that reviewed several LCA studies found that the existing LCA studies could not obtain a unanimous conclusion regarding the positive impact of microalgal fuels over conventional fuels. Similarly, most of the existing studies concluded that GHG emissions of algal biofuels were lower than equivalent fossil fuels. Still, emissions per functional unit of biofuel differed significantly among those studies. Such a significant difference in results was mainly due to the complexity of the microalgae bioenergy production and the lack of industrial or commercialized production (Rajesh Banu et al., 2020).

21.5.2 Techno-Economic Analysis

The TEA studies, too, have shown variable results as different studies used different modeling assumptions. These assumptions include production capacities, biomass productivity, cultivation systems, final products, oil content, and conversion technologies. Studies showed that open raceway ponds were a better choice for microalgae cultivation in countries like Australia because large areas in Australia can be devoted to microalgal ponds without affecting their biodiversity (Campbell et al., 2011). One study compared the PBR and open raceway ponds and found that the cost investment in PBR was four times the open pond system cost. The harvesting process accounts for 20%–30% of the production cost, and the filtration method has a medium price for filtering pollutants with higher efficiency. Biomass productivity with higher lipid content shows a positive impact on final biofuel cost. 10,000 tons of microalgae with 30% lipids were estimated to be $2.80/L, including higher taxes than petroleum which costs $1.10/L. Cultivation in an open raceway pond costs $2.25/L, PBR costs $4.79/L, and upgrading to biodiesel cost in case of open raceway ponds is $2.6/L, and PBR is $5.42/L (Rajesh Banu et al., 2020). Studies suggest that Tubular PBRs have negative Net Energy Ratios. Hence, they are economically unfeasible, while the Flat bed PBRs have the best NERs. However, their operating costs are way higher than open raceway ponds thus are generally used to cultivate microalgae for higher-value products like pharmaceuticals (Aggarwal and Remya, 2022). With the increase in productivity of microalgal biomass, the NER reduces. However, compact PBR for cultivation causes less impact to the environment from a Life Cycle point of view than the raceway ponds. As a result, there is a high-density microalgae biomass production (Silva et al., 2015). Hybrid cultivation systems have proved better than open raceway ponds and PBRs alone because of high productivity and GWP and FER savings by 42% and 38%. Integration of algal biomass waste-based electricity showed a negligible reduction in the energy demand of the cultivation process (Togarcheti et al., 2017). The utilization of waste carbon dioxide from different plants like fossil fuel and ammonia plants can reduce the release of CO_2 into the atmosphere and hence increases the carbon credit (Campbell et al., 2011; Passell et al., 2013).

It was observed that the flocculation method could be a more economically feasible method than centrifugation for dewatering. Moreover, oven drying can be seen as a more sustainable and financially viable method for drying microalgae. Research has shown that the lipid extraction process uses the bulk of the total energy requirement. Thus, the traditional chloroform-Methanol-based extraction can be substituted with more novel types of lipid extraction technology like CO_2 expanded Methanol (CXM), non-expanded methanol inflow through the reactor (NCXM). CXM extraction proves to be the best alternative as it consumes lesser electricity and heat energy inputs. Also, the CXM uses five times less volume of methanol than used in NCXM extraction (Collotta et al., 2017).

If the costly feedstock is used, the contribution to the economics will significantly be higher. Thus, the production of value-added co-products will help drive the economics. However, high economics also depends on geographical area, cultivation variation due to seasons, labor charges, solvents used, and some other factors, thus to improve cost-effectiveness, technologies dealing with these factors have to be taken care of (Dutta et al., 2016).

Many studies found that different conversion systems will have different NER values and GHG emissions, i.e., the NER values were 1.52 (transesterification), 1.25 (HTL-Hydrothermal Liquefaction), 2.46 (pyrolysis), 1.27 (for Anaerobic Digestion without pretreatment), 0.71 (for anaerobic digestion with pretreatment). Whereas the GHG emissions were −10.37 (for transesterification), −6.77 (for HTL), 265.26 (for pyrolysis), −173.02 (for AD without pretreatment), and −60.84 (for AD with pretreatment) g CO_2-eq /MJ biofuel, respectively. On account of this, the AD system with pretreatment was found to be more industrially feasible and ecofriendly (Sun et al., 2019). It can be concluded that the cultivation of microalgae in hybrid systems provides the best results with better productivity of microalgal biomass and value-added co-products. It can help reduce the power supply costs from other sources because these co-products and byproducts can be used to produce renewable energy. Moreover, using nutrient-rich wastewater like municipal wastewater will save costs of additional fertilizers. Setting up the cultivation system near the fossil fuel plants can provide a good source of CO_2 for microalgae while also reducing CO_2 emissions from the atmosphere.

21.6 Policies and Incentives for Algal Biofuel

Currently, there are no concrete EU policies that explicitly support the production of cleaner biofuels. But recently, small but noticeable changes are brought out by some governments that suggest that the trend is slowly shifting towards a greener economy. For example, in countries like Germany and France, biodiesel is excluded from fuel taxes. Similarly, some Asian countries are providing market intensives and bringing out specific energy policies that will go a long way in promoting biofuel production. For example, Indonesia aims to achieve 5% of its total energy production from biofuels by the year 2025. Countries like India, Malaysia, and Pakistan are also aiming to blend petroleum-based fuels with different proportions of biofuels in the near future. Cost incentives are being introduced to the industry to promote the production of biofuels (Batan et al., 2016).

21.7 Conclusions

In conclusion, microalgae have shown over the years that they can be a reliable option to treat different types of wastewaters. There is also no denying that microalgal biofuels can compete with fossil fuels in terms of quality. However, the environmental and economic viability of these biofuels to meet the market demand is still unclear. This is due to the lack of application of microalgae biofuels at an industrial scale. The presence of biofuel production pathways in which the microalgae biofuels are generated leads to uncertain LCA conclusions.

Moreover, different TEA models and assumptions adopted by different authors in their studies make it difficult to compare one study. There needs to be a firm backing by the governments regarding policy and resource allocation to the third-generation biofuel industry. For example, when the US government supported first-generation biofuel production from corn ethanol, the entire biofuel industry boomed. Likewise, similar support will likely create more green jobs and improve energy security.

REFERENCES

Abdel-Raouf N, Al-Homaidan AA, Ibraheem IBM (2012) Microalgae and wastewater treatment. *Saudi J Biol Sci* 19:257–275. https://doi.org/10.1016/j.sjbs.2012.04.005

Aggarwal M, Remya N (2021) Application of native mix algal strain for gray water treatment and biofuel production: Preliminary study. *J Hazard Toxic Radioact Waste* 25:04020070. https://doi.org/10.1061/(ASCE)HZ.2153-5515.0000582

Aggarwal M, Remya N (2022) The state-of-the-art production of biofuel from microalgae with simultaneous wastewater treatment: Influence of process variables on biofuel yield and production cost. *Bioenergy Res* 15:62–76.

Bach QV, Chen WH, Lin SC, et al. (2017) Wet torrefaction of microalga *Chlorella vulgaris* ESP-31 with microwave-assisted heating. *Energy Convers Manag* 141:163–170. https://doi.org/10.1016/j.enconman.2016.07.035

Bach QV, Skreiberg O (2016) Upgrading biomass fuels via wet torrefaction: A review and comparison with dry torrefaction. *Renew Sustain Energy Rev* 54:665–677

Batan LY, Graff GD, Bradley TH (2016) Techno-economic and Monte Carlo probabilistic analysis of microalgae biofuel production system. *Bioresour Technol* 219:45–52. https://doi.org/10.1016/J.BIORTECH.2016.07.085

Bohutskyi P, Chow S, Ketter B, et al. (2016) Phytoremediation of agriculture runoff by filamentous algae polyculture for biomethane production, and nutrient recovery for secondary cultivation of lipid generating microalgae. *Bioresour Technol* 222:294–308. https://doi.org/10.1016/j.biortech.2016.10.013

Campbell PK, Beer T, Batten D (2011) Life cycle assessment of biodiesel production from microalgae in ponds. *Bioresour Technol* 102:50–56. https://doi.org/10.1016/j.biortech.2010.06.048

Chandra, R, Kumar, V 2017a. Detection of *Bacillus* and *Stenotrophomonas* species growing in an organic acid and endocrine-disrupting chemicals rich environment of distillery spent wash and its phytotoxicity. *Environ Monit Assess* 189:26. https://doi.org/10.1007/s10661-016-5746–9

Chandra, R, Kumar, V, 2017b. Detection of androgenic-mutagenic compounds and potential autochthonous bacterial communities during in-situ bioremediation of post methanated distillery sludge. *Front Microbiol* 8:87. https://doi.org/10.3389/fmicb.2017.00887

Cheng F, Cui Z, Chen L, et al. (2017) Hydrothermal liquefaction of high- and low-lipid algae: Bio-crude oil chemistry. *Appl Energy* 206:278–292 https://doi.org/10.1016/j.apenergy.2017.08.105

Cho HU, Kim HG, Kim YM, Park JM (2017) Volatile fatty acid recovery by anaerobic fermentation from blue-green algae: Effect of pretreatment. *Bioresour Technol* 244:1433–1438 https://doi.org/10.1016/J.BIORTECH.2017.05.076

Collotta M, Busi L, Champagne P, et al. (2017) Comparative LCA of three alternative technologies for lipid extraction in biodiesel from microalgae production. *Energy Procedia* 113:244–250 https://doi.org/10.1016/j.egypro.2017.04.061

de Jesus SS, Ferreira GF, Moreira LS, Filho RM (2020) Biodiesel production from microalgae by direct transesterification using green solvents. *Renew Energy* 160:1283–1294 https://doi.org/10.1016/J.RENENE.2020.07.056

Duan P, Bai X, Xu Y, et al. (2013) Catalytic upgrading of crude algal oil using platinum/gamma alumina in supercritical water. *Fuel* 109:225–233 https://doi.org/10.1016/j.fuel.2012.12.074

Dutta S, Neto F, Coelho MC (2016) Microalgae biofuels: A comparative study on techno-economic analysis & life-cycle assessment. *Algal Res* 20:44–52 https://doi.org/10.1016/j.algal.2016.09.018

Gan YY, Ong HC, Show PL, et al. (2018) Torrefaction of microalgal biochar as potential coal fuel and application as bio-adsorbent. *Energy Convers Manag* 165:152–162 https://doi.org/10.1016/J.ENCONMAN.2018.03.046

Goh BHH, Ong HC, Cheah MY, et al. (2019) Sustainability of direct biodiesel synthesis from microalgae biomass: A critical review. *Renew Sustain Energy Rev* 107:59–74 https://doi.org/10.1016/J.RSER.2019.02.012

Jayaraman RS, Gopinath KP, Arun J, et al. (2021) Co-hydrothermal gasification of microbial sludge and algae *Kappaphycus alvarezii* for bio-hydrogen production: Study on aqueous phase reforming. *Int J Hydrogen Energy* 46:16555–16564 https://doi.org/10.1016/J.IJHYDENE.2021.02.038

Li Y, Leow S, Fedders AC, et al. (2017) Quantitative multiphase model for hydrothermal liquefaction of algal biomass. *Green Chem* 19:1163–1174. https://doi.org/10.1039/c6gc03294j

Mälkki H, Alanne K (2017) An overview of life cycle assessment (LCA) and research-based teaching in renewable and sustainable energy education. *Renew Sustain Energy Rev* 69:218–231. https://doi.org/10.1016/j.rser.2016.11.176

Passell H, Dhaliwal H, Reno M, et al. (2013) Algae biodiesel life cycle assessment using current commercial data. *J Environ Manage* 129:103–111. https://doi.org/10.1016/j.jenvman.2013.06.055

Pittman JK, Dean AP, Osundeko O (2011) The potential of sustainable algal biofuel production using wastewater resources. *Bioresour Technol* 102:17–25. https://doi.org/10.1016/j.biortech.2010.06.035

Rajesh Banu J, Kavitha S, Gunasekaran M, et al. (2020) Microalgae based biorefinery promoting circular bioeconomy-techno economic and life-cycle analysis. *Bioresour Technol* 302:122822. https://doi.org/10.1016/J.BIORTECH.2020.122822

Salakkam A, Sittijunda S, Mamimin C, et al. (2021) Valorization of microalgal biomass for biohydrogen generation: A review. *Bioresour Technol* 322:124533. https://doi.org/10.1016/J.BIORTECH.2020.124533

Silva AG, Carter R, Merss FLM, et al. (2015) Life cycle assessment of biomass production in microalgae compact photobioreactors. *GCB Bioenergy* 7. https://doi.org/10.1111/gcbb.12120

Sun C-H, Fu Q, Liao Q, et al. (2019) Life-cycle assessment of biofuel production from microalgae via various bioenergy conversion systems. *Energy* 171:1033–1045. https://doi.org/10.1016/j.energy.2019.01.074

Togarcheti SC, Mediboyina kumar M, Chauhan VS, et al. (2017) Life cycle assessment of microalgae based biodiesel production to evaluate the impact of biomass productivity and energy source. *Resour Conserv Recycl* 122:286–294. https://doi.org/10.1016/j.resconrec.2017.01.008

Waqas M, Rehan M, Daud Khan M, Nizami A-S (2018) Conversion of Food Waste to Fermentation Products. *Encycl Food Secur Sustain* 1–9. https://doi.org/10.1016/B978-0-12-812687-5.22294-6

Wicker RJ, Kumar G, Khan E, Bhatnagar A (2021) Emergent green technologies for cost-effective valorization of microalgal biomass to renewable fuel products under a biorefinery scheme. *Chem Eng J* 415:128932. https://doi.org/10.1016/J.CEJ.2021.128932

Wollmann F, Dietze S, Ackermann J-U, et al. (2019) Microalgae wastewater treatment: Biological and technological approaches. *Eng Life Sci* 19:860–871. https://doi.org/10.1002/ELSC.201900071

Yang C, Li R, Zhang B, et al. (2019) Pyrolysis of microalgae: A critical review. *Fuel Process Technol* 186:53–72. https://doi.org/10.1016/J.FUPROC.2018.12.012

Yu G, Zhang Y, Schideman L, et al. (2011) Distributions of carbon and nitrogen in the products from hydrothermal liquefaction of low-lipid microalgae. *Energy Environ Sci* 4:4587–4595. https://doi.org/10.1039/c1ee01541a

Zanchetta E, Damergi E, Patel B, et al. (2021) Algal cellulose, production and potential use in plastics: Challenges and opportunities. *Algal Res* 56:102288. https://doi.org/10.1016/J.ALGAL.2021.102288

Index

Note: Bold page numbers refer to tables; *italic* page numbers refer to figures.

For Product Safety Concerns and Information please contact our EU
representative GPSR@taylorandfrancis.com
Taylor & Francis Verlag GmbH, Kaufingerstraße 24, 80331 München, Germany

www.ingramcontent.com/pod-product-compliance
Ingram Content Group UK Ltd.
Pitfield, Milton Keynes, MK11 3LW, UK
UKHW052218190926
475702UK00006B/1083
* 9 7 8 1 0 3 2 1 3 7 5 8 2 *